STUDENT SOLUTIONS MANUAL

MULTIVARIABLE CALCULUS

6e

Edwards & Penney

Prentice
Hall

Upper Saddle River, NJ 07458

Acquisitions Editor: Eric Frank
Supplement Editor: Aja Shevelew
Assistant Managing Editor: John Matthews
Production Editor: Wendy A. Perez
Supplement Cover Management/Design: Paul Gourhan
Manufacturing Buyer: Ilene Kahn

 © 2002 by Pearson Education, Inc.
Pearson Education, Inc.
Upper Saddle River, NJ 07458

Printed in the United States of America

10 9 8 7 6 5 4

ISBN 0-13-062023-8

Pearson Education Ltd., *London*
Pearson Education Australia Pty. Ltd., *Sydney*
Pearson Education Singapore, Pte. Ltd.
Pearson Education North Asia Ltd., *Hong Kong*
Pearson Education Canada, Inc., *Toronto*
Pearson Educacíon de Mexico, S.A. de C.V.
Pearson Education—Japan, *Tokyo*
Pearson Education Malaysia, Pte. Ltd.
Pearson Education, *Upper Saddle River, New Jersey*

Contents

Preface

This manual contains the solutions of all odd-numbered problems in Chapters 11 through 15 (and the *Appendices*) of *Calculus: Regular Version*, 6th edition (2002), by C. Henry Edwards and David E. Penney.

Many calculus problems can be solved by more than one method. We have used the most natural method whenever possible, and only in rare instances have we used instead a "clever" method; and then only when its educational value justifies such substitution. With very rare exceptions, we have used only solution methods discussed in the text prior to the statement of any given problem.

We gratefully acknowledge the assistance of our colleagues in helping us to correct errors in this edition of the solutions manual. Special thanks go to the members of the accuracy checking team of M. and N. Toscano (`www.toscano.com`), who verified the solution of almost every problem here and the validity of almost every worked-out example in the text. Any errors that remain are the responsibility of the authors. We need to thank as well

> Gail Suggs and Sharon Southwick, Department of Mathematics, University of Georgia, for their assistance with TEX;

> Brendan Mullen and Jose Malagon, System Support, Department of Mathematics, University of Georgia, for their assistance with hardware and software;

> Ed Azoff, Brian Boe, Frank Lether, Ted Shifrin, and Robert Varley, Department of Mathematics, University of Georgia, for their aid in solving difficult problems.

Most of the solutions were also checked using *Mathematica* 3.0, *Derive* 2.56, or *Maple* V (release 5.1); illustrations were generally created in Adobe Illustrator or in *Mathematica* (and then fine-tuned using Adobe Illustrator). The manual was typeset by $\mathcal{AMS}$-TEX version 2.1 (copyright 1991, American Mathematical Society). If you have constructive suggestions or valid corrections, please send them to us at `hedwards@math.uga.edu` and/or `dpenney@math.uga.edu`. Please include the complete name of the text and the edition (on the front cover) as well as which printing (near the bottom of the copyright page, normally the second [physical] page of the book; it's the last integer in the sequence that begins 10 9 8 7 ...).

C. Henry Edwards

David E. Penney

December 2001

Section 11.2

C11S02.001: The most obvious pattern is that $a_n = n^2$ for $n \geq 1$.

C11S02.003: The most obvious pattern is that $a_n = \dfrac{1}{3^n}$ for $n \geq 1$.

C11S02.005: The most obvious pattern is that $a_n = \dfrac{1}{3n-1}$ for $n \geq 1$.

C11S02.007: Perhaps the most obvious pattern is that $a_n = 1 + (-1)^n$ for $n \geq 1$.

C11S02.009: $\displaystyle\lim_{n\to\infty} \frac{2n}{5n-3} = \lim_{n\to\infty} \frac{2}{5 - \dfrac{3}{n}} = \frac{2}{5-0} = \frac{2}{5}.$

C11S02.011: $\displaystyle\lim_{n\to\infty} \frac{n^2 - n + 7}{2n^3 + n^2} = \lim_{n\to\infty} \frac{\dfrac{1}{n} - \dfrac{1}{n^2} + \dfrac{7}{n^3}}{2 + \dfrac{1}{n}} = \frac{0+0+0}{2+0} = 0.$

C11S02.013: Example 9 tells us that if $|r| < 1$, then $r^n \to 0$ as $n \to +\infty$. Take $r = \frac{9}{10}$ to deduce that

$$\lim_{n\to\infty} \left[1 + \left(\tfrac{9}{10}\right)^n\right] = 1 + 0 = 1.$$

C11S02.015: Given: $a_n = 1 + (-1)^n$ for $n \geq 1$. If n is odd then $a_n = 1 + (-1) = 0$; if n is even then $a_n = 1 + 1 = 2$. Therefore the sequence $\{a_n\}$ diverges. To prove this, we appeal to the definition of limit of a sequence given in Section 11.2. Suppose that $\{a_n\}$ converges to the number L. Let $\epsilon = \frac{1}{2}$ and suppose that N is a positive integer.

Case 1: $L \geq 1$. Then choose $n \geq N$ such that n is odd. Then $a_n = 0$, so

$$|a_n - L| = |0 - L| = L \geq 1 > \epsilon.$$

Case 2: $L < 1$. Then choose $n \geq N$ such that n is even. Then $a_n = 2$, so

$$|a_n - L| = |2 - L| = 2 - L > 1 > \epsilon.$$

No matter what the value of L, it cannot be made to fit the definition of the limit of the sequence $\{a_n\}$. Therefore the sequence $\{a_n\} = \{1 + (-1)^n\}$ has no limit. (We can't even say that it approaches $+\infty$ or $-\infty$; it does not.)

C11S02.017: We use l'Hôpital's rule for sequences (Eq. (9) of Section 11.2):

$$\lim_{n\to\infty} a_n = \lim_{n\to\infty} \frac{1 + (-1)^n\sqrt{n}}{\left(\frac{3}{2}\right)^n} = \lim_{x\to\infty} \frac{1 \pm x^{1/2}}{\left(\frac{3}{2}\right)^x} = \pm\lim_{x\to\infty} \frac{1}{2x^{1/2}\left(\frac{3}{2}\right)^x \ln\left(\frac{3}{2}\right)} = 0.$$

C11S02.019: First we need a lemma.

Lemma: If $r > 0$, then $\displaystyle\lim_{n\to\infty} \frac{1}{n^r} = 0.$

Proof: Suppose that $r > 0$. Given $\epsilon > 0$, let $N = 1 + \left[1/\epsilon^{1/r}\right]$. Then N is a positive integer, and if $n > N$, then $n > 1/\epsilon^{1/r}$, so that $n^r > 1/\epsilon$. Therefore

$$\left| \frac{1}{n^r} - 0 \right| < \epsilon.$$

Thus, by definition, $\lim\limits_{n \to \infty} \dfrac{1}{n^r} = 0.$ ◀

Next, we use the squeeze law for sequences (Theorem 3 of Section 11.2): $-1 \leqq \sin n \leqq 1$ for all integers $n \geqq 1$, and therefore

$$0 \leqq \frac{\sin^2 n}{\sqrt{n}} \leqq \frac{1}{\sqrt{n}}$$

for all integers $n \geqq 1$. But by the preceding lemma, $1/\sqrt{n} \to 0$ as $n \to +\infty$. Therefore

$$\lim_{n \to \infty} \frac{\sin^2 n}{\sqrt{n}} = 0.$$

C11S02.021: If n is a positive integer, then $\sin \pi n = 0$. Therefore $a_n = 0$ for every integer $n \geqq 0$. So $\{a_n\} \to 0$ as $n \to +\infty$.

C11S02.023: Suppose that $a > 0$. Then $f(x) = a^x$ is continuous, so $a^x \to 1 = a^0$ as $x \to 0$. But $-\dfrac{\sin n}{n} \to 0$ as $n \to +\infty$. Therefore $\lim\limits_{n \to \infty} \pi^{-(\sin n)/n} = 1.$

C11S02.025: We use l'Hôpital's rule for sequences (Eq. (9)):

$$\lim_{n \to \infty} \frac{\ln n}{\sqrt{n}} = \lim_{x \to \infty} \frac{\ln x}{x^{1/2}} = \lim_{x \to \infty} \frac{2x^{1/2}}{x} = \lim_{x \to \infty} \frac{2}{x^{1/2}} = 0.$$

C11S02.027: We use l'Hôpital's rule for sequences (Eq. (9)):

$$\lim_{n \to \infty} \frac{(\ln n)^2}{n} = \lim_{x \to \infty} \frac{(\ln x)^2}{x} = \lim_{x \to \infty} \frac{2 \ln x}{x} = \lim_{x \to \infty} \frac{2}{x} = 0.$$

C11S02.029: Because $-\pi/2 < \tan^{-1} x < \pi/2$ for all x,

$$-\frac{\pi}{2n} < \frac{\tan^{-1} n}{n} < \frac{\pi}{2n}$$

for all integers $n \geqq 1$. Therefore, by the squeeze law for limits, $\lim\limits_{n \to \infty} \dfrac{\tan^{-1} n}{n} = 0.$

C11S02.031: We use the squeeze law for limits of sequences:

$$0 < \frac{2^n + 1}{e^n} < \frac{2^n + 2^n}{e^n} = 2 \left(\frac{2}{e} \right)^n.$$

By the result in Example 9, $(2/e)^n \to 0$ as $n \to +\infty$. Therefore $\lim\limits_{n \to \infty} \dfrac{2^n + 1}{e^n} = 0.$

C11S02.033: By Eq. (3) in Section 7.3,

$$\lim_{n \to \infty} \left(1 + \frac{1}{n} \right)^n = e.$$

516

C11S02.035: We use l'Hôpital's rule for sequences:

$$\lim_{n\to\infty}\left(\frac{n-1}{n+1}\right)^n = \lim_{n\to\infty}\exp\left(x\ln\frac{x-1}{x+1}\right) = \exp\left(\lim_{x\to\infty}\frac{\ln(x-1)-\ln(x+1)}{\frac{1}{x}}\right)$$

$$= \exp\left(\lim_{x\to\infty}\frac{\frac{1}{x-1}-\frac{1}{x+1}}{-\frac{1}{x^2}}\right) = \exp\left(\lim_{x\to\infty}\frac{-2x^2}{x^2-1}\right) = e^{-2}.$$

C11S02.037: Let $f(x) = 2^x$ and note that f is continuous on the set of all real numbers. Thus

$$\lim_{n\to\infty} a_n = \lim_{x\to\infty} f\left(\frac{x+1}{x}\right) = f\left(\lim_{x\to\infty}\frac{x+1}{x}\right) = f(1) = 2^1 = 2.$$

It is the continuity of f at $x = 1$ that makes the second equality valid.

C11S02.039: By Example 11, $\lim_{n\to\infty} n^{1/n} = 1$. Thus by Example 7,

$$\lim_{n\to\infty}\left(\frac{2}{n}\right)^{3/n} = \lim_{n\to\infty}\frac{8^{1/n}}{\left(n^{1/n}\right)^3} = \frac{1}{1^3} = 1.$$

C11S02.041: Given: $a_n = \left(\frac{2-n^2}{3+n^2}\right)^n$. First note that

$$\lim_{n\to\infty}\frac{2-n^2}{3+n^2} = \lim_{n\to\infty}\frac{\frac{2}{n^2}-1}{\frac{3}{n^2}+1} = \frac{0-1}{0+1} = -1.$$

Moreover,

$$\lim_{n\to\infty}\left(\frac{n^2-2}{n^2+3}\right)^n = \lim_{x\to\infty}\exp\left(\frac{\ln(x^2-2)-\ln(x^2+3)}{\frac{1}{x}}\right) = \exp\left(\lim_{x\to\infty}\frac{\frac{2x}{x^2-2}-\frac{2x}{x^2+3}}{-\frac{1}{x^2}}\right)$$

$$= \exp\left(\lim_{x\to\infty}\frac{-x^2(2x^3+6x-2x^3+4x)}{(x^2-2)(x^2+3)}\right) = \exp\left(\lim_{x\to\infty}\frac{-10x^3}{(x^2-2)(x^2+3)}\right) = e^0 = 1.$$

Therefore, as $n \to +\infty$ through *even* values, $a_n \to 1$, whereas as $n \to +\infty$ through *odd* values, $a_n \to -1$. So we can now show that the sequence $\{a_n\}$ has no limit as $n \to \infty$.

Let $\epsilon = \frac{1}{10}$. Choose N_1 so large that if $n > N_1$ and n is even, then $|a_n - 1| < \epsilon$. Choose N_2 so large that if $n > N_2$ and n is odd, then $|a_n - (-1)| < \epsilon$. Let N be the maximum of N_1 and N_2. Then if $n > N$,

$$|a_n - 1| < \epsilon \quad \text{if } n \text{ is even};$$

$$|a_n - (-1)| < \epsilon \quad \text{if } n \text{ is odd}.$$

Put another way, if $n > N$ and n is even, then a_n lies in the interval $(0.9, 1.1)$. If $n > N$ and n is odd, then a_n lies in the interval $(-1.1, -0.9)$. It follows that no interval of length 0.2 can contain a_n for all $n > K$, no matter how large K might be. Because every real number L is the midpoint of such an interval, this means that no real number L can be the limit of the sequence $\{a_n\}$. Therefore

$$\lim_{n \to \infty} a_n \quad \text{does not exist.}$$

C11S02.043: Let $f(n) = \dfrac{n-2}{n+13}$. Then the *Mathematica* command

    Table[{ n, N[f[10^n]] }, { n, 1, 7 }]

yielded the response

$$\{\{1, 0.347826\}, \{2, 0.867257\}, \{3, 0.985192\}, \{4, 0.998502\}, \{5, 0.99985\}, \{6, 0.999985\}, \{7, 0.999999\}\}.$$

Indeed,

$$\lim_{n \to \infty} \frac{n-2}{n+13} = \lim_{n \to \infty} \frac{1 - \dfrac{2}{n}}{1 + \dfrac{13}{n}} = \frac{1-0}{1+0} = 1.$$

C11S02.045: Let $f(n) = a_n = \sqrt{\dfrac{4n^2 + 7}{n^2 + 3n}}$. Results of our experiment:

n	$f(n)$
10	1.76940
100	1.97083
1000	1.99701
10000	1.99970
100000	1.99997
1000000	2.00000

By Theorem 4 and l'Hôpital's rule (used twice),

$$\lim_{n \to \infty} \frac{4n^2 + 7}{n^2 + 3n} = \lim_{x \to \infty} \frac{4x^2 + 7}{x^2 + 3x} = \lim_{x \to \infty} \frac{8x}{2x + 3} = \lim_{x \to \infty} \frac{8}{2} = 4.$$

Therefore, by Theorem 2, $\lim_{n \to \infty} a_n = \sqrt{4} = 2$.

C11S02.047: Let $f(n) = a_n = \exp\left(-1/\sqrt{n}\right)$. Results of an experiment:

518

n	$f(n)$
10	0.728893
100	0.904837
1000	0.968872
10000	0.990050
100000	0.996843
1000000	0.999000
10000000	0.999684
100000000	0.999900
1000000000	0.999968

Because $-1/\sqrt{n} \to 0$ as $n \to +\infty$ and $g(x) = e^x$ is continuous at $x = 0$, Theorem 2 implies that

$$\lim_{n \to \infty} a_n = g(0) = 1.$$

C11S02.049: Let $f(n) = a_n = 4 \tan^{-1} \dfrac{n-1}{n+1}$. The results of an experiment:

n	$f(n)$
10	2.74292
100	3.10159
1000	3.13759
10000	3.14119
100000	3.14155
1000000	3.14159
10000000	3.14159

The limit appears to be π. Because

$$\lim_{n \to \infty} \frac{n-1}{n+1} = \lim_{n \to \infty} \left(1 - \frac{2}{n+1} \right) = 1 - 0 = 1$$

and $g(x) = 4 \tan^{-1} x$ is continuous at $x = 1$, Theorem 2 implies that

$$\lim_{n \to \infty} a_n = g(1) = 4 \tan^{-1}(1) = 4 \cdot \frac{\pi}{4} = \pi.$$

C11S02.051: Proof: Suppose that

$$\lim_{n \to \infty} a_n = A \neq 0.$$

Without loss of generality we may suppose that $A > 0$. Let $\epsilon = A/3$ and choose N so large that if $n > N$, then $|a_n - A| < \epsilon$. Then if n is even, $(-1)^n a_n = a_n$; in this case $|(-1)^n a_n - A| < \epsilon$ if $n > N$. If n is odd, then $(-1)^n a_n = -a_n$; in this case $|(-1)^n a_n - (-A)| < \epsilon$ if $n > N$. In other words, $(-1)^n a_n$ lies in the interval $I = (A - \epsilon, A + \epsilon)$ if n is even, whereas $(-1)^n a_n$ lies in the interval $J = (-A - \epsilon, -A + \epsilon)$ if n is odd. This means that no open interval of length 2ϵ can contain every number $(-1)^n a_n$ for which $n > K$, no matter how large the value of K. (Note that no such interval can contains points of both I and J because the distance between their closest endpoints is 4ϵ.) Because every real number is the midpoint of an open interval of length 2ϵ, it now follows that no real number can be the limit of the sequence $\{(-1)^n a_n\}$. This concludes the proof. ◀

C11S02.053: Given: $A > 0$, $x_1 \neq 0$,

$$x_{n+1} = \frac{1}{2}\left(x_n + \frac{A}{x_n}\right) \quad \text{if} \quad n \geq 1, \qquad \text{and} \qquad L = \lim_{n \to \infty} x_n.$$

Then

$$\lim_{n \to \infty} x_{n+1} = L \qquad \text{and} \qquad \lim_{n \to \infty} \frac{1}{2}\left(x_n + \frac{A}{x_n}\right) = \frac{1}{2}\left(L + \frac{A}{L}\right).$$

It follows that

$$L = \frac{1}{2}\left(L + \frac{A}{L}\right); \qquad 2L = \frac{L^2 + A}{L};$$

$$2L^2 = L^2 + A; \qquad L^2 = A.$$

Therefore $L = \pm\sqrt{A}$.

C11S02.055: Part (a): Note first that $F_1 = 1$ and $F_2 = 1$. If $n \geq 3$, then F_{n-1} is the total number of pairs present in the preceding month and F_{n-2} is the total number of productive pairs. Therefore $F_n = F_{n-1} + F_{n-2}$; that is, $F_{n+1} = F_n + F_{n-1}$ for $n \geq 2$. So $\{F_n\}$ is the Fibonacci sequence of Example 2.

Part (b): Note first that $G_1 = G_2 = G_3 = 1$. If $n \geq 4$, then G_{n-1} is the total number of pairs present in the preceding month and G_{n-3} is the total number of productive pairs. Therefore $G_n = G_{n-1} + G_{n-3}$; that is, $G_{n+1} = G_n + G_{n-2}$. The *Mathematica* commands

```
g[1] = 1; g[2] = 1; g[3] = 1;
g[n_] := g[n] = g[n - 1] + g[n - 3]
```

serve as one way to enter the formula for the recursively defined function g. Then the command

```
Table[ {n, g[n]}, {n, 4, 25} ]
```

produces the output

$$\{\{4, 2\}, \{5, 3\}, \{6, 4\}, \{7, 6\}, \{8, 9\}, \{9, 13\}, \{10, 19\}, \{11, 28\}, \{12, 41\},$$
$$\{13, 60\}, \{14, 88\}, \{15, 129\}, \{16, 189\}, \{17, 277\}, \{18, 406\}, \{19, 595\},$$
$$\{20, 872\}, \{21, 1278\}, \{22, 1873\}, \{23, 2745\}, \{24, 4023\}, \{25, 5896\}\}$$

C11S02.057: Part (a): Clearly $a_1 < 4$. Suppose that $a_k < 4$ for some integer $k \geq 1$. Then

$$a_{k+1} = \frac{1}{2}(a_k + 4) < \frac{1}{2}(4 + 4) = 4.$$

Therefore, by induction, $a_n < 4$ for every integer $n \geqq 1$. Next, $a_2 = 3$, so that $a_1 < a_2$. Suppose that $a_k < a_{k+1}$ for some integer $k \geqq 1$. Then

$$a_k + 4 < a_{k+1} + 4; \qquad \frac{1}{2}(a_k + 4) < \frac{1}{2}(a_{k+1} + 4); \qquad a_{k+1} < a_{k+2}.$$

Therefore, by induction, $a_n < a_{n+1}$ for every integer $n \geqq 1$.

Part (b): Part (a) establishes that $\{a_n\}$ is a bounded increasing sequence. Therefore the bounded monotonic sequence property of Section 11.2 implies that the sequence $\{a_n\}$ converges. Let L denote its limit. Then

$$L = \lim_{n \to \infty} a_{n+1} = \lim_{n \to \infty} \frac{1}{2}(a_n + 4) = \frac{1}{2}(L + 4).$$

It now follows immediately that $L = 4$.

C11S02.059: Given the positive real number r, let

$$L = \frac{1 + \sqrt{1 + 4r}}{2}.$$

Define the sequence $\{a_n\}$ recursively as follows: $a_1 = \sqrt{r}$, and for each integer $n \geqq 1$, $a_{n+1} = \sqrt{r + a_n}$. We plan to show first that $\{a_n\}$ is a bounded sequence, then that $\{a_n\}$ is an increasing sequence. Only then will we attempt to evaluate its limit, because our method for doing so depends on knowing that the sequence $\{a_n\}$ converges.

First,

$$a_1 = \sqrt{r} = \frac{\sqrt{4r}}{2} < \frac{1 + \sqrt{1 + 4r}}{2} = L.$$

Suppose that $a_k < L$ for some integer $k \geqq 1$. Then

$$a_k < \frac{1 + \sqrt{1 + 4r}}{2}; \qquad a_k + r < L + r;$$

$$4(a_k + r) < 2 + 2\sqrt{1 + 4r} + 4r; \qquad 4(a_k + r) < \left(1 + \sqrt{1 + 4r}\right)^2;$$

$$a_k + r < L^2; \qquad a_{k+1} = \sqrt{a_k + r} < L.$$

Therefore, by induction, $a_n < L$ for all $n \geqq 1$.

Next, $0 < \sqrt{r}$, so that $r < r + \sqrt{r}$. Thus $\sqrt{r} < \sqrt{r + \sqrt{r}}$. That is, $a_1 < a_2$. Suppose that $a_k < a_{k+1}$ for some integer $k \geqq 1$. Then

$$r + a_k < r + a_{k+1}; \qquad \sqrt{r + a_k} < \sqrt{r + a_{k+1}}; \qquad a_{k+1} < a_{k+2}.$$

Therefore, by induction, $a_n < a_{n+1}$ for all $n \geqq 1$.

Now that we know that the sequence $\{a_n\}$ converges, we may denote its limit by M. Then

$$M = \lim_{n \to \infty} a_{n+1} = \lim_{n \to \infty} \sqrt{r + a_n} = \sqrt{r + M}.$$

It now follows that $M^2 - M - r = 0$, and thus that

$$M = \frac{1 \pm \sqrt{1 + 4r}}{2}.$$

521

Because $M > 0$, we conclude that $\lim_{n \to \infty} a_n = \dfrac{1 + \sqrt{1 + 4r}}{2} = L$. And in conclusion, if we take $r = 20$ we then find that

$$\sqrt{20 + \sqrt{20 + \sqrt{20 + \sqrt{20 + \cdots}}}} = L = 5.$$

C11S02.061: Suppose that $\{a_n\}$ is a bounded monotonic sequence; without loss of generality, suppose that it is an increasing sequence. Because the set of values of a_n is a nonempty set of real numbers with an upper bound, it has a least upper bound λ. We claim that λ is the limit of the sequence $\{a_n\}$.

Let $\epsilon > 0$ be given. Then there exists a positive integer N such that

$$\lambda - \epsilon < a_N \leqq \lambda.$$

For if not, then $\lambda - \epsilon$ would be an upper bound for $\{a_n\}$ smaller than λ, its least upper bound. But because $\{a_n\}$ is an increasing sequence with upper bound λ, it now follows that if $n > N$, then $a_N \leqq a_n \leqq \lambda$. Thus if $n > N$, then $|a_n - \lambda| < \epsilon$. Therefore, by definition, λ is the limit of the sequence $\{a_n\}$.

C11S02.063: For each integer $n \geqq$, let a_n be the largest integral multiple of $1/10^n$ such that $a_n^2 \leqq 2$. (For example, $a_1 = 1.4$, $a_2 = 1.41$, and $a_3 = 1.414$.)

Part (a): First note that the numbers 1 and $\frac{3}{2}$ are multiples of $1/10^n$ (for each $n \geqq 1$) with $1^2 < 2$ and $\left(\frac{3}{2}\right)^2 > 2$. It follows that $1 \leqq a_n \leqq \frac{3}{2}$ for each integer $n \geqq 1$, and therefore the sequence $\{a_n\}$ is bounded. Next, a_n as an integral multiple of $1/10^n$ is also an integral multiple of $1/10^{n+1}$ whose square does not exceed 2. But a_{n+1} is the *largest* multiple of $1/10^{n+1}$ whose square does not exceed 2; it follows that $a_n \leqq a_{n+1}$, and thus the sequence $\{a_n\}$ is also an increasing sequence.

Part (b): Because $\{a_n\}$ is a bounded increasing sequence, it has a limit A. Then the limit laws give

$$A^2 = \left(\lim_{n \to \infty} a_n\right)^2 = \lim_{n \to \infty} (a_n)^2 \leqq \lim_{n \to \infty} 2 = 2,$$

so we see that $A^2 \leqq 2$.

Part (c): Assume that $A^2 < 2$. Then $2 - A^2 > 0$. Choose the integer k so large that $4/10^k \leqq 2 - A^2$. Then

$$\left(a_k + \frac{1}{10^k}\right)^2 = a_k^2 + \frac{2a_k}{10^k} + \frac{1}{10^{2k}}$$

$$< a_k^2 + \frac{4}{10^k} \quad \left(\text{because } a_k < \frac{3}{2} \text{ and } \frac{1}{10^{2k}} < \frac{1}{10^k}\right)$$

$$\leqq A^2 + (2 - A^2) = 2.$$

Thus the assumption that $A^2 < 2$ implies that $(a_k + 1/10^k)^2 < 2$, which contradicts the fact that a_k is, by definition, the largest integral multiple of $1/10^k$ whose square does not exceed 2. It therefore follows that A^2 is *not* less than 2; that is, that $A^2 \geqq 2$.

Part (d): It follows immediately from the results in parts (c) and (d) that $A^2 = 2$.

Section 11.3

C11S03.001: The series is geometric with first term 1 and ratio $\frac{1}{3}$. Therefore it converges to

$$\frac{1}{1 - \frac{1}{3}} = \frac{3}{2}.$$

C11S03.003: This series diverges by the nth-term test. Alternatively, you can show by induction that

$$S_k = \sum_{n=1}^{k} (2n - 1) = k^2,$$

so this series diverges because $\lim_{k \to \infty} S_k = +\infty$.

C11S03.005: This series is geometric but its ratio is -2 and $|-2| > 1$. Therefore the given series diverges. Alternatively, it diverges by the nth-term test for divergence.

C11S03.007: The given series is geometric with first term 4 and ratio $\frac{1}{3}$. Therefore it converges to

$$\frac{4}{1 - \frac{1}{3}} = 6.$$

C11S03.009: This series is geometric with first term 1 and ratio $r = 1.01$. Because $|r| > 1$, the series diverges. Alternatively, you can apply the nth-term test for divergence.

C11S03.011: The given series diverges by the nth-term test:

$$\lim_{n \to \infty} \frac{n}{n + 1} = 1, \quad \text{and therefore} \quad \lim_{n \to \infty} \frac{(-1)^n n}{n + 1} \neq 0.$$

C11S03.013: The given series is geometric with first term 1 and ratio $r = -3/e$. It diverges because $|r| > 1$.

C11S03.015: The given series is geometric with first term 1 and ratio $1/\sqrt{2}$. Therefore its sum is

$$\frac{1}{1 - \frac{1}{\sqrt{2}}} = \frac{\sqrt{2}}{\sqrt{2} - 1} = \frac{\sqrt{2}\,(\sqrt{2} + 1)}{2 - 1} = 2 + \sqrt{2} \approx 3.414213562373.$$

C11S03.017: Because the limit of the nth term is

$$\lim_{n \to \infty} \frac{n}{10n + 17} = \lim_{n \to \infty} \frac{1}{10 + \frac{17}{n}} = \frac{1}{10} \neq 0,$$

the given series diverges.

C11S03.019: The given series is the difference of two convergent geometric series, so its sum is

$$\sum_{n=1}^{\infty} \left(5^{-n} - 7^{-n} \right) = \sum_{n=1}^{\infty} \frac{1}{5^n} - \sum_{n=1}^{\infty} \frac{1}{7^n} = \frac{\frac{1}{5}}{1 - \frac{1}{5}} - \frac{\frac{1}{7}}{1 - \frac{1}{7}} = \frac{1}{4} - \frac{1}{6} = \frac{1}{12}.$$

C11S03.021: The series is geometric with first term e/π and ratio $r = e/\pi$. Because $|r| < 1$, the series converges to

$$\frac{\dfrac{e}{\pi}}{1 - \dfrac{e}{\pi}} = \frac{e}{\pi - e} \approx 6.421479600999.$$

C11S03.023: The given series is geometric with ratio $r = \frac{100}{99}$. But its first term is nonzero and $|r| > 1$, so the series diverges.

C11S03.025: The given series is the sum of three convergent geometric series, and its sum is

$$\sum_{n=0}^{\infty} \frac{1 + 2^n + 3^n}{5^n} = \sum_{n=0}^{\infty} \frac{1}{5^n} + \sum_{n=0}^{\infty} \left(\frac{2}{5}\right)^n + \sum_{n=0}^{\infty} \left(\frac{3}{5}\right)^n$$

$$= \frac{1}{1 - \dfrac{1}{5}} + \frac{1}{1 - \dfrac{2}{5}} + \frac{1}{1 - \dfrac{3}{5}} = \frac{5}{4} + \frac{5}{3} + \frac{5}{2} = \frac{65}{12} \approx 5.41666666667.$$

C11S03.027: We use both parts of Theorem 2:

$$\sum_{n=0}^{\infty} \frac{7 \cdot 5^n + 3 \cdot 11^n}{13^n} = 7 \cdot \left[\sum_{n=0}^{\infty} \left(\frac{5}{13}\right)^n \right] + 3 \cdot \left[\sum_{n=0}^{\infty} \left(\frac{11}{13}\right)^n \right]$$

$$= 7 \cdot \frac{1}{1 - \dfrac{5}{13}} + 3 \cdot \frac{1}{1 - \dfrac{11}{13}} = \frac{91}{8} + \frac{39}{2} = \frac{247}{8} = 30.875.$$

C11S03.029: The given series converges because it is the difference of two convergent geometric series. Its sum is

$$\sum_{n=1}^{\infty} \left[\left(\frac{7}{11}\right)^n - \left(\frac{3}{5}\right)^n \right] = \sum_{n=1}^{\infty} \left(\frac{7}{11}\right)^n - \sum_{n=1}^{\infty} \left(\frac{3}{5}\right)^n = \frac{\dfrac{7}{11}}{1 - \dfrac{7}{11}} - \frac{\dfrac{3}{5}}{1 - \dfrac{3}{5}} = \frac{7}{4} - \frac{3}{2} = \frac{1}{4}.$$

C11S03.031: The given series diverges by the nth-term test for divergence, because

$$\lim_{n \to \infty} \frac{n^2 - 1}{3n^2 - 1} = \lim_{n \to \infty} \frac{1 - \dfrac{1}{n^2}}{3 - \dfrac{1}{n^2}} = \frac{1 - 0}{3 - 0} = \frac{1}{3} \neq 0.$$

C11S03.033: The given series is geometric with nonzero first term and ratio $r = \tan 1 \approx 1.557407724655$. Because $|r| > 1$, this series diverges.

C11S03.035: This is a geometric series with first term $\pi/4$ and ratio $r = \pi/4 \approx 0.785398163397$. Because $|r| < 1$, it converges; its sum is

$$\frac{\dfrac{\pi}{4}}{1 - \dfrac{\pi}{4}} = \frac{\pi}{4 - \pi} \approx 3.659792366325.$$

C11S03.037: A figure similar to Fig. 11.3.4 of the text shows that if n is an integer and $n \geqq 2$, then

$$\frac{1}{n \ln n} \geqq \int_n^{n+1} \frac{1}{x \ln x} \, dx.$$

Therefore the kth partial sum S_k of the given series satisfies the equalities and inequalities

$$S_k = \sum_{n=2}^k \frac{1}{x \ln x} \geqq \int_2^{k+1} \frac{1}{x \ln x} \, dx = \Big[\ln(\ln x) \Big]_2^{k+1} = \ln(\ln(k+1)) - \ln(\ln 2) = \ln \left(\frac{\ln(k+1)}{\ln 2} \right).$$

Hence the given series diverges because $\{S_k\} \to +\infty$ as $k \to +\infty$.

C11S03.039: Here we have

$$0.47\,47\,47\,47 \cdots = \frac{47}{100} + \frac{47}{10000} + \frac{47}{1000000} + \cdots = \frac{\dfrac{47}{100}}{1 - \dfrac{1}{100}} = \frac{47}{99}.$$

C11S03.041: $0.123\,123\,123 \cdots = \dfrac{123}{1000} + \dfrac{123}{1000000} + \dfrac{123}{1000000000} + \cdots = \dfrac{\dfrac{123}{1000}}{1 - \dfrac{1}{1000}} = \dfrac{123}{999} = \dfrac{41}{333}.$

C11S03.043: As in Example 7,

$$3.14159\,14159\,14159 \cdots = 3 + \frac{14159}{10^5} + \frac{14150}{10^{10}} + \frac{14159}{10^{15}} + \cdots = 3 + \frac{\dfrac{14159}{100000}}{1 - \dfrac{1}{100000}}$$

$$= 3 + \frac{14159}{99999} = \frac{299997 + 14159}{99999} = \frac{314156}{99999}.$$

C11S03.045: The series is geometric with ratio $x/3$. Thus it will converge when

$$\left| \frac{x}{3} \right| < 1; \quad \text{that is, when} \quad -3 < x < 3.$$

For such x, we have

$$\sum_{n=1}^\infty \left(\frac{x}{3} \right)^n = \frac{\dfrac{x}{3}}{1 - \dfrac{x}{3}} = \frac{x}{3 - x}.$$

C11S03.047: The given series is geometric with ratio $(x-2)/3$. Hence it will converge when

$$\left| \frac{x-2}{3} \right| < 1; \quad \text{that is, when} \quad -1 < x < 5.$$

For such x, we have

$$\sum_{n=1}^\infty \left(\frac{x-2}{3} \right)^n = \frac{\dfrac{x-2}{3}}{1 - \dfrac{x-2}{3}} = \frac{x-2}{3 - (x-2)} = \frac{x-2}{5 - x}.$$

C11S03.049: This series is geometric with ratio $5x^2/(x^2+16)$. Hence it will converge when

$$\frac{5x^2}{x^2+16} < 1: \quad 5x^2 < x^2 + 16;$$

$$4x^2 < 16;$$

$$x^2 < 4;$$

$$-2 < x < 2.$$

For such x we have

$$\sum_{n=1}^{\infty} \left(\frac{5x^2}{x^2+16} \right)^n = \frac{\dfrac{5x^2}{x^2+16}}{1 - \dfrac{5x^2}{x^2+16}} = \frac{5x^2}{x^2+16-5x^2} = \frac{5x^2}{16-4x^2}.$$

C11S03.051: The method of partial fractions yields

$$\frac{1}{9n^2+3n-2} = \frac{1}{3} \left(\frac{1}{3n-1} + \frac{-1}{3n+2} \right).$$

Therefore the kth partial sum of the given series is

$$S_k = \sum_{n=1}^{k} \frac{1}{9n^2+3n-2} = \frac{1}{3} \left(\frac{1}{2} - \frac{1}{5} + \frac{1}{5} - \frac{1}{8} + \frac{1}{8} - \frac{1}{11} + \cdots - \frac{1}{3k+2} \right) = \frac{1}{3} \left(\frac{1}{2} - \frac{1}{3k+2} \right).$$

Thus

$$\sum_{n=1}^{\infty} \frac{1}{9n^2+3n-2} = \lim_{k \to \infty} S_k = \frac{1}{6}.$$

C11S03.053: The method of partial fractions yields

$$\frac{1}{16n^2-8n-3} = \frac{1}{4} \left(\frac{1}{4n-3} - \frac{1}{4n+1} \right).$$

Thus the kth partial sum of the given series is

$$S_k = \sum_{n=1}^{k} \frac{1}{16n^2-8n-3} = \frac{1}{4} \left(1 - \frac{1}{5} + \frac{1}{5} - \frac{1}{9} + \frac{1}{9} - \frac{1}{13} + \cdots - \frac{1}{4k+1} \right) = \frac{1}{4} \left(1 - \frac{1}{4k+1} \right).$$

Therefore

$$\sum_{n=1}^{\infty} \frac{1}{16n^2-8n-3} = \lim_{k \to \infty} S_k = \frac{1}{4}.$$

C11S03.055: The method of partial fractions yields

$$\frac{1}{n^2-1} = \frac{1}{2} \left(\frac{1}{n-1} - \frac{1}{n+1} \right).$$

So the kth partial sum of the given series is

$$S_k = \sum_{n=2}^{k} \frac{1}{n^2 - 1}$$

$$= \frac{1}{2}\left(1 - \frac{1}{3} + \frac{1}{2} - \frac{1}{4} + \frac{1}{3} - \frac{1}{5} + \frac{1}{4} - \frac{1}{6} + \frac{1}{5} - \frac{1}{7} + \cdots + \frac{1}{k-2} - \frac{1}{k} + \frac{1}{k-1} - \frac{1}{k+1}\right)$$

$$= \frac{1}{2}\left(1 + \frac{1}{2} - \frac{1}{k} - \frac{1}{k+1}\right).$$

Thus the sum of the given series is $\displaystyle\lim_{k\to\infty} S_k = \frac{3}{4}$.

C11S03.057: In *Derive* 2.56 application of the command **Expand** to the nth term of the given series yields the partial fraction decomposition

$$\frac{6n^2 + 2n - 1}{n(n+1)(4n^2 - 1)} = \frac{1}{n} - \frac{1}{n+1} + \frac{1}{2n-1} - \frac{1}{2n+1}.$$

So the kth partial sum of the given series is

$$S_k = \sum_{n=1}^{k} \frac{6n^2 + 2n - 1}{n(n+1)(4n^2 - 1)}$$

$$= \left(1 - \frac{1}{2} + 1 - \frac{1}{3}\right) + \left(\frac{1}{2} - \frac{1}{3} + \frac{1}{3} - \frac{1}{5}\right) + \left(\frac{1}{3} - \frac{1}{4} + \frac{1}{5} - \frac{1}{7}\right) + \left(\frac{1}{4} - \frac{1}{5} + \frac{1}{7} - \frac{1}{9}\right)$$

$$= \left(\frac{1}{5} - \frac{1}{6} + \frac{1}{9} - \frac{1}{11}\right) + \left(\frac{1}{6} - \frac{1}{7} + \frac{1}{11} - \frac{1}{13}\right) + \cdots + \left(\frac{1}{k} - \frac{1}{k+1} + \frac{1}{2k-1} - \frac{1}{2k+1}\right)$$

$$= 1 - \frac{1}{k+1} + 1 - \frac{1}{2k+1}.$$

Therefore the sum of the given series is $\displaystyle\lim_{k\to\infty} S_k = 2$.

C11S03.059: In *Mathematica* 3.0 the command

```
Apart[ 6/(n*(n + 1)*(n + 2)*(n + 3)) ]
```

yields the partial fraction decomposition

$$\frac{6}{n(n+1)(n+2)(n+3)} = \frac{1}{n} - \frac{3}{n+1} + \frac{3}{n+2} - \frac{1}{n+3}.$$

Therefore the kth partial sum of the given series is

$$S_k = \sum_{n=1}^{k} \frac{6}{n(n+1)(n+2)(n+3)} = 1 - \frac{3}{2} + \frac{3}{3} - \frac{1}{4}$$

$$+ \frac{1}{2} - \frac{3}{3} + \frac{3}{4} - \frac{1}{5}$$

$$+ \frac{1}{3} - \frac{3}{4} + \frac{3}{5} - \frac{1}{6}$$

527

$$+ \frac{1}{4} - \frac{3}{5} + \frac{3}{6} - \frac{1}{7}$$

$$+ \frac{1}{5} - \frac{3}{6} + \frac{3}{7} - \frac{1}{8}$$

$$+ \frac{1}{6} - \frac{3}{7} + \frac{3}{8} - \frac{1}{9}$$

$$+ \cdots$$

$$+ \frac{1}{k-4} - \frac{3}{k-3} + \frac{3}{k-2} - \frac{1}{k-1}$$

$$+ \frac{1}{k-3} - \frac{3}{k-2} + \frac{3}{k-1} - \frac{1}{k}$$

$$+ \frac{1}{k-2} - \frac{3}{k-1} + \frac{3}{k} - \frac{1}{k+1}$$

$$+ \frac{1}{k-1} - \frac{3}{k} + \frac{3}{k+1} - \frac{1}{k+2}$$

$$+ \frac{1}{k} - \frac{3}{k+1} + \frac{3}{k+2} - \frac{1}{k+3}$$

Examine the diagonals that run from southwest to northeast. The four fractions with denominator 4 all cancel one another, as do those with denominators 5, 6, ..., $k-1$, and k. Thus

$$S_k = 1 - \frac{2}{2} + \frac{1}{3} - \frac{1}{k+1} + \frac{2}{k+2} - \frac{1}{k+3} = \frac{1}{3} - \frac{1}{k+1} + \frac{2}{k+2} - \frac{1}{k+3}.$$

Therefore the sum of the given series is $\lim\limits_{k \to \infty} S_k = \frac{1}{3}$.

C11S03.061: By part 2 of Theorem 2, if $c \neq 0$ and $\sum ca_n$ converges, then

$$\sum \frac{1}{c} \cdot ca_n = \sum a_n$$

converges. Therefore if $c \neq 0$ and $\sum a_n$ diverges, then $\sum ca_n$ diverges.

C11S03.063: Let

$$S_n = \sum_{i=1}^{n} a_i \quad \text{and} \quad T_n = \sum_{i=1}^{n} b_i,$$

let k be a fixed positive integer, and suppose that $a_j = b_j$ for every integer $j \geqq k$. If $n = k+1$, then

$$S_n - T_n = (S_k + a_n) - (T_k + b_n) = (S_k + a_n) - (T_k + a_n) = S_k - T_k.$$

Assume that $S_n - T_n = S_k - T_k$ for some integer $n \geqq k+1$. Then

$$S_{n+1} - T_{n+1} = (S_n + a_{n+1}) - (T_n + b_{n+1}) = (S_n + a_{n+1}) - (T_n + a_{n+1}) = S_n - T_n = S_k - T_k.$$

Therefore, by induction, $S_n - T_n = S_k - T_k$ for every integer $n \geqq k+1$.

C11S03.065: The total time the ball spends bouncing is

528

$$T = \sqrt{2a/g} + 2\sqrt{2ar/g} + 2\sqrt{2ar^2/g} + 2\sqrt{2ar^3/g} + \cdots$$

$$= -\sqrt{2a/g} + 2\sqrt{2a/g}\left(1 + r^{1/2} + r + r^{3/2} + \cdots\right) = -\sqrt{2a/g} + 2\sqrt{2a/g}\left(\frac{1}{1 - r^{1/2}}\right)$$

$$= \sqrt{2a/g}\left(-1 + \frac{2}{1 - r^{1/2}}\right) = \sqrt{2a/g}\left(\frac{-1 + r^{1/2} + 2}{1 - r^{1/2}}\right) = \sqrt{2a/g}\left(\frac{1 + r^{1/2}}{1 - r^{1/2}}\right).$$

If we take $r = 0.64$, $a = 4$, and $g = 32$, we find the total bounce time to be

$$T = \sqrt{8/32}\left(\frac{1 + 0.8}{1 - 0.8}\right) = \frac{1}{2} \cdot \frac{1.8}{0.2} = 4.5 \quad \text{(seconds)}.$$

C11S03.067: Let $r = 0.95$. Then $M_1 = r M_0$, $M_2 = r M_1 = r^2 M_0$, and so on; in the general case, $M_n = r^n M_0$. Because $-1 < r < 1$, it now follows that

$$\lim_{n \to \infty} M_n = \lim_{n \to \infty} r^n M_0 = 0.$$

C11S03.069: Peter's probability of winning is the sum of:

The probability that he wins in the first round;

The probability that everyone tosses tails in the first round and Peter wins in the second round;

The probability that everyone tosses tails in the first two rounds and Peter wins in the third round;

Et cetera, et cetera, et cetera.

Thus his probability of winning is

$$\frac{1}{2} + \frac{1}{2^4} + \frac{1}{2^7} + \frac{1}{2^{10}} + \cdots = \frac{\dfrac{1}{2}}{1 - \dfrac{1}{8}} = \frac{4}{7}.$$

Similarly, the probability that Paul wins is

$$\frac{1}{2^2} + \frac{1}{2^5} + \frac{1}{2^8} + \frac{1}{2^{11}} + \cdots = \frac{\dfrac{1}{4}}{1 - \dfrac{1}{8}} = \frac{2}{7}$$

and the probability that Mary wins is

$$\frac{1}{2^3} + \frac{1}{2^6} + \frac{1}{2^9} + \frac{1}{2^{12}} + \cdots = \frac{\dfrac{1}{8}}{1 - \dfrac{1}{8}} = \frac{1}{7}.$$

Note that the three probabilities have sum 1, as they should.

C11S03.071: The amount of light transmitted is

$$\frac{I}{2^4} + \frac{I}{2^6} + \frac{I}{2^7} + \frac{I}{2^{10}} + \cdots = I \cdot \frac{\frac{1}{16}}{1 - \frac{1}{4}} = \frac{I}{12},$$

$\frac{1}{12}$ of the incident light.

Section 11.4

C11S04.001: Because $f^{(n)}(x) = (-1)^n e^{-x}$, we see that $f^{(n)}(0) = (-1)^n$ if $n \geqq 0$. Thus

$$P_5(x) = 1 - x + \frac{x^2}{2!} - \frac{x^3}{3!} + \frac{x^4}{4!} - \frac{x^5}{5!} \quad \text{and}$$

$$R_5(x) = \frac{x^6}{6!} e^{-z} \quad \text{for some } z \text{ between } 0 \text{ and } x.$$

C11S04.003: Given $f(x) = \cos x$ and $n = 4$, we have

$$f'(x) = -\sin x \qquad f'(0) = 0$$

$$f''(x) = -\cos x \qquad f''(0) = -1$$

$$f^{(3)}(x) = \sin x \qquad f^{(3)}(0) = 0$$

$$f^{(4)}(x) = \cos x \qquad f^{(4)}(0) = 1$$

$$f^{(5)}(x) = -\sin x$$

Therefore

$$P_4(x) = 1 - \frac{x^2}{2!} + \frac{x^4}{4!} \quad \text{and} \quad R_4(x) = -\frac{x^5}{5!} \sin z$$

for some number z between 0 and x.

C11S04.005: Given $f(x) = (1+x)^{1/2}$ and $n = 3$, we compute

$$f'(x) = \frac{1}{2(1+x)^{1/2}} \qquad f'(0) = \frac{1}{2}$$

$$f''(x) = -\frac{1}{4(1+x)^{3/2}} \qquad f''(0) = -\frac{1}{4}$$

$$f^{(3)}(x) = \frac{3}{8(1+x)^{5/2}} \qquad f^{(3)}(0) = \frac{3}{8}$$

$$f^{(4)}(x) = -\frac{15}{16(1+x)^{7/2}}$$

Therefore

$$P_3(x) = 1 + \frac{x}{2} - \frac{x^2}{8} + \frac{x^3}{16} \quad \text{and} \quad R_3(x) = -\frac{5x^4}{128(1+z)^{7/2}}$$

530

for some number z between 0 and x.

C11S04.007: Given $f(x) = \tan x$ and $n = 3$, we find that

$$f'(x) = \sec^2 x \qquad\qquad\qquad f'(0) = 1$$

$$f''(x) = 2\sec^2 x \tan x \qquad\qquad f''(0) = 0$$

$$f^{(3)}(x) = 2\sec^4 x + 4\sec^2 x \tan^2 x \qquad f^{(3)}(0) = 2$$

$$f^{(4)}(x) = 16\sec^4 x \tan x + 8\sec^2 x \tan^3 x$$

Therefore

$$P_3(x) = x + \frac{x^3}{3} \qquad \text{and} \qquad R_3(x) = \frac{x^4}{4!}(16\sec^4 z \tan z + 8\sec^2 z \tan^3 z)$$

for some number z between 0 and x.

C11S04.009: Given $f(x) = \arcsin x$ and $n = 2$, we compute

$$f'(x) = \frac{1}{\sqrt{1-x^2}} \qquad f'(0) = 1$$

$$f''(x) = \frac{x}{(1-x^2)^{3/2}} \qquad f''(0) = 0$$

$$f^{(3)}(x) = \frac{1+2x^2}{(1-x^2)^{5/2}}$$

Therefore

$$P_2(x) = x \qquad \text{and} \qquad R_2(x) = \frac{x^3(1+2z^2)}{3!(1-z^2)^{5/2}}$$

for some number z between 0 and x.

C11S04.011: Because $f^{(n)}(x) = e^x$ for all $n \geqq 0$, we have $f^{(n)}(1) = e$ for such n. Therefore

$$e^x = e + e(x-1) + \frac{e}{2}(x-1)^2 + \frac{e}{6}(x-1)^3 + \frac{e}{24}(x-1)^4 + \frac{e^z}{120}(x-1)^5$$

for some z between 1 and x.

C11S04.013: Given: $f(x) = \sin x$, $a = \pi/6$, and $n = 3$. We compute

$$f'(x) = \cos x \qquad f'(a) = \frac{\sqrt{3}}{2}$$

$$f''(x) = -\sin x \qquad f''(a) = -\frac{1}{2}$$

$$f^{(3)}(x) = -\cos x \qquad f^{(3)}(a) = -\frac{\sqrt{3}}{2}$$

$$f^{(4)}(x) = \sin x$$

Therefore

$$\sin x = \frac{1}{2} + \frac{\sqrt{3}}{2}\left(x - \frac{\pi}{6}\right) - \frac{1}{4}\left(x - \frac{\pi}{6}\right)^2 - \frac{\sqrt{3}}{12}\left(x - \frac{\pi}{6}\right)^3 + \frac{\sin z}{24}\left(x - \frac{\pi}{6}\right)^4$$

for some number z between $\pi/6$ and x.

C11S04.015: Given $f(x) = (x-4)^{-2}$, $a = 5$, and $n = 5$, we compute

$$f'(x) = -2(x-4)^{-3} \qquad f'(a) = -2$$

$$f''(x) = 6(x-4)^{-4} \qquad f''(a) = 6$$

$$f^{(3)}(x) = -24(x-4)^{-5} \qquad f^{(3)}(a) = -24$$

$$f^{(4)}(x) = 120(x-4)^{-6} \qquad f^{(4)}(a) = 120$$

$$f^{(5)}(x) = -720(x-4)^{-7} \qquad f^{(5)}(a) = -720$$

$$f^{(6)}(x) = 5040(x-4)^{-8}$$

Therefore

$$\frac{1}{(x-4)^2} = 1 - 2(x-5) + 3(x-5)^2 - 4(x-5)^3 + 5(x-4)^4 - 6(x-5)^5 + \frac{(x-5)^6}{720} \cdot \frac{5040}{(z-4)^8}$$

for some number z between 5 and x.

C11S04.017: Given $f(x) = \cos x$, $a = \pi$, and $n = 4$, we compute

$$f(x) = \cos x \qquad f(a) = -1$$

$$f'(x) = -\sin x \qquad f'(a) = 0$$

$$f''(x) = -\cos x \qquad f''(a) = 1$$

$$f^{(3)}(x) = \sin x \qquad f^{(3)}(a) = 0$$

$$f^{(4)}(x) = \cos x \qquad f^{(4)}(a) = -1$$

$$f^{(5)}(x) = -\sin x$$

Therefore

$$\cos x = -1 + \frac{(x-\pi)^2}{2} - \frac{(x-\pi)^4}{24} - \frac{\sin z}{120}(x-\pi)^5$$

for some number z between π and x.

C11S04.019: Given $f(x) = x^{3/2}$, $a = 1$, and $n = 4$, we compute

$$f(x) = x^{3/2} \qquad f(a) = 1$$

$$f'(x) = \frac{3}{2}x^{1/2} \qquad f'(a) = \frac{3}{2}$$

$$f''(x) = \frac{3}{4}x^{-1/2} \qquad f''(a) = \frac{3}{4}$$

$$f^{(3)}(x) = -\frac{3}{8}x^{-3/2} \qquad f^{(3)}(a) = -\frac{3}{8}$$

$$f^{(4)}(x) = \frac{9}{16}x^{-5/2} \qquad f^{(4)}(a) = \frac{9}{16}$$

$$f^{(5)}(x) = -\frac{45}{32}x^{-7/2}$$

Therefore

$$x^{3/2} = 1 + \frac{3}{2}(x-1) + \frac{3}{8}(x-1)^2 - \frac{1}{16}(x-1)^3 + \frac{3}{128}(x-1)^4 - \frac{(x-1)^5}{120} \cdot \frac{45}{32z^{7/2}}$$

for some number z between 1 and x.

C11S04.021: Substitution of $-x$ for x in the series in Eq. (19) yields

$$e^{-x} = 1 - x + \frac{x^2}{2!} - \frac{x^3}{3!} + \frac{x^4}{4!} - \frac{x^5}{5!} + \cdots = \sum_{n=0}^{\infty} \frac{(-1)^n x^n}{n!}.$$

This representation is valid for all x.

C11S04.023: Substitution of $-3x$ for x in the series in Eq. (19) yields

$$e^{-3x} = 1 - 3x + \frac{9x^2}{2!} - \frac{27x^3}{3!} + \frac{81x^4}{4!} - \frac{243x^5}{5!} + \cdots = \sum_{n=0}^{\infty} \frac{(-1)^n 3^n x^n}{n!}.$$

This representation is valid for all x.

C11S04.025: Substitution of $2x$ for x in the series in Eq. (22) yields

$$\sin 2x = 2x - \frac{8x^3}{3!} + \frac{32x^5}{5!} - \frac{128x^7}{7!} + \frac{512x^9}{9!} - \cdots = \sum_{n=0}^{\infty} \frac{(-1)^n (2x)^{2n+1}}{(2n+1)!}.$$

This representation is valid for all x.

C11S04.027: Substitution of x^2 for x in the series in Eq. (22) yields

$$\sin\left(x^2\right) = x^2 - \frac{x^6}{3!} + \frac{x^{10}}{5!} - \frac{x^{14}}{7!} + \frac{x^{18}}{9!} - \cdots = \sum_{n=0}^{\infty} \frac{(-1)^n x^{4n+2}}{(2n+1)!}.$$

This representation is valid for all x.

C11S04.029: Given $f(x) = \ln(1+x)$ and $a = 0$, we compute:

$$f(x) = \ln(1+x) \qquad f(a) = 0$$

$$f'(x) = \frac{1}{1+x} \qquad f'(a) = 1$$

$$f''(x) = -\frac{1}{(1+x)^2} \qquad f''(a) = -1$$

$$f^{(3)}(x) = \frac{2}{(1+x)^3} \qquad f^{(3)}(a) = 2$$

$$f^{(4)}(x) = -\frac{6}{(1+x)^4} \qquad f^{(4)}(a) = -6$$

$$f^{(5)}(x) = \frac{24}{(1+x)^5} \qquad f^{(5)}(a) = 24$$

$$f^{(6)}(x) = -\frac{120}{(1+x)^6} \qquad f^{(6)}(a) = -120$$

Evidently $f^{(n)}(a) = (-1)^{n+1}(n-1)!$ if $n \geq 1$. (For a proof, use proof by induction. We omit the proof to save space.) Therefore the Taylor series for $f(x)$ at $a = 0$ is

$$\sum_{n=1}^{\infty} \frac{(-1)^{n+1}(n-1)!x^n}{n!} = \sum_{n=1}^{\infty} \frac{(-1)^{n+1}x^n}{n} = x - \frac{x^2}{2} + \frac{x^3}{3} - \frac{x^4}{4} + \frac{x^5}{5} - \cdots.$$

This representation of $f(x) = \ln(1 + x)$ is valid if $-1 < x \leq 1$.

C11S04.031: If $f(x) = e^{-x}$, then $f^{(n)}(x) = (-1)^n e^{-x}$ for all $n \geq 0$. With $a = 0$, this implies that $f^{(n)}(a) = (-1)^n$ for all $n \geq 0$. Therefore the Taylor series for $f(x)$ at a is

$$\sum_{n=0}^{\infty} \frac{(-1)^n x^n}{n!} = 1 - x + \frac{x^2}{2!} - \frac{x^3}{3!} + \frac{x^4}{4!} - \frac{x^5}{5!} + \frac{x^6}{6!} - \cdots.$$

This representation of $f(x) = e^{-x}$ is valid for all x.

C11S04.033: Given $f(x) = \ln x$ and $a = 1$, we compute:

$$f(x) = \ln x \qquad f(a) = 0$$

$$f'(x) = \frac{1}{x} \qquad f'(a) = 1$$

$$f''(x) = -\frac{1}{x^2} \qquad f''(a) = -1$$

$$f^{(3)}(x) = \frac{2}{x^3} \qquad f^{(3)}(a) = 2$$

$$f^{(4)}(x) = -\frac{6}{x^4} \qquad f^{(4)}(a) = -6$$

$$f^{(5)}(x) = \frac{24}{x^5} \qquad f^{(5)}(a) = 24$$

$$f^{(6)}(x) = -\frac{120}{x^6} \qquad f^{(6)}(a) = -120$$

We have here convincing evidence that if $n \geq 1$, then $f^{(n)}(a) = (-1)^{n+1}(n-1)!$. (To prove this rigorously, use proof by induction; we omit any proof to save space.) Therefore the Taylor series for $f(x) = \ln x$ at $a = 1$ is

$$\sum_{n=1}^{\infty} \frac{(-1)^{n+1}(n-1)!(x-1)^n}{n!} = \sum_{n=1}^{\infty} \frac{(-1)^{n+1}(x-1)^n}{n}$$

$$= (x-1) - \frac{1}{2}(x-1)^2 + \frac{1}{3}(x-1)^3 - \frac{1}{4}(x-1)^4 + \frac{1}{5}(x-1)^5 - \frac{1}{6}(x-1)^6 + \cdots.$$

This representation of $f(x) = \ln x$ is valid if $0 < x \leqq 2$.

C11S04.035: Given $f(x) = \cos x$ and $a = \pi/4$, we compute:

$$f(x) = \cos x \qquad\qquad f(a) = \frac{\sqrt{2}}{2}$$

$$f'(x) = -\sin x \qquad\qquad f'(a) = -\frac{\sqrt{2}}{2}$$

$$f''(x) = -\cos x \qquad\qquad f''(a) = -\frac{\sqrt{2}}{2}$$

$$f^{(3)}(x) = \sin x \qquad\qquad f^{(3)}(a) = \frac{\sqrt{2}}{2}$$

$$f^{(4)}(x) = \cos x \qquad\qquad f^{(4)}(a) = \frac{\sqrt{2}}{2}$$

$$f^{(5)}(x) = -\sin x \qquad\qquad f^{(5)}(a) = -\frac{\sqrt{2}}{2}$$

$$f^{(6)}(x) = -\cos x \qquad\qquad f^{(6)}(a) = -\frac{\sqrt{2}}{2}$$

It should be clear that

$$f^{(n)}(a) = \frac{\sqrt{2}}{2} \qquad \text{if } n \text{ is of the form } 4k \text{ or } 4k+3, \text{ whereas}$$

$$f^{(n)}(a) = -\frac{\sqrt{2}}{2} \qquad \text{if } n \text{ is of the form } 4k+1 \text{ or } 4k+2.$$

Therefore the Taylor series for $f(x) = \cos x$ at $a = \pi/4$ is

$$\frac{\sqrt{2}}{2} - \frac{\sqrt{2}}{2}\left(x - \frac{\pi}{4}\right) - \frac{\sqrt{2}}{2!\cdot 2}\left(x - \frac{\pi}{4}\right)^2 + \frac{\sqrt{2}}{3!\cdot 2}\left(x - \frac{\pi}{4}\right)^3 + \frac{\sqrt{2}}{4!\cdot 2}\left(x - \frac{\pi}{4}\right)^4 - \cdots.$$

This representation of $f(x) = \cos x$ is valid for all x.

C11S04.037: Given $f(x) = \dfrac{1}{x}$ and $a = 1$, we compute:

$$f(x) = \frac{1}{x} \qquad\qquad f(a) = 1$$

$$f'(x) = -\frac{1}{x^2} \qquad\qquad f'(a) = -1$$

$$f''(x) = \frac{2}{x^3} \qquad\qquad f''(a) = 2$$

$$f^{(3)}(x) = -\frac{6}{x^4} \qquad\qquad f^{(3)}(a) = -6$$

$$f^{(4)}(x) = \frac{24}{x^5} \qquad\qquad f^{(4)}(a) = 24$$

$$f^{(5)}(x) = -\frac{120}{x^6} \qquad\qquad f^{(5)}(a) = -120$$

$$f^{(6)}(x) = \frac{720}{x^7} \qquad\qquad f^{(6)}(a) = 720$$

Clearly $f^{(n)}(a) = (-1)^n \cdot n!$ for $n \geqq 0$. Therefore the Taylor series for $f(x)$ at $a = 1$ is

$$\sum_{n=0}^{\infty} \frac{(-1)^n n! (x-1)^n}{n!} = \sum_{n=0}^{\infty} (-1)^n (x-1)^n$$

$$= 1 - (x-1) + (x-1)^2 - (x-1)^3 + (x-4)^4 - (x-1)^5 + (x-1)^6 - (x-1)^7 + \cdots.$$

This representation of $f(x)$ is valid for $0 < x < 2$.

C11S04.039: Given $f(x) = \sin x$ and $a = \pi/4$, we compute:

$$f(x) = \sin x \qquad\qquad f(a) = \frac{\sqrt{2}}{2}$$

$$f'(x) = \cos x \qquad\qquad f'(a) = \frac{\sqrt{2}}{2}$$

$$f''(x) = -\sin x \qquad\qquad f''(a) = -\frac{\sqrt{2}}{2}$$

$$f^{(3)}(x) = -\cos x \qquad\qquad f^{(3)}(a) = -\frac{\sqrt{2}}{2}$$

$$f^{(4)}(x) = \sin x \qquad\qquad f^{(4)}(a) = \frac{\sqrt{2}}{2}$$

$$f^{(5)}(x) = \cos x \qquad\qquad f^{(5)}(a) = \frac{\sqrt{2}}{2}$$

$$f^{(6)}(x) = -\sin x \qquad\qquad f^{(6)}(a) = -\frac{\sqrt{2}}{2}$$

Therefore the Taylor series for $f(x) = \sin x$ at $a = \pi/4$ is

$$\frac{\sqrt{2}}{2} + \frac{\sqrt{2}}{2}\left(x - \frac{\pi}{4}\right) - \frac{\sqrt{2}}{2! \cdot 2}\left(x - \frac{\pi}{4}\right)^2 - \frac{\sqrt{2}}{3! \cdot 2}\left(x - \frac{\pi}{4}\right)^3 + \frac{\sqrt{2}}{4! \cdot 2}\left(x - \frac{\pi}{4}\right)^4 + \frac{\sqrt{2}}{5! \cdot 2}\left(x - \frac{\pi}{4}\right)^5 - \cdots.$$

This representation of $f(x) = \sin x$ is valid for all x.

C11S04.041: Given $f(x) = \sin x$ and $a = 0$, we compute:

$$f(x) = \sin x \qquad\qquad f(a) = 0$$

$$f'(x) = \cos x \qquad\qquad f'(a) = 1$$

$$f''(x) = -\sin x \qquad\qquad f''(a) = 0$$

$$f^{(3)}(x) = -\cos x \qquad\qquad f^{(3)}(a) = -1$$

$$f^{(4)}(x) = \sin x \qquad\qquad f^{(4)}(a) = 0$$

$$f^{(5)}(x) = \cos x \qquad\qquad f^{(5)}(a) = 1$$

$$f^{(6)}(x) = -\sin x \qquad\qquad f^{(6)}(a) = 0$$

Therefore Taylor's formula for $f(x)$ at $a = 0$ is

$$f(x) = x - \frac{x^3}{3!} + \frac{x^5}{5!} - \cdots + (-1)^n \frac{x^{2n+1}}{(2n+1)!} + (-1)^{n+1} \frac{\sin z}{(2n+3)!} x^{2n+3} \qquad (1)$$

for some number z between 0 and x. Because $|\cos z| \leq 1$ for all z, it follows from Eq. (18) of the text that the remainder term in Eq. (1) approaches zero as $n \to \infty$. Therefore the Taylor series of $f(x) = \sin x$ at $a = 0$ is

$$\sin x = \sum_{n=0}^{\infty} \frac{(-1)^n x^{2n+1}}{(2n+1)!} = x - \frac{x^3}{3!} + \frac{x^5}{5!} - \frac{x^7}{7!} + \cdots,$$

and this representation is valid for all x.

C11S04.043: Given $f(x) = \cosh x$, $g(x) = \sinh x$, and $a = 0$, we compute:

$$f(x) = \cosh x \qquad\qquad f(a) = 1$$

$$f'(x) = \sinh x \qquad\qquad f'(a) = 0$$

$$f''(x) = \cosh x \qquad\qquad f''(a) = 1$$

$$f^{(3)}(x) = \sinh x \qquad\qquad f^{(3)}(a) = 0$$

$$f^{(4)}(x) = \cosh x \qquad\qquad f^{(4)}(a) = 1$$

$$f^{(5)}(x) = \sinh x \qquad\qquad f^{(5)}(a) = 0$$

$$f^{(6)}(x) = \cosh x \qquad\qquad f^{(6)}(a) = 1$$

Evidently $f^{(n)}(a) = 1$ if n is even and $f^{(n)}(a) = 0$ if n is odd. Therefore the Maclaurin series for $f(x) = \cosh x$ is

$$\sum_{n=0}^{\infty} \frac{x^{2n}}{(2n)!} = 1 + \frac{x^2}{2!} + \frac{x^4}{4!} + \frac{x^6}{6!} + \frac{x^8}{8!} + \cdots. \qquad (1)$$

The remainder term in Taylor's formula is

$$\frac{\sinh z}{(2n+1)!} x^{2n+1}$$

where z is between 0 and x. The remainder term approaches zero as $n \to +\infty$ by Eq. (18) of the text. Therefore the series in Eq. (1) converges to $f(x) = \cosh x$ for all x. Similarly,

$$g(x) = \sinh x \qquad\qquad g(a) = 0$$

$$g'(x) = \cosh x \qquad\qquad g'(a) = 1$$

$$g''(x) = \sinh x \qquad\qquad g''(a) = 0$$

$$g^{(3)}(x) = \cosh x \qquad\qquad g^{(3)}(a) = 1$$

$$g^{(4)}(x) = \sinh x \qquad\qquad g^{(4)}(a) = 0$$

$$g^{(5)}(x) = \cosh x \qquad\qquad g^{(5)}(a) = 1$$

$$g^{(6)}(x) = \sinh x \qquad\qquad g^{(6)}(a) = 0$$

It is clear that $g^{(n)}(a) = 0$ if n is even, whereas $g^{(n)} = 1$ if n is odd. Therefore the Maclaurin series for $g(x) = \sinh x$ is

$$\sum_{n=0}^{\infty} \frac{x^{2n+1}}{(2n+1)!} = x + \frac{x^3}{3!} + \frac{x^5}{5!} + \frac{x^7}{7!} + \frac{x^9}{9!} + \cdots . \tag{2}$$

This series converges to $g(x) = \sinh x$ for all x by an argument very similar to that given for the hyperbolic cosine series.

Next, substitution of ix for x yields

$$\cosh ix = 1 + \frac{(ix)^2}{2!} + \frac{(ix)^4}{4!} + \frac{(ix)^6}{6!} + \cdots = 1 - \frac{x^2}{2!} + \frac{x^4}{4!} - \frac{x^6}{6!} + \cdots = \cos x.$$

Similarly, $\sinh ix = \sin x$. This is one way to describe the relationship of the hyperbolic functions to the circular functions. A more prosaic response to the concluding question in Problem 43 would be that if the signs in the Maclaurin series for the cosine function are changed so that they are all plus signs, you get the Maclaurin series for the hyperbolic cosine function; the same relation hold for the sine and hyperbolic sine series.

C11S04.045: Given $f(x) = e^{-x}$, its plot together with that of

$$P_3(x) = 1 - x + \frac{x^2}{2!} - \frac{x^3}{3!}$$

are shown next.

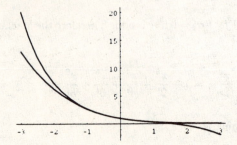

The graphs of $f(x) = e^{-x}$ and

$$P_6(x) = 1 - x + \frac{x^2}{2!} - \frac{x^3}{3!} + \frac{x^4}{4!} - \frac{x^5}{5!} + \frac{x^6}{6!}$$

are shown together next.

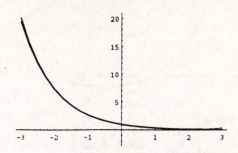

C11S04.047: Given $f(x) = \cos x$, two of its Taylor polynomials are

$$P_4(x) = 1 - \frac{x^2}{2!} + \frac{x^4}{4!} \quad \text{and} \quad P_8(x) = 1 - \frac{x^2}{2!} + \frac{x^4}{4!} - \frac{x^6}{6!} + \frac{x^8}{8!}.$$

The graphs of f and P_4 are shown next, on the left; the graph of f and P_8 are on the right.

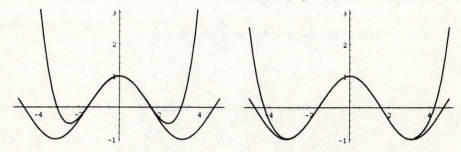

C11S04.049: Given $f(x) = \dfrac{1}{1 + x}$, two of its Taylor polynomials are

$$P_3(x) = 1 - x + x^2 - x^3 \quad \text{and} \quad P_4(x) = 1 - x + x^2 - x^3 + x^4.$$

The graphs of f and P_3 are shown together next, on the left; the graphs of f and P_4 are on the right.

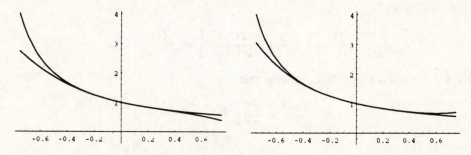

C11S04.051: The graph of the Taylor polynomial

$$P_4(x) = 1 - \frac{x}{2!} + \frac{x^2}{4!} - \frac{x^3}{6!} + \frac{x^4}{8!}$$

of $f(x)$ and the graph of $g(x)$ are shown together, next.

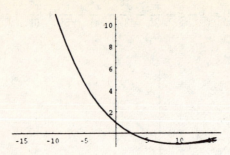

C11S04.053: We begin with the formula

$$\tan(A+B) = \frac{\tan A + \tan B}{1 - \tan A \tan B}.$$

Let $A = \arctan x$ and $B = \arctan y$. Thus

$$\tan(\arctan x + \arctan y) = \frac{x+y}{1-xy}, \quad \text{so that} \quad \arctan x + \arctan y = \arctan \frac{x+y}{1-xy}$$

(if $xy < 1$). Thus

$$\arctan \frac{1}{2} + \arctan \frac{1}{5} = \arctan \frac{\frac{7}{10}}{\frac{9}{10}} = \arctan \frac{7}{9}.$$

Therefore

$$\arctan \frac{1}{2} + \arctan \frac{1}{5} + \arctan \frac{1}{8} = \arctan \frac{7}{9} + \arctan \frac{1}{8} = \arctan \frac{\frac{65}{72}}{\frac{65}{72}} = \arctan 1 = \frac{\pi}{4}.$$

C11S04.055: Prove that

$$\lim_{n \to \infty} \frac{x^n}{n!} = 0$$

for every real number x.

Proof: Suppose that x is a real number. Choose the integer k so large that $k > |2x|$. Let $L = |x|^k/k!$. Suppose that $n = k+1$. Then

$$\frac{|x|^n}{n!} = \frac{|x|^{k+1}}{(k+1)!} = \frac{|x|^k}{k!} \cdot \frac{|x|}{k+1} < \frac{L}{2} = \frac{L}{2^{n-k}}$$

because $|2x| < k < k+1$ and $n - k = 1$. Next, assume that

$$\frac{|x|^m}{m!} < \frac{L}{2^{m-k}}$$

for some integer $m > k$. Then

$$\frac{|x|^{m+1}}{(m+1)!} = \frac{|x|^m}{m!} \cdot \frac{|x|}{m+1} < \frac{L}{2^{m-k}} \cdot \frac{1}{2} = \frac{L}{2^{m+1-k}}$$

because $|2x| < k < m$. Therefore, by induction,

$$\frac{|x|^n}{n!} < \frac{L}{2^{n-k}}$$

for every integer $n > k$. Now let $n \to +\infty$ to conclude that

$$\lim_{n \to \infty} \frac{x^n}{n!} = 0.$$

C11S04.057: By Theorem 4 of Section 11.3, S is not a number. Thus attempts to do "arithmetic" with S are meaningless and may lead to all sorts of absurd results.

C11S04.059: Results: With $x = 1$ in the Maclaurin series in Problem 56, we find that

$$a = \sum_{n=1}^{50} \frac{(-1)^{n+1}}{n} \approx 0.68324716057591818842565811649.$$

With $x = \frac{1}{3}$ in the second series in Problem 58, we find that

$$b = \sum_{\substack{n=1 \\ n \text{ odd}}}^{49} \frac{2}{n \cdot 3^n} \approx 0.69314718055994530941723210107.$$

Because $|a - \ln 2| \approx 0.0099000019984$, whereas $|b - \ln 2| \approx 2.039 \times 10^{-26}$, it is clear that the second series of Problem 58 is far superior to the series of Problem 56 for the accurate approximation of $\ln 2$.

Section 11.5

C11S05.001: $\displaystyle\int_0^\infty \frac{x}{x^2+1}\,dx = \left[\frac{1}{2}\ln(x^2+1)\right]_0^\infty = +\infty.$ Therefore $\displaystyle\sum_{n=1}^\infty \frac{n}{n^2+1}$ diverges.

C11S05.003: $\displaystyle\int_0^\infty (x+1)^{-1/2}\,dx = \left[2(x+1)^{1/2}\right]_0^\infty = +\infty.$ Therefore $\displaystyle\sum_{n=1}^\infty \frac{1}{\sqrt{n+1}}$ diverges.

C11S05.005: $\displaystyle\int_0^\infty \frac{1}{x^2+1}\,dx = \left[\arctan x\right]_0^\infty = \frac{\pi}{2} < +\infty.$ Therefore $\displaystyle\sum_{n=1}^\infty \frac{1}{n^2+1}$ converges.

This is a special case of series 6.1.32 in Eldon R. Hansen, *A Table of Series and Products*, Prentice-Hall Inc. (Englewood Cliffs, N.J.), 1975. According to Hansen, its sum is

$$-\frac{1}{2} + \frac{\pi}{2}\coth \pi \approx 1.0766740474685811741340507947500.$$

Mathematica 3.0 reports that the sum of the first 1,000,000 terms of this series is approximately 1.07667.

C11S05.007: $\displaystyle\int_2^\infty \frac{1}{x \ln x}\,dx = \Big[\ln(\ln x)\Big]_2^\infty = +\infty.$ Therefore $\displaystyle\sum_{n=2}^\infty \frac{1}{n \ln n}$ diverges.

C11S05.009: $\displaystyle\int_0^\infty 2^{-x}\,dx = \left[-\frac{1}{2^x \ln 2}\right]_0^\infty = \frac{1}{\ln 2} < +\infty.$ Therefore $\displaystyle\sum_{n=1}^\infty \frac{1}{2^n}$ converges (to 1).

C11S05.011: For each positive integer n, let

541

$$I_n = \int x^n e^{-x} \, dx.$$

Let $u = x^n$ and $dv = e^{-x} \, dx$. Then $du = nx^{n-1} \, dx$; choose $v = -e^{-x}$. Then

$$I_n = -x^n e^{-x} + n \int x^{n-1} e^{-x} \, dx = -x^n e^{-x} + n I_{n-1}.$$

Therefore

$$\int x^2 e^{-x} \, dx = I_2 = -x^2 e^{-x} + 2I_1 = -x^2 e^{-x} + 2 \left(-xe^{-x} + \int e^{-x} \, dx \right) = -(x^2 + 2x + 2)e^{-x} + C.$$

Hence

$$\int_0^\infty x^2 e^{-x} \, dx = -\left[(x^2 + 2x + 2)e^{-x} \right]_0^\infty = 2 < +\infty,$$

and so $\displaystyle\sum_{n=1}^\infty \frac{n^2}{e^n}$ converges. It can be shown that its sum is

$$\frac{e(e+1)}{(e-1)^3} \approx 1.992294767125.$$

C11S05.013: Choose $u = \ln x$ and $dv = \dfrac{1}{x^2} \, dx$. Then $du = \dfrac{1}{x} \, dx$; choose $v = -\dfrac{1}{x}$. Then

$$\int \frac{\ln x}{x^2} \, dx = -\frac{\ln x}{x} + \int \frac{1}{x^2} \, dx = -\frac{\ln x}{x} - \frac{1}{x} + C.$$

Thus

$$\int_1^\infty \frac{\ln x}{x^2} \, dx = \left[-\frac{1 + \ln x}{x} \right]_1^\infty = \frac{1+0}{1} - \lim_{x \to \infty} \frac{1 + \ln x}{x} = 1 - \lim_{x \to \infty} \frac{1}{x} = 1 - 0 = 1 < +\infty$$

(we used l'Hôpital's rule to find the limit). Therefore $\displaystyle\sum_{n=1}^\infty \frac{\ln n}{n^2}$ converges.

C11S05.015: Because

$$\int_0^\infty \frac{x}{x^4 + 1} \, dx = \left[\frac{1}{2} \arctan \left(x^2 \right) \right]_0^\infty = \frac{\pi}{4} < +\infty,$$

the series $\displaystyle\sum_{n=1}^\infty \frac{n}{n^4 + 1}$ converges.

With the aid of *Mathematica 3.0* and Theorem 2, we find that the sum of this series is approximately 0.694173022150715 (only the last digit shown here is in doubt; it may round to 6 instead of 5).

C11S05.017: Because

$$\int_1^\infty \frac{2x + 5}{x^2 + 5x + 17} \, dx = \left[\ln(x^2 + 5x + 17) \right]_1^\infty = +\infty,$$

542

the series $\displaystyle\sum_{n=1}^{\infty} \frac{2n+5}{n^2+5n+17}$ diverges.

C11S05.019: Choose $u = \ln(1 + x^{-2})$ and $dv = dx$. Then

$$du = \frac{-2x^{-3}}{1+x^{-2}}\,dx = -\frac{2}{x^3+x}\,dx;$$

choose $v = x$. Thus

$$\int_1^{\infty} \ln\left(1 + \frac{1}{x^2}\right)\,dx = \left[x\ln(1+x^{-2})\right]_1^{\infty} + 2\int_1^{\infty} \frac{1}{1+x^2}\,dx$$

$$= \lim_{x\to\infty} \frac{\ln(1+x^{-2})}{1/x} - \ln 2 + \left[2\arctan x\right]_1^{\infty}$$

$$= \left[\lim_{z\to 0^+} \frac{\ln(1+z^2)}{z}\right]_1^{\infty} - \ln 2 + \pi - \frac{\pi}{2} = \frac{\pi}{2} - \ln 2 < +\infty$$

(use l'Hôpital's rule to evaluate the last limit). Therefore $\displaystyle\sum_{n=1}^{\infty} \ln\left(1 + \frac{1}{n^2}\right)$ converges.

C11S05.21: Because

$$\int_1^{\infty} \frac{x}{4x^2+5}\,dx = \left[\frac{1}{8}\ln(4x^2+5)\right]_1^{\infty} = +\infty,$$

the series $\displaystyle\sum_{n=1}^{\infty} \frac{n}{4n^2+5}$ diverges.

C11S05.023: Because

$$\int_2^{\infty} \frac{1}{x\sqrt{\ln x}}\,dx = \int_2^{\infty} \frac{(\ln x)^{-1/2}}{x}\,dx = \left[2(\ln x)^{1/2}\right]_2^{\infty} = +\infty,$$

the series $\displaystyle\sum_{n=2}^{\infty} \frac{1}{n\sqrt{\ln n}}$ diverges.

C11S05.025: The substitution $u = 2x$ and integral formula 17 of the endpapers of the text yields

$$\int \frac{1}{4x^2+9}\,dx = \frac{1}{2}\int \frac{1}{u^2+9}\,du = \frac{1}{2}\cdot\frac{1}{3}\arctan\left(\frac{u}{3}\right) + C = \frac{1}{6}\arctan\left(\frac{2x}{3}\right) + C.$$

Therefore

$$\int_1^{\infty} \frac{1}{4x^2+9}\,dx = \left[\frac{1}{6}\arctan\left(\frac{2x}{3}\right)\right]_1^{\infty} = \frac{1}{6}\cdot\frac{\pi}{2} - \frac{1}{6}\arctan\left(\frac{2}{3}\right) < +\infty,$$

and thus $\displaystyle\sum_{n=1}^{\infty} \frac{1}{4n^2+9}$ converges.

This series is a special case of Eq. (6.1.32) of Eldon R. Hansen's *A Table of Series and Products*, Prentice-Hall, Inc. (Englewood Cliffs, N.J.), 1975. *Mathematica* 3.0 summed this series in a fraction of a second and obtained the same answer as Hansen, viz.,

$$-\frac{1}{18} + \frac{\pi}{12} \coth\left(\frac{3\pi}{2}\right) \approx 0.20628608982235128529.$$

C11S05.027: Because

$$\int_1^\infty \frac{x}{(x^2+1)^2}\, dx = \left[-\frac{1}{2(x^2+1)}\right]_1^\infty = \frac{1}{4} < +\infty,$$

the series $\displaystyle\sum_{n=1}^\infty \frac{n}{n^4 + 2n^2 + 1}$ converges.

The *Mathematica* command

```
NSum[ n/(n^4 + 2*n^2 + 1), { n, 1, Infinity }, WorkingPrecision -> 28 ] // Timing
```

when executed on a Power Macintosh 7600/120 yielded the approximate sum 0.39711677137965943 in 3.45 seconds.

C11S05.029: Because

$$\int_1^\infty \frac{\arctan x}{x^2+1}\, dx = \left[\frac{1}{2}\left(\arctan x\right)^2\right]_1^\infty = \frac{\pi^2}{8} - \frac{\pi^2}{32} = \frac{3}{32}\pi^2 < +\infty,$$

the series $\displaystyle\sum_{n=1}^\infty \frac{\arctan n}{n^2+1}$ converges.

C11S05.031: The integral test cannot be applied because this is not a positive-term series. In Section 11.7 you will see how to prove that it converges, and in Problem 61 there you will see that its sum is $-\ln 2$.

C11S05.033: The terms of this series are not monotonically decreasing. For specific examples, if we let $a_n = (2 + \sin n)/(n^2)$, then

$$a_5 \approx 0.0416430290 < 0.0477940139 \approx a_6, \qquad a_{11} \approx 0.0082645437 < 0.0101626881 \approx a_{12},$$

$$a_{18} \approx 0.0038549776 < 0.0059553385 \approx a_{19}, \qquad a_{24} \approx 0.0019000376 < 0.0029882372 \approx a_{25},$$

$$a_{30} \approx 0.0011244093 < 0.0016607309 \approx a_{31}, \qquad a_{37} \approx 0.0009908414 < 0.0015902829 \approx a_{38}.$$

In Section 11.6 you will see how to use the *comparison test* to prove that this series converges.

C11S05.035: There is some implication that we are to use the integral test to solve this problem. Hence we consider only the case in which $p > 0$. And if $p \neq 1$, then

$$\int_1^\infty p^{-x}\, dx = \left[-\frac{p^{-x}}{\ln p}\right]_{x=1}^\infty = \frac{1}{p\ln p} - \lim_{x\to\infty} \frac{1}{p^x \ln p}. \tag{1}$$

If $0 < p < 1$, then the limit in (1) is $+\infty$. If $p > 1$ then the limit in (1) is 0. If $p = 1$ then the series diverges because

$$\int_1^\infty \frac{1}{p^x}\, dx = \int_1^\infty 1\, dx = +\infty.$$

Answer: The series diverges if $0 < p \leqq 1$ and converges if $p > 1$.

C11S05.037: If $p = 1$ then

$$\int_2^\infty \frac{1}{x(\ln x)^p}\, dx = \int_2^\infty \frac{1}{x \ln x}\, dx = \left[\ln(\ln x)\right]_2^\infty = +\infty.$$

Otherwise,

$$\int_2^\infty \frac{1}{x(\ln x)^p}\, dx = \int_2^\infty \frac{(\ln x)^{-p}}{x}\, dx = \left[\frac{(\ln x)^{1-p}}{1-p}\right]_2^\infty = \left[-\frac{1}{(p-1)(\ln x)^{p-1}}\right]_2^\infty.$$

So this improper integral diverges if $p < 1$. If $p > 1$ then it converges to

$$\frac{1}{(p-1)(\ln 2)^{p-1}} < +\infty.$$

Therefore the series $\displaystyle\sum_{n=2}^\infty \frac{1}{n(\ln n)^p}$ diverges if $p \leqq 1$ and converges if $p > 1$.

C11S05.039: We require $R_n < 0.0001$. This will hold provided that

$$\int_n^\infty \frac{1}{x^2}\, dx < 0.0001$$

because R_n cannot exceed the integral. So we require

$$\left[-\frac{1}{x}\right]_n^\infty < 0.0001;$$

that is, that $n > 10000$.

C11S05.041: We require $R_n < 0.00005$. This will hold provided that

$$\int_n^\infty \frac{1}{x^3}\, dx < 0.00005$$

because R_n cannot exceed the integral. So we require

$$\left[-\frac{1}{2x^2}\right]_n^\infty < 0.00005;$$

$$\frac{1}{2n^2} < 0.00005;$$

$$2n^2 > 20000;$$

thus we require that $n > 100$.

C11S05.043: We require $R_n < 0.005$. This will hold provided that

$$\int_n^\infty \frac{1}{x^{3/2}}\, dx < 0.005;$$

that is, provided that

$$\left[-\frac{2}{x^{1/2}}\right]_n^\infty < 0.005,$$

545

so that $n^{1/2} > 400$, and thus $n > N = 160000$. *Mathematica 3.0* reports that

$$S_N = \sum_{n=1}^{N} \frac{1}{n^{3/2}} \approx 2.607375356498 \quad \text{and that} \quad S = \sum_{n=1}^{\infty} \frac{1}{n^{3/2}} \approx 2.612375348685.$$

Note that $S - S_N \approx 0.004999992187 < 0.005$.

C11S05.045: We require $R_n < 0.000005$. This will hold provided that

$$\int_n^{\infty} \frac{1}{x^5} \, dx = \frac{1}{4n^4} < 0.000005,$$

so that $n > 14.9535$. Choose $N = 15$. Then *Mathematica 3.0* reports that

$$S_N = \sum_{n=1}^{N} \frac{1}{n^5} \approx 1.036923438841 \quad \text{and that} \quad S = \sum_{n=1}^{\infty} \frac{1}{n^5} \approx 1.036927755143.$$

Note that $S - S_N \approx 0.000004316302 < 0.000005$.

C11S05.047: If $p = 1$, then

$$\int_1^{\infty} \frac{\ln x}{x^p} \, dx = \int_1^{\infty} \frac{\ln x}{x} \, dx = \left[\frac{1}{2} (\ln x)^2 \right]_1^{\infty} = +\infty,$$

so in this case the given series $\displaystyle\sum_{n=1}^{\infty} \frac{\ln n}{n^p}$ diverges. Otherwise (with the aid of *Mathematica 3.0* for the antiderivative)

$$\int_1^{\infty} \frac{\ln x}{x^p} \, dx = \left[\frac{x^{1-p} \ln x - x^{1-p}}{(1-p)^2} \right]_1^{\infty} = \left[\frac{-1 + \ln x}{(p-1)^2 x^{p-1}} \right]_1^{\infty}.$$

Thus if $p < 1$ the given series diverges, whereas if $p > 1$ it converges. Answer: $p > 1$.

C11S05.049: From the proof of Theorem 1 (the integral test), we see that if

$$a_n = \frac{1}{n}, \quad f(x) = \frac{1}{x}, \quad \text{and} \quad S_n = \sum_{k=1}^{n} a_n$$

for each integer $n \geqq 1$, then

$$S_n \geqq \int_1^{n+1} \frac{1}{x} \, dx = \left[\ln x \right]_1^{n+1} = \ln(n+1)$$

and

$$S_n - a_1 \leqq \int_1^{n} \frac{1}{x} \, dx = \left[\ln x \right]_1^{n} = \ln n.$$

Therefore

$$\ln n < \ln(n+1) \leqq S_n \leqq 1 + \ln n;$$

put another way,

$$\ln n \leqq 1 + \frac{1}{2} + \frac{1}{3} + \frac{1}{4} + \cdots + \frac{1}{n} \leqq 1 + \ln n$$

for every integer $n \geqq 1$. So if a computer adds a million terms of the harmonic series every second, the partial sum S_n will first reach 50 when $n \leqq e^{50} \leqq e \cdot n$. This means that n must satisfy the inequalities

$$1 + \left[\!\left[e^{49} \right]\!\right] \leqq n \leqq \left[\!\left[e^{50} \right]\!\right];$$

that is,

$$19073465724950099690526 \leqq n \leqq 51847055285870724640 87.$$

Divide the smaller of these bounds by one million (additions the computer carries out each second), then by 3600 to convert to hours, by 24 and then by 365.242199 to convert to years, and finally by 100 to convert to an answer: It will require over 604414 centuries. For a more precise answer, if $N = 2911002088526872100231$, then *Mathematica* 3.0 reports that

$$\sum_{n=1}^{N-1} \frac{1}{n} \approx 49.9999999999999999999999713 \qquad \text{and}$$

$$\sum_{n=1}^{N} \frac{1}{n} \approx 50.00000000000000000000000057.$$

After converting to centuries as before, we finally get the "right" answer: It will require a little over 922460 centuries.

C11S05.051: Suppose that f is continuous and $f(x) > 0$ for all $x \geqq 1$. For each positive integer n, let

$$b_n = \int_1^n f(x)\, dx.$$

Part (a): Note that the sequence $\{b_n\}$ is increasing. Suppose that it is bounded, so that

$$B = \lim_{n \to \infty} b_n$$

exists. The definition of the value of an improper integral then implies that

$$\int_1^\infty f(x)\, dx = \lim_{a \to \infty} \int_1^a f(x)\, dx. \qquad (1)$$

Therefore, by Theorem 4 in Section 11.2,

$$\int_1^\infty f(x)\, dx = \lim_{n \to \infty} \int_1^n f(x)\, dx = \lim_{n \to \infty} b_n = B.$$

Part (b): If the increasing sequence $\{b_n\}$ is not bounded, then by Problem 52 of Section 11.2,

$$\lim_{n \to \infty} b_n = +\infty.$$

Then Eq. (1) implies that

547

$$\int_1^\infty f(x)\, dx = +\infty$$

because $\int_1^\alpha f(x)\, dx$ is an increasing function of α.

Section 11.6

C11S06.001: The series

$$\sum_{n-1}^\infty \frac{1}{n^2+n+1} \quad \text{is dominated by} \quad \sum_{n=1}^\infty \frac{1}{n^2},$$

which converges because it is a p-series with $p = 2 > 1$. Therefore the dominated series also converges.

C11S06.003: The series

$$\sum_{n=1}^\infty \frac{1}{n+n^{1/2}}$$

diverges by limit-comparison with the harmonic series, demonstrated by the computation

$$\frac{\dfrac{1}{n+n^{1/2}}}{\dfrac{1}{n}} = \frac{n}{n+n^{1/2}} = \frac{1}{1+\dfrac{1}{n^{1/2}}} \longrightarrow \frac{1}{1+0} = 1$$

as $n \to +\infty$.

C11S06.005: The series

$$\sum_{n=1}^\infty \frac{1}{1+3^n} \quad \text{is dominated by} \quad \sum_{n=1}^\infty \frac{1}{3^n},$$

and the latter converges because it is a geometric series with ratio $\frac{1}{3} < 1$. Therefore the dominated series also converges.

C11S06.007: The series

$$\sum_{n=2}^\infty \frac{10n^2}{n^3-1}$$

diverges by limit-comparison with the harmonic series, demonstrated by the computation

$$\frac{\dfrac{10n^2}{n^3-1}}{\dfrac{1}{n}} = \frac{10n^3}{n^3-1} = \frac{10}{1-\dfrac{1}{n^3}} \longrightarrow \frac{10}{1-0} = 10$$

as $n \to +\infty$.

C11S06.009: First note that if $n \geqq 1$, then $\dfrac{1}{\sqrt{37n^3+3}} \leqq \dfrac{1}{\sqrt{n^3}} = \dfrac{1}{n^{3/2}}$. Therefore the series

$$\sum_{n=1}^{\infty} \frac{1}{\sqrt{37n^3 + 3}} \quad \text{is dominated by} \quad \sum_{n=1}^{\infty} \frac{1}{n^{3/2}},$$

and the latter series converges because it is a p-series with $p = \frac{3}{2} > 1$. Therefore the dominated series also converges.

C11S06.011: Because $\dfrac{\sqrt{n}}{n^2 + n} \leqq \dfrac{n^{1/2}}{n^2} = \dfrac{1}{n^{3/2}}$, the series

$$\sum_{n=1}^{\infty} \frac{\sqrt{n}}{n^2 + n} \quad \text{is dominated by} \quad \sum_{n=1}^{\infty} \frac{1}{n^{3/2}},$$

which converges because it is a p-series with $p = \frac{3}{2} > 1$. Therefore the dominated series also converges.

C11S06.013: First we need a lemma: $\ln x < x$ if $x > 0$.

Proof: Let $f(x) = x - \ln x$. Then

$$f'(x) = 1 - \frac{1}{x}.$$

Because $f'(x) < 0$ if $0 < x < 1$, $f'(1) = 0$, and $f'(x) > 0$ if $1 < x$, the graph of $y = f(x)$ has a global minimum value at $x = 1$. Its minimum is $f(1) = 1 - \ln 1 = 1 > 0$, and $f(x) \geqq f(1)$ if $x > 0$. Therefore $f(x) > 0$ for all $x > 0$; that is, $\ln x < x$ if $x > 0$. ◀

Therefore the series

$$\sum_{n=2}^{\infty} \frac{1}{\ln n} \quad \text{dominates} \quad \sum_{n=2}^{\infty} \frac{1}{n}.$$

The latter series diverges because it is "eventually the same" as the harmonic series, and therefore the dominating series also diverges.

C11S06.015: Because $0 \leqq \sin^2 n \leqq 1$ for every integer $n \geqq 1$, the series

$$\sum_{n=1}^{\infty} \frac{\sin^2 n}{n^2 + 1} \quad \text{is dominated by} \quad \sum_{n=1}^{\infty} \frac{1}{n^2}.$$

The latter series converges because it is a p-series with $p = 2 > 1$. Therefore the dominated series also converges.

C11S06.017: First we need a lemma: If n is a positive integer, then $n < 2^n$.

Proof: The lemma is true for $n = 1$ because $1 < 2$, so that $1 < 2^1$. Suppose that $k < 2^k$ for some integer $k \geqq 1$. Then

$$2^{k+1} = 2 \cdot 2^k \geqq 2 \cdot k = k + k \geqq k + 1.$$

Thus whenever the lemma holds for the integer $k \geqq 1$, it also hold for $k + 1$. Therefore, by induction, $n < 2^n$ for every integer $n \geqq 1$. ◀

Next, note that as a consequence of the lemma,

$$\frac{n+2^n}{n+3^n} \leq \frac{2^n + 2^n}{3^n} = \frac{2 \cdot 2^n}{3^n} = 2 \cdot \left(\frac{2}{3}\right)^n.$$

Therefore the series

$$\sum_{n=1}^{\infty} \frac{n+2^n}{n+3^n} \quad \text{is dominated by} \quad \sum_{n=1}^{\infty} 2 \cdot \left(\frac{2}{3}\right)^n.$$

The latter series converges because it is geometric with ratio $\frac{2}{3} < 1$. Therefore the dominated series also converges by the comparison test.

C11S06.019: Because $\dfrac{1}{n^2 \ln n} \leq \dfrac{1}{n^2}$ if $n \geq 3$, the given series

$$\sum_{n=2}^{\infty} \frac{1}{n^2 \ln n} \quad \text{is eventually dominated by} \quad \sum_{n=2}^{\infty} \frac{1}{n^2}.$$

The latter series converges because it is eventually the same as the p-series with $p = 2 > 1$. Therefore the dominated series converges by the comparison test. See the discussion of "eventual domination" following the proof of Theorem 1 (the comparison test) in Section 11.6 of the text.

C11S06.021: First, a lemma: There is a positive integer K such that $\ln n \leq \sqrt{n}$ if n is a positive integer and $n \geq K$.

Proof: We use l'Hôpital's rule:

$$\lim_{n \to \infty} \frac{\ln n}{\sqrt{n}} = \lim_{x \to \infty} \frac{\ln x}{x^{1/2}} = \lim_{x \to \infty} \frac{2x^{1/2}}{x} = \lim_{x \to \infty} \frac{2}{x^{1/2}} = 0.$$

Therefore there exists a positive integer K such that $\ln n \leq \sqrt{n}$ if $n \geq K$. ◀

Alternatively, let $f(x) = x^{1/2} - \ln x$. Apply methods of calculus to show that $f'(x) < 0$ if $0 < x < 4$, $f'(4) = 0$, and $f'(x) > 0$ if $x > 4$. It follows that $f(x) \geq f(4) = 2 - \ln 2 > 0$ for all $x > 0$, and hence $x^{1/2} > \ln x$ for all $x > 0$. Thus the integer K of the preceding proof may be chosen to be 1. But relying only on the lemma, we now conclude that

$$\sum_{n=1}^{\infty} \frac{\ln n}{n^2} \quad \text{is eventually dominated by} \quad \sum_{n=1}^{\infty} \frac{n^{1/2}}{n^2} = \sum_{n=1}^{\infty} \frac{1}{n^{3/2}}.$$

The last series converges because it is the p-series with $p = \frac{3}{2} > 1$. Therefore the dominated series converges by the comparison test.

C11S06.023: Because $0 \leq \sin^2(1/n) \leq 1$ for every positive integer n, the given series

$$\sum_{n=1}^{\infty} \frac{\sin^2(1/n)}{n^2} \quad \text{is dominated by} \quad \sum_{n=1}^{\infty} \frac{1}{n^2}.$$

The latter series converges because it is the p-series with $p = 2 > 1$. Therefore the dominated series also converges by the comparison test.

C11S06.025: We showed in the solution of Problem 13 that $\ln n \leq n$ for every positive integer n. We showed in the solution of Problem 17 that $n \leq 2^n$ for every positive integer n. Therefore

550

$$\sum_{n=1}^{\infty} \frac{\ln n}{e^n} \quad \text{is dominated by} \quad \sum_{n=1}^{\infty} \frac{2^n}{e^n} = \sum_{n=1}^{\infty} \left(\frac{2}{e}\right)^n.$$

The last series converges because it is geometric with ratio $2/e < 1$. Therefore the dominated series converges by the comparison test.

C11S06.027: The given series

$$\sum_{n=1}^{\infty} \frac{n^{3/2}}{n^2 + 4} \quad \text{diverges by limit-comparison with} \quad \sum_{n=1}^{\infty} \frac{1}{n^{1/2}},$$

shown by the computation

$$\lim_{n \to \infty} \frac{\dfrac{n^{3/2}}{n^2 + 4}}{\dfrac{1}{n^{1/2}}} = \lim_{n \to \infty} \frac{n^2}{n^2 + 4} = \lim_{n \to \infty} \frac{1}{1 + \dfrac{4}{n^2}} = \frac{1}{1 + 0} = 1;$$

note that $\displaystyle\sum_{n=1}^{\infty} \frac{1}{n^{1/2}}$ diverges because it is a p-series with $p = \frac{1}{2} \leqq 1$.

C11S06.029: First note that the series

$$\sum_{n=1}^{\infty} \frac{1}{n^{1/2}} \tag{1}$$

diverges because it is a p-series with $p = \frac{1}{2} \leqq 1$. Therefore the given series

$$\sum_{n=1}^{\infty} \frac{3}{4 + \sqrt{n}}$$

diverges by limit-comparison with the series in (1), as shown by the computation

$$\lim_{n \to \infty} \frac{\dfrac{3}{4 + n^{1/2}}}{\dfrac{1}{n^{1/2}}} = \lim_{n \to \infty} \frac{3n^{1/2}}{4 + n^{1/2}} = \lim_{n \to \infty} \frac{3}{\dfrac{4}{n^{1/2}} + 1} = \frac{3}{0 + 1} = 3.$$

C11S06.031: First note that

$$\frac{2n^2 - 1}{n^2 \cdot 3^n} \leqq \frac{2n^2}{n^2 \cdot 3^n} = \frac{2}{3^n}$$

for each positive integer n. Therefore the given series

$$\sum_{n=1}^{\infty} \frac{2n^2 - 1}{n^2 \cdot 3^n} \quad \text{is dominated by} \quad \sum_{n=1}^{\infty} \frac{2}{3^n}.$$

The latter series converges because it is geometric with ratio $\frac{1}{3} < 1$. Therefore the series of Problem 31 converges by the comparison test.

C11S06.033: Because $1 \leqq 2 + \sin n \leqq 3$ for each integer $n \geqq 1$, the given series

$$\sum_{n=1}^{\infty} \frac{2 + \sin n}{n^2} \quad \text{is dominated by} \quad \sum_{n=1}^{\infty} \frac{3}{n^2}.$$

The latter series converges because it is a constant multiple of the p-series with $p = 2 > 1$. Therefore the dominated series converges as well by the comparison test.

C11S06.035: The given series

$$\sum_{n=1}^{\infty} \frac{(n+1)^n}{n^{n+1}}$$

diverges by limit-comparison with the harmonic series, demonstrated by the following computation:

$$\lim_{n \to \infty} \frac{\dfrac{(n+1)^n}{n^{n+1}}}{\dfrac{1}{n}} = \lim_{n \to \infty} \frac{(n+1)^n}{n^n} = \lim_{n \to \infty} \left(1 + \frac{1}{n}\right)^n = e.$$

C11S06.037: The sum of the first ten terms of the given series is

$$S_{10} = \sum_{n=1}^{10} \frac{1}{n^2 + 1} = \frac{1662222227}{1693047850} \approx 0.981792822335.$$

The error in using S_{10} to approximate the sum S of the infinite series is

$$S - S_{10} = R_{10} \leqq \int_{10}^{\infty} \frac{1}{x^2 + 1} \, dx = \left[\arctan x\right]_{10}^{\infty} = \frac{\pi}{2} - \arctan 10 \approx 0.099668652491.$$

Because $S \approx 1.076674047469$, the true value of the error is approximately 0.09488123.

C11S06.039: The sum of the first ten terms of the given series is

$$\sum_{n=1}^{10} \frac{\cos^2 n}{n^2} \approx 0.528869678057.$$

The error in using S_{10} to approximate the sum S of the infinite series is

$$S - S_{10} = R_{10} \leqq \int_{10}^{\infty} \frac{1}{x^2} \, dx = \left[-\frac{1}{x}\right]_{10}^{\infty} = \frac{1}{10} = 0.1.$$

Because $S \approx 0.574137740053$, the true value of the error is approximately 0.04526806.

C11S06.041: The sum of the series is

$$S = \sum_{n=1}^{\infty} \frac{1}{n^3 + 1} \approx 0.686503342339,$$

and $S - 0.005 \approx 0.681503342339$. Because

$$\sum_{n=1}^{9} \frac{1}{n^3 + 1} \approx 0.680981 < S - 0.005 < 0.681980 = \sum_{n=1}^{10} \frac{1}{n^3 + 1},$$

the smallest positive integer n such that $R_n < 0.005$ is $n = 10$. Without advance knowledge of the sum of the given series, you can obtain a conservative overestimate of n in the following way. We know that

$$R_n \leq \int_n^\infty \frac{1}{x^3}\,dx = \left[-\frac{1}{2x^2}\right]_n^\infty = \frac{1}{2n^2}.$$

So it will be sufficient if

$$\frac{1}{2n^2} < 0.005; \qquad 2n^2 > 200; \qquad n > 10;$$

that is, if $n = 11$. More accuracy, and a smaller value of n, might be obtained had we used instead the better estimate

$$R_n \leq \int_n^\infty \frac{1}{x^3 + 1}\,dx,$$

and if $n = 10$ the value of this integral is approximately 0.004998001249, but one must question whether the extra work in evaluating the antiderivative and solving the resulting inequality would be worth the trouble.

C11S06.043: The sum of the series is

$$S = \sum_{n=1}^\infty \frac{\cos^4 n}{n^4} \approx 0.100714442927,$$

and $S - 0.005 \approx 0.095714442927$. Because

$$\sum_{n=1}^2 \frac{\cos^4 n}{n^4} \approx 0.087095 < S - 0.005 < 0.098954 \approx \sum_{n=1}^3 \frac{\cos^4 n}{n^4},$$

the smallest positive integer n such that $R_n < 0.005$ is $n = 3$. Without advance knowledge of the sum of the given series, you can obtain a conservative overestimate of n in the following way. We know that

$$R_n \leq \int_n^\infty \frac{1}{x^4}\,dx = \left[-\frac{1}{3x^3}\right]_n^\infty = \frac{1}{3n^3}.$$

So it will be sufficient if

$$\frac{1}{3n^3} < 0.005; \qquad 3n^3 > 200; \qquad n^3 > 67;$$

that is, $n = 5$. There are ways to lower this estimate but they are highly technical and probably not worth the extra trouble.

C11S06.045: We suppose that $\sum a_n$ is a convergent positive-term series. Apply the mean value theorem to $f(t) = (\sin t)/t$ on the interval $[0, x]$ to show that $\sin x < x$ for all $x > 0$. Moreover, the converse of Theorem 3 in Section 11.3 implies that $a_n \to 0$ as $n \to +\infty$. Thus there exists a positive integer K such that if $n \geq K$, then $a_n < \pi$. Therefore

$$0 < \sin(a_n) < a_n \qquad \text{if} \qquad n \geq K.$$

Consequently $\sum a_n$ eventually dominates the eventually positive-term series $\sum \sin(a_n)$. Therefore the latter series converges because the values of its terms for $1 \leq n < K$ cannot affect its convergence or divergence.

C11S06.047: If $\sum a_n$ is a convergent positive-term series, then we may assume that $n \geq 1$, and hence

$$0 < \frac{a_n}{n} \leqq a_n \quad \text{for all} \quad n.$$

Therefore $\sum a_n$ dominates the positive-term series $\sum (a_n/n)$, so by the comparison test the latter series converges as well.

C11S06.049: Convergence of $\sum b_n$ implies that $\{b_n\} \rightarrow 0$ (the converse of Theorem 3 of Section 10.3). Therefore $\sum a_n b_n$ converges by Problem 48.

C11S06.051: By Problem 50 in Section 10.5, if n is a positive integer then

$$0 \leqq 1 + \frac{1}{2} + \frac{1}{3} + \frac{1}{4} + \cdots + \frac{1}{n} - \ln n \leqq 1.$$

Therefore

$$1 + \frac{1}{2} + \frac{1}{3} + \frac{1}{4} + \cdots + \frac{1}{n} \leqq 1 + \ln n$$

for every positive integer n. Hence

$$\sum_{n=1}^{\infty} \frac{1}{1 + \frac{1}{2} + \frac{1}{3} + \frac{1}{4} + \cdots + \frac{1}{n}} \quad \text{dominates} \quad \sum_{n=1}^{\infty} \frac{1}{1 + \ln n}.$$

But we showed in the solution of Problem 13 of this section that $\ln n < n$ for all $n \geqq 1$. So the last series dominates

$$\sum_{n=1}^{\infty} \frac{1}{1 + n},$$

which diverges because it is eventually the same as the harmonic series. Therefore the series of Problem 51 diverges.

Section 11.7

C11S07.001: The sequence $\{1/n^2\}$ is monotonically decreasing with limit zero. So the given series meets both criteria of the alternating series test and therefore converges. It is known that

$$\sum_{n=1}^{\infty} \frac{(-1)^{n+1}}{n^2} = \frac{\pi^2}{12} \approx 0.822467033424;$$

this can be derived from results in Problem 68 of Section 11.8.

C11S07.003: Because

$$\lim_{n \to \infty} \frac{n}{3n + 2} = \frac{1}{3} \neq 0, \qquad \lim_{n \to \infty} \frac{(-1)^n n}{3n + 2} \quad \text{does not exist.}$$

Therefore the given series diverges by the nth-term test for divergence.

C11S07.005: Because

554

$$\lim_{n\to\infty} \frac{n}{\sqrt{n^2+2}} = \lim_{n\to\infty} \frac{1}{\left(1+\dfrac{2}{n^2}\right)^{1/2}} = 1 \neq 0, \qquad \lim_{n\to\infty} \frac{(-1)^{n+1}n}{\sqrt{n^2+2}} \quad \text{does not exist.}$$

Therefore the given series diverges by the nth-term test for divergence.

C11S07.007: We showed in the solution of Problem 13 of Section 11.6 that $n > \ln n$ for every integer $n \geq 1$. Also, by l'Hôpital's rule,

$$\lim_{n\to\infty} \frac{\ln n}{n} = 0.$$

Therefore

$$\lim_{n\to\infty} \frac{n}{\ln n} = +\infty,$$

so the given series diverges by the nth-term test for divergence.

C11S07.009: First we claim that if n is a positive integer, then

$$\frac{n}{2^n} \geq \frac{n+1}{2^{n+1}}. \tag{1}$$

This assertion is true if $n = 1$ because

$$\frac{1}{2} \geq \frac{2}{4}, \quad \text{and thus} \quad \frac{1}{2^1} \geq \frac{2}{2^2}.$$

Suppose that the inequality in (1) holds for some integer $k \geq 1$. Then

$$\frac{k}{2^k} \geq \frac{k+1}{2^{k+1}}; \qquad \frac{k}{2^k} + \frac{1}{2^k} \geq \frac{k+1}{2^{k+1}} + \frac{2}{2^{k+1}};$$

$$\frac{k+1}{2^k} \geq \frac{k+3}{2^{k+1}}; \qquad \frac{k+1}{2^{k+1}} > \frac{k+2}{2^{k+2}}.$$

Therefore, by induction, the inequality in (1) holds for every integer $n \geq 1$; indeed, strict inequality holds if $n \geq 2$. Therefore if $a_n = n/2^n$ for $n \geq 1$, then the sequence $\{a_n\}$ is monotonically decreasing. Its limit is zero by l'Hôpital's rule:

$$\lim_{n\to\infty} a_n = \lim_{x\to\infty} \frac{x}{2^x} = \lim_{x\to\infty} \frac{1}{2^x \ln 2} = 0.$$

Therefore the given series satisfies both criteria of the alternating series test and thus it converges. To find its sum, note that

$$\sum_{n=1}^{\infty} \frac{(-1)^n n}{2^n} = f\left(\frac{1}{2}\right) \quad \text{where} \quad f(x) = \sum_{n=1}^{\infty} (-1)^{n+1} n x^n = x \sum_{n=1}^{\infty} (-1)^n n x^{n-1} = xg(x)$$

where $g(x) = h'(x)$ if we let

$$h(x) = \sum_{n=1}^{\infty} (-1)^n x^n = -x + x^2 - x^3 + x^4 - x^5 + \cdots = -\frac{x}{1+x}.$$

Thus

$$g(x) = h'(x) = -\frac{1}{(1+x)^2}, \quad \text{so that} \quad f(x) = -\frac{x}{(1+x)^2}.$$

It can be shown that all these computations are valid if $-1 < x < 1$, and therefore

$$\sum_{n=1}^{\infty} \frac{(-1)^n n}{2^n} = f\left(\frac{1}{2}\right) = -\frac{2}{9} \approx -0.222222222222.$$

C11S07.011: Given the series

$$\sum_{n=1}^{\infty} \frac{(-1)^n n}{\sqrt{2^n + 1}}, \tag{1}$$

first observe that, by l'Hôpital's rule,

$$\lim_{x \to \infty} \frac{x}{(2^x+1)^{1/2}} = \lim_{x \to \infty} \frac{2(2^x+1)^{1/2}}{2^x \ln 2} = \lim_{x \to \infty} \frac{2(2^x+1)^{1/2}}{(2^{2x})^{1/2} \ln 2}$$

$$= \lim_{x \to \infty} \frac{2}{\ln 2} \left(\frac{2^x+1}{2^{2x}}\right)^{1/2} = \lim_{x \to \infty} \frac{2}{\ln 2} \left(\frac{1}{2^x} + \frac{1}{2^{2x}}\right)^{1/2} = 0.$$

Next,

$$\lim_{n \to \infty} \frac{2^{n+1}+1}{2^n+1} = \lim_{n \to \infty} \frac{2+2^{-n}}{1+2^{-n}} = \frac{2+0}{1+0} = 2$$

and

$$\lim_{n \to \infty} \frac{(n+1)^2}{n^2} = \lim_{n \to \infty} \left(\frac{n+1}{n}\right)^2 = 1^2 = 1.$$

Therefore there exists a positive integer K such that, if $n \geqq K$, then

$$\frac{2^{n+1}+1}{2^n+1} > \frac{3}{2} > \frac{(n+1)^2}{n^2}.$$

For such n, it follows that

$$\frac{n^2}{2^n+1} > \frac{(n+1)^2}{2^{n+1}+1}; \quad \text{thus}$$

$$\frac{n}{\sqrt{2^n+1}} > \frac{n+1}{\sqrt{2^{n+1}+1}}.$$

This shows that the terms of the series in (1) are monotonically decreasing for $n \geq K$, and so both criteria of the alternating series test are met for $n \geq K$. Altering the terms for $n < K$ cannot change the convergence or divergence of a series, so the series in (1) converges. Its sum is approximately -0.178243455603. (By the way, the least value of K that "works" in this proof is $K = 5$, although the terms of the series begin to decrease in magnitude after $n = 3$.)

C11S07.013: The values of $\sin(n\pi/2)$ for $n = 1, 2, 3, \ldots$ are $1, 0, -1, 0, 1, 0, -1, 0, 1 \cdots$. So we rewrite the given series in the form

$$\sum_{n=1}^{\infty} \frac{(-1)^{n+1}}{(2n-1)^{2/3}}$$

to present it as an alternating series in the strict sense of the definition. Because the sequence $\{1/(2n-1)^{2/3}\}$ clearly meets the criteria of the alternating series test, this series converges. Its sum is approximately 0.711944418056.

C11S07.015: Because

$$\lim_{n\to\infty} \sin\left(\frac{1}{n}\right) = \lim_{u\to 0^+} \sin u = 0$$

and because $\sin u$ decreases monotonically through positive values as $u \to 0^+$, this series converges by the alternating series test. Its sum is approximately -0.550796848134.

C11S07.017: By Example 7 of Section 11.2, $2^{1/n} \to 1$ as $n \to +\infty$. So the given series diverges by the nth-term test for divergence.

C11S07.019: By the result in Example 11 of Section 11.2 of the text, $n^{1/n} \to 1$ as $n \to +\infty$. Therefore the given series diverges by the nth-term test for divergence.

C11S07.021: The ratio test yields

$$\rho = \lim_{n\to\infty} \frac{2^n}{2^{n+1}} = \frac{1}{2} < 1,$$

so the series

$$\sum_{n=1}^{\infty} \frac{(-1)^{n+1}}{2^n}$$

converges absolutely. Because it is geometric with ratio $r = -\frac{1}{2}$ and first term $\frac{1}{2}$, its sum is $\frac{1}{3}$.

C11S07.023: If

$$f(x) = \frac{\ln x}{x}, \quad \text{then} \quad f'(x) = \frac{1 - \ln x}{x^2},$$

so the sequence $\{(\ln n)/n\}$ is monotonically decreasing if $n \geqq 3$. By l'Hôpital's rule,

$$\lim_{n\to\infty} \frac{\ln n}{n} = \lim_{x\to\infty} \frac{\ln x}{x} = \lim_{x\to\infty} \frac{1}{x} = 0.$$

Therefore the series

$$\sum_{n=1}^{\infty} \frac{(-1)^n \ln n}{n}$$

converges by the alternating series test. Because

$$\int_1^{\infty} \frac{\ln x}{x}\, dx = \left[\frac{1}{2}(\ln x)^2 \right]_1^{\infty} = +\infty,$$

the given series converges conditionally rather than absolutely. Its sum is approximately 0.159868903742.

C11S07.025: The series

$$\sum_{n=1}^{\infty} \left(\frac{10}{n}\right)^n$$

converges absolutely by the root test, because

$$\rho = \lim_{n\to\infty} \left[\left(\frac{10}{n}\right)^n\right]^{1/n} = \lim_{n\to\infty} \frac{10}{n} = 0 < 1.$$

Its sum is approximately 186.724948614024. In spite of the relatively large sum, this series converges extremely rapidly; for example, the sum of its first 25 terms is approximately 186.724948614005. The sum is large because the first ten terms of the series are each at least 10; the largest is the fourth term, 39.0625. But the 25th term is less than 1.126×10^{-10}.

C11S07.027: Given the infinite series $\sum_{n=0}^{\infty} \frac{(-10)^n}{n!}$, the ratio test yields

$$\rho = \lim_{n\to\infty} \frac{n! \, 10^{n+1}}{(n+1)! \, 10^n} = \lim_{n\to\infty} \frac{10}{n+1} = 0.$$

Therefore this series converges absolutely. It is the result of substitution of -10 for x in the Maclaurin series for $f(x) = e^x$ (see Eq. (19) in Section 11.4); therefore its sum is $e^{-10} \approx 0.0000453999297625$.

C11S07.029: The series $\sum_{n=1}^{\infty} (-1)^{n+1} \left(\frac{n}{n+1}\right)^n$ diverges by the nth-term test for divergence because

$$\lim_{n\to\infty} \left(\frac{n}{n+1}\right)^n = \frac{1}{e} \neq 0.$$

The evaluation of the limit is made easy by Eq. (3) in Section 7.3.

C11S07.031: Given the series $\sum_{n=1}^{\infty} \left(\frac{\ln n}{n}\right)^n$, the root test yields

$$\rho = \lim_{n\to\infty} \left[\left(\frac{\ln n}{n}\right)^n\right]^{1/n} = \lim_{n\to\infty} \frac{\ln n}{n} = \lim_{x\to\infty} \frac{\ln x}{x} = \lim_{x\to\infty} \frac{1}{x} = 0$$

(with the aid of l'Hôpital's rule). Because $\rho < 1$, the given series converges absolutely. Its sum is approximately 0.187967875056.

C11S07.033: First note that

$$\sqrt{n+1} - \sqrt{n} = \frac{n+1-n}{\sqrt{n+1} + \sqrt{n}} = \frac{1}{\sqrt{n+1} + \sqrt{n}} \to 0$$

as $n \to +\infty$. Moreover, the sequence $\{\sqrt{n+1} - \sqrt{n}\}$ is monotonically decreasing. Proof: Suppose that n is a positive integer. Then

$$n^2 + 2n + 1 > n^2 + 2n; \qquad n+1 > \sqrt{n^2 + 2n}$$

$$2n+2 > 2\sqrt{n^2 + 2n}; \qquad 4(n+1) > n + 2\sqrt{n^2 + 2n} + n + 2;$$

$$2\sqrt{n+1} > \sqrt{n} + \sqrt{n+2}; \qquad \sqrt{n+1} - \sqrt{n} > \sqrt{n+2} - \sqrt{n+1}.$$

Therefore the series $\displaystyle\sum_{n=0}^{\infty} (-1)^n \left(\sqrt{n+1} - \sqrt{n}\right)$ converges by the alternating series test.

Its sum is approximately 0.760209625219. It converges conditionally, not absolutely. The reason is that

$$\sum_{n=1}^{\infty} \frac{1}{\sqrt{n+1} + \sqrt{n}} \qquad \text{dominates} \qquad \sum_{n=1}^{\infty} \frac{1}{2\sqrt{n+1}},$$

which diverges because it is a constant multiple of a series eventually the same as the p-series with $p = \frac{1}{2} \leqq 1$.

C11S07.035: Because

$$\lim_{n \to \infty} \left(\ln \frac{1}{n} \right)^n = \lim_{n \to \infty} (-\ln n)^n$$

does not exist, the series $\displaystyle\sum_{n=1}^{\infty} \left(\ln \frac{1}{n} \right)^n$ diverges by the nth-term test for divergence.

C11S07.037: First, for each positive integer n,

$$\frac{3^n}{n(2^n + 1)} \geqq \frac{3^n}{n(2^n + 2^n)} = \frac{3^n}{2n \cdot 2^n} = \frac{1}{2n} \cdot \left(\frac{3}{2} \right)^n.$$

Next, using l'Hôpital's rule,

$$\lim_{x \to \infty} \frac{\left(\frac{3}{2}\right)^x}{2x} = \lim_{x \to \infty} \frac{\left(\frac{3}{2}\right)^x \ln\left(\frac{3}{2}\right)}{2} = +\infty$$

because $\ln\left(\frac{3}{2}\right) > 0$ and because, if $a > 1$, then $a^x \to +\infty$ as $x \to +\infty$. Therefore the series

$$\sum_{n=1}^{\infty} \frac{(-1)^{n+1} 3^n}{n(2^n + 1)}$$

diverges by the nth-term test for divergence.

C11S07.0:39 Given the series $\displaystyle\sum_{n=1}^{\infty} \frac{(-1)^{n+1} n!}{1 \cdot 3 \cdot 5 \cdots (2n-1)}$, the ratio test yields

$$\rho = \lim_{n \to \infty} \frac{(n+1)! \cdot 1 \cdot 3 \cdot 5 \cdots (2n-1)}{n! \cdot 1 \cdot 3 \cdot 5 \cdots (2n-1) \cdot (2n+1)} = \lim_{n \to \infty} \frac{n+1}{2n+1} = \frac{1}{2}.$$

Because $\rho < 1$, the series in question converges absolutely. Its sum is approximately 0.586781998767.

C11S07.041: Given the series $\displaystyle\sum_{n=1}^{\infty} \frac{(n+2)!}{3^n (n!)^2}$, the ratio test yields

$$\rho = \lim_{n \to \infty} \frac{(n+3)! 3^n (n!)^2}{(n+2)! 3^{n+1} [(n+1)!]^2} = \lim_{n \to \infty} \frac{n+3}{3(n+1)^2} = \lim_{n \to \infty} \frac{\frac{1}{n} + \frac{3}{n^2}}{3\left(1 + \frac{1}{n}\right)^2} = \frac{0+0}{3 \cdot 1} = 0.$$

Because $\rho < 1$, the given series converges absolutely. Its sum can be computed exactly, as follows. Note first that

$$\sum_{n=1}^{\infty} \frac{(n+2)!}{3^n (n!)^2} = \sum_{n=1}^{\infty} \frac{(n+2)(n+1)}{3^n \cdot (n!)} = f\left(\frac{1}{3}\right)$$

where

$$f(x) = \sum_{n=1}^{\infty} \frac{(n+2)(n+1)x^n}{n!}.$$

But $f(x) = g'(x)$ where

$$g(x) = \sum_{n=1}^{\infty} \frac{(n+2)x^{n+1}}{n!},$$

and $g(x) = h'(x)$ where

$$h(x) = \sum_{n=1}^{\infty} \frac{x^{n+2}}{n!} = x^2 \sum_{n=1}^{\infty} \frac{x^n}{n!} = x^2(e^x - 1).$$

But then, $f(x) = h''(x) = (x^2 + 4x + 2)e^x - 2$, so the sum of the series in this problem is

$$f\left(\frac{1}{3}\right) = \frac{31e^{1/3} - 18}{9} \approx 2.807109464185.$$

This is confirmed by *Mathematica* 3.0, which in response to the command

    NSum[((n + 2)*(n + 1))/((3^n)*(n!)), { n, 1, Infinity }]

returns the approximate sum 2.80711.

C11S07.043: The sum of the first five terms of the given series is

$$S_5 = \sum_{n=1}^{5} \frac{(-1)^{n+1}}{n^3} = \frac{195353}{216000} \approx 0.904412037037.$$

The sixth term of the series is

$$-\frac{1}{216} \approx -0.004629629629.$$

Thus S_5 approximates the sum S of the series with error less than 0.005. Indeed, we can conclude that $S_6 \approx 0.899782 < S < 0.904412 \approx S_5$. To two decimal places, $S \approx 0.90$. *Mathematica* 3.0 reports that $S \approx 0.901542677370$.

C11S07.045: The sum of the first six terms of the given series is

$$S_6 = \sum_{n=1}^{6} \frac{(-1)^{n+1}}{n!} = \frac{91}{144} \approx 0.631944444444.$$

The seventh term of the series is

$$\frac{1}{5040} \approx 0.000198412698.$$

Thus S_6 approximates the sum S of the series with error less than 0.0002. Indeed, we can conclude that $S_6 \approx 0.631945 < S < 0.632142 \approx S_7$ (here we round *down* lower bounds and round *up* upper bounds). To three places, $S \approx 0.632$. *Mathematica* 3.0 reports that $S \approx 0.632120558829$. Using Eq. (19) in Section 11.4, we see that the exact value of the sum is

$$S = 1 - \frac{1}{e}.$$

We have in this problem another example of a series that *Mathematica* 3.0 can sum exactly (using the command Sum instead of NSum); the command

    Sum[((-1)∧(n+1))/(n!), { n, 1, Infinity }]

produces the exact answer in the form $-\dfrac{1-e}{e}$.

C11S07.047: The sum of the first 12 terms of the series is

$$S_{12} = \sum_{n=1}^{12} \frac{(-1)^{n+1}}{n} = \frac{18107}{27720} \approx 0.653210678211.$$

The 13th term of the series is

$$\frac{1}{13} \approx 0.076923076923.$$

Thus S_{12} approximates the sum S of the series with error less than 0.08. Indeed, we may conclude that $S_{12} \approx 0.653211 < S < 0.730133 \approx S_{13}$ (we round *down* lower bounds and round *up* upper bounds). Thus to one decimal place, $S \approx 0.7$. This is a series that *Mathematica* 3.0 can sum exactly; the command

    Sum[((-1)∧(n+1))/n, { n, 1, Infinity }]

produces the response $\ln 2$.

C11S07.049: The condition

$$\frac{1}{n^4} < 0.0005 \quad \text{leads to} \quad n > 6.69,$$

so the sum of the terms through $n = 6$ will provide three-place accuracy. The sum of the first six terms of the series is

$$\sum_{n=1}^{6} \frac{(-1)^{n+1}}{n^4} = \frac{4090037}{4320000} \approx 0.946767824074,$$

so to three places, the sum of the infinite series is 0.947. The exact value of the sum of this series is

$$\frac{7\pi^4}{720} \approx 0.947032829497.$$

C11S07.051: The condition

$$\frac{1}{n! \cdot 2^n} < 0.00005 \quad \text{leads to} \quad 5 < n < 6,$$

so the sum of the terms through $n = 5$ will provide four-place accuracy. The sum of the first six terms of the series is

$$\sum_{n=0}^{5} \frac{(-1)^n}{n! \cdot 2^n} = \frac{2329}{3840} \approx 0.606510416667,$$

so to four places, the sum of the infinite series is 0.6065. The exact value of the sum of this series is

$$\sum_{n=0}^{\infty} \frac{(-1)^n}{n! \cdot 2^n} = e^{-1/2} \approx 0.606530659713.$$

C11S07.053: The condition

$$\frac{1}{(2n+1)!} \cdot \left(\frac{\pi}{3}\right)^{2n+1} < 0.000005 \quad \text{leads to} \quad 4 < n < 5,$$

so the sum of the terms through $n = 4$ will provide five-place accuracy. The sum of the first five terms of the series is

$$\sum_{n=0}^{4} \frac{(-1)^n}{(2n+1)!} \cdot \left(\frac{\pi}{3}\right)^{2n+1} \approx 0.866025445100,$$

so to five places, the sum of the infinite series is 0.86603. The exact value of the sum of the infinite series is

$$\sum_{n=0}^{\infty} \frac{(-1)^n}{(2n+1)!} \cdot \left(\frac{\pi}{3}\right)^{2n+1} = \sin\left(\frac{\pi}{3}\right) = \frac{\sqrt{3}}{2} \approx 0.866025403784.$$

C11S07.055: Because

$$0 < a_n \leqq \frac{1}{n} \quad \text{for all} \quad n \geqq 1,$$

$a_n \to 0$ as $n \to +\infty$ by the squeeze law for limits (Section 2.3 of the text). The alternating series test does not apply because the sequence $\{a_n\}$ is not monotonically decreasing. The series $\sum a_n$ diverges because its $2n$th partial sum S_{2n} satisfies the inequality

$$S_{2n} > 1 + \frac{1}{3} + \frac{1}{5} + \frac{1}{7} + \cdots + \frac{1}{2n-1} > \frac{1}{2} + \frac{1}{4} + \frac{1}{6} + \frac{1}{8} + \cdots + \frac{1}{2n},$$

and the last expression is half the nth partial sum of the harmonic series. Similar remarks hold for S_{2n+1}, and hence $S_n \to +\infty$ as $n \to +\infty$. Therefore $\sum a_n$ diverges.

C11S07.057: Let

$$a_n = b_n = \frac{(-1)^n}{\sqrt{n}} \quad \text{for} \quad n \geqq 1.$$

Then $\sum a_n$ and $\sum b_n$ converge by the alternating series test. But

$$\sum_{n=1}^{\infty} a_n b_n = \sum_{n=1}^{\infty} \frac{(-1)^{2n}}{n} = \sum_{n=1}^{\infty} \frac{1}{n}$$

diverges because it is the harmonic series.

C11S07.059: Let $b = |a|$. Then the ratio test applied to $\sum (a^n/n!)$ yields

$$\rho = \lim_{n \to \infty} \frac{n! b^{n+1}}{(n+1)! b^n} = \lim_{n \to \infty} \frac{b}{n+1} = 0 < 1.$$

Therefore the series

$$\sum_{n=0}^{\infty} \frac{a^n}{n!} \tag{1}$$

converges for every real number a. Thus by the nth-term test for divergence, it follows that

$$\lim_{n \to \infty} \frac{a^n}{n!} = 0$$

for every real number a. The sum of the series in (1) is e^a.

C11S07.061: We are given

$$H_n = \sum_{k=1}^{n} \frac{1}{k} \quad \text{and} \quad S_n = \sum_{k=1}^{n} \frac{(-1)^{k+1}}{k}.$$

Part (a): Note first that

$$S_2 = 1 - \frac{1}{2} \quad \text{and} \quad H_2 - H_1 = 1 + \frac{1}{2} - 1,$$

so $S_{2n} = H_{2n} - H_n$ if $n = 1$. Assume that $S_{2m} = H_{2m} - H_m$ for some integer $m \geq 1$. Then

$$S_{2(m+1)} = S_{2m} + \frac{1}{2m+1} - \frac{1}{2m+2} = H_{2m} - H_m + \frac{1}{2m+1} - \frac{1}{2m+2}$$

$$= H_{2m} + \frac{1}{2m+1} + \frac{1}{2m+2} - H_m - \frac{2}{2m+2} = H_{2(m+1)} - H_{m+1}.$$

Therefore, by induction, $S_{2n} = H_{2n} - H_n$ for every positive integer n.

Part (b): Let $m = 2n$. Then

$$\lim_{n \to \infty} (H_{2n} - \ln 2n) = \lim_{m \to \infty} (H_m - \ln m) = \gamma$$

by Problem 50 in Section 11.5.

Part (c): By the results in parts (a) and (b),

$$\lim_{n \to \infty} (H_{2n} - \ln 2n - H_n + \ln n) = 0;$$

$$\lim_{n \to \infty} (S_{2n} - \ln 2 - \ln n + \ln n) = 0;$$

$$\lim_{n \to \infty} S_{2n} = \ln 2.$$

Thus the "even" partial sums of the alternating harmonic series converge to $\ln 2$. But the alternating harmonic series converges by the alternating series test. Therefore the sequence of *all* of its partial sums converges to $\ln 2$; that is,

$$\sum_{k=1}^{\infty} \frac{1}{k} = 1 - \frac{1}{2} + \frac{1}{3} - \frac{1}{4} + \frac{1}{5} - \frac{1}{6} + \cdots = \ln 2.$$

But see the solution of Problem 28 in Section 11.6 for a better way to approximate $\ln 2$.

C11S07.063: The answer consists of the first twelve terms of the following series:

$$1 + \frac{1}{3} - \frac{1}{2} + \frac{1}{5} - \frac{1}{4} + \frac{1}{7} + \frac{1}{9} - \frac{1}{6} + \frac{1}{11} + \frac{1}{13} - \frac{1}{8} + \frac{1}{15} + \frac{1}{17} - \frac{1}{10} + \frac{1}{19} + \frac{1}{21} - \frac{1}{12} + \frac{1}{23} + \frac{1}{25} - \frac{1}{14}$$

$$+ \frac{1}{27} - \frac{1}{16} + \frac{1}{29} + \frac{1}{31} - \frac{1}{18} + \frac{1}{33} + \frac{1}{35} - \frac{1}{20} + \frac{1}{37} + \frac{1}{39} - \frac{1}{22} + \frac{1}{41} + \frac{1}{43} - \frac{1}{24} + \frac{1}{45} + \frac{1}{47} - \frac{1}{26} + \cdots.$$

The 12th partial sum of the series shown here is

$$\frac{353201}{360360} \approx 0.9801337551337551$$

and the 13th is

$$\frac{6364777}{6126120} \approx 1.0389572845455198,$$

so the convergence to the sum 1 is quite slow (as might be expected when dealing with variations of the harmonic series). To generate and view many more partial sums, enter the following commands in *Mathematica* 3.0 (or modify them to use in another computer algebra program):

```
u = Table[ 1/(2*n - 1), { n, 1, 2 + 1000 } ]
```

$$\left\{ 1, \frac{1}{3}, \frac{1}{5}, \frac{1}{7}, \cdots, \frac{1}{2003} \right\}$$

(Of course, the ellipsis is ours, not *Mathematica's*. And you may replace 1000 in the first two commands with as large a positive integer as you and your computer will tolerate.)

```
v = Table[ 1/(2*n), { n, 1, 2 + 1000 } ]
```

$$\left\{ \frac{1}{2}, \frac{1}{4}, \frac{1}{6}, \frac{1}{8}, \cdots, \frac{1}{2004} \right\}$$

```
x = 0;    i = 0;    j = 0;
```

(Here x denotes the running sum of the first k terms of the series; i and j are merely subscripts to be used in the arrays u and v, respectively.)

```
While[ i < 1000, {
    While[ x <= 1, { i = i + 1, x = x + u[[i]], Print[ { i, u[[i]], N[x,40] } ] } ],
    While[ x >= 1, { j = j + 1, x = x - v[[j]], Print[ { j, v[[j]], N[x,40] } ] } ] } ]
```

If you execute these commands, be prepared for 1543 lines of output, concluding with

$$\left\{ 1001, \frac{1}{2001}, 1.0003529867391675223067581695773251551 87 \right\}$$

$$\left\{ 542, \frac{1}{1084}, 0.9994304775140752138424894448507423267 8 \right\}$$

There is evidence that the series is converging to 1 but still stronger evidence that the convergence is painfully slow.

C11S07.065: The sum of the first 50 terms of the rearrangement is $S_{50} \approx -0.00601599$. Also,

$$S_{500} \approx -0.000622656, \qquad S_{5000} \approx -0.0000624766,$$

$$S_{50000} \approx -0.0000062497656, \qquad S_{500000} \approx -0.00000062499766,$$

$$S_{5000000} \approx -0.000000062499977, \quad \text{and} \quad S_{50000000} \approx -0.0000000062499998.$$

We have strong circumstantial evidence here that the sum of the series is 0. (It is.)

Section 11.8

C11S08.001: Given the series $\displaystyle\sum_{n=1}^{\infty} nx^n$, the ratio test yields

$$\lim_{n\to\infty} \frac{(n+1)|x|^{n+1}}{n|x|^n} = |x|,$$

so the series converges if $-1 < x < 1$. It clearly diverges at both endpoints of this interval, so its interval of convergence is $(-1, 1)$. To find its sum, note that

$$\sum_{n=1}^{\infty} nx^n = x \sum_{n=1}^{\infty} nx^{n-1} = xf'(x)$$

where

$$f(x) = \sum_{n=1}^{\infty} x^n = \frac{x}{1-x}, \quad \text{so that} \quad f'(x) = \frac{1}{(1-x)^2}.$$

Therefore $\displaystyle\sum_{n=1}^{\infty} nx^n = \frac{x}{(1-x)^2}$ if $-1 < x < 1$.

C11S08.003: Given the series $\displaystyle\sum_{n=1}^{\infty} \frac{nx^n}{2^n}$, the ratio test yields

$$\lim_{n\to\infty} \frac{(n+1)2^n|x|^{n+1}}{n2^{n+1}|x|^n} = \frac{|x|}{2},$$

so the series converges if $-2 < x < 2$. It diverges at each endpoint of this interval by the nth-term test for divergence, so its interval of convergence is $(-2, 2)$.

C11S08.005: Given the series $\displaystyle\sum_{n=1}^{\infty} n!x^n$, the ratio test yields

$$\lim_{n\to\infty} \frac{(n+1)!|x|^{n+1}}{n!|x|^n} = \lim_{n\to\infty} n|x|.$$

This limit is zero if $x = 0$ but is $+\infty$ otherwise. Therefore the series converges only at the real number $x = 0$. Thus its interval of convergence is $[0, 0]$. If you prefer the strict interpretation of the word "interval," the interval $[a, b]$ is defined only if $a < b$ according to Appendix A. If so, we must say that this series has no interval of convergence and that it converges only if $x = 0$.

C11S08.007: Given the series $\displaystyle\sum_{n=1}^{\infty} \frac{3^n x^n}{n^3}$, the ratio test yields

565

$$\lim_{n \to \infty} \frac{n^3 3^{n+1} |x|^{n+1}}{(n+1)^3 3^n |x|^n} = 3|x|,$$

so the series converges if $-\frac{1}{3} < x < \frac{1}{3}$. When $x = \frac{1}{3}$ it is the p-series with $p = 3 > 1$, and thus it converges. When $x = -\frac{1}{3}$ the series converges by the alternating series test. Therefore its interval of convergence is $\left[-\frac{1}{3}, \frac{1}{3} \right]$.

C11S08.009: Given the series $\displaystyle\sum_{n=1}^{\infty} (-1)^n n^{1/2} (2x)^n$, the ratio test yields

$$\lim_{n \to \infty} \frac{(n+1)^{1/2} 2^{n+1} |x|^{n+1}}{n^{1/2} 2^n |x|^n} = 2|x|,$$

so the series converges if $-\frac{1}{2} < x < \frac{1}{2}$. It diverges at each endpoint of this interval by the nth-term test for divergence, and therefore its interval of convergence is $\left(-\frac{1}{2}, \frac{1}{2} \right)$.

C11S08.011: Given the series $\displaystyle\sum_{n=1}^{\infty} \frac{(-1)^n n x^n}{2^n (n+1)^3}$, the ratio test yields

$$\lim_{n \to \infty} \frac{(n+1)^4 2^n |x|^{n+1}}{n(n+2)^3 2^{n+1} |x|^n} = \frac{|x|}{2},$$

so this series converges if $-2 < x < 2$. If $x = 2$ it becomes

$$\sum_{n=1}^{\infty} \frac{(-1)^n n}{(n+1)^3},$$

which converges by the alternating series test. If $x = -2$ it becomes

$$\sum_{n=1}^{\infty} \frac{n}{(n+1)^3},$$

which converges because it is dominated by the p-series with $p = 2 > 1$. Therefore its interval of convergence is $[-2, 2]$.

C11S08.013: Given the series $\displaystyle\sum_{n=1}^{\infty} \frac{(\ln n) x^n}{3^n}$, the ratio test yields

$$\lim_{n \to \infty} \frac{[\ln(n+1)] \, 3^n |x|^{n+1}}{[\ln n] \, 3^{n+1} |x|^n} = \frac{|x|}{3},$$

so the series converges if $-3 < x < 3$. At the two endpoints of this interval it diverges by the nth-term test for divergence. Hence its interval of convergence is $(-3, 3)$. Note:

$$\lim_{n \to \infty} \frac{\ln(n+1)}{\ln n} = \lim_{x \to \infty} \frac{\ln(x+1)}{\ln x} = \lim_{x \to \infty} \frac{x}{x+1} = 1$$

by l'Hôpital's rule.

C11S08.015: Given the series $\displaystyle\sum_{n=0}^{\infty} (5x - 3)^n$, the ratio test yields

$$\lim_{n \to \infty} \frac{|5x - 3|^{n+1}}{|5x - 3|^n} = |5x - 3|,$$

and we solve $|5x - 3| < 1$ as follows:

$$-1 < 5x - 3 < 1; \qquad 2 < 5x < 4; \qquad \frac{2}{5} < x < \frac{4}{5}.$$

So this series converges on the interval $I = \left(\frac{2}{5}, \frac{4}{5}\right)$. It diverges at each endpoint by the nth-term test for divergence, so I is its interval of convergence. On this interval it is a geometric series with ratio in $(-1, 1)$, so its sum is

$$\sum_{n=0}^{\infty} (5x - 3)^n = \frac{1}{1 - (5x - 3)} = \frac{1}{4 - 5x}$$

provided that x is in I.

C11S08.017: Given the series $\displaystyle\sum_{n=1}^{\infty} \frac{2^n(x - 3)^n}{n^2}$, the ratio test yields

$$\lim_{n \to \infty} \frac{n^2 2^{n+1} |x - 3|^{n+1}}{(n+1)^2 2^n |x - 3|^n} = 2|x - 3|,$$

so this series converges if $2|x - 3| < 1$:

$$|x - 3| < \frac{1}{2}; \qquad -\frac{1}{2} < x - 3 < \frac{1}{2}; \qquad \frac{5}{2} < x < \frac{7}{2}.$$

If $x = \frac{5}{2}$, the series converges by the alternating series test. If $x = \frac{7}{2}$, it converges because it is the p-series with $p = 2 > 1$. Thus its interval of convergence is $\left[\frac{5}{2}, \frac{7}{2}\right]$.

C11S08.019: Given the series $\displaystyle\sum_{n=1}^{\infty} \frac{(2n)!}{n!} x^n$, the ratio test yields

$$\lim_{n \to \infty} \frac{n!(2n+2)! |x|^{n+1}}{(n+1)!(2n)! |x|^n} = \lim_{n \to \infty} \frac{(2n+2)(2n+1)|x|}{n+1} = \lim_{n \to \infty} 2(2n+1)|x|.$$

The last limit is $+\infty$ if $x \neq 0$ but zero if $x = 0$. Therefore this series converges only at the single point $x = 0$.

C11S08.021: Given the series $\displaystyle\sum_{n=1}^{\infty} \frac{n^3(x + 1)^n}{3^n}$, the ratio test yields

$$\lim_{n \to \infty} \frac{(n+1)^3 3^n |x + 1|^{n+1}}{n^3 3^{n+1} |x + 1|^n} = \frac{|x + 1|}{3},$$

so the given series converges if

$$-1 < \frac{x + 1}{3} < 1; \qquad -3 < x + 1 < 3; \qquad -4 < x < 2.$$

At the endpoints of this interval, the series diverges by the nth-term test for divergence. Thus its interval of convergence is $(-4, 2)$. To find its sum in closed form, let

$$f(x) = \sum_{n=1}^{\infty} \frac{n^3(x + 1)^n}{3^n}.$$

Then

567

$$f(x) = (x+1) \sum_{n=1}^{\infty} \frac{n^3 (x+1)^{n-1}}{3^n} = (x+1)g'(x)$$

where

$$g(x) = \sum_{n=1}^{\infty} \frac{n^2 (x+1)^n}{3^n} = (x+1) \sum_{n=1}^{\infty} \frac{n^2 (x+1)^{n-1}}{3^n}.$$

But $g(x) = (x+1)h'(x)$ where

$$h(x) = \sum_{n=1}^{\infty} \frac{n(x+1)^n}{3^n} = (x+1) \sum_{n=1}^{\infty} \frac{n(x+1)^{n-1}}{3^n} = (x+1)k'(x)$$

where

$$k(x) = \sum_{n=1}^{\infty} \frac{(x+1)^n}{3^n} = \frac{x+1}{2-x}$$

if $-4 < x < 2$ because the last series is geometric and convergent for such x. It now follows that

$$k'(x) = \frac{3}{(x-2)^2}; \qquad h(x) = \frac{3(x+1)}{(x-2)^2};$$

$$h'(x) = -\frac{3(x+4)}{(x-2)^3}; \qquad g(x) = -\frac{3(x+1)(x+4)}{(x-2)^3};$$

$$g'(x) = \frac{3(x^2 + 14x + 22)}{(x-2)^4}; \qquad f(x) = \frac{3(x+1)(x^2 + 14x + 22)}{(x-2)^4}.$$

C11S08.023: Given the series $\sum_{n=1}^{\infty} \frac{(3-x)^n}{n^3}$, the ratio test yields

$$\lim_{n \to \infty} \frac{n^3 |3-x|^{n+1}}{(n+1)^3 |3-x|^n} = |3-x| = |x-3|,$$

so this series converges if $-1 < x - 3 < 1$; that is, if $2 < x < 4$. It also converges at $x = 2$ because it is the p-series with $p = 3 > 1$ and converges at $x = 4$ by the alternating series test. Therefore its interval of convergence is $[2, 4]$.

C11S08.025: Given the series $\sum_{n=1}^{\infty} \frac{n!}{2^n} (x-5)^n$, the ratio test yields

$$\lim_{n \to \infty} \frac{(n+1)! 2^n |x-5|^{n+1}}{n! 2^{n+1} |x-5|^n} = \lim_{n \to \infty} \frac{n+1}{2} |x-5| = +\infty$$

unless $x = 5$, in which case the limit is zero. So this series converges only at the single point $x = 5$; its radius of convergence is zero.

C11S08.027: Given the series $\sum_{n=0}^{\infty} x^{(2^n)}$, the ratio test yields

$$\lim_{n \to \infty} \frac{|x|^{(2^{n+1})}}{|x|^{(2^n)}} = \lim_{n \to \infty} |x|^{(2^n)}.$$

This limit is zero if $-1 < x < 1$, is 1 if $x = \pm 1$, and is $+\infty$ if $|x| > 1$. The series diverges if $x = \pm 1$ by the nth-term test for divergence, and hence its interval of convergence is $(-1, 1)$.

C11S08.029: Given the series $\displaystyle\sum_{n=1}^{\infty} \frac{(-1)^n x^n}{1 \cdot 3 \cdot 5 \cdots (2n-1)}$, the ratio test yields

$$\lim_{n \to \infty} \frac{1 \cdot 3 \cdot 5 \cdots (2n-1) \cdot |x|^{n+1}}{1 \cdot 3 \cdot 5 \cdots (2n-1) \cdot (2n+1) \cdot |x|^n} = \lim_{n \to \infty} \frac{|x|}{2n+1} = 0$$

for all x. Hence the interval of convergence of this series is $(-\infty, +\infty)$.

C11S08.031: The function is the sum of a geometric series with first term and ratio x, and hence

$$f(x) = \frac{x}{1-x} = x + x^2 + x^3 + x^4 + x^5 + \cdots .$$

This series has radius of convergence 1 and interval of convergence $(-1, 1)$.

C11S08.033: Substitute $-3x$ for x in the Maclaurin series for e^x in Eq. (2), then multiply by x^2 to obtain

$$f(x) = x^2 e^{-3x} = x^2 \left(1 - \frac{3x}{1!} + \frac{9x^2}{2!} - \frac{27x^3}{3!} + \frac{81x^4}{4!} - \frac{243x^5}{5!} + \cdots \right)$$

$$= x^2 - \frac{3x^3}{1!} + \frac{3^2 x^4}{2!} - \frac{3^3 x^5}{3!} + \frac{3^4 x^6}{4!} - \frac{3^5 x^7}{5!} + \cdots .$$

The ratio test gives radius of convergence $+\infty$, so the interval of convergence of this series is $(-\infty, +\infty)$.

C11S08.035: Substitute x^2 for x in the Maclaurin series in (4) to obtain

$$f(x) = \sin x^2 = x^2 - \frac{x^6}{3!} + \frac{x^{10}}{5!} - \frac{x^{14}}{7!} + \frac{x^{18}}{9!} - \cdots .$$

The ratio test yields radius of convergence $+\infty$, so the interval of convergence of this series is $(-\infty, +\infty)$.

C11S08.037: Substitution of $\alpha = \frac{1}{3}$ in the binomial series in Eq. (14) yields

$$(1+x)^{1/3} = 1 + \frac{1}{3}x - \frac{1}{3} \cdot \frac{2}{3} \cdot \frac{x^2}{2!} + \frac{1}{3} \cdot \frac{2}{3} \cdot \frac{5}{3} \cdot \frac{x^3}{3!} - \frac{1}{3} \cdot \frac{2}{3} \cdot \frac{5}{3} \cdot \frac{8}{3} \cdot \frac{x^4}{4!} + \cdots$$

$$= 1 + \frac{1}{3}x - \frac{2}{3^2} \cdot \frac{x^2}{2!} + \frac{2 \cdot 5}{3^3} \cdot \frac{x^3}{3!} - \frac{2 \cdot 5 \cdot 8}{3^4} \cdot \frac{x^4}{4!} + \cdots .$$

Next, replacement of x with $-x$ yields

$$f(x) = (1-x)^{1/3} = 1 - \frac{1}{3}x - \frac{2}{3^2} \cdot \frac{x^2}{2!} - \frac{2 \cdot 5}{3^3} \cdot \frac{x^3}{3!} - \frac{2 \cdot 5 \cdot 8}{3^4} \cdot \frac{x^4}{4!} - \frac{2 \cdot 5 \cdot 8 \cdot 11}{3^5} \cdot \frac{x^5}{5!} - \cdots .$$

The radius of convergence of this series is 1.

C11S08.039: Substitution of $\alpha = -3$ in the binomial series in Eq. (14) yields

$$f(x) = (1+x)^{-3} = 1 - 3x + 3 \cdot 4 \cdot \frac{x^2}{2!} - 3 \cdot 4 \cdot 5 \cdot \frac{x^3}{3!} + 3 \cdot 4 \cdot 5 \cdot 6 \cdot \frac{x^4}{4!} - \cdots .$$

The radius of convergence of this series is 1.

C11S08.041: Let $g(x) = \ln(1 + x)$. Then

$$g'(x) = \frac{1}{1+x} = 1 - x + x^2 - x^3 + x^4 - x^5 + \cdots, \quad -1 < x < 1.$$

So

$$g(x) = C + x - \frac{x^2}{2} + \frac{x^3}{3} - \frac{x^4}{4} + \frac{x^5}{5} - \frac{x^6}{6} + \cdots$$

by Theorem 3. Also $0 = g(0) = \ln 1 = C$, so that

$$g(x) = x - \frac{x^2}{2} + \frac{x^3}{3} - \frac{x^4}{4} + \frac{x^5}{5} - \frac{x^6}{6} + \cdots.$$

Therefore

$$f(x) = \frac{g(x)}{x} = 1 - \frac{x}{2} + \frac{x^2}{3} - \frac{x^3}{4} + \frac{x^4}{5} - \frac{x^5}{6} + \cdots.$$

The ratio test tells us that the radius of convergence is 1; the interval of convergence is $(-1, 1]$.

C11S08.043: Termwise integration yields

$$f(x) = \int_0^x \sin t^3 \, dt = \int_0^x \left(t^3 - \frac{t^9}{3!} + \frac{t^{15}}{5!} - \frac{t^{21}}{7!} + \cdots \right) dt$$

$$= \left[\frac{t^4}{4} - \frac{t^{10}}{3!10} + \frac{t^{16}}{5!16} - \frac{t^{22}}{7!22} + \cdots \right]_0^x = \frac{x^4}{4} - \frac{x^{10}}{3!10} + \frac{x^{16}}{5!16} - \frac{x^{22}}{7!22} + \cdots.$$

This representation is valid for all real x.

C11S08.045: Termwise integration yields

$$f(x) = \int_0^x \exp\left(-t^3\right) dt = \int_0^x \left(1 - t^3 + \frac{t^6}{2!} - \frac{t^9}{3!} + \frac{t^{12}}{4!} - \cdots \right) dt$$

$$= \left[t - \frac{t^4}{4} + \frac{t^7}{2!7} - \frac{t^{10}}{3!10} + \frac{t^{13}}{4!13} - \cdots \right]_0^x = x - \frac{x^4}{4} + \frac{x^7}{2!7} - \frac{x^{10}}{3!10} + \frac{x^{13}}{4!13} - \cdots.$$

This representation is valid for all real x.

C11S08.047: First,

$$1 - \exp\left(-t^2\right) = 1 - \left(1 - t^2 + \frac{t^4}{2!} - \frac{t^6}{3!} + \frac{t^8}{4!} - \cdots \right) = t^2 - \frac{t^4}{2!} + \frac{t^6}{3!} - \frac{t^8}{4!} + \frac{t^{10}}{5!} - \cdots.$$

Then termwise integration yields

$$f(x) = \int_0^x \frac{1 - \exp\left(-t^2\right)}{t^2} \, dt = \int_0^x \left(1 - \frac{t^2}{2!} + \frac{t^4}{3!} - \frac{t^6}{4!} + \frac{t^8}{5!} - \cdots \right) dt$$

$$= \left[t - \frac{t^3}{2! \cdot 3} + \frac{t^5}{3! \cdot 5} - \frac{t^7}{4! \cdot 7} + \frac{t^9}{5! \cdot 9} - \cdots \right]_0^x = x - \frac{x^3}{2! \cdot 3} + \frac{x^5}{3! \cdot 5} - \frac{x^7}{4! \cdot 7} + \frac{x^9}{5! \cdot 9} - \cdots.$$

This representation is valid for all real x.

C11S08.049: We begin with

$$f(x) = \sum_{n=0}^{\infty} x^n = 1 + x + x^2 + x^3 + x^4 + x^5 + \cdots = \frac{1}{1-x}, \quad -1 < x < 1.$$

Then termwise differentiation yields

$$f'(x) = \sum_{n=1}^{\infty} nx^{n-1} = \frac{1}{(1-x)^2}; \quad \text{thus}$$

$$xf'(x) = \sum_{n=1}^{\infty} nx^n = \frac{x}{(1-x)^2}, \quad -1 < x < 1.$$

C11S08.051: We found in the solution of Problem 49 that if

$$f(x) = \sum_{n=0}^{\infty} x^n = \frac{1}{1-x}, \quad -1 < x < 1,$$

then

$$xf'(x) = \sum_{n=1}^{\infty} nx^n = \frac{x}{(1-x)^2}, \quad -1 < x < 1.$$

Therefore

$$D_x\left[xf'(x)\right] = \sum_{n=1}^{\infty} n^2 x^{n-1} = \frac{1+x}{(1-x)^3},$$

and hence

$$\sum_{n=1}^{\infty} n^2 x^n = \frac{x + x^2}{(1-x)^3}, \quad -1 < x < 1.$$

C11S08.053: If

$$y = e^x = 1 + x + \frac{x^2}{2!} + \frac{x^3}{3!} + \cdots + \frac{x^n}{n!} + \frac{x^{n+1}}{(n+1)!} + \cdots, \quad \text{then}$$

$$\frac{dy}{dx} = 0 + 1 + \frac{2x}{2!} + \frac{3x^2}{3!} + \cdots + \frac{nx^{n-1}}{n!} + \frac{(n+1)x^n}{(n+1)!} + \cdots$$

$$= 1 + x + \frac{x^2}{2!} + \frac{x^3}{3!} + \cdots + \frac{x^{n-1}}{(n-1)!} + \frac{x^n}{n!} + \cdots = e^x = y.$$

C11S08.055: If

$$y = \sinh x = x + \frac{x^3}{3!} + \frac{x^5}{5!} + \frac{x^7}{7!} + \frac{x^9}{9!} + \frac{x^{11}}{11!} + \cdots, \quad \text{then}$$

$$\frac{dy}{dx} = 1 + \frac{x^2}{2!} + \frac{x^4}{4!} + \frac{x^6}{6!} + \frac{x^8}{8!} + \frac{x^{10}}{10!} + \cdots = \cosh x \quad \text{and}$$

$$\frac{d^2y}{dx^2} = x + \frac{x^3}{3!} + \frac{x^5}{5!} + \frac{x^7}{7!} + \frac{x^9}{9!} + \cdots = \sinh x.$$

Therefore both the hyperbolic sine and hyperbolic cosine functions satisfy the differential equation

$$\frac{d^2y}{dx^2} - y = 0.$$

C11S08.057: From Example 7 we have

$$J_0(x) = \sum_{n=0}^{\infty} \frac{(-1)^n x^{2n}}{2^{2n}(n!)^2}; \quad \text{we are also given} \quad J_1(x) = \sum_{n=0}^{\infty} \frac{(-1)^n x^{2n+1}}{2^{2n+1} n!(n+1)!}.$$

We apply the ratio test to the series for $J_1(x)$ with the following result:

$$\lim_{n \to \infty} \frac{2^{2n+1} n!(n+1)! |x|^{2n+3}}{2^{2n+3}(n+1)!(n+2)! |x|^{2n+1}} = \lim_{n \to \infty} \frac{x^2}{4(n+1)(n+2)} = 0$$

for all x. Therefore this series converges for all x. Next,

$$J_0'(x) = \sum_{n=1}^{\infty} \frac{(-1)^n 2n x^{2n-1}}{2^{2n}(n!)^2} = \sum_{n=1}^{\infty} \frac{(-1)^n x^{2n-1}}{2^{2n-1} n!(n-1)!}$$

$$= \sum_{n=0}^{\infty} \frac{(-1)^{n+1} x^{2n+1}}{2^{2n+1}(n+1)! n!} = -\sum_{n=0}^{\infty} \frac{(-1)^n x^{2n+1}}{2^{2n+1}(n+1)! n!} = -J_1(x).$$

C11S08.059: We begin with

$$y(x) = J_0(x) = \sum_{n=0}^{\infty} \frac{(-1)^n x^{2n}}{2^{2n}(n!)^2}. \quad \text{Then:}$$

$$y'(x) = \sum_{n=1}^{\infty} \frac{(-1)^n 2n x^{2n-1}}{2^{2n}(n!)^2} = \sum_{n=0}^{\infty} \frac{(-1)^{n+1}(2n+2) x^{2n+1}}{2^{2n+2}[(n+1)!]^2} = \sum_{n=0}^{\infty} \frac{(-1)^{n+1} x^{2n+1}}{2^{2n+1} n!(n+1)!};$$

$$y''(x) = \sum_{n=0}^{\infty} \frac{(-1)^{n+1}(2n+1) x^{2n}}{2^{2n+1} n!(n+1)!};$$

$$x^2 y''(x) = \sum_{n=0}^{\infty} \frac{(-1)^{n+1}(2n+1) x^{2n+2}}{2^{2n+1} n!(n+1)!};$$

$$x y'(x) = \sum_{n=0}^{\infty} \frac{(-1)^{n+1} x^{2n+2}}{2^{2n+1} n!(n+1)!};$$

$$x^2 y(x) = \sum_{n=0}^{\infty} \frac{(-1)^n x^{2n+2}}{2^{2n}(n!)^2}.$$

Note that the coefficient n in Bessel's equation is zero. Therefore

$$x^2 y''(x) + xy'(x) + x^2 y(x) = \sum_{n=0}^{\infty} \frac{(-1)^n x^{2n+2}}{2^{2n}(n!)^2} \left[-\frac{2n+1}{2(n+1)} - \frac{1}{2(n+1)} + 1 \right]$$

$$= \sum_{n=0}^{\infty} \frac{(-1)^n x^{2n+2}}{2^{2n}(n!)^2} \left[-\frac{2n+2}{2n+2} + 1 \right] \equiv 0.$$

C11S08.061: The Taylor series of f centered at $a = 0$ is

$$f(x) = \frac{\sin x}{x} = 1 - \frac{x^2}{3!} + \frac{x^4}{5!} - \frac{x^6}{7!} + \frac{x^8}{9!} - \cdots.$$

This representation is valid for all real x. We plotted the Taylor polynomials for $f(x)$ with center $a = 0$ of degree 4, 6, and 8 and the graph of $y = f(x)$ simultaneously, with the following result.

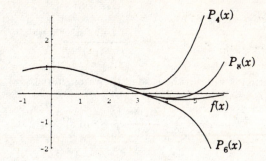

C11S08.063: Equation 20 is

$$\arctan x = x - \frac{x^3}{3} + \frac{x^5}{5} - \frac{x^7}{7} + \frac{x^9}{9} - \cdots, \quad -1 < x < 1.$$

Thus

$$\arctan x = \sum_{n=0}^{\infty} \frac{(-1)^n x^{2n+1}}{2n+1} = x \sum_{n=0}^{\infty} \frac{(-1)^n x^{2n}}{2n+1}, \quad -1 < x < 1.$$

Consequently,

$$\pi = 6 \cdot \frac{\pi}{6} = 6 \cdot \arctan \frac{1}{\sqrt{3}} = \frac{6}{\sqrt{3}} \sum_{n=0}^{\infty} \frac{(-1)^n}{2n+1} \cdot \left(\frac{1}{\sqrt{3}} \right)^{2n} = \frac{6}{\sqrt{3}} \sum_{n=0}^{\infty} \frac{(-1)^n}{2n+1} \cdot \frac{1}{3^n}.$$

Mathematica 3.0 reports that

$$\frac{6}{\sqrt{3}} \sum_{n=0}^{10} \frac{(-1)^n}{2n+1} \cdot \frac{1}{3^n} \approx 3.1415933045030815,$$

$$\frac{6}{\sqrt{3}} \sum_{n=0}^{20} \frac{(-1)^n}{2n+1} \cdot \frac{1}{3^n} \approx 3.1415926535956350,$$

$$\frac{6}{\sqrt{3}} \sum_{n=0}^{30} \frac{(-1)^n}{2n+1} \cdot \frac{1}{3^n} \approx 3.1415926535897932, \quad \text{and}$$

$$\frac{6}{\sqrt{3}} \sum_{n=0}^{40} \frac{(-1)^n}{2n+1} \cdot \frac{1}{3^n} \approx 3.1415926535897932.$$

C11S08.065: Part (a): From the Maclaurin series for the natural exponential function in Eq. (2) of this section, we derive the fact that

$$e^{-t} = \frac{1}{e^t} = \sum_{n=0}^{\infty} \frac{(-1)^n}{n!} t^n$$

for all real t. Substitute $t = x \ln x = \ln(x^x)$ to obtain

$$\frac{1}{x^x} = \sum_{n=0}^{\infty} \frac{(-1)^n}{n!} (x \ln x)^n$$

if $x > 0$. Part (b): The formula in Problem 53 of Section 8.8 states that if m and n are fixed positive integers, then

$$\int_0^1 x^m (\ln x)^n \, dx = \frac{(-1)^n n!}{(m+1)^{n+1}}. \tag{1}$$

Moreover, term-by-term integration is valid in the case of the result in part (a), so

$$\int_0^1 \frac{1}{x^x} \, dx = \sum_{n=0}^{\infty} \frac{(-1)^n}{n!} \int_0^1 (x \ln x)^n \, dx.$$

Thus Eq. (1) here yields

$$\int_0^1 \frac{1}{x^x} \, dx = \sum_{n=0}^{\infty} \frac{(-1)^n}{n!} \cdot \frac{(-1)^n n!}{(n+1)^{n+1}} = \sum_{n=0}^{\infty} \frac{1}{(n+1)^{n+1}} = \sum_{n=1}^{\infty} \frac{1}{n^n}.$$

C11S08.067: Part (a):

$$f(x) = \sum_{n=0}^{\infty} \frac{\alpha(\alpha-1)(\alpha-2)\cdots(\alpha-n+1)}{n!} x^n$$

$$= 1 + \alpha x + \frac{\alpha(\alpha-1)}{2!} x^2 + \frac{\alpha(\alpha-1)(\alpha-2)}{3!} x^3 + \cdots + \frac{\alpha(\alpha-1)(\alpha-2)\cdots(\alpha-n+1)}{n!} x^n + \cdots.$$

$$f'(x) = \alpha + \frac{\alpha(\alpha-1)}{1!} x + \frac{\alpha(\alpha-1)(\alpha-2)}{2!} x^2 + \frac{\alpha(\alpha-1)(\alpha-2)(\alpha-3)}{3!} x^3 + \cdots$$

$$+ \frac{\alpha(\alpha-1)(\alpha-2)\cdots(\alpha-n+1)}{(n-1)!} x^{n-1} + \frac{\alpha(\alpha-1)(\alpha-2)\cdots(\alpha-n)}{n!} x^n + \cdots.$$

$$x f'(x) = \alpha x + \frac{\alpha(\alpha-1)}{1!} x^2 + \frac{\alpha(\alpha-1)(\alpha-2)}{2!} x^3 + \frac{\alpha(\alpha-1)(\alpha-2)(\alpha-3)}{3!} x^4 + \cdots$$

$$+ \frac{\alpha(\alpha-1)(\alpha-2)\cdots(\alpha-n+1)}{(n-1)!} x^n + \frac{\alpha(\alpha-1)(\alpha-2)\cdots(\alpha-n)}{n!} x^{n+1} + \cdots.$$

Therefore

$$(1 + x)f'(x) = f'(x) + xf'(x)$$

$$= \alpha + (\alpha^2 - \alpha + \alpha)x + \frac{1}{2!}\left[\alpha(\alpha - 1)(\alpha - 2 + 2)\right]x^2 + \frac{1}{3!}\left[\alpha(\alpha - 1)(\alpha - 2)(\alpha - 3 + 3)\right]x^3 + \cdots$$

$$+ \frac{1}{n!}\left[\alpha(\alpha - 1)(\alpha - 2)\cdots(\alpha - n + 1)(\alpha - n + n)\right]x^n + \cdots$$

$$= \alpha + \alpha^2 x + \frac{\alpha^2(\alpha - 1)}{2!}x^2 + \frac{\alpha^2(\alpha - 1)(\alpha - 2)}{3!}x^3 + \cdots$$

$$+ \frac{\alpha^2(\alpha - 1)(\alpha - 2)\cdots(\alpha - n + 1)}{n!}x^n + \cdots = \alpha f(x).$$

Part (b): From the result $(1 + x)f'(x) = \alpha f(x)$ in part (a), we derive

$$\frac{f'(x)}{f(x)} = \frac{\alpha}{1 + x}; \qquad \ln(f(x)) = C + \alpha \ln(1 + x);$$

$$f(x) = K(1 + x)^{\alpha}; \qquad 1 = f(0) = K \cdot 1^{\alpha}: \quad K = 1.$$

Therefore $f(x) = (1 + x)^{\alpha}$, $-1 < x < 1$.

C11S08.069: Assume that the power series

$$\sum_{n=0}^{\infty} a_n x^n$$

converges for some $x = x_0 \neq 0$. Then $\{a_n x_0^n\} \to 0$ as $n \to +\infty$. Thus there exists a positive integer K such that, if $n \geq K$ and $|x| < |x_0|$, then

$$|a_n x_0^n| \leq 1; \qquad |a_n x^n| \leq \left|\frac{x^n}{x_0^n}\right|; \qquad |a_n x^n| \leq \left|\frac{x}{x_0}\right|^n.$$

This implies that the series

$$\sum_{n=0}^{\infty} |a_n x^n| \quad \text{is eventually dominated by} \quad \sum_{n=0}^{\infty} \left|\frac{x}{x_0}\right|^n,$$

a convergent geometric series. Therefore $\sum a_n x^n$ converges absolutely if $|x| < |x_0|$.

Section 11.9

C11S09.001: To estimate $65^{1/3}$ using the binomial series, first write

$$65^{1/3} = (4^3 + 1)^{1/3} = 4\left(1 + \frac{1}{64}\right)^{1/3}.$$

According to Eq. (14) in Section 11.8, the binomial series is

$$(1 + x)^{\alpha} = 1 + \alpha x + \frac{\alpha(\alpha - 1)}{2!}x^2 + \frac{\alpha(\alpha - 1)(\alpha - 2)}{3!}x^3 + \cdots;$$

it has radius of convergence $R = 1$. Therefore

575

$$4(1+x)^{1/3} = 4 + \frac{4}{3}x - \frac{4 \cdot 2}{3^2} \cdot \frac{x^2}{2!} + \frac{4 \cdot 2 \cdot 5}{3^3} \cdot \frac{x^3}{3!} - \frac{4 \cdot 2 \cdot 5 \cdot 8}{3^4} \cdot \frac{x^4}{4!} + \cdots.$$

With $x = \frac{1}{64}$, this series is alternating after the first two terms. For three digits correct to the right of the decimal in our approximation, we note that if $a = \frac{1}{64}$, then

$$\frac{4 \cdot 2}{3^2} \cdot \frac{a^2}{2!} < 0.00011.$$

Therefore

$$65^{1/3} \approx 4 + \frac{4}{3} \cdot \frac{1}{64} \approx 4.02083333 \approx 4.021.$$

For more accuracy, the sum of the first seven terms of the series is approximately 4.02072575858904; compare this with $65^{1/3} \approx 4.02072575858906$.

C11S09.003: The Maclaurin series for the sine function yields

$$\sin(0.5) = \frac{1}{2} - \frac{1}{3! \cdot 2^3} + \frac{1}{5! \cdot 2^5} - \frac{1}{7! \cdot 2^7} - \cdots.$$

The alternating series remainder estimate (Theorem 2 in Section 11.7) tells us that because

$$\frac{1}{5! \cdot 2^5} \approx 0.000260417 < 0.0003,$$

the error in approximating $\sin(0.5)$ using only the first two terms of this series will be no greater than 0.0003. Thus

$$\sin(0.5) \approx \frac{1}{2} - \frac{1}{3! \cdot 2^3} \approx 0.47916667.$$

Therefore, to three places, $\sin(0.5) \approx 0.479$. As a separate check, the sum of the first three terms of the series is approximately 0.47942708, the sum of its first six terms is approximately 0.479425538604, and this agrees with the true value of $\sin(0.5)$ to the number of digits shown.

C11S09.005: The Maclaurin series of the inverse tangent function is given in Eq. (20) of Section 11.8; it is

$$\arctan x = x - \frac{1}{3}x^3 + \frac{1}{5}x^5 - \frac{1}{7}x^7 + \frac{1}{9}x^9 - \cdots.$$

When $x = \frac{1}{2}$, the sum of the first four terms of this series is approximately 0.4634672619. With five terms we get 0.4636842758 and with six terms we get 0.4636398868. To three places, $\arctan(0.5) \approx 0.464$.

C11S09.007: When $x = \pi/10$ is substituted in the Maclaurin series for the sine function (Eq. (4) in Section 11.8), we obtain

$$\sin\left(\frac{\pi}{10}\right) = \frac{\pi}{10} - \frac{\pi^3}{3! \cdot 10^3} + \frac{\pi^5}{5! \cdot 10^5} - \frac{\pi^7}{7! \cdot 10^7} + \cdots.$$

Because

$$\frac{\pi^5}{5! \cdot 10^5} \approx 0.0000255011641 < 0.00003,$$

the sum of the first two terms of the series should yield three-place accuracy. Thus because

$$\sin\left(\frac{\pi}{10}\right) \approx \frac{\pi}{10} - \frac{\pi^3}{3! \cdot 10^3} \approx 0.30899155,$$

we may conclude that $\sin(\pi/10) \approx 0.309$ to three places. In fact, the sum of the first three terms of the series is approximately 0.30901699 and $\sin(\pi/10) \approx 0.30901699$ to the number of digits shown.

C11S09.009: First we convert $10°$ into $\pi/18$ radians, then use the Maclaurin series for the sine function (Eq. (4) of Section 11.8):

$$\sin 10° = \sin\left(\frac{\pi}{18}\right) = \frac{\pi}{18} - \frac{\pi^3}{3! \cdot 18^3} + \frac{\pi^5}{5! \cdot 18^5} - \frac{\pi^7}{7! \cdot 18^7} + \cdots.$$

Because

$$\frac{\pi^3}{3! \cdot 18^3} \approx 0.000886096,$$

the first term of the Maclaurin series alone may not give three-place accuracy (it doesn't). But

$$\frac{\pi^5}{5! \cdot 18^5} \approx 0.0000013496016,$$

so we will certainly obtain three-place accuracy by adding the first two terms of the series:

$$\sin\left(\frac{\pi}{18}\right) \approx \frac{\pi}{18} - \frac{\pi^3}{3! \cdot 18^3} \approx 0.17364683.$$

Thus, to three places, $\sin 10° \approx 0.174$. To check, the sum of the first three terms of the series is approximately 0.1736481786 and the true value of $\sin 10°$ is approximately 0.1736481777 (to the number of digits shown).

C11S09.011: Four-place accuracy demands that the error not exceed 0.00005. Here we have

$$I = \int_0^1 \frac{\sin x}{x}\,dx = \int_0^1 \left(1 - \frac{x^2}{3!} + \frac{x^4}{5!} - \frac{x^6}{7!} + \cdots\right)dx$$

$$= 1 - \frac{1}{3!3} + \frac{1}{5!5} - \frac{1}{7!7} + \frac{1}{9!9} - \frac{1}{11!11} + \cdots.$$

The alternating series remainder estimate (Theorem 2 of Section 11.7), when applied, yields

$$\frac{1}{7!7} \approx 0.0000566893 \quad \text{(not good enough) and} \quad \frac{1}{9!9} \approx 0.000000008748 \quad \text{(great accuracy)},$$

so that

$$I \approx 1 - \frac{1}{3!3} + \frac{1}{5!5} - \frac{1}{7!7} \approx 0.946082766;$$

to four places, $I \approx 0.9461$. To check, the sum of the first five terms of the series is approximately 0.9460830726 and the true value of the integral is 0.9460830703671830, correct to the number of digits shown here.

C11S09.013: The Maclaurin series for the inverse tangent function (Eq. (20) in Section 11.8) leads to

$$K = \int_0^{1/2} \frac{\arctan x}{x}\, dx = \int_0^{1/2} \left(1 - \frac{x^2}{3} + \frac{x^4}{5} - \frac{x^6}{7} + \cdots\right) dx$$

$$= \frac{1}{2} - \frac{1}{2^3 \cdot 3^2} + \frac{1}{2^5 \cdot 5^2} - \frac{1}{2^7 \cdot 7^2} + \frac{1}{2^9 \cdot 9^2} - \cdots.$$

Now

$$\frac{1}{2^9 \cdot 9^2} \approx 0.000024112654 < 0.00003,$$

so by the alternating series error estimate, the sum of the first four terms of the numerical series should yield four-place accuracy. Thus

$$K \approx \frac{1}{2} - \frac{1}{2^3 \cdot 3^2} + \frac{1}{2^5 \cdot 5^2} - \frac{1}{2^7 \cdot 7^2} \approx 0.4872016723;$$

that is, to four places $K \approx 0.4872$. To check this result, the sum of the first five terms of the series is approximately 0.487225784990 and the approximate value of the integral is 0.487222358295 (to the number of digits shown).

C11S09.015: The Maclaurin series for $\ln(1+x)$, in Eq. (19) of Section 11.8, yields

$$I = \int_0^{1/10} \frac{\ln(1+x)}{x}\, dx = \int_0^{1/10} \left(1 - \frac{x}{2} + \frac{x^2}{3} - \frac{x^3}{4} + \frac{x^4}{5} - \frac{x^5}{6} + \cdots\right) dx$$

$$= \left[x - \frac{x^2}{4} + \frac{x^3}{9} - \frac{x^4}{16} + \frac{x^5}{25} - \frac{x^6}{36} + \cdots\right]_0^{1/10} = \frac{1}{10} - \frac{1}{4 \cdot 10^2} + \frac{1}{9 \cdot 10^3} - \frac{1}{16 \cdot 10^4} + \frac{1}{25 \cdot 10^5} - \cdots.$$

Because

$$\frac{1}{16 \cdot 10^4} = 0.00000625,$$

the alternating series remainder estimate tells us that the sum of the first three terms of the numerical series should give four-place accuracy. Because

$$I \approx \frac{1}{10} - \frac{1}{4 \cdot 10^2} + \frac{1}{9 \cdot 10^3} \approx 0.097611111,$$

the four-place approximation we seek is $I \approx 0.0976$. To check this result, the sum of the first four terms of the series is approximately 0.097604861111 and the value of the integral, to the number of digits shown here, is 0.0976052352293216.

C11S09.017: The Maclaurin series for the natural exponential function—Eq. (2) in Section 11.8—yields

$$J = \int_0^{1/2} \frac{1 - e^{-x}}{x}\, dx = \int_0^{1/2} \left(1 - \frac{x}{2!} + \frac{x^2}{3!} - \frac{x^3}{4!} + \frac{x^4}{5!} - \cdots\right) dx$$

$$= \left[x - \frac{x^2}{2! 2} + \frac{x^3}{3! 3} - \frac{x^4}{4! 4} + \frac{x^5}{5! 5} - \cdots\right]_0^{1/2} = \frac{1}{2} - \frac{1}{2! \cdot 2 \cdot 2^2} + \frac{1}{3! \cdot 3 \cdot 2^3} - \frac{1}{4! \cdot 4 \cdot 2^4} + \cdots.$$

Because

$$\frac{1}{6! \cdot 6 \cdot 2^6} \approx 0.00000361690,$$

578

the alternating series remainder estimate assures us that the sum of the first five terms of the series, which is approximately 0.443845486111, will give us four-place accuracy: $J \approx 0.4438$. The value of the integral, accurate to the number of digits shown here, is 0.4438420791177484.

C11S09.019: The Maclaurin series for the natural exponential function—Eq. (2) in Section 11.8—yields

$$\exp(-x^2) = 1 - x^2 + \frac{x^4}{2!} - \frac{x^6}{3!} + \frac{x^8}{4!} - \frac{x^{10}}{5!} + \cdots .$$

Then termwise integration gives

$$\int_0^1 \exp(-x^2)\, dx = 1 - \frac{1}{3} + \frac{1}{2! \cdot 5} - \frac{1}{3! \cdot 7} + \frac{1}{4! \cdot 9} - \frac{1}{5! \cdot 11} + \cdots$$

Next, the sum of the first six terms of the numerical series is approximately 0.746729196729, the sum of the first seven terms is approximately 0.746836034336, and the sum of the first eight terms is approximately 0.746822806823. To four places, the value of the integral is 0.7468. The actual value of the integral, to the number of digits shown here, is 0.7468241328124270.

C11S09.021: The binomial series (Example 8 in Section 11.8) yields

$$(1 + x^2)^{1/3} = 1 + \frac{x^2}{3} - \frac{2x^4}{2! \cdot 3^2} + \frac{2 \cdot 5 x^6}{3! \cdot 3^3} - \frac{2 \cdot 5 \cdot 8 x^8}{4! \cdot 3^4} + \frac{2 \cdot 5 \cdot 8 \cdot 11 x^{10}}{5! \cdot 3^5} - \cdots .$$

Then term-by-term integration yields

$$I = \int_0^{1/2} (1 + x^2)^{1/3}\, dx = \left[x + \frac{x^3}{3 \cdot 3} - \frac{2x^5}{2! \cdot 3^2 \cdot 5} + \frac{2 \cdot 5 x^7}{3! \cdot 3^3 \cdot 7} - \frac{2 \cdot 5 \cdot 8 x^9}{4! \cdot 3^4 \cdot 9} + \frac{2 \cdot 5 \cdot 8 \cdot 11 x^{11}}{5! \cdot 3^5 \cdot 11} - \cdots \right]_0^{1/2}$$

$$= \frac{1}{2} + \frac{1}{3 \cdot 3 \cdot 2^3} - \frac{2}{2! \cdot 3^2 \cdot 5 \cdot 2^5} + \frac{2 \cdot 5}{3! \cdot 3^3 \cdot 7 \cdot 2^7} - \frac{2 \cdot 5 \cdot 8}{4! \cdot 3^4 \cdot 9 \cdot 2^9} + \frac{2 \cdot 5 \cdot 8 \cdot 11}{5! \cdot 3^5 \cdot 11 \cdot 2^{11}} - \cdots .$$

The sum of the first five terms of the numerical series is approximately 0.513254407130 and the sum of its first six terms is approximately 0.513255746722. So to four places, $I \approx 0.5133$. The actual value of the integral, accurate to the number of digits shown here, is 0.5132555590033423.

C11S09.023: The Maclaurin series for the natural exponential function (Eq. (2) in Section 11.8) yields

$$\lim_{x \to 0} \frac{1 + x - e^x}{x^2} = \lim_{x \to 0} \frac{1}{x^2} \left(-\frac{x^2}{2!} - \frac{x^3}{3!} - \frac{x^4}{4!} - \cdots \right) = \lim_{x \to 0} \left(-\frac{1}{2} - \frac{x}{6} - \frac{x^2}{24} - \cdots \right) = -\frac{1}{2} .$$

C11S09.025: The series in Eqs. (2) and (3) of Section 11.8 yield

$$\lim_{x \to 0} \frac{1 - \cos x}{x(e^x - 1)} = \lim_{x \to 0} \frac{\dfrac{x^2}{2!} - \dfrac{x^4}{4!} + \dfrac{x^6}{6!} - \cdots}{x^2 + \dfrac{x^3}{2!} + \dfrac{x^4}{3!} + \cdots} = \lim_{x \to 0} \frac{\dfrac{1}{2!} - \dfrac{x^2}{4!} + \dfrac{x^4}{6!} - \cdots}{1 + \dfrac{x}{2!} + \dfrac{x^2}{3!} + \cdots} = \frac{1}{2} .$$

C11S09.027: The Maclaurin series for the sine function in Eq. (4) of Section 11.8 yields

$$\lim_{x \to 0} \left(\frac{1}{x} - \frac{1}{\sin x} \right) = \lim_{x \to 0} \frac{(\sin x) - x}{x \sin x}$$

$$= \lim_{x \to 0} \frac{-\dfrac{x^3}{3!} + \dfrac{x^5}{5!} - \dfrac{x^7}{7!} + \cdots}{x^2 - \dfrac{x^4}{3!} + \dfrac{x^6}{5!} - \cdots} = \lim_{x \to 0} \frac{-\dfrac{x}{3!} + \dfrac{x^3}{5!} - \dfrac{x^5}{7!} + \cdots}{1 - \dfrac{x^2}{3!} + \dfrac{x^4}{5!} - \cdots} = \frac{0}{1} = 0.$$

C11S09.029: The Taylor series with center $\pi/2$ for the sine function is

$$\sin x = 1 - \frac{1}{2!}\left(x - \frac{\pi}{2}\right)^2 + \frac{1}{4!}\left(x - \frac{\pi}{2}\right)^4 - \frac{1}{6!}\left(x - \frac{\pi}{4}\right)^6 + \cdots.$$

We convert $80°$ into $x = 4\pi/9$ and substitute:

$$\sin 80° = 1 - \frac{1}{2!}\left(\frac{\pi}{18}\right)^2 + \frac{1}{4!}\left(\frac{\pi}{18}\right)^4 - \frac{1}{6!}\left(\frac{\pi}{18}\right)^6 + \cdots. \qquad (1)$$

For four-place accuracy, we need

$$\frac{1}{n!}\left(\frac{\pi}{18}\right)^n < 0.00005,$$

and the smallest even positive integer for which this inequality holds is $n = 4$. Thus only the first two terms of the series in (1) are needed to show that, to four places, $\sin 80° \approx 0.9848$. The sum of the first two terms is approximately 0.984769129011, the sum of the first three terms is approximately 0.984807792249, and $\sin 80° \approx 0.984807753012$ (all digits given here are correct).

C11S09.031: The Taylor series with center $\pi/4$ for the cosine function is

$$\cos x = \frac{\sqrt{2}}{2}\left[1 - \left(x - \frac{\pi}{4}\right) - \frac{1}{2!}\left(x - \frac{\pi}{4}\right)^2 + \frac{1}{3!}\left(x - \frac{\pi}{4}\right)^3 + \frac{1}{4!}\left(x - \frac{\pi}{4}\right)^4 - \frac{1}{5!}\left(x - \frac{\pi}{4}\right)^5 - \cdots\right].$$

We convert $47°$ to radians and substitute to find that

$$\cos 47° = \frac{\sqrt{2}}{2}\left[1 - \frac{\pi}{90} - \frac{1}{2!}\left(\frac{\pi}{90}\right)^2 + \frac{1}{3!}\left(\frac{\pi}{90}\right)^3 + \frac{1}{4!}\left(\frac{\pi}{90}\right)^4 - \frac{1}{5!}\left(\frac{\pi}{90}\right)^5 - \cdots\right].$$

This series is absolutely convergent, so rearrangement will not change its convergence or its sum. We make it into an alternating series by grouping terms 2 and 3, terms 4 and 5, and so on. We seek six-place accuracy here. If $x = \pi/90$, then

$$\frac{x^3}{3!} + \frac{x^4}{4!} \approx 0.000007 \quad \text{and that} \quad \frac{x^5}{5!} + \frac{x^6}{6!} \approx 0.00000000043,$$

so the first five terms of the [ungrouped] series—those through exponent 4—yield the required six-place accuracy: $\cos 47° \approx 0.681998$. The actual value of $\cos 47°$ is approximately 0.6819983600624985 (the digits shown here are all correct or correctly rounded).

C11S09.033: Note that $e^{0.1} < 1.2 = \frac{6}{5}$, and if $|x| \leq 0.1$, then the Taylor series remainder estimate yields

$$\left|\frac{e^z}{120}x^5\right| \leq \frac{6}{600}\left(\frac{1}{10}\right)^5 = 10^{-7} < 0.5 \times 10^{-6},$$

so six-place accuracy is assured.

C11S09.035: The Taylor series remainder estimate is difficult to work with; we use instead the cruder alternating series remainder estimate:

$$\frac{(0.1)^5}{5} < 0.5 \times 10^{-5},$$

so five-place accuracy is assured.

C11S09.037: Clearly $|e^z| < \frac{5}{3}$ if $|z| < 0.5$. Hence the Taylor series remainder estimate yields

$$\left| \frac{e^z}{120} x^5 \right| \leqq \frac{5}{3 \cdot 120} \left(\frac{1}{2} \right)^5 \approx 0.434 \times 10^{-3},$$

so two-place accuracy will be obtained if $|x| \leqq 0.5$. In particular,

$$e^{1/3} \approx 1 + \frac{1}{3} + \frac{1}{18} + \frac{1}{486} + \frac{1}{1944} \approx 1.39.$$

In fact, to the number of digits shown here, $e^{1/3} \approx 1.395612425086$.

C11S09.039: The Taylor series remainder estimate is

$$|R_3(x)| = \frac{\sqrt{2}}{2} \cdot \frac{\cos z}{4!} \left(x - \frac{\pi}{4} \right)^4.$$

Part (a): If $40^\circ \leqq x^\circ \leqq 50^\circ$, then

$$\frac{2\pi}{9} \leqq x \leqq \frac{5\pi}{18} \quad \text{and} \quad \cos z \leqq \cos\left(\frac{\pi}{6} \right) = \frac{\sqrt{3}}{2},$$

so

$$|R_3(x)| \leqq \frac{\sqrt{2}}{2} \cdot \frac{1}{24} \cdot \left(\frac{\pi}{36} \right)^4 \cdot \frac{\sqrt{3}}{2} \approx 0.0000014797688 < 0.000002,$$

thereby giving five-place accuracy. Part (b): If $44^\circ \leqq x^\circ \leqq 46^\circ$, then

$$\frac{44\pi}{180} \leqq x \leqq \frac{46\pi}{180},$$

so that

$$|R_3(x)| \leqq \frac{\sqrt{2}}{2} \cdot \frac{1}{24} \cdot \left(\frac{\pi}{180} \right)^4 \cdot \frac{\sqrt{3}}{2} \approx 0.0000000023676302 < 0.000000003,$$

thereby giving eight-place accuracy.

C11S09.041: The volume of revolution around the x-axis is

$$V = 2 \int_0^\pi \pi \frac{\sin^2 x}{x^2} \, dx = 2\pi \int_0^\pi \frac{1 - \cos 2x}{2x^2} \, dx$$

$$= \pi \int_0^\pi \frac{1}{x^2} \left(\frac{(2x)^2}{2!} - \frac{(2x)^4}{4!} + \frac{(2x)^6}{6!} - \frac{(2x)^8}{8!} + \cdots \right) dx = \pi \int_0^\pi \left(\frac{2^2}{2!} - \frac{2^4 x^2}{4!} + \frac{2^6 x^4}{6!} - \frac{2^8 x^6}{8!} + \cdots \right) dx$$

$$= \pi \left[\frac{2^2 x}{2!} - \frac{2^4 x^3}{4! \cdot 3} + \frac{2^6 x^5}{6! \cdot 5} - \frac{2^8 x^7}{8! \cdot 7} + \cdots \right]_0^\pi = \frac{(2\pi)^2}{2!} - \frac{(2\pi)^4}{4! \cdot 3} + \frac{(2\pi)^6}{6! \cdot 5} - \frac{(2\pi)^8}{8! \cdot 7} + \cdots.$$

This series converges rapidly after the first 10 or 15 terms. For example, the sum of the first seven terms is about 8.927353886225. The *Mathematica 3.0* command

```
NSum[ ((-1)∧(k+1))*((2*Pi)∧(2*k))/(((2*k)!)*(2*k-1)),

   { k, 1, Infinity }, WorkingPrecision → 28 ]
```

returns the approximate sum 8.910509146510103807178167 8. The *Mathematica* 3.0 command

```
2*Integrate[ Pi*(Sin[x]/x)∧2, {x, 0, Pi} ]
```

produces the exact value of the volume:

$$V = -1 + \text{HypergeometricPFQ}\left[\left\{-\frac{1}{2}\right\}, \left\{\frac{1}{2}, \frac{1}{2}\right\}, -\pi^2\right]$$

$$\approx 8.9105091465101038071781677928811594135107930070735323609643.$$

The *Mathematica* function `HypergeometricPFQ` is the generalized hypergeometric function $_pF_q$. Space prohibits further discussion; we've mentioned this only to give you a reference in case you're interested in further details.

C11S09.043: The volume is

$$V = \int_0^{2\pi} 2\pi x \frac{1 - \cos x}{x^2} \, dx = 2\pi \int_0^{2\pi} \left(\frac{x}{2!} - \frac{x^3}{4!} + \frac{x^5}{6!} - \frac{x^7}{8!} + \frac{x^9}{10!} - \cdots\right) dx$$

$$= 2\pi \left[\frac{x^2}{2! \cdot 2} - \frac{x^4}{4! \cdot 4} + \frac{x^6}{6! \cdot 6} - \frac{x^8}{8! \cdot 8} + \frac{x^{10}}{10! \cdot 10} - \cdots\right]_0^{2\pi}$$

$$= \frac{(2\pi)^3}{2! \cdot 2} - \frac{(2\pi)^5}{4! \cdot 4} + \frac{(2\pi)^7}{6! \cdot 6} - \frac{(2\pi)^9}{8! \cdot 8} + \frac{(2\pi)^{11}}{10! \cdot 10} - \cdots = \sum_{n=1}^{\infty} \frac{(-1)^{n+1}(2\pi)^{2n+1}}{(2n)! \cdot 2n}.$$

This alternating series converges slowly at first—its ninth term is approximately 0.0127—but its 21st term is less than 4×10^{-19}, so the sum of its first 20 terms is a very accurate estimate of its value. That partial sum is approximately 15.3162279832536178, and all the digits shown here are accurate.

The *Mathematica* 3.0 command

```
Integrate[ 2*Pi*x*(1 - Cos[x])/(x*x), {x, 0, 2*Pi} ]
```

produces the exact value of the volume:

$$V = 2\pi \left[\text{EulerGamma} - \text{CosIntegral}(2\pi) + \ln(2\pi)\right]$$

$$\approx 15.316227983253617819314890707705969367325235855602999990575827.$$

Here, `EulerGamma` is Euler's constant $\gamma \approx 0.577216$, which first appears in the textbook in Problem 50 of Section 11.5; `CosIntegral` is defined to be

$$\text{CosIntegral}(x) = \gamma + \ln x + \int_0^x \frac{(\cos t) - 1}{t} \, dt.$$

Again, the reference is provided only for your convenience if you care to pursue further study of this topic.

C11S09.045: The long division is shown next.

$$
\begin{array}{r}
1 + x + x^2 + x^3 + \cdots \\
1 - x \overline{\smash{\big)}\ 1 } \\
\underline{1 - x} \\
x \\
\underline{x - x^2} \\
x^2 \\
\underline{x^2 - x^3} \\
x^3 \\
\cdots
\end{array}
$$

C11S09.047: The equation (actually, the *identity*)

$$(1 - x)(a_0 + a_1 x + a_2 x^2 + a_3 x^3 + \cdots + a_n x^n + \cdots) = 1$$

leads to

$$a_0 + (a_1 - a_0)x + (a_2 - a_1)x^2 + (a_3 - a_2)x^3 + (a_4 - a_3)x^4 + \cdots = 1.$$

It now follows that $a_0 = 1$ and that $a_{n+1} = a_n$ if $n \geqq 0$, and therefore $a_n = 1$ for every integer $n \geqq 0$. Consequently

$$\frac{1}{1 - x} = 1 + x + x^2 + x^3 + x^4 + \cdots + x^n + \cdots, \quad -1 < x < 1.$$

C11S09.049: The method of Example 3 uses the identity

$$\sec x \cos x = 1$$

and begins by assuming the existence of coefficients $\{a_i\}$ such that

$$\sec x = a_0 + a_1 x + a_2 x^2 + a_3 x^3 + a_4 x^4 + \cdots.$$

Thus

$$(a_0 + a_1 x + a_2 x^2 + a_3 x^3 + a_4 x^4 + \cdots)\left(1 - \frac{x^2}{2} + \frac{x^4}{24} - \frac{x^6}{720} + \cdots\right) = 1,$$

so that

$$a_0 + a_1 x + \left(a_2 - \frac{1}{2}a_0\right)x^2 + \left(a_3 - \frac{1}{2}a_1\right)x^3 + \left(a_4 - \frac{1}{2}a_2 + \frac{1}{24}a_0\right)x^4 + \cdots = 1.$$

It now follows that

$$a_0 = 1; \qquad a_1 = 0;$$

$$a_2 = \frac{1}{2}a_0 = \frac{1}{2}; \qquad a_3 = \frac{1}{2}a_1 = 0;$$

$$a_4 = \frac{1}{2}a_2 - \frac{1}{24}a_0 = \frac{5}{24}.$$

Therefore

$$\sec x = 1 + \frac{1}{2}x^2 + \frac{5}{24}x^4 + \frac{61}{720}x^6 + \frac{277}{8064}x^8 + \frac{50521}{3628800}x^{10} + \frac{540553}{95800320}x^{12} + \frac{199360981}{87178291200}x^{14} + \cdots.$$

We had *Mathematica 3.0* compute a few extra terms in case you did as well and want to check your work. (We used the command

```
Series[ Sec[x], { x, 0, 20 } ] // Normal
```

but to save space we show here only the first eight terms of the resulting Taylor polynomial.)

C11S09.051: Example 10 in Section 11.8 shows how to derive the series

$$\ln(1+x) = x - \frac{x^2}{2} + \frac{x^3}{3} - \frac{x^4}{4} + \frac{x^5}{5} - \cdots, \quad -1 < x < 1.$$

Hence

$$1 + x = \exp(\ln(1+x)) = \sum_{n=0}^{\infty} a_n \left(x - \frac{x^2}{2} + \frac{x^3}{3} - \frac{x^4}{4} + \frac{x^5}{5} - \cdots \right)^n$$

$$= a_0 + a_1 \left(x - \frac{x^2}{2} + \frac{x^3}{3} - \frac{x^4}{4} + \frac{x^5}{5} - \cdots \right) + a_2 \left(x^2 - x^3 + \left[\frac{1}{4} + \frac{2}{3} \right] x^4 - \cdots \right)$$

$$+ a_3 \left(x^3 + \left[-\frac{1}{2} - 1 \right] x^4 + \cdots \right) + \cdots$$

$$= a_0 + a_1 x + \left(a_2 - \frac{1}{2}a_1 \right) x^2 + \left(a_3 - a_2 + \frac{1}{3}a_1 \right) x^3 + \cdots.$$

Therefore

$$a_0 = 1, \qquad a_1 = 1, \qquad a_2 = \frac{1}{2}a_1 = \frac{1}{2}, \qquad \text{and} \qquad a_3 = a_2 - \frac{1}{3}a_1 = \frac{1}{6}.$$

C11S09.053: Long division of the finite power series $1 + x + x^2$ into the finite power series $2 + x$ proceeds as shown next.

$$
\begin{array}{r}
2 - x - x^2 + \cdots \\
1 + x + x^2 \overline{\smash{\big)}\, 2 + x } \\
\underline{2 + 2x + 2x^2 } \\
-x - 2x^2 \\
\underline{-x - x^2 - x^3} \\
-x^2 + x^3 \\
\underline{-x^2 - x^3 - x^4} \\
2x^3 + x^4 \\
\cdots
\end{array}
$$

The next dividend is the original dividend with exponents increased by 3, so the next three terms in the numerator can be obtained by multiplying the first three by x^3. Thus we obtain the series representation

$$\frac{2+x}{1+x+x^2} = 2 - x - x^2 + 2x^3 - x^4 - x^5 + 2x^6 - x^7 - x^8 + \cdots.$$

584

This series is the sum of three geometric series each with ratio x^3, so it converges on the interval $(-1, 1)$. Summing the three geometric series separately, we obtain

$$\frac{2}{1-x^3} - \frac{x}{1-x^3} - \frac{x^2}{1-x^3} = \frac{2-x-x^2}{1-x^3} = \frac{(1-x)(2+x)}{(1-x)(1+x+x^2)},$$

thus verifying our computations.

C11S09.055: The first two steps in the long division of $1 + x^2 + x^4$ into 1 give quotient $1 - x^2$ and remainder (and new dividend) x^6. So the process will repeat with exponents increased by 6, and thus

$$\frac{1}{1+x^2+x^4} = 1 - x^2 + x^6 - x^8 + x^{12} - x^{14} + x^{18} - x^{20} + \cdots.$$

Therefore

$$\int_0^{1/2} \frac{1}{1+x^2+x^4}\, dx = \left[x - \frac{x^3}{3} + \frac{x^7}{7} - \frac{x^9}{9} + \frac{x^{13}}{13} - \frac{x^{15}}{15} \cdots \right]_0^{1/2}$$

$$= \frac{1}{2} - \frac{1}{2^3 \cdot 3} + \frac{1}{2^7 \cdot 7} - \frac{1}{2^9 \cdot 9} + \frac{1}{2^{13} \cdot 13} - \frac{1}{2^{15} \cdot 15} + \cdots.$$

The sum of the first five terms of the last series is approximately 0.459239824988 and the sum of the first six terms is approximately 0.459239825000. The *Mathematica* 3.0 command

```
NIntegrate[ 1/(1 + x∧2 + x∧4), { x, 0, 1/2 }, WorkingPrecision → 28 ]
```

returns 0.4592398249998759. The computer algebra system *Derive* 2.56 yields

$$\int_0^{1/2} \frac{1}{1+x^2+x^4}\, dx$$

$$= \left[\frac{\sqrt{3}}{6} \left\{ \arctan\left(\frac{\sqrt{3}}{3}(2x+1) \right) + \arctan\left(\frac{\sqrt{3}}{3}(2x-1) \right) \right\} + \frac{1}{4} \ln\left(\frac{x^2+x+1}{x^2-x+1} \right) \right]_0^{1/2}$$

$$= \frac{\sqrt{3}}{6} \arctan\left(\frac{2\sqrt{3}}{3} \right) + \frac{1}{4} \ln\left(\frac{7}{3} \right) \approx 0.45923982499987591403.$$

C11S09.057: See the solution of Problem 61 in Section 11.8. We plotted the Taylor polynomials with center zero for $f(x)$,

$$P_n(x) = \sum_{k=1}^{n} \frac{(-1)^{k+1} x^{2k-2}}{(2k-1)!},$$

for $n = 3, \ 6,$ and 9. Their graphs, together with the graph of f, are shown next.

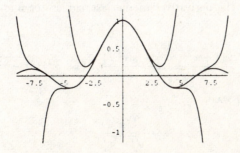

585

C11S09.059: The four Maclaurin series we need are in Eq. (4) of Section 11.8, Eq. (7) of Section 11.9, Eq. (21) of Section 11.8, and Eq. (20) of Section 11.8. They are

$$\sin x = x - \frac{x^3}{6} + \frac{x^5}{120} - \frac{x^7}{5040} + \cdots,$$

$$\tan x = x + \frac{x^3}{3} + \frac{2x^5}{15} + \frac{17x^7}{315} + \cdots,$$

$$\arcsin x = x + \frac{x^3}{6} + \frac{3x^5}{40} + \frac{5x^7}{112} + \cdots, \quad \text{and}$$

$$\arctan x = x - \frac{x^3}{3} + \frac{x^5}{5} - \frac{x^7}{7} + \cdots.$$

Therefore

$$\frac{\sin x - \tan x}{\arcsin x - \arctan x} = \frac{-\frac{1}{2}x^3 - \frac{1}{8}x^5 - \cdots}{\frac{1}{2}x^3 + \frac{1}{8}x^5 + \cdots} = \frac{-\frac{1}{2} - \frac{1}{8}x^2 - \cdots}{\frac{1}{2} + \frac{1}{8}x^2 + \cdots} \rightarrow \frac{-\frac{1}{2} - 0 - 0 - \cdots}{\frac{1}{2} + 0 + 0 + \cdots} = -1$$

as $x \to 0$.

Use of l'Hôpital's rule to solve this problem—even with the aid of a computer algebra program—is troublesome. Using *Mathematica* 3.0, we first defined

$$f(x) = \sin x - \tan x \quad \text{and} \quad g(x) = \arcsin x - \arctan x.$$

Then the command

```
Limit[ f[x]/g[x], x → 0 ]
```

elicited the disappointing response "Indeterminate." But when we computed the quotient of the derivatives using

```
f'[x]/g'[x]
```

we obtained the expected fraction

$$\frac{\cos x - \sec^2 x}{\dfrac{1}{\sqrt{1 - x^2}} - \dfrac{1}{1 + x^2}},$$

and *Mathematica* reported that the limit of this fraction, as $x \to 0$, was -1. To avoid *Mathematica's* invocation of l'Hôpital's rule (with the intent of obtaining a fraction that was *not* indeterminate), we had *Mathematica* find the quotient of the derivatives of the last numerator and denominator; the resulting numerator was

$$-2x\sqrt{1 - x^2}\,(\cos x - \sec^2 x) + \frac{x(1 + x^2)(\cos x - \sec^2 x)}{\sqrt{1 - x^2}} + (1 + x^2)\sqrt{1 - x^2}\,(\sin x + 2\sec^2 x \tan x)$$

and the corresponding denominator was

$$-2x - \frac{x}{\sqrt{1 - x^2}}.$$

The quotient is still indeterminate, but repeating the process—you really don't want to see the results—next (and finally) led to a form not indeterminate, whose value at $x = 0$ was (still) -1.

C11S09.061: Part (a): Assume that $a \geq b > 0$. The parametrization $x = a\cos t$, $y = b\sin t$ yields arc length element

$$ds = (a^2 \sin^2 t + b^2 \cos^2 t)^{1/2} \, dt = [a^2 \sin^2 t + a^2 \cos^2 t + (b^2 - a^2)\cos^2 t]^{1/2} \, dt$$

$$= [a^2 + (b^2 - a^2)\cos^2 t]^{1/2} \, dt = a\left[1 - \frac{a^2 - b^2}{a^2}\cos^2 t\right]^{1/2} \, dt = a(1 - \epsilon^2 \cos^2 t)^{1/2} \, dt$$

where

$$\epsilon = \sqrt{\frac{a^2 - b^2}{a^2}} = \sqrt{1 - (b/a)^2}$$

is the eccentricity of the ellipse; recall that $0 \leq \epsilon < 1$ for the case of an ellipse. We multiply the length of the part of the ellipse in the first quadrant by 4 to find its total arc length is

$$p = 4a \int_0^{\pi/2} \sqrt{1 - \epsilon^2 \cos^2 t} \, dt.$$

Part (b): The binomial formula yields

$$(1 - x)^{1/2} = 1 - \frac{1}{2}x - \frac{1}{2^2} \cdot \frac{x^2}{2!} - \frac{3}{2^3} \cdot \frac{x^3}{3!} - \frac{3 \cdot 5}{2^4} \cdot \frac{x^4}{4!} - \frac{3 \cdot 5 \cdot 7}{2^5} \cdot \frac{x^5}{5!} - \cdots.$$

Consequently,

$$\sqrt{1 - \epsilon^2 \cos^2 t} = 1 - \frac{1}{2}\epsilon^2 \cos^2 t - \frac{1}{2^2 \cdot 2!}\epsilon^4 \cos^4 t - \frac{3}{2^3 \cdot 3!}\epsilon^6 \cos^6 t$$

$$- \frac{3 \cdot 5}{2^4 \cdot 4!}\epsilon^8 \cos^8 t - \frac{3 \cdot 5 \cdot 7}{2^5 \cdot 5!}\epsilon^{10} \cos^{10} t - \cdots.$$

Then formula 113 in the Table of Integrals in the text yields

$$p = 4a \int_0^{\pi/2} \sqrt{1 - \epsilon^2 \cos^2 t} \, dt$$

$$= 4a\left(\left[t\right]_0^{\pi/2} - \frac{1}{2}\epsilon^2 \cdot \frac{1}{2} \cdot \frac{\pi}{2} - \frac{1}{2^2 \cdot 2!}\epsilon^4 \cdot \frac{1}{2} \cdot \frac{3}{4} \cdot \frac{\pi}{2}\right.$$

$$\left. - \frac{3}{2^3 \cdot 3!}\epsilon^6 \cdot \frac{1}{2} \cdot \frac{3}{4} \cdot \frac{5}{6} \cdot \frac{\pi}{2} - \frac{3 \cdot 5}{2^4 \cdot 4!}\epsilon^8 \cdot \frac{1}{2} \cdot \frac{3}{4} \cdot \frac{5}{6} \cdot \frac{7}{8} \cdot \frac{\pi}{2} - \cdots\right)$$

$$= 2\pi a\left(1 - \frac{1}{4}\epsilon^2 - \frac{3}{64}\epsilon^4 - \frac{5}{256}\epsilon^6 - \frac{175}{16384}\epsilon^8 - \cdots\right).$$

Section 11.10

C11S10.001: We use series methods to solve $\dfrac{dy}{dx} = y$. Assume that

587

$$y(x) = \sum_{n=0}^{\infty} a_n x^n,$$

so that

$$y'(x) = \sum_{n=1}^{\infty} n a_n x^{n-1} = \sum_{n=0}^{\infty} (n+1) a_{n+1} x^n.$$

Substitution in the given differential equation yields

$$\sum_{n=0}^{\infty} \left[(n+1) a_{n+1} - a_n \right] x^n = 0,$$

and hence

$$a_{n+1} = \frac{a_n}{n+1} \quad \text{for} \quad n \geqq 0.$$

Thus

$$a_1 = a_0, \qquad a_2 = \frac{a_1}{2} = \frac{a_0}{2},$$

$$a_3 = \frac{a_2}{3} = \frac{a_0}{3 \cdot 2}, \qquad a_4 = \frac{a_3}{4} = \frac{a_0}{4!},$$

$$a_5 = \frac{a_4}{5} = \frac{a_0}{5!}, \qquad \cdots .$$

In general, $a_n = \dfrac{a_0}{n!}$ if $n \geqq 0$. Hence

$$y(x) = \sum_{n=0}^{\infty} \frac{a_0}{n!} x^n = a_0 \sum_{n=0}^{\infty} \frac{x^n}{n!} = a_0 e^x.$$

Finally,

$$\lim_{n \to \infty} \left| \frac{n! x^{n+1}}{(n+1)! x^n} \right| = |x| \cdot \left(\lim_{n \to \infty} \frac{1}{n+1} \right) = 0,$$

so the radius of convergence of the series we found is $+\infty$.

C11S10.003: We use series methods to solve $2\dfrac{dy}{dx} + 3y = 0$. Assume that

$$y(x) = \sum_{n=0}^{\infty} a_n x^n,$$

so that

$$y'(x) = \sum_{n=1}^{\infty} n a_n x^{n-1} = \sum_{n=0}^{\infty} (n+1) a_{n+1} x^n.$$

Substitution in the given differential equation yields

$$\sum_{n=0}^{\infty} \left[2(n+1)a_{n+1} + 3a_n \right] x^n = 0,$$

and thus

$$a_{n+1} = -\frac{3a_n}{2(n+1)} \quad \text{if} \quad n \geqq 0.$$

Therefore

$$a_1 = -\frac{3}{2}a_0, \qquad a_2 = -\frac{3}{2} \cdot \frac{a_1}{2} = \left(\frac{3}{2} \right)^2 \cdot \frac{a_0}{2},$$

$$a_3 = -\frac{3}{2} \cdot \frac{a_2}{3} = -\left(\frac{3}{2} \right)^3 \cdot \frac{a_0}{3!}, \qquad \ldots .$$

In general,

$$a_n = (-1)^n \left(\frac{3}{2} \right)^n \cdot \frac{a_0}{n!} \quad \text{for} \quad n \geqq 1.$$

Therefore

$$y(x) = a_0 \sum_{n=0}^{\infty} (-1)^n \left(\frac{3}{2} \right)^n \cdot \frac{x_n}{n!} = a_0 \sum_{n=0}^{\infty} \frac{(-1)^n}{n!} \left(\frac{3x}{2} \right)^n = a_0 e^{-3x/2}.$$

By computations quite similar to those in the solution of Problem 1, this series has radius of convergence $+\infty$.

C11S10.005: We use series methods to solve $\dfrac{dy}{dx} = x^2 y$. Assume that

$$y(x) = \sum_{n=0}^{\infty} a_n x^n,$$

so that

$$y'(x) = \sum_{n=1}^{\infty} n a_n x^{n-1} = \sum_{n=0}^{\infty} (n+1)a_{n+1} x^n.$$

Substitution in the given differential equation yields

$$\sum_{n=0}^{\infty} (n+1)a_{n+1} x^n = \sum_{n=0}^{\infty} a_n x^{n+2} = \sum_{n=2}^{\infty} a_{n-2} x^n.$$

Therefore $a_1 = 0$, $a_2 = 0$, and $(n+1)a_{n+1} = a_{n-2}$ if $n \geqq 2$; that is,

$$a_1 = a_2 = a_4 = a_5 = a_7 = a_8 = \cdots 0 \quad \text{and} \quad a_{n+3} = \frac{a_n}{n+3}$$

if $n \geqq 0$. Hence

589

$$a_3 = \frac{a_0}{3} = \frac{a_0}{1! \cdot 3}, \qquad a_6 = \frac{a_3}{6} = \frac{a_0}{6 \cdot 3} = \frac{a_0}{2! \cdot 3^2},$$

$$a_9 = \frac{a_6}{9} = \frac{a_0}{9 \cdot 6 \cdot 3} = \frac{a_0}{3! \cdot 3^3}, \qquad \cdots \;;$$

in general,

$$a_{3n} = \frac{a_0}{n! \cdot 3^n} \quad \text{if } n \geqq 1.$$

Therefore

$$y(x) = a_0 \left[1 + \frac{1}{1!} \cdot \frac{x^3}{3} + \frac{1}{2!} \left(\frac{x^3}{3} \right)^2 + \frac{1}{3!} \left(\frac{x^3}{3} \right)^3 + \cdots \right] = a_0 \exp \left(\frac{x^3}{3} \right).$$

As in previous solutions, the radius of convergence of this series is $+\infty$.

C11S10.007: We use series methods to solve $(2x - 1)\dfrac{dy}{dx} + 2y = 0$. Assume that

$$y(x) = \sum_{n=0}^{\infty} a_n x^n,$$

so that

$$y'(x) = \sum_{n=1}^{\infty} n a_n x^{n-1} = \sum_{n=0}^{\infty} (n+1)a_{n+1} x^n.$$

Substitution in the given differential equation yields

$$\sum_{n=1}^{\infty} 2n a_n x^n - \sum_{n=0}^{\infty} (n+1)a_{n+1} x^n + \sum_{n=0}^{\infty} 2a_n x^n = 0.$$

If $n = 0$, we find that $-a_1 + 2a_0 = 0$, and thus that $a_1 = 2a_0$. If $n \geqq 1$, then

$$2n a_n - (n+1)a_{n+1} + 2a_n = 0; \qquad (n+1)a_{n+1} = 2(n+1)a_n; \qquad a_{n+1} = 2a_n.$$

Hence $a_1 = 2a_0$, $a_2 = 2^2 a_0$, $a_3 = 2^3 a_0$, etc.; in general, $a_n = 2^n a_0$ if $n \geqq 1$. Therefore

$$y(x) = a_0 \sum_{n=0}^{\infty} 2^n x^n = a_0 \sum_{n=0}^{\infty} (2x)^n = \frac{a_0}{1 - 2x}$$

because the series is geometric; for the same reason, its radius of convergence is $R = \frac{1}{2}$.

C11S10.009: We use series methods to solve $(x - 1)\dfrac{dy}{dx} + 2y = 0$. Assume that

$$y(x) = \sum_{n=0}^{\infty} a_n x^n,$$

so that

$$y'(x) = \sum_{n=1}^{\infty} n a_n x^{n-1} = \sum_{n=0}^{\infty} (n+1)a_{n+1} x^n.$$

Substitution in the given differential equation yields

$$\sum_{n=1}^{\infty} na_n x^n - \sum_{n=0}^{\infty} (n+1)a_{n+1} x^n + \sum_{n=0}^{\infty} 2a_n x^n = 0.$$

When $n = 0$, we have $-a_1 + 2a_0 = 0$, so that $a_1 = 2a_0$. If $n \geq 1$, then

$$na_n - (n+1)a_{n+1} + 2a_n = 0 : (n+1)a_{n+1} = (n+2)a_n,$$

and hence

$$a_{n+1} = \frac{n+2}{n+1} a_n \quad \text{if} \quad n \geq 0.$$

Therefore

$$a_2 = \frac{3}{2} a_1 = 3a_0, \qquad a_3 = \frac{4}{3} a_2 = 4a_0,$$

$$a_4 = \frac{5}{4} a_3 = 5a_0, \qquad \dots ;$$

in general, $a_n = (n+1)a_0$ if $n \geq 1$. Therefore

$$y(x) = a_0 \sum_{n=0}^{\infty} (n+1)x^n.$$

Now $y(x) = F'(x)$ where

$$F(x) = a_0 \sum_{n=0}^{\infty} x^{n+1} = \frac{a_0 x}{1 - x}.$$

Consequently,

$$y(x) = F'(x) = \frac{a_0(1 - x + x)}{(1 - x)^2} = \frac{a_0}{(1 - x)^2}.$$

The radius of convergence of the series for $y(x)$ is $R = 1$.

C11S10.011: We use series methods to solve the differential equation $y'' = y$. Assume the existence of a solution of the form

$$y(x) = \sum_{n=0}^{\infty} a_n x^n.$$

Then

$$y'(x) = \sum_{n=1}^{\infty} na_n x^{n-1} = \sum_{n=0}^{\infty} (n+1)a_{n+1} x^n \qquad \text{and}$$

$$y''(x) = \sum_{n=2}^{\infty} n(n-1)a_n x^{n-2} = \sum_{n=0}^{\infty} (n+2)(n+1)a_{n+2} x^n.$$

591

Then substitution in the given differential equation yields

$$a_{n+2} = \frac{a_n}{(n+2)(n+1)} \quad \text{for} \quad n \geqq 0.$$

Therefore

$$a_2 = \frac{a_0}{2 \cdot 1}, \qquad a_3 = \frac{a_1}{3 \cdot 2},$$

$$a_4 = \frac{a_0}{4!}, \qquad a_5 = \frac{a_1}{5!},$$

and so on. Hence

$$y(x) = a_0 \left(1 + \frac{x^2}{2!} + \frac{x^4}{4!} + \frac{x^6}{6!} + \cdots \right) + a_1 \left(x + \frac{x^3}{3!} + \frac{x^5}{5!} + \frac{x^7}{7!} + \cdots \right)$$

$$= a_0 \cosh x + a_1 \sinh x.$$

The radius of convergence of all series here is $R = +\infty$. The solution may also be expressed in the form $y(x) = c_1 e^x + c_2 e^{-x}$.

C11S10.013: We use series methods to solve the differential equation $y'' + 9y = 0$. Assume the existence of a solution of the form

$$y(x) = \sum_{n=0}^{\infty} a_n x^n.$$

Then

$$y'(x) = \sum_{n=1}^{\infty} n a_n x^{n-1} = \sum_{n=0}^{\infty} (n+1) a_{n+1} x^n \qquad \text{and}$$

$$y''(x) = \sum_{n=2}^{\infty} n(n-1) a_n x^{n-2} = \sum_{n=0}^{\infty} (n+2)(n+1) a_{n+2} x^n.$$

Then substitution in the given differential equation leads—as in the solution of Problem 11—to the recursion formula $(n+2)(n+1)a_{n+2} + 9a_n = 0$, and thus

$$a_{n+2} = -\frac{9}{(n+2)(n+1)} a_n \quad \text{for} \quad n \geqq 0.$$

Hence

$$a_2 = -\frac{9}{2!} a_0, \qquad a_3 = -\frac{9}{3!} a_1,$$

$$a_4 = \frac{9^2}{4!} a_0, \qquad a_5 = \frac{9^2}{5!} a_1,$$

$$a_6 = -\frac{9^3}{6!} a_0, \qquad a_7 = -\frac{9^3}{7!} a_1,$$

and so on. Hence

592

$$y(x) = a_0 \left(1 - \frac{9x^2}{2!} + \frac{9^2 x^4}{4!} - \frac{9^3 x^6}{6!} + \cdots \right) + a_1 \left(x - \frac{9x^3}{3!} + \frac{9^2 x^5}{5!} - \frac{9^3 x^7}{7!} + \cdots \right)$$

$$= a_0 \left(1 - \frac{(3x)^2}{2!} + \frac{(3x)^4}{4!} - \frac{(3x)^6}{6!} + \cdots \right) + \frac{a_1}{3} \left(3x - \frac{(3x)^3}{3!} + \frac{(3x)^5}{5!} - \frac{(3x)^7}{7!} + \cdots \right)$$

$$= a_0 \cos 3x + \frac{a_1}{3} \sin 3x = c_1 \cos 3x + c_2 \sin 3x.$$

The radius of convergence of each series here is $R = +\infty$.

C11S10.015: Given the differential equation $x \frac{dy}{dx} + y = 0$, substitution of the series

$$y(x) = \sum_{n=0}^{\infty} a_n x^n \tag{1}$$

as in earlier solutions in this section yields

$$\sum_{n=1}^{\infty} n a_n x^n + \sum_{n=0}^{\infty} a_n x^n = 0.$$

It then follows that $a_0 = 0$ and that $n a_n + a_n = 0$ if $n \geq 1$. The latter equation implies that $a_n = 0$ if $n \geq 1$. Thus we obtain only the trivial solution $y(x) \equiv 0$, which is not part of the general solution because it contains no arbitrary constant and is not independent of any other solution. Part of the reason that the series has no solution of the form in (1) is that a general solution is

$$y(x) = \frac{C}{x}.$$

This solution is undefined at $x = 0$ and, of course, has no power series expansion with center $c = 0$. Here's an experiment for you: Assume a solution of the form

$$\sum_{n=0}^{\infty} b_n (x-1)^n$$

and see what happens. Then assume a solution of the form

$$\sum_{n=-1}^{\infty} c_n x^n$$

and see what happens. You can learn more about these ideas, and their consequences, in a standard course in differential equations (make sure that the syllabus includes the topic of series solution of ordinary differential equations).

C11S10.017: Given the differential equation $x \frac{dy}{dx} + y = 0$, substitution of the series

$$y(x) = \sum_{n=0}^{\infty} a_n x^n \tag{1}$$

as in earlier solutions in this section yields

$$\sum_{n=1}^{\infty} n a_n x^{n+1} + \sum_{n=0}^{\infty} a_n x^n = 0;$$

593

$$\sum_{n=2}^{\infty} (n-1)a_{n-1}x^n + \sum_{n=0}^{\infty} a_n x^n = 0.$$

Examination of the cases $n = 0$ and $n = 1$ yields $a_0 = a_1 = 0$. If $n \geq 2$ we see that $(n-1)a_{n-1} + a_n = 0$, and hence that $a_n = 0$ for all $n \geq 0$. Thus the series method using the form in (1) uncovers only the trivial solution $y(x) \equiv 0$, not a general solution of the given differential equation. Part of the reason is that a general solution of the differential equation is

$$y(x) = C \exp\left(\frac{1}{x}\right).$$

C11S10.019: Given the initial value problem

$$y'' + 4y = 0; \qquad y(0) = 0, \quad y'(0) = 3,$$

we assume the existence of a series solution of the form

$$y(x) = \sum_{n=0}^{\infty} a_n x^n.$$

Then

$$y'(x) = \sum_{n=1}^{\infty} n a_n x^{n-1} = \sum_{n=0}^{\infty} (n+1)a_{n+1}x^n \qquad \text{and}$$

$$y''(x) = \sum_{n=2}^{\infty} n(n-1)a_n x^{n-2} = \sum_{n=0}^{\infty} (n+2)(n+1)a_{n+2}x^n.$$

Substitution in the given differential equation yields $(n+2)(n+1)a_{n+2} + 4a_n = 0$, from which we obtain the recurrence relation

$$a_{n+2} = -\frac{4}{(n+2)(n+1)}a_n \qquad \text{for} \quad n \geq 0.$$

Thus we may choose a_0 and a_1 to be arbitrary constants, and find that

$$a_2 = -\frac{4}{2!}a_0, \qquad a_3 = -\frac{4}{3!}a_1,$$

$$a_4 = -\frac{4}{4\cdot 3}a_2 = \frac{4^2}{4!}a_0, \qquad a_5 = \frac{4^2}{5!}a_1,$$

$$a_6 = -\frac{4^3}{6!}a_0, \qquad a_7 = -\frac{4^3}{7!}a_1,$$

and so on. Therefore the general solution of the given differential equation may be written in the form

$$y(x) = a_0\left(1 - \frac{4x^2}{2!} + \frac{4^2 x^4}{4!} - \frac{4^3 x^6}{6!} + \frac{4^4 x^8}{8!} - \cdots\right) + a_1\left(x - \frac{4x^3}{3!} + \frac{4^2 x^5}{5!} - \frac{4^3 x^7}{7!} + \frac{4^4 x^9}{9!} - \cdots\right)$$

$$= a_0 \cos 2x + \frac{a_1}{2}\sin 2x = A\cos 2x + B\sin 2x.$$

594

Substitution of the initial conditions yields $a_0 = y(0) = 0$ and $a_1 = y'(0) = 3$, so the particular solution of the differential equation is

$$y(x) = \frac{3}{2} \sin 2x.$$

C11S10.021: Given the initial value problem

$$y'' - 2y' + y = 0; \qquad y(0) = 0, \quad y'(0) = 1,$$

we assume the existence of a series solution of the form

$$y(x) = \sum_{n=0}^{\infty} a_n x^n.$$

Then

$$y'(x) = \sum_{n=1}^{\infty} n a_n x^{n-1} = \sum_{n=0}^{\infty} (n+1) a_{n+1} x^n \qquad \text{and}$$

$$y''(x) = \sum_{n=2}^{\infty} n(n-1) a_n x^{n-2} = \sum_{n=0}^{\infty} (n+2)(n+1) a_{n+2} x^n.$$

Substitution in the given differential equation then yields

$$(n+2)(n+1) a_{n+2} - 2(n+1) a_{n+1} + a_n = 0 \qquad \text{for} \quad n \geq 0,$$

so that

$$a_{n+2} = \frac{2(n+1) a_{n+1} - a_n}{(n+2)(n+1)}, \qquad n \geq 0.$$

At this point it would be easier to use the information that $a_0 = 0$ and $a_1 = 1$ to help find the general form of the coefficient a_n, but we choose to demonstrate that it is not necessary and, instead, find the general solution of the differential equation in terms of a_0 and a_1 as yet unspecified. Using the recursion formula just derived, we find that

$$a_2 = \frac{2a_1 - a_0}{2 \cdot 1} = \frac{2a_1 - a_0}{2!},$$

$$a_3 = \frac{4a_2 - a_1}{3 \cdot 2} = \frac{4a_1 - 2a_0 - a_1}{3 \cdot 2} = \frac{3a_1 - 2a_0}{3!},$$

$$a_4 = \frac{6a_3 - a_2}{4 \cdot 3} = \frac{3a_1 - 2a_0 - a_1 + \frac{1}{2} a_0}{4 \cdot 3} = \frac{4a_1 - 3a_0}{4!}, \qquad \text{and}$$

$$a_5 = \frac{8a_4 - a_3}{5 \cdot 4} = \frac{\frac{4}{3} a_1 - a_0 - \frac{1}{2} a_1 + \frac{1}{3} a_0}{5 \cdot 4} = \frac{8a_1 - 6a_0 - 3a_1 + 2a_0}{5!} = \frac{5a_1 - 4a_0}{5!}.$$

At this point one might conjecture that

$$a_n = \frac{n a_1 - (n-1) a_0}{n!} \qquad \text{if} \quad n \geq 2,$$

and this can be established using a proof by induction on n. That granted, it follows that

595

$$y(x) = a_0 + a_1 x + \frac{2a_1 - a_0}{2!} x^2 + \frac{3a_1 - 2a_0}{3!} x^3 + \frac{4a_1 - 3a_0}{4!} x^4 + \frac{5a_1 - 4a_0}{5!} x^5 + \cdots$$

$$= a_0 \left(1 - \frac{x^2}{2!} - \frac{2x^3}{3!} - \frac{3x^4}{4!} - \frac{4x^5}{5!} - \cdots \right) + a_1 \left(x + x^2 + \frac{x^3}{2!} + \frac{x^4}{3!} + \frac{x^5}{4!} + \cdots \right)$$

$$= a_0 \left(1 - \frac{x^2}{2!} - \frac{2x^3}{3!} - \frac{3x^4}{4!} - \frac{4x^5}{5!} - \cdots \right) + a_1 x \left(1 + x + \frac{x^2}{2!} + \frac{x^3}{3!} + \frac{x^4}{4!} + \cdots \right).$$

Let

$$F(x) = 1 - \frac{x^2}{2!} - \frac{2x^3}{3!} - \frac{3x^4}{4!} - \frac{4x^5}{5!} - \cdots .$$

Then

$$F'(x) = -x - x^2 - \frac{x^3}{2!} - \frac{x^4}{3!} - \frac{x^5}{4!} - \cdots$$

$$= -x \left(1 + x + \frac{x^2}{2!} + \frac{x^3}{3!} + \frac{x^4}{4!} + \cdots \right) = -xe^x.$$

Therefore $F(x) = (1 - x)e^x + C$. Moreover, $F(0) = 1$, so that $C = 0$. Consequently,

$$y(x) = a_0(1 - x)e^x + a_1 x e^x = a_0 e^x + (a_1 - a_0)x e^x = A e^x + B x e^x.$$

Finally, the given initial conditions imply that $A = y(0) = 0$ and that $B = y'(0) = 1$. Therefore the particular solution of the original initial value problem is $y(x) = x e^x$.

C11S10.023: Suppose that the differential equation

$$x^2 y'' + x^2 y' + y = 0$$

has a series solution of the form

$$y(x) = \sum_{n=0}^{\infty} c_n x^n.$$

Then

$$y'(x) = \sum_{n=1}^{\infty} n c_n x^{n-1} = \sum_{n=0}^{\infty} (n+1)c_{n+1} x^n \qquad \text{and}$$

$$y''(x) = \sum_{n=2}^{\infty} n(n-1)c_n x^{n-2} = \sum_{n=0}^{\infty} (n+2)(n+1)c_{n+2} x^n.$$

Substitution in the given differential equation then yields

$$\sum_{n=2}^{\infty} n(n-1)c_n x^n + \sum_{n=1}^{\infty} n c_n x^{n+1} + \sum_{n=0}^{\infty} c_n x^n = 0;$$

$$\sum_{n=2}^{\infty} n(n-1)c_n x^n + \sum_{n=2}^{\infty} (n-1)c_{n-1} x^n + \sum_{n=0}^{\infty} c_n x^n = 0.$$

It then follows that $c_0 = c_1 = 0$ and that, if $n \geq 2$,

$$n(n-1)c_n + (n-1)c_{n-1} + c_n = 0; \quad \text{that is,} \quad c_n = -\frac{n-1}{n^2 - n + 1}c_{n-1}.$$

Thus

$$c_2 = -\frac{1}{3}c_1 = 0, \quad c_3 = -\frac{2}{7}c_2 = 0, \quad c_4 = -\frac{3}{13}c_3 = 0,$$

and so on: $c_n = 0$ for all $n \geq 0$. Therefore the only solution discovered by the series method used here is the trivial solution $y(x) \equiv 0$. Not only do we not find two linearly independent solutions, there is not even one because the trivial solution is neither independent of any solution nor has it the form of a general solution.

C11S10.025: Part (a): The method of separation of variables yields

$$\frac{1}{1+y^2}\, dy = 1\, dx; \qquad \arctan y = x + C;$$

$$y(x) = \tan(x+C). \qquad 0 = y(0) = \tan C :$$

$$C = n\pi \quad (n \text{ is an integer}); \qquad y(x) = \tan(x + n\pi) = \tan x.$$

Part (b): If

$$y(x) = x + c_3 x^3 + c_5 x^5 + c_7 x^7 + c_9 x^9 + \cdots, \qquad \text{then}$$

$$y'(x) = 1 + 3c_3 x^2 + 5c_5 x^4 + 7c_7 x^6 + 9c_9 x^8 + \cdots. \qquad \text{Hence}$$

$$1 + [y(x)]^2 = 1 + x^2 + 2c_3 x^4 + (c_3^2 + 2c_5)x^6 + (2c_3 c_5 + 2c_7)x^8$$

$$+ (2c_3 c_7 + c_5^2 + 2c_9)x^{10} + (2c_3 c_9 + 2c_5 c_7 + 2c_{11})x^{12} + (2c_3 c_{11} + 2c_5 c_9 + c_7^2 + 2c_{13})x^{14} + \cdots$$

$$= y'(x) = 1 + 3c_3 x^2 + 5c_5 x^4 + 7c_7 x^6 + 9c_9 x^8 + 11c_{11}x^{10} + \cdots.$$

It follows that

$$3c_3 = 1 : \qquad c_3 = \frac{1}{3}.$$

$$5c_5 = 2c_3 = \frac{2}{3} : \qquad c_5 = \frac{2}{15}.$$

$$7c_7 = c_3^2 + 2c_5 = \frac{1}{9} + \frac{4}{15} = \frac{17}{45} : \qquad c_7 = \frac{17}{315}.$$

$$9c_9 = 2c_3 c_5 + 2c_7 = \frac{62}{315} : \qquad c_9 = \frac{62}{2835}.$$

$$11c_{11} = 2c_3 c_7 + c_5^2 + 2c_9 = \frac{1382}{14175} : \qquad c_{11} = \frac{1382}{155925}.$$

Part (c): Continuing in this manner, we find that

$$\tan x = x + \frac{1}{3}x^3 + \frac{2}{15}x^5 + \frac{17}{315}x^7 + \frac{62}{2835}x^9 + \frac{1382}{155925}x^{11} + \frac{21844}{6081075}x^{13} + \frac{929569}{638512875}x^{15}$$

597

$$+ \frac{6404582}{10854518875}x^{17} + \frac{443861162}{1856156927825}x^{19} + \frac{18888466084}{194896477400625}x^{21}$$

$$+ \frac{113927491862}{2900518163668125}x^{23} + \frac{58870668456604}{3698160658676859375}x^{25} + \cdots.$$

Chapter 11 Miscellaneous Problems

C11S0M.001: Divide each term in numerator and denominator by n^2:

$$\lim_{n \to \infty} \frac{n^2 + 1}{n^2 + 4} = \lim_{n \to \infty} \frac{1 + \dfrac{1}{n^2}}{1 + \dfrac{4}{n^2}} = \frac{1 + 0}{1 + 0} = 1.$$

C11S0M.003: In Example 9 of Section 11.2, it is shown that if $|r| < 1$, then $r^n \to 0$ as $n \to +\infty$. Therefore

$$\lim_{n \to \infty} \left[10 - (0.99)^n \right] = 10 - 0 = 10.$$

C11S0M.005: Because

$$0 \le |a_n| = \frac{|1 + (-1)^n \sqrt{n}\,|}{n + 1} \le \frac{2\sqrt{n}}{n} = \frac{2}{\sqrt{n}} \to 0$$

as $n \to +\infty$, the sequence with the given general term converges to 0 by the squeeze law for limits. This problem is the result of a typograhical error; it was originally intended to be the somewhat more challenging problem in which

$$a_n = \frac{1 + (-1)^n n^{1/n}}{n + 1}$$

for $n \ge 1$.

C11S0M.007: Because $-1 \le \sin 2n \le 1$ for every positive integer n,

$$-\frac{1}{n} \le \frac{\sin 2n}{n} \le \frac{1}{n} \quad .$$

for every positive integer n. Therefore, by the squeeze law for limits (Theorem 3 of Section 11.2),

$$\lim_{n \to \infty} \frac{\sin 2n}{n} = 0.$$

C11S0M.009: If n is an even positive integer, then $\sin(n\pi/2) = 0$. Therefore $a_n = (-1)^0 = 1$ for arbitrarily large values of n. Hence if the sequence $\{a_n\}$ has a limit, it must be 1. But if n is an odd positive integer, then $\sin(n\pi/2) = \pm 1$. Therefore $a_n = -1$ for arbitrarily large values of n. So if the sequence $\{a_n\}$ has a limit, it must be -1. Because $1 \ne -1$, the sequence $\{a_n\}$ has no limit as $n \to +\infty$.

C11S0M.011: Because $-1 \le \sin x \le 1$ for all x,

$$-\frac{1}{n} \le \frac{1}{n}\sin\frac{1}{n} \le \frac{1}{n}$$

for every positive integer n. Therefore by the squeeze law for sequences, $\lim\limits_{n\to\infty} \dfrac{1}{n}\sin\dfrac{1}{n} = 0$.

C11S0M.013: Here we have

$$\lim_{n\to\infty}\frac{\sinh n}{n} = \lim_{n\to\infty}\frac{e^n - e^{-n}}{2n} = \lim_{n\to\infty}\frac{1 - e^{-2n}}{2ne^{-n}}.$$

The numerator in the last fraction is approaching 1 as $n \to +\infty$, but the denominator is approaching zero through positive values (by Eq. (8) of Section 7.2). Therefore $a_n \to +\infty$ as $n \to +\infty$. Alternatively, you may say that the limit in question does not exist.

C11S0M.015: Example 7 in Section 11.2 implies that $3^{1/n} \to 1$ as $n \to +\infty$. Example 11 in Section 11.2 shows that $n^{1/n} \to 1$ as $n \to +\infty$. Moreover, if n is a positive integer, then

$$n^{1/n} \leqq (2n^2 + 1)^{1/n} \leqq (3n^2)^{1/n} = 3^{1/n} \cdot \left(n^{1/n}\right)^2,$$

and

$$\lim_{n\to\infty} 3^{1/n} \cdot \left(n^{1/n}\right)^2 = \left(\lim_{n\to\infty} 3^{1/n}\right) \cdot \left(\lim_{n\to\infty} n^{1/n}\right)^2 = 1 \cdot 1 = 1.$$

Therefore, by the squeeze law for limits,

$$\lim_{n\to\infty} (2n^2 + 1)^{1/n} = 1.$$

C11S0M.017: By l'Hôpital's rule,

$$\lim_{x\to\infty}\frac{\ln x}{x} = \lim_{x\to\infty}\frac{1}{x} = 0,$$

so $(\ln n)/n \to 0$ as $n \to +\infty$ by Theorem 4 of Section 11.2. Also, if

$$f(x) = \frac{\ln x}{x}, \quad\text{then}\quad f'(x) = \frac{1 - \ln x}{x^2},$$

which is negative for $x > e$, and so the sequence $\{(\ln n)/n\}$ is monotonically decreasing if $n \geqq 3$. Therefore the given series converges by the alternating series test. Its sum is approximately 0.080357603217. The *Mathematica* 3.0 command

```
Sum[ ((-1)^(n+1))*(Log[n])/n, {n, 2, Infinity} ]
```

almost immediately produces the exact value of the sum of the series; it is

$$(-\gamma + \ln 2)\ln 2 \approx 0.0803576032166697405766033928384159153690544520408140507626608$$

(Euler's constant γ is first discussed in Problem 50 of Section 10.5 of the text).

C11S0M.019: This series converges because the ratio test yields

$$\rho = \lim_{n\to\infty}\frac{(n+1)!\exp(n^2)}{n!\exp([n+1]^2)} = \lim_{n\to\infty}(n+1)\exp\left(n^2 - [n+1]^2\right) = \lim_{n\to\infty}\frac{n+1}{\exp(2n+1)} = 0$$

by a result from Chapter 7 and the squeeze law for limits:

$$0 \leq \frac{n+1}{e^{2n+1}} \leq \frac{2n+1}{e^{2n+1}}$$

for every positive integer n. The sum of the series is approximately 1.405253880284.

C11S0M.021: For every positive integer n,

$$\left| \frac{(-2)^n}{3^n + 1} \right| \leq \frac{2^n}{3^n} = \left(\frac{2}{3} \right)^n.$$

Therefore the series

$$\sum_{n=0}^{\infty} \left| \frac{(-2)^n}{3^n + 1} \right| \quad \text{is dominated by} \quad \sum_{n=0}^{\infty} \left(\frac{2}{3} \right)^n.$$

The latter series converges because it is geometric with ratio $\frac{2}{3}$, and therefore the dominated series converges. Hence the original series of Problem 21 converges absolutely, and therefore it converges by Theorem 3 of Section 11.7. Its sum is approximately 0.230836643803.

C11S0M.023: Three applications of l'Hôpital's rule yield

$$\lim_{x \to \infty} \frac{x}{(\ln x)^3} = \lim_{x \to \infty} \frac{x}{3(\ln x)^2} = \lim_{x \to \infty} \frac{x}{6 \ln x} = \lim_{x \to \infty} \frac{x}{6} = +\infty.$$

Therefore $\displaystyle\sum_{n=2}^{\infty} \frac{(-1)^n \cdot n}{(\ln n)^3}$ diverges by the nth-term test for divergence (Theorem 3 of Section 11.3).

C11S0M.025: For every positive integer n,

$$0 \leq \frac{n^{1/2} + n^{1/3}}{n^2 + n^3} \leq \frac{2n^{1/2}}{n^3} = \frac{2}{n^{5/2}}.$$

Therefore

$$\sum_{n=1}^{\infty} \frac{n^{1/2} + n^{1/3}}{n^2 + n^3} \quad \text{is dominated by} \quad \sum_{n=1}^{\infty} \frac{2}{n^{5/2}}.$$

The latter series converges because it is a constant multiple of the p-series with $p = \frac{5}{2} > 1$. Therefore the dominated series converges by the comparison test (Theorem 1 of Section 11.6). The sum of the given series is approximately 1.459973884376.

C11S0M.027: Given: The alternating series

$$\sum_{n=1}^{\infty} \frac{(-1)^{n+1} \arctan n}{\sqrt{n}}. \tag{1}$$

We plan to show that this series meets the criterion for convergence stated in the alternating series test (Theorem 1 of of Section 11.7). First,

$$0 \leq \arctan n \leq \frac{\pi}{2}$$

for every positive integer n. Thus for such n,

$$0 \leq \frac{\arctan n}{\sqrt{n}} \leq \frac{\pi}{2\sqrt{n}}.$$

Therefore, by the squeeze law for limits,

$$\lim_{n \to \infty} \frac{\arctan n}{\sqrt{n}} = 0.$$

Now let

$$f(x) = \frac{\arctan x}{\sqrt{x}} \quad \text{for} \quad x \geq 1. \quad \text{Then:}$$

$$f'(x) = \frac{1}{x} \cdot \left(\frac{x^{1/2}}{1+x^2} - \frac{\arctan x}{2x^{1/2}} \right) = \frac{1}{x^{3/2}} \left(\frac{x}{1+x^2} - \frac{\arctan x}{2} \right)$$

$$= \frac{1}{2x^{3/2}} \left(\frac{2x}{1+x^2} - \arctan x \right) = \frac{2x - (1+x^2) \arctan x}{2x^{3/2}(1+x^2)}.$$

Now if $x \geq 2$, then $1 \leq \arctan x$. Therefore

$$1 + x^2 \leq (1+x^2) \arctan x; \qquad -(1+x^2) \arctan x \leq -(1+x^2);$$

$$2x - (1+x^2) \arctan x \leq 2x - (1+x^2); \qquad 2x - (1+x^2) \arctan x \leq -(x-1)^2;$$

$$2x - (1+x^2) \arctan x < 0 \quad \text{if} \quad x > 1; \qquad f'(x) < 0 \quad \text{if} \quad x > 1.$$

Consequently the sequence of terms of the series in (1) is (after the first term) monotonically decreasing in magnitude. Because they alternate in sign, the series in (1) converges by the alternating series test. Its sum is approximately 0.378868816198.

C11SOM.029: We use the integral test (Theorem 1 of Section 11.5):

$$\int_3^\infty \frac{1}{x(\ln x)(\ln \ln x)} \, dx = \left[\ln(\ln \ln x) \right]_3^\infty = +\infty,$$

and therefore $\displaystyle\sum_{n=3}^\infty \frac{1}{n(\ln n)(\ln \ln n)}$ diverges.

C11SOM.031: The ratio test yields

$$\rho = \lim_{n \to \infty} \frac{2^{n+1} \cdot n! \cdot |x|^{n+1}}{2^n \cdot (n+1)! \cdot |x|^n} = \lim_{n \to \infty} \frac{2|x|}{n+1} = 0$$

for every real number x. Therefore the given series converges for all x; its interval of convergence is $(-\infty, +\infty)$. Its sum is e^{2x}.

C11SOM.033: The ratio test yields

$$\rho = \lim_{n \to \infty} \frac{n \cdot 3^n \cdot |x-1|^{n+1}}{(n+1) \cdot 3^{n+1} \cdot |x-1|^n} = \lim_{n \to \infty} \frac{n \cdot |x-1|}{3(n+1)} = \frac{|x-1|}{3}.$$

601

So the given series converges if $-3 < x - 1 < 3$; that is, if $-2 < x < 4$. It diverges if $x = 4$ (because it becomes the harmonic series). It converges if $x = -2$ by the alternating series test. Thus its interval of convergence is $[-2, 4)$.

C11S0M.035: The ratio test yields

$$\rho = \lim_{n \to \infty} \frac{(4n^2 - 1) \cdot |x|^{n+1}}{[4(n+1)^2 - 1] \cdot |x|^n} = \lim_{n \to \infty} \frac{(4n^2 - 1) \cdot |x|}{4n^2 + 8n + 3} = \lim_{n \to \infty} \frac{\left(4 - \dfrac{1}{n^2}\right) \cdot |x|}{4 + \dfrac{8}{4} + \dfrac{3}{n^2}} = |x|.$$

Thus the series converges if $-1 < x < 1$. If $x = \pm 1$ then the given series converges absolutely because it is dominated by the p-series with $p = 2 > 1$. Hence the interval of convergence of the given series is $[-1, 1]$. The *Mathematica 3.0* command

```
Sum[ ((-1)^n)*(x^n)/(4*n*n - 1), {n, 1, Infinity} ]
```

quickly returns the value of the sum of this series on *part* of its interval of convergence; the response is

$$\frac{\sqrt{x} - \arctan\left(\sqrt{x}\right) - x\arctan\left(\sqrt{x}\right)}{2\sqrt{x}}.$$

This result raises some intriguing new questions concerning the behavior of the series, and particularly of its sum, for $-1 \leqq x < 0$.

C11S0M.037: The ratio test yields

$$\rho = \lim_{n \to \infty} \frac{(n+1)! \cdot 10^n \cdot |x|^{2n+2}}{n! \cdot 10^{n+1} \cdot |x|^{2n}} = \lim_{n \to \infty} \frac{(n+1)x^2}{10} = +\infty$$

if $x \neq 0$. Therefore the given series converges only if $x = 0$.

C11S0M.039: Note that

$$\sum_{n=0}^{\infty} \frac{1 + (-1)^n}{n! \cdot 2} x^n = 1 + \frac{x^2}{2!} + \frac{x^4}{4!} + \frac{x^6}{6!} + \cdots = 1 + \sum_{n=1}^{\infty} \frac{x^{2n}}{(2n)!}. \tag{1}$$

So the ratio test yields

$$\rho = \lim_{n \to \infty} \frac{(2n)! \cdot x^{2n+2}}{(2n+2)! \cdot x^{2n}} = \lim_{n \to \infty} \frac{x^2}{(2n+2)(2n+1)} = 0$$

for all real x. Hence the interval of convergence of the series in (1) is $(-\infty, +\infty)$. This series converges to $f(x) = \cosh x$ (see Example 6 in Section 11.8).

C11S0M.041: The given series diverges for every real number x by the nth-term test for divergence.

C11S0M.043: The ratio test yields

$$\rho = \lim_{n \to \infty} \frac{n! \cdot e^{(n+1)x}}{(n+1)! \cdot e^{nx}} = \lim_{n \to \infty} \frac{e^x}{n+1} = 0$$

for every real number x, so the given series converges for all x. Its sum is $\exp(e^x)$.

C11S0M.045: Let

$$a_n = b_n = \frac{(-1)^{n+1}}{\sqrt{n}} \quad \text{for} \quad n \geq 1.$$

Then $\sum a_n$ and $\sum b_n$ converge by the alternating series test, but $\sum a_n b_n$ diverges because it is the harmonic series.

C11S0M.047: Assuming that A exists, we have

$$A = \lim_{n \to \infty} a_n = \lim_{n \to \infty} \left(1 + \frac{1}{1 + a_n}\right) = 1 + \frac{1}{1 + A}$$

because $A \neq -1$. Therefore $A + A^2 = 2 + A$, and it follows that $A^2 = 2$. Because $A \geq 0$ (the limit of a sequence of positive numbers cannot be negative), $A = \sqrt{2}$.

C11S0M.049: If $a_n = \dfrac{1}{n}$, then the series

$$\sum_{n=1}^{\infty} \ln(1 + a_n) = \sum_{n=1}^{\infty} \ln\left(1 + \frac{1}{n}\right)$$

diverges because

$$S_k = \sum_{n=1}^{k} \ln\left(1 + \frac{1}{n}\right) = \sum_{n=1}^{k} [\ln(n+1) - \ln n]$$

$$= \ln 2 - \ln 1 + \ln 3 - \ln 2 + \ln 4 - \ln 3 + \cdots + \ln(k+1) - \ln k = \ln(k+1),$$

and therefore $S_k \to +\infty$ as $k \to +\infty$. Alternatively, using the integral test,

$$J = \int_1^{\infty} \ln\left(1 + \frac{1}{x}\right) dx = \int_1^{\infty} [\ln(x+1) - \ln x]\, dx = \left[(x+1)\ln(x+1) - x \ln x\right]_1^{\infty} = +\infty$$

because

$$\lim_{x \to \infty} \left[(x+1)\ln(x+1) - x \ln x\right] \geq \lim_{x \to \infty} \left[(x+1)\ln x - x \ln x\right] = \lim_{x \to \infty} \ln x = +\infty$$

and, at the lower limit $x = 1$ of integration, we have $(x+1)\ln(x+1) - x \ln x = \ln 4$. Therefore, because

$$J = \int_1^{\infty} \ln\left(1 + \frac{1}{x}\right) dx = +\infty,$$

the infinite product $\displaystyle\prod_{n=1}^{\infty} \left(1 + \frac{1}{n}\right)$ diverges.

C11S0M.051: The binomial series is

$$(1 + x)^{1/5} = 1 + \frac{x}{5} - \frac{4}{5^2} \cdot \frac{x^2}{2!} + \frac{4 \cdot 9}{5^3} \cdot \frac{x^4}{3!} - \frac{4 \cdot 9 \cdot 14}{5^4} \cdot \frac{x^4}{4!} + \cdots.$$

Substitution of $x = \frac{1}{2}$ and summing the first five terms of this series yields 1.0839 (exactly); summing the first six terms yields 1.0843788 (exactly). So to three places, $\left(1 + \frac{1}{2}\right)^{1/5} \approx 1.084$. The true value of the expression is closer to 1.084471771198.

C11S0M.053: We substitute $-x^2$ for x in the Maclaurin series for the natural exponential function. Thus we find that

$$\int_0^{1/2} \exp(-x^2)\,dx = \int_0^{1/2} \left(1 - x^2 + \frac{x^4}{2!} - \frac{x^6}{3!} + \frac{x^8}{4!} - \cdots\right) dx$$

$$= \left[x - \frac{x^3}{3} + \frac{x^5}{2!\cdot 5} - \frac{x^7}{3!\cdot 7} + \frac{x^9}{4!\cdot 9} - \cdots\right]_0^{1/2}$$

$$= \frac{1}{2} - \frac{1}{3\cdot 2^3} + \frac{1}{2!\cdot 5\cdot 2^5} - \frac{1}{3!\cdot 7\cdot 2^7} + \frac{1}{4!\cdot 9\cdot 2^9} - \cdots = \sum_{n=0}^{\infty} \frac{(-1)^n}{n!\cdot(2n+1)\cdot 2^{2n+1}}.$$

The sum of the first two terms and the sum of the first three terms of this series are

$$\frac{443}{960} \approx 0.461458333333 \quad \text{and} \quad \frac{4133}{8960} \approx 0.461272321429,$$

respectively. Thus the value of the integral to three places is approximately 0.461. A closer approximation is 0.4612810064127924 (to the number of digits shown).

C11S0M.055: The Maclaurin series for the natural exponential function yields

$$\frac{1}{x}(1 - e^{-x}) = \frac{1}{x}\left(x - \frac{x^2}{2!} + \frac{x^3}{3!} - \frac{x^4}{4!} + \frac{x^5}{5!} - \cdots\right) = 1 - \frac{x}{2!} + \frac{x^2}{3!} - \frac{x^3}{4!} + \frac{x^4}{5!} - \frac{x^5}{6!} + \cdots.$$

Then termwise integration produces

$$\int_0^1 \frac{1 - e^{-x}}{x}\,dx = \left[x - \frac{x^2}{2!\cdot 2} + \frac{x^3}{3!\cdot 3} - \frac{x^4}{4!\cdot 4} + \frac{x^5}{5!\cdot 5} - \frac{x^6}{6!\cdot 6} + \cdots\right]_0^1$$

$$= 1 - \frac{1}{2!\cdot 2} + \frac{1}{3!\cdot 3} - \frac{1}{4!\cdot 4} + \frac{1}{5!\cdot 5} - \frac{1}{6!\cdot 6} + \cdots.$$

The sum of the first five terms of the last series and the sum of its first six terms are

$$\frac{5737}{7200} \approx 0.796805555556 \quad \text{and} \quad \frac{8603}{10800} \approx 0.796574074074,$$

respectively. So to three places, the value of the integral is 0.797. A more accurate approximation is 0.7965995992970531.

C11S0M.057: We will need both the recursion formula

$$\int_0^{\infty} t^{2n} \exp(-t^2)\,dt = \frac{2n-1}{2} \int_0^{\infty} t^{2n-2} \exp(-t^2)\,dt \quad (n \geqq 1),$$

which follows from the formula in Problem 50 of Section 8.3, and the famous formula

$$\int_0^{\infty} \exp(-t^2)\,dt = \frac{\sqrt{\pi}}{2},$$

which is derived in Example 5 of Section 14.4 (it is Eq. (9) there). We begin with the Maclaurin series of the cosine function.

$$\int_0^{\infty} \exp(-t^2)\cos 2xt\,dt = \int_0^{\infty} \exp(-t^2)\left(1 - \frac{2^2 x^2 t^2}{2!} + \frac{2^4 x^4 t^4}{4!} - \frac{2^6 x^6 t^6}{6!} + \frac{2^8 x^8 t^8}{8!} - \cdots\right) dt$$

$$= \int_0^\infty \left(\exp(-t^2) - \frac{2^2 x^2}{2!} t^2 \exp(-t^2) + \frac{2^4 x^4}{4!} t^4 \exp(-t^2) - \frac{2^6 x^6}{6!} t^6 \exp(-t^2) + \cdots \right) dt$$

$$= \int_0^\infty e^{-t^2}\, dt - \frac{2^2 x^2}{2!} \int_0^\infty t^2 e^{-t^2}\, dt + \frac{2^4 x^4}{4!} \int_0^\infty t^4 e^{-t^2}\, dt - \frac{2^6 x^6}{6!} \int_0^\infty t^6 e^{-t^2}\, dt + \cdots$$

$$= \int_0^\infty e^{-t^2}\, dt - \frac{2^2 x^2}{2!} \cdot \frac{1}{2} \int_0^\infty e^{-t^2}\, dt + \frac{2^4 x^4}{4!} \cdot \frac{3}{2} \cdot \frac{1}{2} \int_0^\infty e^{-t^2}\, dt - \frac{2^6 x^6}{6!} \cdot \frac{5}{2} \cdot \frac{3}{2} \cdot \frac{1}{2} \int_0^\infty e^{-t^2}\, dt + \cdots$$

$$= \left(\int_0^\infty e^{-t^2}\, dt \right) \left(1 - \frac{2^2 x^2}{2!} \cdot \frac{1}{2} + \frac{2^4 x^4}{4!} \cdot \frac{3 \cdot 1}{2^2} - \frac{2^6 x^6}{6!} \cdot \frac{5 \cdot 3 \cdot 1}{2^3} + \frac{2^8 x^8}{8!} \cdot \frac{7 \cdot 5 \cdot 3 \cdot 1}{2^4} - \cdots \right).$$

The typical term in the last infinite series is

$$\frac{2^{2n} x^{2n}}{(2n)!} \cdot \frac{(2n-1)(2n-3) \cdots 5 \cdot 3 \cdot 1}{2^n} = \frac{2^{2n} x^{2n}}{(2n)!} \cdot \frac{(2n)!}{2^n \cdot (2n)(2n-2) \cdots 6 \cdot 4 \cdot 2} = \frac{2^{2n} x^{2n}}{n! \cdot 2^n \cdot 2^n} = \frac{(x^2)^n}{n!}.$$

Consequently,

$$\int_0^\infty e^{-t^2} \cos 2xt\, dt = \left(\int_0^\infty e^{-t^2}\, dt \right) \left(1 - \frac{x^2}{1!} + \frac{(x^2)^2}{2!} - \frac{(x^2)^3}{3!} + \frac{(x^2)^4}{4!} - \cdots \right)$$

$$= \left(\int_0^\infty e^{-t^2}\, dt \right) e^{-x^2} = \frac{\sqrt{\pi}}{2} e^{-x^2}.$$

C11S0M.059: The binomial series takes the form

$$(1 + t^2)^{-1/2} = 1 - \frac{1}{2} t^2 + \frac{1 \cdot 3}{2^2} \cdot \frac{t^4}{2!} - \frac{1 \cdot 3 \cdot 5}{2^3} \cdot \frac{t^6}{3!} + \frac{1 \cdot 3 \cdot 5 \cdot 7}{2^4} \cdot \frac{t^8}{4!} - \cdots.$$

Thus

$$\sinh^{-1} x = \int_0^x (1 + t^2)^{-1/2}\, dt = x - \frac{x^3}{3} + \frac{1 \cdot 3}{2^2 \cdot 5} \cdot \frac{x^5}{2!} - \frac{1 \cdot 3 \cdot 5}{2^3 \cdot 7} \cdot \frac{x^7}{3!} + \frac{1 \cdot 3 \cdot 5 \cdot 7}{2^4 \cdot 9} \cdot \frac{x^9}{4!} - \cdots$$

$$= \sum_{n=0}^\infty \frac{1 \cdot 3 \cdot 5 \cdots (2n-1)}{2^n \cdot n!} \cdot \frac{x^{2n+1}}{2n+1} = \sum_{n=0}^\infty \frac{1 \cdot 3 \cdot 5 \cdots (2n-1)}{2 \cdot 4 \cdot 6 \cdots (2n)} \cdot \frac{x^{2n+1}}{2n+1}$$

$$= \sum_{n=0}^\infty \frac{(2n)!}{2^n \cdot n! \cdot 2 \cdot 4 \cdot 6 \cdots (2n)} \cdot \frac{x^{2n+1}}{2n+1} = \sum_{n=0}^\infty \frac{(2n)!}{2^{2n} \cdot (n!)^2} \cdot \frac{x^{2n+1}}{2n+1}$$

provided that $|x| < 1$.

C11S0M.061: We let *Mathematica 3.0* do this problem. First we defined

```
mu = 1/(12*n) - 1/(360*n^3) + 1/(1260*n^5)
```

Then the command

```
Series[ Exp[x], { x, 0, 10 } ] // Normal
```

produced the 10th-degree Taylor polynomial with center zero for e^x:

$$1 + x + \frac{x^2}{2} + \frac{x^3}{6} + \frac{x^4}{24} + \frac{x^5}{120} + \frac{x^6}{720} + \frac{x^7}{5040} + \frac{x^8}{40320} + \frac{x^9}{362880} + \frac{x^{10}}{3628800}.$$

Recall that % means "last output" to *Mathematica*. Thus the next command

    % /.   x → mu

tells *Mathematica* to substitute mu for x in the Taylor polynomial, producing the response

$$1 + \left(\frac{1}{1260n^5} - \frac{1}{360n^3} + \frac{1}{12n}\right) + \frac{1}{2}\left(\frac{1}{1260n^5} - \frac{1}{360n^3} + \frac{1}{12n}\right)^2 + \frac{1}{6}\left(\frac{1}{1260n^5} - \frac{1}{360n^3} + \frac{1}{12n}\right)^3$$

$$+ \frac{1}{24}\left(\frac{1}{1260n^5} - \frac{1}{360n^3} + \frac{1}{12n}\right)^4 + \cdots + \frac{1}{3628800}\left(\frac{1}{1260n^5} - \frac{1}{360n^3} + \frac{1}{12n}\right)^{10}.$$

Finally, the **Expand** command resulted in almost a full page of output, including the answer:

$$\exp(\mu(n)) = 1 + \frac{1}{12n} + \frac{1}{288n^2} - \frac{139}{51840n^3} - \frac{571}{2488320n^4} + \frac{163879}{209018880n^5} + \cdots.$$

C11S0M.063: Proof: Assume that e is a rational number. Then $e = p/q$ where p and q are positive integers and $q > 1$ (because e is not an integer). Thus

$$\frac{p}{q} = e = 1 + \frac{1}{1!} + \frac{1}{2!} + \frac{1}{3!} + \cdots + \frac{1}{q!} + R_q$$

where

$$R_q = \frac{1}{(q+1)!} + \frac{1}{(q+2)!} + \frac{1}{(q+3)!} + \frac{1}{(q+4)!} + \cdots$$

$$= \frac{1}{q!} \cdot \left(\frac{1}{q+1} + \frac{1}{(q+1)(q+2)} + \frac{1}{(q+1)(q+2)(q+3)} + \cdots\right)$$

$$< \frac{1}{q!} \cdot \left(\frac{1}{q+1} + \frac{1}{(q+1)^2} + \frac{1}{(q+1)^3} + \cdots\right) = \frac{1}{q!} \cdot \frac{\frac{1}{q+1}}{1 - \frac{1}{q+1}} = \frac{1}{q! \cdot q}.$$

Thus

$$1 + \frac{1}{1!} + \frac{1}{2!} + \frac{1}{3!} + \cdots + \frac{1}{q!} < \frac{p}{q} < 1 + \frac{1}{1!} + \frac{1}{2!} + \frac{1}{3!} + \cdots + \frac{1}{q!} + \frac{1}{q! \cdot q}.$$

If the left member of the last inequality is multiplied by $q!$, the product is an integer; call it M. Thus when all three members of the last inequality are multiplied by $q!$, the result is

$$M < (q-1)! \cdot p < M + \frac{1}{q} < M + 1.$$

This is a contradiction because it asserts that the *integer* $(q-1)! \cdot p$ lies strictly between the *consecutive* integers M and $M+1$. Therefore e is irrational. ◄

C11S0M.065: Suppose that $x^2 = 5$. Then

$$x^2 - 4 = 1; \quad x - 2 = \frac{1}{2+x}; \quad x = 2 + \frac{1}{2+x}.$$

Now substitute the last expression for the last x. The result is

$$x = 2 + \cfrac{1}{4 + \cfrac{1}{2 + x}}.$$

Repeat: Substitute the right-hand side of the last equation for the last x. Thus

$$x = 2 + \cfrac{1}{4 + \cfrac{1}{4 + \cfrac{1}{2 + x}}}.$$

Continue this process. It follows that $a_0 = 2$ and that $a_n = 4$ for all $n \geq 1$.

C10S0M.Extra: Curious about the Riemann zeta function? Questions about its behavior are currently the deepest and most important unsolved problems in mathematics; some of the answers have important consequences in the theory of the distribution of prime numbers. Some of those consequences are related to a remarkable identity discovered by Leonhard Euler:

Theorem: If $s > 1$ then

$$\zeta(s) = \prod_{p \text{ prime}} \frac{1}{1 - p^{-s}}.$$

Note that the product is taken over all *primes* p.

Recall that if s is a real number and $s > 1$, then

$$\zeta(s) = \sum_{n=1}^{\infty} \frac{1}{n^s}$$

and that the function ζ may be extended to most other numbers, including most complex numbers, by the condition that it is required to be infinitely differentiable. We'll have no need for its values at complex numbers here; we are mostly concerned with its values when s is an integer and $s > 1$. In the text we have seen a few of the values of the zeta function; for example,

$$\zeta(2) = \frac{\pi^2}{6}, \qquad \zeta(4) = \frac{\pi^4}{90}, \qquad \zeta(6) = \frac{\pi^6}{945}, \qquad \text{and} \qquad \zeta(8) = \frac{\pi^8}{9450}.$$

It is known that $\zeta(2n)$ is a rational multiple of π^{2n} if n is a positive integer; much less is known about $\zeta(n)$ if n is odd and $n \geq 3$. The values of $\zeta(2n)$ continue the preceding list as follows:

$$\frac{\pi^{10}}{93555}, \quad \frac{691\pi^{12}}{638512875}, \quad \frac{2\pi^{14}}{18243225}, \quad \frac{3617\pi^{16}}{325641566250}, \quad \frac{43867\pi^{18}}{38979295480125}, \quad \text{and} \quad \frac{174611\pi^{20}}{1531329465290625}.$$

The pattern of the coefficients is related to the *Bernoulli numbers* $\{B_n\}$, the values of which may be defined as follows. Write the Taylor series with center zero for

$$g(t) = \frac{t}{e^t - 1}.$$

(Note that $g(0)$ may be defined by the usual requirement that g be continuous at $t = 0$.) The resulting series is

$$g(t) = 1 - \frac{t}{2} + \frac{t^2}{12} - \frac{t^4}{720} + \frac{t^6}{30240} - \frac{t^8}{1209600} + \frac{t^{10}}{47900160} - \cdots. \tag{1}$$

Then for n an even nonnegative integer, the nth Bernoulli number B_n may be defined to be the product of $n!$ and the coefficient of t^n in the series in (1). Finally, if n is an integer and $n \geq 1$, then the coefficient of π^{2n} in the expression for $\zeta(2n)$ is of the form

$$\frac{2^j |B_{2n}|}{(2k)!}$$

where j and k are integers very closely related to n. We leave it to you to discover that simple relationship —extrapolation from the data given here will yield a valid result. Finally, if you need more numbers, the *Mathematica* commands `Zeta[ n ]` and `BernoulliB[ n ]`, or the *Maple* commands `Zeta(n)` and `bernoulli(n)`, will provide you with more values of the zeta function and more Bernoulli numbers.

Section 12.1

C12S01.001: $\mathbf{v} = \overrightarrow{RS} = \langle 3 - 1, 5 - 2 \rangle = \langle 2, 3 \rangle$. The position vector of the point $P(2, 3)$ and $\overrightarrow{RS}$ are shown next.

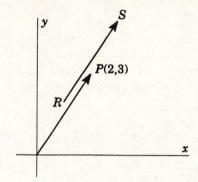

C12S01.003: $\mathbf{v} = \overrightarrow{RS} = \langle -5 - 5, -10 - 10 \rangle = \langle -10, -20 \rangle$. The position vector of the point P and $\overrightarrow{RS}$ are shown next.

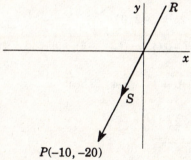

C12S01.005: $\mathbf{w} = \mathbf{u} + \mathbf{v} = \langle 1, -2 \rangle + \langle 3, 4 \rangle = \langle 1 + 3, -2 + 4 \rangle = \langle 4, 2 \rangle$. The next figure illustrates this computation in the form of the triangle law for vector addition (see Fig. 12.1.6 of the text).

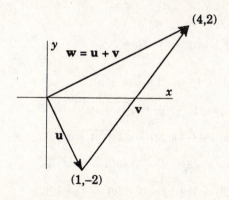

C12S01.007: Given: $\mathbf{u} = 3\mathbf{i} + 5\mathbf{j}$, $\mathbf{v} = 2\mathbf{i} - 7\mathbf{j}$:

$$\mathbf{u} + \mathbf{v} = 3\mathbf{i} + 5\mathbf{j} + 2\mathbf{i} - 7\mathbf{j} = (3 + 2)\mathbf{i} + (5 - 7)\mathbf{j} = 5\mathbf{i} - 2\mathbf{j}.$$

609

The next figure illustrates the triangle law for vector addition using **u** and **v**.

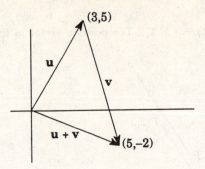

C12S01.009: Given: $\mathbf{a} = \langle 1, -2 \rangle$ and $\mathbf{b} = \langle -3, 2 \rangle$. Then:

$$|\mathbf{a}| = \sqrt{(1)^2 + (-2)^2} = \sqrt{5}\,,$$

$$|-2\mathbf{b}| = |\langle -6, 4 \rangle| = \sqrt{36 + 16} = 2\sqrt{13}\,,$$

$$|\mathbf{a} - \mathbf{b}| = |\langle 1 - (-3), -2 - 2 \rangle| = \sqrt{16 + 16} = 4\sqrt{2}\,,$$

$$\mathbf{a} + \mathbf{b} = \langle 1 - 3, -2 + 2 \rangle = \langle -2, 0 \rangle,$$

$$3\mathbf{a} - 2\mathbf{b} = \langle 3, -6 \rangle - \langle -6, 4 \rangle = \langle 3 - (-6), -6 - 4 \rangle = \langle 9, -10 \rangle.$$

C12S01.011: Given: $\mathbf{a} = \langle -2, -2 \rangle$ and $\mathbf{b} = \langle -3, -4 \rangle$. Then:

$$|\mathbf{a}| = \sqrt{4 + 4} = \sqrt{8} = 2\sqrt{2}\,,$$

$$|-2\mathbf{b}| = |\langle 6, 8 \rangle| = \sqrt{36 + 64} = 10,$$

$$|\mathbf{a} - \mathbf{b}| = |\langle -2 - (-3), -2 - (-4) \rangle| = |\langle 1, 2 \rangle| = \sqrt{1 + 4} = \sqrt{5}\,,$$

$$\mathbf{a} + \mathbf{b} = \langle -2 - 3, -2 - 4 \rangle = \langle -5, -6 \rangle,$$

$$3\mathbf{a} - 2\mathbf{b} = \langle -6, -6 \rangle - \langle -6, -8 \rangle = \langle -6 - (-6), -6 - (-8) \rangle = \langle 0, 2 \rangle.$$

C12S01.013: Given: $\mathbf{a} = \mathbf{i} + 3\mathbf{j}$ and $\mathbf{b} = 2\mathbf{i} - 5\mathbf{j}$. Then:

$$|\mathbf{a}| = |\mathbf{i} + 3\mathbf{j}| = \sqrt{1 + 9} = \sqrt{10}\,,$$

$$|-2\mathbf{b}| = |4\mathbf{i} - 10\mathbf{j}| = \sqrt{16 + 100} = 2\sqrt{29}\,,$$

$$|\mathbf{a} - \mathbf{b}| = |-\mathbf{i} + 8\mathbf{j}| = \sqrt{1 + 64} = \sqrt{65}\,,$$

$$\mathbf{a} + \mathbf{b} = \mathbf{i} + 3\mathbf{j} + 2\mathbf{i} - 5\mathbf{j} = 3\mathbf{i} - 2\mathbf{j},$$

$$3\mathbf{a} - 2\mathbf{b} = 3\mathbf{i} + 9\mathbf{j} - 4\mathbf{i} + 10\mathbf{j} = -\mathbf{i} + 19\mathbf{j}.$$

C12S01.015: Given: $\mathbf{a} = 4\mathbf{i}$ and $\mathbf{b} = -7\mathbf{j}$. Then:

$$|\mathbf{a}| = |4\mathbf{i}| = \sqrt{16} = 4,$$

$$|-2\mathbf{b}| = |14\mathbf{j}| = \sqrt{(14)^2} = 14,$$

$$|\mathbf{a} - \mathbf{b}| = |4\mathbf{i} + 7\mathbf{j}| = \sqrt{16 + 49} = \sqrt{65},$$

$$\mathbf{a} + \mathbf{b} = 4\mathbf{i} - 7\mathbf{j},$$

$$3\mathbf{a} - 2\mathbf{b} = 12\mathbf{i} + 14\mathbf{j}.$$

C12S01.017: Because $|\mathbf{a}| = \sqrt{9 + 16} = 5,$

$$\mathbf{u} = \frac{1}{5}\mathbf{a} = -\frac{3}{5}\mathbf{i} - \frac{4}{5}\mathbf{j} \quad \text{and} \quad \mathbf{v} = -\frac{1}{5}\mathbf{a} = \frac{3}{5}\mathbf{i} + \frac{4}{5}\mathbf{j}.$$

C12S01.019: Because $|\mathbf{a}| = \sqrt{64 + 225} = \sqrt{289} = 17,$

$$\mathbf{u} = \frac{1}{17}\mathbf{a} = \frac{8}{17}\mathbf{i} + \frac{15}{17}\mathbf{j} \quad \text{and} \quad \mathbf{v} = -\frac{1}{17}\mathbf{a} = -\frac{8}{17}\mathbf{i} - \frac{15}{17}\mathbf{j}.$$

C12S01.021: $\mathbf{a} = \overrightarrow{PQ} = \langle 3 - 3, -2 - 2 \rangle = \langle 0, -4 \rangle = -4\mathbf{j}.$

C12S01.023: $\mathbf{a} = \overrightarrow{PQ} = \langle 4 - (-4), -7 - 7 \rangle = \langle 8, -14 \rangle = 8\mathbf{i} - 14\mathbf{j}.$

C12S01.025: Given $\mathbf{a} = \langle 6, 0 \rangle$ and $\mathbf{b} = \langle 0, -7 \rangle$, $\mathbf{c} = \mathbf{b} - \mathbf{a} = \langle -6, -7 \rangle$. Then

$$|\mathbf{c}|^2 = 36 + 49 = |\mathbf{a}|^2 + |\mathbf{b}|^2.$$

Therefore the vectors $\mathbf{a}$ and $\mathbf{b}$ are perpendicular because of the (true) converse of the Pythagorean theorem. See Example 5.

C12S01.027: Given $\mathbf{a} = \langle 2, -1 \rangle$ and $\mathbf{b} = \langle 4, 8 \rangle$, $\mathbf{c} = \mathbf{b} - \mathbf{a} = \langle 2, 9 \rangle$. Then

$$|\mathbf{c}|^2 = 4 + 81 = 85 = 5 + 80 = |\mathbf{a}|^2 + |\mathbf{b}|^2.$$

Therefore the vectors $\mathbf{a}$ and $\mathbf{b}$ are perpendicular by the (true) converse of the Pythagorean theorem. See Example 5.

C12S01.029: Given $\mathbf{a} = 2\mathbf{i} + 3\mathbf{j}$ and $\mathbf{b} = 3\mathbf{i} + 4\mathbf{j}$, we have

$$3\mathbf{a} - 2\mathbf{b} = 6\mathbf{i} + 9\mathbf{j} - 6\mathbf{i} - 8\mathbf{j} = \mathbf{j} \quad \text{and}$$

$$4\mathbf{a} - 3\mathbf{b} = 8\mathbf{i} + 12\mathbf{j} - 9\mathbf{i} - 12\mathbf{j} = -\mathbf{i}.$$

Therefore $\mathbf{i} = -4\mathbf{a} + 3\mathbf{b}$ and $\mathbf{j} = 3\mathbf{a} - 2\mathbf{b}.$

C12S01.031: Given $\mathbf{a} = \mathbf{i} + \mathbf{j}$ and $\mathbf{b} = \mathbf{i} - \mathbf{j}$, we have

$$\mathbf{c} = 2\mathbf{i} - 3\mathbf{j} = r\mathbf{a} + s\mathbf{b} = r\mathbf{i} + r\mathbf{j} + s\mathbf{i} - s\mathbf{j} = (r + s)\mathbf{i} + (r - s)\mathbf{j}.$$

It follows that $r + s = 2$ and that $r - s = -3$. These simultaneous equations are easily solved, and thereby we find that $\mathbf{c} = -\frac{1}{2}\mathbf{a} + \frac{5}{2}\mathbf{b}.$

C12S01.033: Part (a): $3(5\mathbf{i} - 7\mathbf{j}) = 15\mathbf{i} - 21\mathbf{j}.$ Part (b): $\frac{1}{3}(5\mathbf{i} - 7\mathbf{j}) = \frac{5}{3}\mathbf{i} - \frac{7}{3}\mathbf{j}.$

C12S01.035: Part (a): If $\mathbf{a} = 7\mathbf{i} - 3\mathbf{j}$, then $|\mathbf{a}| = \sqrt{58}$. Thus a unit vector with the same direction as $\mathbf{a}$ is

$$\mathbf{u} = \frac{\mathbf{a}}{\sqrt{58}}. \quad \text{Answer:} \quad 5\mathbf{u} = \frac{5\sqrt{58}}{58}(7\mathbf{i} - 3\mathbf{j}).$$

Part (b): If $\mathbf{b} = 8\mathbf{i} + 5\mathbf{j}$, then $|\mathbf{b}| = \sqrt{89}$. Thus a unit vector with the same direction as $\mathbf{b}$ is

$$\mathbf{v} = \frac{\mathbf{b}}{\sqrt{89}}. \quad \text{Answer:} \quad -5\mathbf{v} = -\frac{5\sqrt{89}}{89}(8\mathbf{i} + 5\mathbf{j}).$$

C12S01.037: Given $\mathbf{a} = \langle 2c, -4 \rangle$ and $\mathbf{b} = \langle 3, c \rangle$, let $\mathbf{w} = \mathbf{b} - \mathbf{a} = \langle 3 - 2c, c + 4 \rangle$. Perpendicularity of $\mathbf{a}$ and $\mathbf{b}$ is equivalent to the Pythagorean relation

$$|\mathbf{a}|^2 + |\mathbf{b}|^2 = |\mathbf{w}|^2;$$

that is, $4c^2 + 16 + c^2 + 9 = 5c^2 - 4c + 25$, and thus the unique solution is $c = 0$.

C12S01.039: With $\mathbf{a} = \langle a_1, a_2 \rangle$, $\mathbf{b} = \langle b_1, b_2 \rangle$, and $\mathbf{c} = \langle c_1, c_2 \rangle$, we have

$$\mathbf{a} + (\mathbf{b} + \mathbf{c}) = \langle a_1, a_2 \rangle + (\langle b_1, b_2 \rangle + \langle c_1, c_2 \rangle) = \langle a_1, a_2 \rangle + \langle b_1 + c_1, b_2 + c_2 \rangle$$

$$= \langle a_1 + (b_1 + c_1), a_2 + (b_2 + c_2) \rangle = \langle (a_1 + b_1) + c_1, (a_2 + b_2) + c_2 \rangle$$

$$= \langle a_1 + b_1, a_2 + b_2 \rangle + \langle c_1, c_2 \rangle = (\langle a_1, a_2 \rangle + \langle b_1, b_2 \rangle) + \langle c_1, c_2 \rangle = (\mathbf{a} + \mathbf{b}) + \mathbf{c}.$$

C12S01.041: With $\mathbf{a} = \langle a_1, a_2 \rangle$ and scalars r and s, we have

$$(rs)\mathbf{a} = (rs)\langle a_1, a_2 \rangle = \langle (rs)a_1, (rs)a_2 \rangle = \langle r(sa_1), r(sa_2) \rangle = r\langle sa_1, sa_2 \rangle = r(s\langle a_1, a_2 \rangle) = r(s\mathbf{a}).$$

C12S01.043: See Fig. 12.1.13. The angle each cable makes with the horizontal is $30°$, so

$$T_1 \cos 30° = T_2 \cos 30°. \quad \text{Therefore} \quad T_1 = T_2.$$

Also

$$T_1 \sin 30° + T_2 \sin 30° = 100; \qquad 2 \cdot \frac{1}{2} \cdot T_1 = 100; \qquad T_1 = T_2 = 100.$$

C12S01.045: See Fig. 12.1.15; let T_1 be the tension in the right-hand cable and T_2 the tension in the left-hand cable. The equations that balance the horizontal and vertical forces, respectively, are

$$T_1 \cos 40° = T_2 \cos 55° \quad \text{and} \quad T_1 \sin 40° + T_2 \sin 55° = 125.$$

A *Mathematica* 3.0 command for solving these equations simultaneously is

```
Solve[ { t1*Cos[40*Pi/180] == t2*Cos[55*Pi/180],
         t1*Sin[40*Pi/180] + t2*Sin[55*Pi/180] == 125 }, {t1,t2} ]
```

and it yields the solution

$$T_1 = \frac{125\cos(11\pi/15)}{\cos(11\pi/36)\sin(2\pi/9) + \cos(2\pi/9)\sin(11\pi/36)} \approx 71.970925644575 \quad \text{(lb)},$$

$$T_2 = \frac{125\cos(2\pi/9)}{\cos(11\pi/36)\sin(2\pi/9) + \cos(2\pi/9)\sin(11\pi/36)} \approx 96.121326055335 \quad \text{(lb)}.$$

C12S01.047: $\mathbf{v}_a = \mathbf{v}_g - \mathbf{w} = \langle 500, 0 \rangle - \langle -25\sqrt{2}, -25\sqrt{2} \rangle = \langle 500 + 25\sqrt{2}, 25\sqrt{2} \rangle$. This corresponds to a compass bearing of about $86° 13'$ (about $3.778°$ north of east) with airspeed approximately 536.52 mi/h (about 467 knots).

C12S01.049: $\mathbf{v}_a = \mathbf{v}_g - \mathbf{w} = \langle -250\sqrt{2}, 250\sqrt{2} \rangle - \langle -25\sqrt{2}, -25\sqrt{2} \rangle = \langle -225\sqrt{2}, 275\sqrt{2} \rangle$. This corresponds to a compass bearing of about $320° 43'$ (about $50.71°$ north of west) with airspeed approximately 502 mi/h (about 437 knots).

C12S01.051: Denote the origin by O. Then $\mathbf{a} = \overrightarrow{OP}$ and $\mathbf{b} = \overrightarrow{OQ}$. Let M denote the midpoint of the line segment PQ. Then $\overrightarrow{PQ} = \mathbf{b} - \mathbf{a}$, so that

$$\overrightarrow{PM} = \frac{1}{2}(\mathbf{b} - \mathbf{a}).$$

Hence the position vector of M is

$$\overrightarrow{OP} + \overrightarrow{MP} = \mathbf{a} + \frac{1}{2}(\mathbf{b} - \mathbf{a}) = \frac{1}{2}(\mathbf{a} + \mathbf{b}).$$

C12S01.053: Let A, B, C, and D be the vertices of a parallelogram, in order, counterclockwise. Then its diagonals are AC and BD. Let M be the point where the diagonals cross. Let $\mathbf{u} = \overrightarrow{AB}$ and let $\mathbf{v} = \overrightarrow{BC}$. Then

$$\overrightarrow{AC} = \mathbf{u} + \mathbf{v} \quad \text{and} \quad \overrightarrow{BD} = \mathbf{v} - \mathbf{u}.$$

Then $\overrightarrow{AM} = r(\mathbf{u} + \mathbf{v})$ and $\overrightarrow{MC} = (1 - r)(\mathbf{u} + \mathbf{v})$ for some scalar r. Similarly, $\overrightarrow{BM} = s(\mathbf{v} - \mathbf{u})$ and $\overrightarrow{MD} = (1 - s)(\mathbf{v} - \mathbf{u})$ for some scalar s. Therefore

$$\overrightarrow{AB} + \overrightarrow{BM} = \overrightarrow{AM}; \qquad \mathbf{u} + s(\mathbf{v} - \mathbf{u}) = r(\mathbf{u} + \mathbf{v});$$

$$(1 - r - s)\mathbf{u} = (r - s)\mathbf{v}; \qquad 1 - r - s = 0 = r - s$$

because $\mathbf{u}$ and $\mathbf{v}$ are not parallel. It follows that $r = s = \frac{1}{2}$, so that $|AM| = |MC|$ and $|BM| = |MD|$. Therefore the diagonals of the arbitrary parallelogram $ABCD$ bisect each other.

C12S01.055: See Fig. 12.1.18, which makes it easy to see that $a_1 = r\cos\theta$ and $a_2 = r\sin\theta$. Then

$$\mathbf{a}_\perp = \left[r\cos\left(\theta + \frac{\pi}{2}\right)\right]\mathbf{i} + \left[r\sin\left(\theta + \frac{\pi}{2}\right)\right]\mathbf{j} = (-r\sin\theta)\mathbf{i} + (r\cos\theta)\mathbf{j} = -a_2\mathbf{i} + a_1\mathbf{j}.$$

Section 12.2

C12S02.001: Given $\mathbf{a} = \langle 2, 5, -4 \rangle$ and $\mathbf{b} = \langle 1, -2, -3, \rangle$, we find that

(a): $2\mathbf{a} + \mathbf{b} = \langle 5, 8, -11 \rangle$, (b): $3\mathbf{a} - 4\mathbf{b} = \langle 2, 23, 0 \rangle$,

(c): $\mathbf{a} \cdot \mathbf{b} = 2 \cdot 1 - 5 \cdot 2 + 4 \cdot 3 = 4$,

(d): $|\mathbf{a} - \mathbf{b}| = |\langle 1, 7, -1 \rangle| = \sqrt{51}$, and

(e): $\dfrac{\mathbf{a}}{|\mathbf{a}|} = \dfrac{1}{\sqrt{4 + 25 + 16}} \langle 2, 5, -4 \rangle = \dfrac{\sqrt{5}}{15} \langle 2, 5, -4 \rangle = \left\langle \dfrac{2\sqrt{5}}{15}, \dfrac{\sqrt{5}}{3}, -\dfrac{4\sqrt{5}}{15} \right\rangle$.

C12S02.003: Given $\mathbf{a} = \langle 1, 1, 1 \rangle$ and $\mathbf{b} = \langle 0, 1, -1 \rangle$, we find that

(a): $2\mathbf{a} + \mathbf{b} = \langle 2, 3, 1 \rangle$,

(b): $3\mathbf{a} - 4\mathbf{b} = \langle 3, -1, 7 \rangle$,

(c): $\mathbf{a} \cdot \mathbf{b} = 1 \cdot 0 + 1 \cdot 1 - 1 \cdot 1 = 0$,

(d): $|\mathbf{a} - \mathbf{b}| = |\langle 1, 0, 2 \rangle| = \sqrt{5}$, and

(e): $\dfrac{\mathbf{a}}{|\mathbf{a}|} = \dfrac{1}{\sqrt{3}} \langle 1, 1, 1 \rangle = \left\langle \dfrac{\sqrt{3}}{3}, \dfrac{\sqrt{3}}{3}, \dfrac{\sqrt{3}}{3} \right\rangle$.

C12S02.005: Given $\mathbf{a} = \langle 2, -1, 0 \rangle$ and $\mathbf{b} = \langle 0, 1, -3 \rangle$, we find that

(a): $2\mathbf{a} + \mathbf{b} = \langle 4, -1, -3 \rangle$,

(b): $3\mathbf{a} - 4\mathbf{b} = \langle 6, -7, 12 \rangle$,

(c): $\mathbf{a} \cdot \mathbf{b} = 2 \cdot 0 - 1 \cdot 1 + 0 \cdot (-3) = -1$,

(d): $|\mathbf{a} - \mathbf{b}| = |\langle 2, -2, 3 \rangle| = \sqrt{17}$, and

(e): $\dfrac{\mathbf{a}}{|\mathbf{a}|} = \dfrac{1}{\sqrt{5}} \langle 2, -1, 0 \rangle = \left\langle \dfrac{2\sqrt{5}}{5}, -\dfrac{\sqrt{5}}{5}, 0 \right\rangle$.

C12S02.007: If θ is the angle between $\mathbf{a}$ and $\mathbf{b}$, then

$$\cos\theta = \frac{\mathbf{a} \cdot \mathbf{b}}{|\mathbf{a}| \cdot |\mathbf{b}|} = \frac{4}{\sqrt{45}\,\sqrt{14}} = \frac{2\sqrt{70}}{105} \approx 0.15936371,$$

and therefore $\theta \approx 80.830028° \approx 81°$.

C12S02.009: If θ is the angle between $\mathbf{a}$ and $\mathbf{b}$, then

$$\cos\theta = \frac{\mathbf{a} \cdot \mathbf{b}}{|\mathbf{a}| \cdot |\mathbf{b}|} = 0 \quad \text{(exactly)},$$

and therefore $\theta = 90°$ (exactly).

C12S02.011: If θ is the angle between $\mathbf{a}$ and $\mathbf{b}$, then

$$\cos\theta = \frac{\mathbf{a} \cdot \mathbf{b}}{|\mathbf{a}| \cdot |\mathbf{b}|} = \frac{-1}{5\sqrt{2}} = -\frac{\sqrt{2}}{10} \approx -0.14142136,$$

and therefore $\theta \approx 98.130102° \approx 98°$.

C12S02.013: Refer to Problem 1;

$$\text{comp}_{\mathbf{a}}\mathbf{b} = \frac{\mathbf{a} \cdot \mathbf{b}}{|\mathbf{a}|} = \frac{4\sqrt{5}}{15} \quad \text{and} \quad \text{comp}_{\mathbf{b}}\mathbf{a} = \frac{\mathbf{a} \cdot \mathbf{b}}{|\mathbf{b}|} = \frac{2\sqrt{14}}{7}.$$

C12S02.015: Refer to Problem 3;

$$\text{comp}_{\mathbf{a}}\mathbf{b} = \frac{\mathbf{a} \cdot \mathbf{b}}{|\mathbf{a}|} = 0 = \text{comp}_{\mathbf{b}}\mathbf{a} = \frac{\mathbf{a} \cdot \mathbf{b}}{|\mathbf{b}|}.$$

C12S02.017: Refer to Problem 5;

$$\text{comp}_{\mathbf{a}}\mathbf{b} = \frac{\mathbf{a} \cdot \mathbf{b}}{|\mathbf{a}|} = -\frac{\sqrt{5}}{5} \quad \text{and} \quad \text{comp}_{\mathbf{b}}\mathbf{a} = \frac{\mathbf{a} \cdot \mathbf{b}}{|\mathbf{b}|} = -\frac{\sqrt{10}}{10}.$$

C12S02.019: An equation of the sphere is $(x-3)^2 + (y-1)^2 + (z-2)^2 = 25$; that is,

$$x^2 - 6x + y^2 - 2y + z^2 - 4z = 11.$$

C12S02.021: The center of the sphere is at

$$\left(\frac{3+7}{2}, \frac{5+3}{2}, \frac{-3+1}{2} \right) = (5,\, 4,\, -1)$$

and its radius is

$$\frac{1}{2}\sqrt{(7-3)^2 + (3-5)^2 + (1+3)^2} = \frac{1}{2}\sqrt{16 + 4 + 16} = 3.$$

Therefore an equation of this sphere is $(x-5)^2 + (y-4)^2 + (z+1)^2 = 9$; that is,

$$x^2 - 10x + y^2 - 8y + z^2 + 2z + 33 = 0.$$

C12S02.023: The sphere has radius 2, thus equation $x^2 + y^2 + (z-2)^2 = 4$; that is, $x^2 + y^2 + z^2 - 4z = 0$.

C12S02.025: We complete the square in x and y:

$$x^2 + 4x + 4 + y^2 - 6y + 9 + z^2 = 13;$$
$$(x+2)^2 + (y-3)^2 + (z-0)^2 = 13.$$

This sphere has center at $(-2,\, 3,\, 0)$ and radius $\sqrt{13}$.

C12S02.027: We complete the square in z:

$$x^2 + y^2 + z^2 - 6z + 9 = 16 + 9 = 25;$$
$$(x-0)^2 + (y-0)^2 + (z-3)^2 = 5^2.$$

This sphere has center at $(0,\, 0,\, 3)$ and radius 5.

C12S02.029: The equation $z = 0$ is an equation of the xy-plane.

C12S02.031: The equation $z = 10$ is an equation of the plane parallel to the xy-plane and passing through the point $(0, 0, 10)$.

C12S02.033: The equation $xyz = 0$ holds when any one (or more) of the three variables is zero. Hence its graph is the union of the three coordinate planes.

C12S02.035: Given the equation $x^2 + y^2 + z^2 = 0$, the sum of the three nonnegative numbers x^2, y^2, and z^2 can be zero only if none of x, y, and z is nonzero. Therefore the graph of the given equation consists of the single point $(0, 0, 0)$.

C12S02.037: Complete the square in x and y:

$$x^2 + y^2 + z^2 - 6x + 8y + 25 = 0;$$

$$x^2 - 6x + 9 + y^2 + 8y + 16 + z^2 + 25 = 25;$$

$$(x - 3)^2 + (y + 4)^2 + z^2 = 0.$$

The sum of three nonnegative real numbers can be zero only if none is positive. So the graph of the given equation consists of the single point $(3, -4, 0)$.

C12S02.039: If $\mathbf{a} = \langle 4, -2, 6 \rangle$ and $\mathbf{b} = \langle 6, -3, 9 \rangle$, then $\mathbf{b} = \frac{3}{2}\mathbf{a}$. Therefore $\mathbf{a}$ and $\mathbf{b}$ are parallel. Because they are nonzero and parallel, they are not perpendicular; alternatively, they are not perpendicular because $\mathbf{a} \cdot \mathbf{b} = 84 \neq 0$.

C12S02.041: Because

$$\mathbf{b} = -9\mathbf{i} + 15\mathbf{j} - 12\mathbf{k} = -\frac{3}{4}(12\mathbf{i} - 20\mathbf{j} + 16\mathbf{k}) = -\frac{3}{4}\mathbf{a},$$

the vectors $\mathbf{a}$ and $\mathbf{b}$ are parallel. Because they are nonzero as well, they are not perpendicular; alternatively, $\mathbf{a} \cdot \mathbf{b} = -600 \neq 0$, and hence they are not perpendicular.

C12S02.043: The distinct points P, Q, and R lie on a single straight line if and only if the vectors representing $\overrightarrow{PQ}$ and $\overrightarrow{QR}$ are parallel. Here we have

$$3\overrightarrow{PQ} = 3(\langle 1, -3, 5 \rangle - \langle 0, -2, 4 \rangle) = 3\langle 1, -1, 1 \rangle = \langle 3, -3, 3 \rangle = \langle 4, -6, 8 \rangle - \langle 1, -3, 5 \rangle = \overrightarrow{QR}.$$

Therefore $\overrightarrow{PQ}$ and $\overrightarrow{QR}$ are parallel, and so the three points P, Q, and R do lie on a single straight line.

C12S02.045: The sides of triangle ABC are represented by the three vectors $\mathbf{u} = \overrightarrow{AB} = \langle -1, 1, 0 \rangle$, $\mathbf{v} = \overrightarrow{BC} = \langle 0, -1, 1 \rangle$, and $\mathbf{w} = \overrightarrow{AC} = \langle -1, 0, 1 \rangle$. Let A denote the angle of the triangle at vertex A. Then

$$\cos A = \frac{\mathbf{u} \cdot \mathbf{w}}{|\mathbf{u}| \cdot |\mathbf{w}|} = \frac{(-1)(-1)}{\sqrt{2}\,\sqrt{2}} = \frac{1}{2},$$

and therefore $A = 60°$ (exactly). Similarly,

$$\cos B = \frac{-\mathbf{u} \cdot \mathbf{v}}{|-\mathbf{u}| \cdot |\mathbf{v}|} = \frac{(-1)(-1)}{\sqrt{2}\,\sqrt{2}} = \frac{1}{2},$$

and so $B = 60°$ as well. Finally,

$$\cos C = \frac{-\mathbf{v} \cdot (-\mathbf{w})}{|-\mathbf{v}| \cdot |-\mathbf{w}|} = \frac{(-1)(-1)}{\sqrt{2}\,\sqrt{2}} = \frac{1}{2},$$

and it follows that $C = 60°$.

C12S02.047: The sides of triangle ABC are represented by the three vectors $\mathbf{u} = \overrightarrow{AB} = \langle 2, -3, 2 \rangle$, $\mathbf{v} = \overrightarrow{BC} = \langle 0, 6, 3 \rangle$, and $\mathbf{w} = \overrightarrow{AC} = \langle 2, 3, 5 \rangle$. Let A denote the angle of the triangle at vertex A. Then

$$\cos A = \frac{\mathbf{u} \cdot \mathbf{w}}{|\mathbf{u}| \cdot |\mathbf{w}|} = \frac{4 - 9 + 10}{\sqrt{17}\sqrt{38}} = \frac{5\sqrt{646}}{646} \approx 0.19672237,$$

and therefore $A \approx 78.654643° \approx 79°$. Similarly,

$$\cos B = \frac{-\mathbf{u} \cdot \mathbf{v}}{|-\mathbf{u}| \cdot |\mathbf{v}|} = \frac{18 - 6}{\sqrt{17}\sqrt{45}} = \frac{4\sqrt{85}}{85} \approx 0.43386082,$$

and so $B \approx 64.287166° \approx 64°$. Finally,

$$\cos C = \frac{-\mathbf{v} \cdot (-\mathbf{w})}{|-\mathbf{v}| \cdot |-\mathbf{w}|} = \frac{18 + 15}{\sqrt{38}\sqrt{45}} = \frac{11\sqrt{190}}{190} \approx 0.79802388,$$

and it follows that $C \approx 37.058191° \approx 37°$.

C12S02.049: If $\mathbf{a} = \overrightarrow{PQ} = \langle 2, 5, 5 \rangle$ and its direction angles are α, β, and γ (as in Fig. 11.2.14), then

$$\cos \alpha = \frac{2}{|\mathbf{a}|} = \frac{2}{\sqrt{54}} = \frac{\sqrt{6}}{9} \approx 0.27216553,$$

so that $\alpha \approx 74.206830951736°$. Also

$$\cos \beta = \cos \gamma = \frac{5}{|\mathbf{a}|} = \frac{5}{\sqrt{54}} = \frac{5\sqrt{6}}{18} \approx 0.68041382,$$

and therefore $\beta = \gamma \approx 47.124011333364°$.

C12S02.051: If $\mathbf{a} = \overrightarrow{PQ} = \langle 6, 8, 10 \rangle$ and its direction angles are α, β, and γ (as in Fig. 12.2.14), then

$$\cos \alpha = \frac{6}{|\mathbf{a}|} = \frac{6}{\sqrt{200}} = \frac{3\sqrt{2}}{10} \approx 0.42426407,$$

so that $\alpha \approx 64.895909749779°$. Next,

$$\cos \beta = \frac{8}{|\mathbf{a}|} = \frac{8}{\sqrt{200}} = \frac{2\sqrt{2}}{5} \approx 0.56568543,$$

and therefore $\beta \approx 55.5500098012047°$. Finally,

$$\cos \gamma = \frac{10}{|\mathbf{a}|} = \frac{10}{\sqrt{200}} = \frac{\sqrt{2}}{2} \approx 0.70710678,$$

and so $\gamma = 45°$ (exactly).

C12S02.053: The displacement vector is $\mathbf{D} = \overrightarrow{PQ} = 3\mathbf{i} + \mathbf{j}$, and consequently the work done is simply $\mathbf{F} \cdot \mathbf{D} = (\mathbf{i} - \mathbf{k}) \cdot (3\mathbf{i} + \mathbf{j}) = 3$.

C12S02.055: $40 \cdot (\cos 40°) \cdot 1000 \cdot (0.239) \approx 7323.385$ (cal; less than 8 Cal).

C12S02.057: Set up a coordinate system in which the lowest point of the inclined plane is at the origin in the xy-plane and its highest point is at (x, h). Then

$$\frac{h}{x} = \tan \alpha, \quad \text{so that} \quad x = \frac{h}{\tan \alpha} = h \cot \alpha.$$

Therefore the displacement vector in this problem is $\mathbf{D} = \langle h \cot \alpha, h \rangle$. A unit vector parallel to the inclined plane is $\langle \cos \alpha, \sin \alpha \rangle$, so a unit vector in the direction of $\mathbf{N}$ is $\langle -\sin \alpha, \cos \alpha \rangle$ (by Problem 55 in Section 12.1). Therefore $\mathbf{N} = \lambda \langle -\sin \alpha, \cos \alpha \rangle$ where λ is a positive scalar. If we denote by F the magnitude $|\mathbf{F}|$ of $\mathbf{F}$, then

$$\mathbf{F} = \langle F \cos \alpha, F \sin \alpha \rangle.$$

The horizontal forces acting on the weight must balance, as must the vertical forces, yielding (respectively) the scalar equations

$$\lambda \sin \alpha = F \cos \alpha \quad \text{and} \quad \lambda \cot \alpha + F \sin \alpha = mg.$$

From the first of these equations we see that $\lambda = F \cos \alpha$, and substitution of this in the second equation yields

$$F \cdot \left(\frac{\cos^2 \alpha}{\sin \alpha} \right) + F \cdot \left(\frac{\sin^2 \alpha}{\sin \alpha} \right) = mg,$$

and it follows that $F = mg \sin \alpha$. Therefore $\mathbf{F} = \langle mg \sin \alpha \cos \alpha, mg \sin^2 \alpha \rangle$. So the work done by $\mathbf{F}$ in moving the weight from the bottom to the top of the inclined plane is (if there's no friction)

$$W = \mathbf{F} \cdot \mathbf{D} = mgh \cos^2 \alpha + mgh \sin^2 \alpha = mgh.$$

C12S02.059: By the first equation in (9), we have

$$|\mathbf{a} + \mathbf{b}|^2 = (\mathbf{a} + \mathbf{b}) \cdot (\mathbf{a} + \mathbf{b}).$$

Then we use the fact that $x \leqq |x|$ for each real number x (in the first line) and the Cauchy-Schwarz inequality of Problem 58 (in the second line):

$$|\mathbf{a} + \mathbf{b}|^2 = \mathbf{a} \cdot \mathbf{a} + 2\mathbf{a} \cdot \mathbf{b} + \mathbf{b} \cdot \mathbf{b} = |\mathbf{a}|^2 + 2\mathbf{a} \cdot \mathbf{b} + |\mathbf{b}|^2 \leqq |\mathbf{a}|^2 + 2|\mathbf{a} \cdot \mathbf{b}| + |\mathbf{b}|^2$$

$$\leqq |\mathbf{a}|^2 + 2|\mathbf{a}| \cdot |\mathbf{b}| + |\mathbf{b}|^2 = (|\mathbf{a}| + |\mathbf{b}|)^2.$$

All these expression are nonnegative. Therefore, for any two vectors $\mathbf{a}$ and $\mathbf{b}$, $|\mathbf{a} + \mathbf{b}| \leqq |\mathbf{a}| + |\mathbf{b}|$.

C12S02.061: Suppose that $\mathbf{u} = \langle u_1, u_2, u_3 \rangle$ and that $\mathbf{v} = \langle v_1, v_2, v_3 \rangle$. If both are the zero vector, let $\mathbf{w} = \mathbf{i}$. Otherwise we may suppose without loss of genrality that $\mathbf{u} \neq \mathbf{0}$. Indeed, we may further suppose without loss of generality that $u_1 \neq 0$. We need to choose $\mathbf{w} = \langle w_1, w_2, w_3 \rangle$ such that

$$\mathbf{u} \cdot \mathbf{w} = u_1 w_1 + u_2 w_2 + u_3 w_3 = 0 \quad \text{and}$$

$$\mathbf{v} \cdot \mathbf{w} = v_1 w_1 + v_2 w_2 + v_3 w_3 = 0.$$

This will obtain provided that

$$u_1 v_1 w_1 + u_2 v_1 w_2 + u_3 v_1 w_3 = 0 \quad \text{and}$$

$$u_1 v_1 w_1 + u_1 v_2 w_2 + u_1 v_3 w_3 = 0.$$

Subtraction of the first of these equations from the second yields the sufficient condition

618

$$(u_1 v_2 - u_2 v_1)w_2 + (u_1 v_3 - u_3 v_1)w_3 = 0.$$

Thus our goal will be accomplished if we let

$$w_2 = u_1 v_3 - u_3 v_1 \quad \text{and}$$

$$w_3 = -(u_1 v_2 - u_2 v_1) = u_2 v_1 - u_1 v_2.$$

Now let

$$w_1 = -\frac{u_2 w_2 + u_3 w_3}{u_1},$$

and we have reached our goal. In particular, if $\mathbf{u} = \langle 1, 2, -3 \rangle$ and $\mathbf{v} = \langle 2, 0, 1 \rangle$, then we obtain

$$w_2 = 1 \cdot 1 + 3 \cdot 2 = 7,$$

$$w_3 = 2 \cdot 2 - 1 \cdot 0 = 4, \quad \text{and}$$

$$w_1 = -(2 \cdot 7 - 3 \cdot 4) = -2,$$

and thus any nonzero multiple of $\mathbf{w} = \langle -2, 7, 4 \rangle$ is a correct answer to this problem. We will see a more efficient construction of a nonzero vector perpendicular to two given nonparallel vectors in Section 12.3.

C12S02.063: Given: The points $P_1(x_1, y_1, z_1)$, $P_2(x_2, y_2, z_2)$, and

$$M\left(\frac{x_1 + x_2}{2}, \frac{y_1 + y_2}{2}, \frac{z_1 + z_2}{2} \right).$$

Let $\mathbf{a} = \overrightarrow{P_1 P_2} = \langle x_2 - x_1, y_2 - y_1, z_2 - z_1 \rangle$. Then $\mathbf{a}$ coincides with the line segment joining P_1 with P_2. Hence if the initial point of $\frac{1}{2}\mathbf{a}$ is placed at P_1, then its terminal point will coincide with the midpoint of the segment $P_1 P_2$. That is, $\overrightarrow{OP_1} + \frac{1}{2}\mathbf{a}$ will be the position vector of the midpoint of $P_1 P_2$. But

$$\overrightarrow{OP_1} + \frac{1}{2}\mathbf{a} = \langle x_1, y_1, z_1 \rangle + \frac{1}{2} \langle x_2 - x_1, y_2 - y_1, z_2 - z_1 \rangle = \left\langle \frac{x_1 + x_2}{2}, \frac{y_1 + y_2}{2}, \frac{z_1 + z_2}{2} \right\rangle = \overrightarrow{OM}.$$

This is enough to establish that M is the midpoint of the segment $P_1 P_2$.

C12S02.065: Suppose that the vectors $\mathbf{a}$ and $\mathbf{b}$ in the plane are nonzero and nonparallel. We will provide a proof in the case that the first components of $\mathbf{a}$ and $\mathbf{b}$ are nonzero. Let $\mathbf{a} = \langle a_1, a_2, \rangle$, $\mathbf{b} = \langle b_1, b_2 \rangle$, and $\mathbf{c} = \langle c_1, c_2 \rangle$. To find α and β such that $\mathbf{c} = \alpha \mathbf{a} + \beta \mathbf{b}$, solve

$$c_1 = \alpha a_1 + \beta b_1,$$
$$c_2 = \alpha a_2 + \beta b_2;$$

$$a_2 c_1 = \alpha a_1 a_2 + \beta a_2 b_1,$$
$$a_1 c_2 = \alpha a_1 a_2 + \beta a_1 b_2.$$

Subtract the next-to-last equation from the last:

$$a_1c_2 - a_2c_1 = \beta(a_1b_2 - a_2b_1);$$

$$\beta = \frac{a_1c_2 - a_2c_1}{a_1b_2 - a_2b_1}.$$

But what if the last denominator is zero? Then $a_1b_2 = a_2b_1$, so that

$$\frac{a_2}{a_1} = \frac{b_2}{b_1} = k,$$

and thus $a_2 = ka_1$ and $b_2 = kb_1$. Therefore

$$\mathbf{a} = \langle a_1, ka_1 \rangle = a_1 \langle 1, k \rangle \quad \text{and}$$

$$\mathbf{b} = \langle b_1, kb_1 \rangle = b_1 \langle 1, k \rangle,$$

so that $\mathbf{a}$ and $\mathbf{b}$ are parallel. Hence $a_1b_2 - a_2b_1 \neq 0$.

In a similar fashion we find that

$$\alpha = \frac{b_2c_1 - b_1c_2}{a_1b_2 - a_2b_1}.$$

All that remains is the verification that α and β perform as advertised:

$$\alpha\mathbf{a} + \beta\mathbf{b} = \frac{b_2c_1 - b_1c_2}{a_1b_2 - a_2b_1} \langle a_1, a_2 \rangle + \frac{a_1c_2 - a_2c_1}{a_1b_2 - a_2b_1} \langle b_1, b_2 \rangle$$

$$= \frac{1}{a_1b_2 - a_2b_1} \left(\langle a_1b_2c_1 - a_1b_1c_2, a_2b_2c_1 - a_2b_1c_2 \rangle + \langle a_1b_1c_2 - a_2b_1c_1, a_1b_2c_2 - a_2b_2c_1 \rangle \right)$$

$$= \frac{1}{a_1b_2 - a_2b_1} \langle (a_1b_2 - a_2b_1)c_1, (a_1b_2 - a_2b_1)c_2 \rangle = \langle c_1, c_2 \rangle = \mathbf{c}.$$

C12S02.067: The equation

$$(x - 3)^2 + (y + 2)^2 + (z - 4)^2 = (x - 5)^2 + (y - 7)^2 + (z + 1)^2$$

can be simplified to $2x + 9y - 5z = 23$. This is an equation of the plane that bisects the segment AB and is perpendicular to it.

C12S02.069: The distance between the last two points listed in Problem 69 is

$$\sqrt{(1 - 0)^2 + (0 - 1)^2 + (1 - 1)^2} = \sqrt{2},$$

and the other five distance computations also yield the same result. Hence the given points are the vertices of a regular tetrahedron. Let $\mathbf{a}$ be the position vector of $(1, 1, 0)$ and let $\mathbf{b}$ be the position vector of $(1, 0, 1)$. Then the angle θ between $\mathbf{a}$ and $\mathbf{b}$ satisfies the equation

$$\cos\theta = \frac{\mathbf{a} \cdot \mathbf{b}}{|\mathbf{a}| \cdot |\mathbf{b}|} = \frac{1 \cdot 1 + 1 \cdot 0 + 0 \cdot 1}{\sqrt{2}\,\sqrt{2}} = \frac{1}{2}.$$

Therefore $\theta = \dfrac{\pi}{3}$.

C12S03.001: If $\mathbf{a} = \langle 5, -1, -2 \rangle$ and $\mathbf{b} = \langle -3, 2, 4 \rangle$, then

$$\mathbf{a} \times \mathbf{b} = \begin{vmatrix} \mathbf{i} & \mathbf{j} & \mathbf{k} \\ 5 & -1 & -2 \\ -3 & 2 & 4 \end{vmatrix} = \langle -4 + 4, \ 6 - 20, \ 10 - 3 \rangle = \langle 0, -14, 7 \rangle.$$

C12S03.003: If $\mathbf{a} = \mathbf{i} - \mathbf{j} + 3\mathbf{k}$ and $\mathbf{b} = -2\mathbf{i} + 3\mathbf{j} + \mathbf{k}$, then

$$\mathbf{a} \times \mathbf{b} = \begin{vmatrix} \mathbf{i} & \mathbf{j} & \mathbf{k} \\ 1 & -1 & 3 \\ -2 & 3 & 1 \end{vmatrix} = (-1 - 9)\mathbf{i} + (-6 - 1)\mathbf{j} + (3 - 2)\mathbf{k} = -10\mathbf{i} - 7\mathbf{j} + \mathbf{k}.$$

C12S03.005: If $\mathbf{a} = \langle 2, -3, 0 \rangle$ and $\mathbf{b} = \langle 4, 5, 0 \rangle$, then

$$\mathbf{a} \times \mathbf{b} = \begin{vmatrix} \mathbf{i} & \mathbf{j} & \mathbf{k} \\ 2 & -3 & 0 \\ 4 & 5 & 0 \end{vmatrix} = \langle 0 - 0, \ 0 - 0, \ 10 + 12 \rangle = \langle 0, 0, 22 \rangle.$$

C12S03.007: If $\mathbf{a} = \langle 3, 12, 0 \rangle$ and $\mathbf{b} = \langle 0, 4, 3 \rangle$, then

$$\mathbf{a} \times \mathbf{b} = \begin{vmatrix} \mathbf{i} & \mathbf{j} & \mathbf{k} \\ 3 & 12 & 0 \\ 0 & 4 & 3 \end{vmatrix} = \langle 36 - 0, \ 0 - 9, \ 12 - 0 \rangle = \langle 36, -9, 12 \rangle.$$

The magnitude of $\mathbf{a} \times \mathbf{b}$ is

$$|\mathbf{a} \times \mathbf{b}| = \sqrt{36^2 + 9^2 + 12^2} = \sqrt{1521} = 39,$$

so the two unit vectors perpendicular to both $\mathbf{a}$ and $\mathbf{b}$ are

$$\mathbf{u} = \frac{1}{39} \langle 36, -9, 12 \rangle = \left\langle \frac{12}{13}, -\frac{3}{13}, \frac{4}{13} \right\rangle \quad \text{and} \quad \mathbf{v} = -\frac{1}{39} \langle 36, -9, 12 \rangle = \left\langle -\frac{12}{13}, \frac{3}{13}, -\frac{4}{13} \right\rangle.$$

C12S03.009: The definition of the cross product in Eq. (5) yields

$$\mathbf{i} \times \mathbf{j} = \begin{vmatrix} \mathbf{i} & \mathbf{j} & \mathbf{k} \\ 1 & 0 & 0 \\ 0 & 1 & 0 \end{vmatrix} = (0 - 0)\mathbf{i} + (0 - 0)\mathbf{j} + (1 - 0)\mathbf{k} = \mathbf{k},$$

$$\mathbf{j} \times \mathbf{k} = \begin{vmatrix} \mathbf{i} & \mathbf{j} & \mathbf{k} \\ 0 & 1 & 0 \\ 0 & 0 & 1 \end{vmatrix} = (1-0)\mathbf{i} + (0-0)\mathbf{j} + (0-0)\mathbf{k} = \mathbf{i}, \quad \text{and}$$

$$\mathbf{k} \times \mathbf{i} = \begin{vmatrix} \mathbf{i} & \mathbf{j} & \mathbf{k} \\ 0 & 0 & 1 \\ 1 & 0 & 0 \end{vmatrix} = (0-0)\mathbf{i} + (1-0)\mathbf{j} + (0-0)\mathbf{k} = \mathbf{j}.$$

C12S03.011: If $\mathbf{a} = \mathbf{i}$, $\mathbf{b} = \mathbf{i} + \mathbf{j}$, and $\mathbf{c} = \mathbf{i} + \mathbf{j} + \mathbf{k}$, then

$$\mathbf{a} \times (\mathbf{b} \times \mathbf{c}) = \mathbf{a} \times \begin{vmatrix} \mathbf{i} & \mathbf{j} & \mathbf{k} \\ 1 & 1 & 0 \\ 1 & 1 & 1 \end{vmatrix} = \mathbf{a} \times (\mathbf{i} - \mathbf{j}) = \begin{vmatrix} \mathbf{i} & \mathbf{j} & \mathbf{k} \\ 1 & 0 & 0 \\ 1 & -1 & 0 \end{vmatrix} = -\mathbf{k},$$

whereas

$$(\mathbf{a} \times \mathbf{b}) \times \mathbf{c} = \begin{vmatrix} \mathbf{i} & \mathbf{j} & \mathbf{k} \\ 1 & 0 & 0 \\ 1 & 1 & 0 \end{vmatrix} \times \mathbf{c} = \mathbf{k} \times \mathbf{c} = \begin{vmatrix} \mathbf{i} & \mathbf{j} & \mathbf{k} \\ 0 & 0 & 1 \\ 1 & 1 & 1 \end{vmatrix} = -\mathbf{i} + \mathbf{j}.$$

Thus the vector product is not associative.

C12S03.013: If $\mathbf{a}$, $\mathbf{b}$, and $\mathbf{c}$ are mutually perpendicular, note that $\mathbf{b} \times \mathbf{c}$ is perpendicular to both $\mathbf{b}$ and $\mathbf{c}$, hence $\mathbf{b} \times \mathbf{c}$ is parallel to $\mathbf{a}$. Therefore $\mathbf{a} \times (\mathbf{b} \times \mathbf{c}) = \mathbf{0}$.

C12S03.015: By Eq. (10) (and the discussion that precedes and follows it), the area of triangle PQR is $A = \frac{1}{2} | \overrightarrow{PQ} + \overrightarrow{PR} |$. But

$$\overrightarrow{PQ} \times \overrightarrow{PR} = \begin{vmatrix} \mathbf{i} & \mathbf{j} & \mathbf{k} \\ 1 & 1 & 7 \\ -4 & -5 & 4 \end{vmatrix} = 39\mathbf{i} - 32\mathbf{j} - \mathbf{k}.$$

The magnitude of $39\mathbf{i} - 32\mathbf{j} - \mathbf{k}$ is

$$|39\mathbf{i} - 32\mathbf{j} - \mathbf{k}| = \sqrt{1521 + 1024 + 1} = \sqrt{2546}.$$

Therefore the area of triangle PQR is $A = \frac{1}{2}\sqrt{2546} \approx 25.2289516231$.

C12S03.017: Let $\mathbf{a} = \overrightarrow{OP} = \langle 1, 3, -2 \rangle$, $\mathbf{b} = \overrightarrow{OQ} = \langle 2, 4, 5 \rangle$, and $\mathbf{c} = \overrightarrow{OR} = \langle -3, -2, 2 \rangle$. Part (a): By Theorem 4 and part (4) of Theorem 3, the volume of the parallelepiped with adjacent edges these three vectors is given by $V = |(\mathbf{a} \times \mathbf{b}) \cdot \mathbf{c}|$. Here we have

$$\mathbf{a} \times \mathbf{b} = \begin{vmatrix} \mathbf{i} & \mathbf{j} & \mathbf{k} \\ 1 & 3 & -2 \\ 2 & 4 & 5 \end{vmatrix} = \langle 23, -9, -2 \rangle,$$

and therefore $V = |\langle 23, -9, -2 \rangle \cdot \langle -3, -2, 2 \rangle| = |-55| = 55$.

Part (b): By Example 7 of Section 12.3, the volume of the pyramid is $\dfrac{55}{6}$.

C12S03.019: Here we have $\overrightarrow{AB} = \langle 2, 1, 3 \rangle$, $\overrightarrow{AC} = \langle 1, -3, 0 \rangle$, and $\overrightarrow{AD} = \langle 3, 5, 6 \rangle$. By Theorem 4 and Eq. (17), the volume of the parallelepiped having these vectors as adjacent sides is the absolute value of

$$\begin{vmatrix} 2 & 1 & 3 \\ 1 & -3 & 0 \\ 3 & 5 & 6 \end{vmatrix} = (-1) \cdot (6 - 15) - 3 \cdot (12 - 9) = 9 - 9 = 0$$

(we expanded the determinant using its second row; see the discussion following Example 1 of the text). By the reasoning given in Example 8, the four given points are coplanar.

C12S03.021: Here we have $\overrightarrow{AB} = \langle 1, 2, 3 \rangle$, $\overrightarrow{AC} = \langle 2, 3, 4 \rangle$, and $\overrightarrow{AD} = \langle 9, 12, 21 \rangle$. By Theorem 4 and Eq. (17), the volume of the parallelepiped having these vectors as adjacent sides is the absolute value of

$$\begin{vmatrix} 1 & 2 & 3 \\ 2 & 3 & 4 \\ 9 & 12 & 21 \end{vmatrix} = 1 \cdot (63 - 48) - 2 \cdot (42 - 36) + 3 \cdot (24 - 27) = 15 - 12 - 9 = -6$$

(we expanded the determinant using its first row; see the discussion following Example 1 of the text). Therefore the four given points are not coplanar. The parallelepiped having the three vectors as adjacent sides has volume 6 and the tetrahedron (pyramid) with the four given points as its vertices has volume 1.

C12S03.023: Name the vertices of the polygon counterclockwise, beginning at the origin: O, A, and B. Then

$$\mathbf{a} = \overrightarrow{OA} = \langle 176 \cos 15°, 176 \sin 15°, 0 \rangle \approx \langle 170.002945, 45.552152, 0 \rangle \quad \text{and}$$

$$\mathbf{b} = \overrightarrow{OB} = \langle 83 \cos 52°, 83 \sin 52°, 0 \rangle \approx \langle 51.099902, 65.404893, 0 \rangle.$$

Using the exact rather than the approximate values, we entered the *Mathematica* 3.0 commands

```
c = Cross[a,b]    (* to compute the vector product of a and b *)
area = (1/2)*Sqrt[c.c]    (* to compute half the magnitude of c *)
N[area,20]    (* to print a 20-place approximation to the area *)
```

The response to the last command rounds to the answer, 4395.6569291026 (m²).

C12S03.025: Name the vertices of the polygon counterclockwise, beginning at the origin: O, A, B, and C. Then

$$\mathbf{a} = \overrightarrow{OA} = \langle 220\cos 25°,\, 220\sin 25°,\, 0 \rangle \approx \langle 227.206664,\, 115.767577,\, 0 \rangle,$$

$$\mathbf{b} = \overrightarrow{OB} = \mathbf{a} + \langle -210\cos 40°,\, 210\sin 40°,\, 0 \rangle \approx \langle 38.518380,\, 227.961416,\, 0 \rangle, \quad \text{and}$$

$$\mathbf{c} = \overrightarrow{OC} = \langle -150\cos 63°,\, 150\sin 63°,\, 0 \rangle \approx \langle -68.098575,\, 133.650979,\, 0 \rangle.$$

Using the exact rather than the approximate values, we entered the *Mathematica 3.0* commands

```
p = Cross[a,b]    (* to compute the vector product of a and b *)
area1 = (1/2)*Sqrt[p.p]    (* to compute half the magnitude of p *)
q = Cross[b,c]
area2 = (1/2)*Sqrt[q.q]
N[area1 + area2, 20]    (* to print a 20-place approximation to the total area *)
```

The response to the last command rounds to the answer, 31271.643253 (ft^2).

C12S03.027: Let $\mathbf{a} = \langle a_1,\, a_2,\, a_3 \rangle$ and $\mathbf{b} = \langle b_1,\, b_2,\, b_3 \rangle$. Then

$$\mathbf{a} \times \mathbf{b} = \begin{vmatrix} \mathbf{i} & \mathbf{j} & \mathbf{k} \\ a_1 & a_2 & a_3 \\ b_1 & b_2 & b_3 \end{vmatrix} = \langle a_2 b_3 - a_3 b_2,\, a_3 b_1 - a_1 b_3\ a_1 b_2 - a_2 b_1 \rangle \quad \text{and}$$

$$\mathbf{b} \times \mathbf{a} = \begin{vmatrix} \mathbf{i} & \mathbf{j} & \mathbf{k} \\ b_1 & b_2 & b_3 \\ a_1 & a_2 & a_3 \end{vmatrix} = \langle b_2 a_3 - b_3 a_2,\, b_3 a_1 - b_1 a_3\ b_1 a_2 - b_2 a_1 \rangle$$

$$= -\langle a_2 b_3 - a_3 b_2,\, a_3 b_1 - a_1 b_3,\, a_1 b_2 - a_2 b_1 \rangle = -(\mathbf{a} \times \mathbf{b}).$$

C12S03.029: Part (a): Please refer to Fig. 12.3.13 of the text. Because the segment of length d is perpendicular to the segment PQ, we know from elementary geometry that the area of the triangle is $\frac{1}{2}|\overrightarrow{PQ}| \cdot d$. Therefore

$$\frac{1}{2}|\overrightarrow{AP} \times \overrightarrow{AQ}| = \frac{1}{2}|\overrightarrow{PQ}| \cdot d, \quad \text{and thus} \quad d = \frac{|\overrightarrow{AP} \times \overrightarrow{AQ}|}{|\overrightarrow{PQ}|}. \tag{1}$$

Part (b): With $A = (1, 0, 1)$, $P = (2, 3, 1)$, and $Q = (-3, 1, 4)$, we have

$$\overrightarrow{AP} = \langle 1, 3, 0 \rangle, \quad \overrightarrow{AQ} = \langle -4, 1, 3 \rangle, \quad \text{and} \quad \overrightarrow{PQ} = \langle -5, -2, 3 \rangle.$$

To use the formula in (1), we first compute

$$\overrightarrow{AP} \times \overrightarrow{AQ} = \begin{vmatrix} \mathbf{i} & \mathbf{j} & \mathbf{k} \\ 1 & 3 & 0 \\ -4 & 1 & 3 \end{vmatrix} = \langle 9 - 0,\ 0 - 3,\ 1 - (-12) \rangle = \langle 9,\ -3,\ 13 \rangle.$$

Then the formula in (1) implies that the distance from the given point to the given line is

$$d = \frac{|\overrightarrow{AP} \times \overrightarrow{AQ}|}{|\overrightarrow{PQ}|} = \frac{|\langle 9,\ -3,\ 13 \rangle|}{\sqrt{25 + 4 + 9}} = \frac{\sqrt{259}}{\sqrt{38}} = \frac{\sqrt{9842}}{38} \approx 2.61070670.$$

C12S03.031: Let $\mathbf{u} = \overrightarrow{P_1 Q_1}$, $\mathbf{v} = \overrightarrow{P_2 Q_2}$, and $\mathbf{w} = \overrightarrow{P_1 P_2}$. A vector $\mathbf{n}$ that is perpendicular to both lines is one perpendicular to both $\mathbf{u}$ and $\mathbf{v}$, and an obvious choice is $\mathbf{n} = \mathbf{u} \times \mathbf{v}$ ($\mathbf{n}$ will be nonzero because the two lines are not parallel). The projection of $\mathbf{w}$ in the direction of $\mathbf{n}$ will be $\pm d$ where d is the (perpendicular) distance between the two lines. Thus

$$d = \frac{|\mathbf{w} \cdot \mathbf{n}|}{|\mathbf{n}|} = \frac{|\overrightarrow{P_1 P_2} \cdot (\overrightarrow{P_1 Q_1} \times \overrightarrow{P_2 Q_2})|}{|\overrightarrow{P_1 Q_1} \times \overrightarrow{P_2 Q_2}|}.$$

Comment: Rather than memorizing formulas such as those in Problems 29, 30, and 31, one should learn the general technique for finding the distance between two objects (points, lines, or planes) in space. First find a vector $\mathbf{n}$ normal to both objects. Then construct another vector $\mathbf{c}$ "connecting" the two objects; that is, with initial point on one and terminal point on the other. Then the projection of $\mathbf{c}$ in the direction of $\mathbf{n}$ is the (perpendicular) distance between the two objects (or its negative); that is, the distance d is given by the simple formula

$$d = \frac{|\mathbf{c} \cdot \mathbf{n}|}{|\mathbf{n}|}.$$

C12S03.033: Given the vectors $\mathbf{a}$, $\mathbf{b}$, and $\mathbf{c}$ in space, we have

$$\mathbf{a} \times (\mathbf{b} \times \mathbf{c}) = -(\mathbf{b} \times \mathbf{c}) \times \mathbf{a} \quad \text{(by Eq. (12))}$$

$$= -[(\mathbf{b} \cdot \mathbf{a})\mathbf{c} - (\mathbf{c} \cdot \mathbf{a})\mathbf{b}] \quad \text{(by the result in Problem 32)}$$

$$= (\mathbf{a} \cdot \mathbf{c})\mathbf{b} - (\mathbf{a} \cdot \mathbf{b})\mathbf{c} \quad \text{(this establishes Eq. (16))}.$$

C12S03.035: Given: The triangle in the plane with vertices at $P(x_1, y_1, 0)$, $Q(x_2, y_2, 0)$, and $R(x_3, y_3, 0)$. Let

$$\mathbf{u} = \overrightarrow{PQ} = \langle x_2 - x_1,\ y_2 - y_1,\ 0 \rangle \quad \text{and} \quad \mathbf{v} = \overrightarrow{PR} = \langle x_3 - x_1,\ y_3 - y_1,\ 0 \rangle.$$

The area of the triangle is $A = \frac{1}{2}|\mathbf{u} \times \mathbf{v}|$. Now

$$\mathbf{u} \times \mathbf{v} = \begin{vmatrix} \mathbf{i} & \mathbf{j} & \mathbf{k} \\ x_2 - x_1 & y_2 - y_1 & 0 \\ x_3 - x_1 & y_3 - y_1 & 0 \end{vmatrix} = \langle 0,\ 0,\ (x_2 - x_1)(y_3 - y_1) - (y_2 - y_1)(x_3 - x_1) \rangle,$$

so that

$$|\mathbf{u} \times \mathbf{v}| = |x_2 y_3 - x_2 y_1 - x_1 y_3 + x_1 y_1 - x_3 y_2 + x_3 y_1 + x_1 y_2 - x_1 y_1|$$

$$= |x_2 y_3 - x_2 y_1 - x_1 y_3 - x_3 y_2 + x_3 y_1 + x_1 y_2|$$

$$= |(x_2 y_3 - x_3 y_2) - (x_1 y_3 - x_3 y_1) + (x_1 y_2 - x_2 y_1)|,$$

which is the *absolute value* of the determinant

$$\begin{vmatrix} 1 & x_1 & y_1 \\ 1 & x_2 & y_2 \\ 1 & x_3 & y_3 \end{vmatrix} \tag{1}$$

(expanded using its first column—see the discussion following Eq. (3) in the text). Thus the area of triangle PQR is half the absolute value of the determinant in (1).

Section 12.4

C12S04.001: Solving the vector equation $\langle x, y, z \rangle = t\mathbf{v} + \overrightarrow{OP}$ yields

$$\langle x, y, z \rangle = t\langle 1, 2, 3 \rangle + \langle 0, 0, 0 \rangle = \langle t, 2t, 3t \rangle,$$

so that the parametric equations of the line are $x = t$, $y = 2t$, $z = 3t$.

C12S04.003: Solving the vector equation $\langle x, y, z \rangle = t\mathbf{v} + \overrightarrow{OP}$ yields

$$\langle x, y, z \rangle = t\langle 2, 0, -3 \rangle + \langle 4, 13, -3 \rangle = \langle 2t + 4, 13, -3t - 3 \rangle,$$

and thus the parametric equations of the line are $x = 2t + 4$, $y = 13$, $z = -3t - 3$.

C12S04.005: Solving the vector equation $\langle x, y, z \rangle = \overrightarrow{OP_1} + t\,\overrightarrow{P_1 P_2}$ yields

$$\langle x, y, z \rangle = \langle 0, 0, 0 \rangle + t\langle -6, 3, 5 \rangle,$$

so the line has parametric equations $x = -6t$, $y = 3t$, $z = 5t$.

C12S04.007: Solving the vector equation $\langle x, y, z \rangle = \overrightarrow{OP_1} + t\,\overrightarrow{P_1 P_2}$ yields

$$\langle x, y, z \rangle = \langle 3, 5, 7 \rangle + t\langle 3, 0, -3 \rangle = \langle 3 + 3t, 5, 7 - 3t \rangle$$

so the line has parametric equations $x = 3t + 3$, $y = 5$, $z = -3t + 7$.

C12S04.009: Solving the vector equation $\langle x, y, z \rangle = \overrightarrow{OP} + t\mathbf{v}$ yields

$$\langle x, y, z \rangle = \langle 2, 3, -4 \rangle + t\langle 1, -1, -2 \rangle = \langle 2 + t, 3 - t, -4 - 2t \rangle$$

so the line has parametric equations $x = t + 2$, $y = -t + 3$, $z = -2t - 4$ and symmetric equations

$$x - 2 = -y + 3 = -\frac{z + 4}{2}.$$

C12S04.011: Solving the vector equation $\langle x, y, z \rangle = \overrightarrow{OP} + t\mathbf{k}$ yields

$$\langle x, y, z \rangle = \langle 1, 1, 1 \rangle + t\langle 0, 0, 1 \rangle = \langle 1, 1, 1+t \rangle$$

so the line has parametric equations $x = 1$, $y = 1$, $z = t + 1$. In a strict sense, the line doesn't have symmetric equations, but its Cartesian equations are

$$x = 1, \quad y = 1.$$

The fact that z is not mentioned means that z is *arbitrary*.

C12S04.013: A line normal to the plane with Cartesian equation $2x - y + 3z = 4$ is parallel to its normal vector $\mathbf{n} = \langle 2, -1, 3 \rangle$. The line of this problem also passes through the point $P(2, -3, 4)$ and thus has vector equation $\langle x, y, z \rangle = t\langle 2, -1, 3 \rangle + \langle 2, -3, 4 \rangle$, parametric equations $x = 2t + 2$, $y = -t - 3$, $z = 3t + 4$, and symmetric equations

$$\frac{x-2}{2} = -(y+3) = \frac{z-4}{3}.$$

C12S04.015: Given: The lines L_1 and L_2 with symmetric equations

$$x - 2 = \frac{y+1}{2} = \frac{z-3}{3} \quad \text{and} \quad \frac{x-5}{3} = \frac{y-1}{2} = z - 4, \tag{1}$$

respectively. Points on L_1 include $P_1(2, -1, 3)$ and $Q_1(3, 1, 6)$, and thus L_1 is parallel to the vector $\overrightarrow{P_1Q_1} = \langle 1, 2, 3 \rangle$. Points on L_2 include $P_2(5, 1, 4)$ and $Q_2(8, 3, 5)$, and thus L_2 is parallel to the vector $\overrightarrow{P_2Q_2} = \langle 3, 2, 1 \rangle$. Clearly L_1 and L_2 are not parallel. To determine whether they intersect, the *Mathematica* 3.0 command for solving the equations in (1) simultaneously is

```
Solve[ {x - 2 == (y + 1)/2, x - 2 == (z - 3)/3,
       (x - 5)/3 == (y - 1)/2, (x - 5)/3 == z - 4 }, { x, y, z } ]
```

and this command yields the response $x = 2$, $y = -1$, $z = 3$. Hence the two lines meet at the single point $(2, -1, 3)$.

C12S04.017: When we try to solve the equations of L_1 and L_2 simultaneously, we find that there is no solution. Thus the lines do not intersect. Two points on L_1 are $P_1(6, 5, 7)$ and $Q_1(8, 7, 10)$, so L_1 is parallel to the vector $\mathbf{v}_1 = \overrightarrow{P_1Q_1} = \langle 2, 2, 3 \rangle$. Two points on L_2 are $P_2(7, 5, 10)$ and $Q_2(10, 8, 15)$, so L_2 is parallel to the vector $\mathbf{v}_2 = \overrightarrow{P_2Q_2} = \langle 3, 3, 5 \rangle$. The equation $\lambda \mathbf{v}_1 = \mathbf{v}_2$ has no solution, so $\mathbf{v}_1$ and $\mathbf{v}_2$ are not parallel. Therefore L_1 and L_2 are skew lines.

C12S04.019: We solve the equations

$$\frac{x-7}{6} = \frac{y+5}{4}, \quad \frac{x-7}{6} = \frac{9-z}{8}, \quad \frac{11-x}{9} = \frac{7-y}{6}, \quad \text{and} \quad \frac{7-y}{6} = \frac{z-13}{12}$$

simultaneously and find that there is no solution. So L_1 and L_2 do not intersect. Two points on L_1 are $P_1(7, -5, 9)$ and $Q_1(13, -1, 1)$, so L_1 is parallel to the vector $\mathbf{v}_1 = \overrightarrow{P_1Q_1} = 6\mathbf{i} + 4\mathbf{j} - 8\mathbf{k}$. Two points of L_2 and $P_2(11, 7, 13)$ and $Q_2(2, 1, 25)$, so L_2 is parallel to the vector $\mathbf{v}_2 = \overrightarrow{P_2Q_2} = -9\mathbf{i} - 6\mathbf{j} + 12\mathbf{k}$. Because $-\frac{3}{2}\mathbf{v}_1 = \mathbf{v}_2$, the vectors $\mathbf{v}_1$ and $\mathbf{v}_2$ are parallel, and therefore L_1 and L_2 are parallel as well.

C12S04.021: The vector equation of the plane is $\mathbf{n} \cdot \langle x, y, z \rangle = \mathbf{n} \cdot \overrightarrow{OP}$, and because $\overrightarrow{OP} = 0$ it follows that a Cartesian equation of the plane is $x + 2y + 3z = 0$.

C12S04.023: The vector equation of the plane is $\mathbf{n} \cdot \langle x, y, z \rangle = \mathbf{n} \cdot \overrightarrow{OP}$; that is,

$$\langle 1, 0, -1 \rangle \cdot \langle x, y, z \rangle = \langle 1, 0, -1 \rangle \cdot \langle 5, 12, 13 \rangle,$$

and thus a Cartesian equation of the plane is $x - z + 8 = 0$.

C12S04.025: A plane parallel to the xz-plane has an equation of the form $y = c$ where c is a constant. The plane of Problem 25 also passes through the point $(5, 7, -6)$, so $c = 7$. Thus a Cartesian equation of this plane is $y = 7$.

C12S04.027: The vector equation of the plane is $\mathbf{n} \cdot \langle x, y, z \rangle = \mathbf{n} \cdot \overrightarrow{OP}$; that is,

$$\langle 7, 11, 0 \rangle \cdot \langle x, y, z \rangle = \langle 7, 11, 0 \rangle \cdot \langle 10, 4, -3 \rangle.$$

Thus a Cartesian equation of this plane is $7x + 11y = 114$.

C12S04.029: A plane parallel to the plane with Cartesian equation $3x + 4y - z = 10$ has normal vector $\mathbf{n} = 3\mathbf{i} + 4\mathbf{j} - \mathbf{k}$, thus a Cartesian equation of the form $3x + 4y - z = c$ for some constant c. The plane of Problem 29 also passes through the origin, so that $c = 0$. Hence it has Cartesian equation $3x + 4y - z = 0$.

C12S04.031: The plane through the origin $O(0, 0, 0)$ and the two points $P(1, 1, 1)$ and $Q(1, -1, 3)$ has normal vector

$$\mathbf{n} = \overrightarrow{OP} \times \overrightarrow{OQ} = \begin{vmatrix} \mathbf{i} & \mathbf{j} & \mathbf{k} \\ 1 & 1 & 1 \\ 1 & -1 & 3 \end{vmatrix} = \langle 4, -2, -2 \rangle.$$

Therefore this plane has a Cartesian equation of the form $4x - 2y - 2z = c$ where c is a constant. Because the plane of Problem 31 passes through the origin, $c = 0$. Therefore a Cartesian equation of this plane is $2x - y - z = 0$.

C12S04.033: The plane $\mathcal{P}$ that contains $P(2, 4, 6)$ and the line L with parametric equations $x = 7 - 3t$, $y = 3 + 4t$, $z = 5 + 2t$ also contains the two points $Q(7, 3, 5)$ (set $t = 0$) and $R(4, 7, 4)$ (set $t = 1$) of L. So the vectors $\overrightarrow{PQ} = 5\mathbf{i} - \mathbf{j} - \mathbf{k}$ amd $\overrightarrow{PR} = 2\mathbf{i} + 3\mathbf{j} + \mathbf{k}$ are parallel to $\mathcal{P}$, and thus $\mathcal{P}$ has normal vector

$$\mathbf{n} = \overrightarrow{PQ} \times \overrightarrow{PR} = \begin{vmatrix} \mathbf{i} & \mathbf{j} & \mathbf{k} \\ 5 & -1 & -1 \\ 2 & 3 & 1 \end{vmatrix} = 2\mathbf{i} - 7\mathbf{j} + 17\mathbf{k}.$$

So $\mathcal{P}$ has vector equation $\mathbf{n} \cdot (x\mathbf{i} + y\mathbf{j} + z\mathbf{k}) = \mathbf{n} \cdot \overrightarrow{OP} = 78$ and thus Cartesian equation $2x - 7y + 17z = 78$.

C12S04.035: There are several ways to solve this problem.

First solution: Two points of L are $P(7, 3, 9)$ and $Q(3, 9, 14)$, so the vector $\overrightarrow{PQ} = \langle -4, 6, 5 \rangle$ is parallel to L. The vector $\mathbf{n} = \langle 4, 1, 2 \rangle$ is normal to the plane $\mathcal{P}$. But

$$\mathbf{n} \cdot \overrightarrow{PQ} = \langle 4, 1, 2 \rangle \cdot \langle -4, 6, 5 \rangle = -16 + 6 + 10 = 0,$$

so that $\mathbf{n}$ and $\overrightarrow{PQ}$ are perpendicular. Therefore L is parallel to $\mathcal{P}$. Moreover, P does not satisfy the equation of $\mathcal{P}$ because

$$4 \cdot 7 + 1 \cdot 3 + 2 \cdot 9 = 28 + 3 + 18 = 49 \neq 17.$$

Because L is parallel to $\mathcal{P}$ and contains a point not on $\mathcal{P}$, the line and the plane cannot coincide.

Second solution: If the line and the plane both contain the point (x, y, z), then there exists a scalar t such that all four equations in Problem 35 are simultaneously true. The *Mathematica* 3.0 command

```
Solve[ { x == 7 - 4*t, y == 3 + 6*t, z == 9 + 5*t,
         4*x + y + 2*z == 17 }, { x, y, z, t } ]
```

returns { }, the "empty set," telling us that these four equations have no simultaneous solution. So no point of L lies in the plane $\mathcal{P}$; the line and the plane are parallel and do not coincide.

Third solution: If $x = 7 - 4t$, $y = 3 + 6t$, and $z = 9 + 5t$ are substituted in the equation $4x + y + 2z = 17$ of the plane, the result is $49 = 17$. This is impossible, so no point of L lies in the plane $\mathcal{P}$. We reach the same conclusion as in the previous two solutions.

C12S04.037: Simultaneous solution of the four equations given in the statement of Problem 37 yields the unique solution $x = \frac{9}{2}$, $y = \frac{9}{4}$, $z = \frac{17}{4}$, and $t = \frac{3}{4}$. So the line and the plane are not parallel and meet at the single point $\left(\frac{9}{2}, \frac{9}{4}, \frac{17}{4}\right)$. The easiest way to solve these equations by hand is to substitute the three parametric equations into the equation of the plane and solve for t:

$$3(3 + 2t) + 2(6 - 5t) - 4(2 + 3t) = 1; \qquad 9 + 6t + 12 - 10t - 8 - 12t = 1;$$

$$-16t = -12; \qquad t = \frac{3}{4}.$$

Then substitution in the parametric equations yields the values of x, y, and z given here.

C12S04.039: The vector $\mathbf{n}_1 = \langle 1, 0, 0 \rangle$ is normal to the first plane; the vector $\mathbf{n}_2 = \langle 1, 1, 1 \rangle$ is normal to the second. If θ is the angle between the normals (this is, by definition, the angle between the planes), then

$$\cos \theta = \frac{\mathbf{n}_1 \cdot \mathbf{n}_2}{|\mathbf{n}_1| \cdot |\mathbf{n}_2|} = \frac{1}{\sqrt{3}} = \frac{\sqrt{3}}{3},$$

so $\theta \approx 54.735610317245°$.

C12S04.041: The vector $\mathbf{n}_1 = \langle 1, -1, -2 \rangle$ is normal to the first plane; the vector $\mathbf{n}_2 = \langle 1, -1, -2 \rangle$ is normal to the second. If θ is the angle between the normals (this is, by definition, the angle between the planes), then $\theta = 0$ because $\mathbf{n}_1$ and $\mathbf{n}_2$ are parallel.

C12S04.043: By inspection, two points that lie on both planes are $P(10, 0, -10)$ and $Q(10, 1, -11)$. Hence a vector parallel to their line of intersection L is $\mathbf{v} = \overrightarrow{PQ} = \langle 0, 1, -1 \rangle$. So a vector equation of L is

$$\langle x, y, z \rangle = \overrightarrow{OP} + t\mathbf{v} = \langle 10, 0, -10 \rangle + t\langle 0, 1, -1 \rangle = \langle 10, t, -10 - t \rangle,$$

its parametric equations are $x = 10$, $y = t$, $z = -10 - t$, and its Cartesian equations are $x = 10$, $y = -10 - z$.

C12S04.045: The planes of Problem 41 are parallel, so there is no line of intersection.

C12S04.047: Substitute $z = 0$, then solve the equations of the planes simultaneously to find that one point on their line of intersection L is $Q\left(\frac{7}{5}, \frac{6}{5}, 0\right)$. Repeat with $z = 1$ to find that another such point is $R\left(\frac{7}{5}, \frac{1}{5}, 1\right)$. Hence the vector $\mathbf{v} = \overrightarrow{QR} = \langle 0, -1, 1 \rangle$ is parallel to L. Therefore the parallel line through the point $P(3, 3, 1)$ has vector equation

$$\langle x, y, z \rangle = \overrightarrow{OP} + t\mathbf{v} = \langle 3, 3, 1 \rangle + t\langle 0, -1, 1 \rangle.$$

Thus L has parametric equations $x = 3$, $y = 3 - t$, $z = 1 + t$ and Cartesian equations $x = 3$, $-y + 3 = z - 1$.

C12S04.049: Because the xy-plane has equation $z = 0$, the plane with equation $3x + 2y - z = 6$ intersects the xy-plane in the line with equation $3x + 2y = 6$. So three points on the plane $\mathcal{P}$ we seek are $P(1, 1, 1)$, $Q(2, 0, 0)$, and $R(0, 3, 0)$. Thus a normal to $\mathcal{P}$ is

$$\mathbf{n} = \overrightarrow{PQ} \times \overrightarrow{PR} = \begin{vmatrix} \mathbf{i} & \mathbf{j} & \mathbf{k} \\ 1 & -1 & -1 \\ -1 & 2 & -1 \end{vmatrix} = \langle 3, 2, 1 \rangle.$$

Therefore $\mathcal{P}$ has vector equation $\mathbf{n} \cdot \langle x, y, z \rangle = \mathbf{n} \cdot \overrightarrow{OP}$; that is,

$$\langle 3, 2, 1 \rangle \cdot \langle x, y, z \rangle = \langle 3, 2, 1 \rangle \cdot \langle 1, 1, 1 \rangle;$$

$$3x + 2y + z = 6.$$

C12S04.051: The plane $\mathcal{P}$ whose equation we seek passes through $P(1, 0, -1)$ and $Q(2, 1, 0)$, and is thus parallel to $\overrightarrow{PQ} = \langle 1, 1, 1 \rangle$. To find two points in the line of intersection of the other two planes, set $z = 1$ and solve their equations simultaneously to find that one such point is $S(2, 2, 1)$. Repeat with $z = 5$ to find that another such point is $R(1, -1, 5)$. Hence another vector parallel to $\mathcal{P}$ is $\overrightarrow{RS} = \langle 1, 3, -4 \rangle$. If $\overrightarrow{RS}$ had turned out to be parallel to $\overrightarrow{PQ}$, then there would have been insufficient information to solve the problem, but they are not parallel. Hence a normal to $\mathcal{P}$ is

$$\mathbf{n} = \overrightarrow{PQ} \times \overrightarrow{RS} = \begin{vmatrix} \mathbf{i} & \mathbf{j} & \mathbf{k} \\ 1 & 1 & 1 \\ 1 & 3 & -4 \end{vmatrix} = \langle -7, 5, 2 \rangle.$$

Hence $\mathcal{P}$ has vector equation $\mathbf{n} \cdot \langle x, y, z \rangle = \mathbf{n} \cdot \overrightarrow{OP}$; that is,

$$\langle -7, 5, 2 \rangle \cdot \langle x, y, z \rangle = \langle -7, 5, 2 \rangle \cdot \langle 1, 0, -1 \rangle; \qquad -7x + 5y + 2z = -9; \qquad 7x - 5y - 2z = 9.$$

C12S04.053: Set $z = 1$ and solve the equations of the two planes simultaneously to find that one point on their line of intersection is $P(1, 1, 1)$. Repeat with $z = 5$ to find that another such point is $Q(-5, 6, 5)$. Substitute $t = 0$ to find that a point on the given line is $R(1, 3, 2)$; substitute $t = 1$ to find that another such point is $S(7, -2, -2)$. Then $\overrightarrow{PQ} = \langle -6, 5, 4 \rangle$ and $\overrightarrow{RS} = \langle 6, -5, -4 \rangle$, so it's clear that the two lines are parallel. Note that R does not lie on the first given plane, so the given line and the line of intersection of the two given planes do not coincide. To obtain two nonparallel vectors in the plane $\mathcal{P}$ that contains both lines, use $\overrightarrow{PQ}$ and $\overrightarrow{PR} = \langle 0,\ 2,\ 1 \rangle$. Then a normal to $\mathcal{P}$ will be

$$\mathbf{n} = \overrightarrow{PQ} \times \overrightarrow{PR} = \begin{vmatrix} \mathbf{i} & \mathbf{j} & \mathbf{k} \\ -6 & 5 & 4 \\ 0 & 2 & 1 \end{vmatrix} = \langle -3,\ 6,\ -12 \rangle.$$

Replace $\mathbf{n}$ with the simpler $\langle 1,\ -2,\ 4 \rangle$ and write a vector equation of $\mathcal{P}$ in the form

$$\langle 1,\ -2,\ 4 \rangle \cdot \langle x,\ y,\ z \rangle = \langle 1,\ -2,\ 4 \rangle \cdot \langle 1,\ 1,\ 1 \rangle,$$

from which we read the Cartesian equation $x - 2y + 4z = 3$.

C12S04.055: The formula in Problem 54 with $x_0 = y_0 = z_0 = 0$, $a = b = c = 1$, and $d = 10$ yields distance

$$D = \frac{10}{\sqrt{3}} = \frac{10\sqrt{3}}{3} \approx 5.773502691896.$$

C12S04.057: If L_1 and L_2 are skew lines, choose two points P_1 and Q_1 in L_1 and two points P_2 and Q_2 in L_2. Let $\mathbf{n} = \overrightarrow{P_1Q_1} \times \overrightarrow{P_2Q_2}$. Let $\mathcal{P}_1$ be the plane through P_1 with normal vector $\mathbf{n}$ and let $\mathcal{P}_2$ be the plane through P_2 with normal vector $\mathbf{n}$. Clearly $\mathcal{P}_1$ contains L_1, $\mathcal{P}_2$ contains L_2, and $\mathcal{P}_1$ and $\mathcal{P}_2$ are parallel.

C12S04.059: See the *Comment* in the solution of Problem 31 of Section 12.3 for our strategy in the following solution. By inspection, $P_1(1,\ -1,\ 4)$ and $Q_1(3,\ 0,\ 4)$ lie in L_1 and we are given the two points $P_2(2,\ 1,\ -3)$ and $Q_2(0,\ 8,\ 4)$ in L_2. Hence $\mathbf{v}_1 = \overrightarrow{P_1Q_1} = \langle 2,\ 1,\ 0 \rangle$ is parallel to L_1 and $\mathbf{v}_2 = \overrightarrow{P_2Q_2} = \langle -2,\ 7,\ 7 \rangle$ is parallel to L_2.

Part (a): The line L_2 has vector equation

$$\langle x,\ y,\ z \rangle = t\mathbf{v}_2 + \langle 2,\ 1,\ -3 \rangle = \langle 2 - 2t,\ 1 + 7t,\ -3 + 7t \rangle,$$

and thus symmetric equations

$$\frac{-x + 2}{2} = \frac{y - 1}{7} = \frac{z + 3}{7}.$$

It is clear that the two lines are not parallel because $\mathbf{v}_2$ is not a scalar multiple of $\mathbf{v}_1$. The *Mathematica 3.0* command

```
Solve[ { (2 - x)/2 == (y - 1)/7, (2 - x)/2 == (z + 3)/7,
      x - 1 == 2*y + 2, z == 4 }, { x, y, z } ]
```

for solving the equations of the two lines simultaneously returns the information that there is no solution. Therefore the two lines are skew lines.

Part (b): Let $\mathbf{c} = \overrightarrow{P_1P_2}$ be a "connector" from L_1 to L_2 and note that $\mathbf{n} = \mathbf{v}_1 \times \mathbf{v}_2 = \langle 7,\ -14,\ 16 \rangle$ is normal to both lines. Therefore the distance between the lines is

$$D = \frac{|\mathbf{n} \cdot \mathbf{c}|}{|\mathbf{n}|} = \frac{133}{\sqrt{501}} = \frac{133\sqrt{501}}{501} \approx 5.942001786397.$$

Alternative Part (b): If we follow the method required in the statement of Problem 59, we find that the first line lies in the plane with equation $7x - 14y + 16z = 85$ and the second line lies in the plane with equation $7x - 14y + 16z = -48$. Therefore, by the formula in Problem 58, the distance between them is

$$D = \frac{|85 - (-48)|}{\sqrt{49 + 196 + 256}} = \frac{133}{\sqrt{501}} = \frac{133\sqrt{501}}{501}.$$

Section 12.5

C12S05.001: Because $y^2 + z^2 = 1$ while x is arbitrary, the graph lies on the cylinder of radius 1 with axis the x-axis. A small part of the graph is shown in Fig. 12.5.17.

C12S05.003: Because $x^2 + y^2 = t^2 = z^2$, the graph lies on the cone with axis the z-axis and equation $z^2 = x^2 + y^2$. A small part of the graph is shown in Fig. 12.5.16.

C12S05.005: If $\mathbf{r}(t) = 3\mathbf{i} - 2\mathbf{j}$, then $\mathbf{r}'(t) = \mathbf{0} = \mathbf{r}''(t)$, and hence $\mathbf{r}'(1) = \mathbf{0} = \mathbf{r}''(1)$.

C12S05.007: If $\mathbf{r}(t) = e^{2t}\mathbf{i} + e^{-t}\mathbf{j}$, then $\mathbf{r}'(t) = 2e^{2t}\mathbf{i} - e^{-t}\mathbf{j}$ and $\mathbf{r}''(t) = 4e^{2t}\mathbf{i} + e^{-t}\mathbf{j}$. Therefore $\mathbf{r}'(0) = 2\mathbf{i} - \mathbf{j}$ and $\mathbf{r}''(0) = 4\mathbf{i} + \mathbf{j}$.

C12S05.009: If $\mathbf{r}(t) = 3\mathbf{i}\cos 2\pi t + 3\mathbf{j}\sin 2\pi t$, then

$$\mathbf{r}'(t) = -6\pi\mathbf{i}\sin 2\pi t + 6\pi\mathbf{j}\cos 2\pi t \quad \text{and} \quad \mathbf{r}''(t) = -12\pi^2\mathbf{i}\cos 2\pi t - 12\pi^2\mathbf{j}\sin 2\pi t.$$

Therefore

$$\mathbf{r}'\left(\frac{3}{4}\right) = 6\pi\mathbf{i} \quad \text{and} \quad \mathbf{r}''\left(\frac{3}{4}\right) = 12\pi^2\mathbf{j}.$$

C12S05.011: If $\mathbf{r}(t) = t\mathbf{i} + t^2\mathbf{j} + t^3\mathbf{k}$, then

$$\mathbf{v}(t) = \langle 1, 2t, 3t^2 \rangle,$$
$$v(t) = \sqrt{1 + 4t^2 + 9t^4}, \quad \text{and}$$
$$\mathbf{a}(t) = \langle 0, 2, 6t \rangle.$$

C12S05.013: If $\mathbf{r}(t) = \langle t, 3e^t, 4e^t \rangle$, then

$$\mathbf{v}(t) = \langle 1, 3e^t, 4e^t \rangle, \quad v(t) = \sqrt{1 + 25e^{2t}}, \quad \text{and} \quad \mathbf{a}(t) = \langle 0, 3e^t, 4e^t \rangle.$$

C12S05.015: If $\mathbf{r}(t) = \langle 3\cos t, 3\sin t, -4t \rangle$, then

$$\mathbf{v}(t) = \langle -3\sin t, 3\cos t, -4 \rangle, \quad v(t) = \sqrt{9\sin^2 t + 9\cos^2 t + 16} = 5, \quad \text{and} \quad \mathbf{a}(t) = \langle -3\cos t, -3\sin t, 0 \rangle.$$

C12S05.017: By Eq. (16), we have

$$\int_0^{\pi/4} \langle \sin t, 2\cos t \rangle \, dt = \Big[\langle -\cos t, 2\sin t \rangle \Big]_0^{\pi/4} = \left\langle -\frac{\sqrt{2}}{2}, \sqrt{2} \right\rangle - \langle -1, 0 \rangle = \left\langle \frac{2 - \sqrt{2}}{2}, \sqrt{2} \right\rangle.$$

C12S05.019: By Eq. (16), we have

$$\int_0^2 \langle t^2(1+t^3)^{3/2}, 0 \rangle \, dt = \left[\left\langle \frac{2}{15}(1+t^3)^{5/2}, 0 \right\rangle \right]_0^2 = \left(\frac{162}{5} - \frac{2}{15} \right) \mathbf{i} = \frac{484}{15} \mathbf{i}.$$

C12S05.021: Given $\mathbf{u}(t) = \langle 3t, -1 \rangle$ and $\mathbf{v}(t) = \langle 2, -5t \rangle$, Theorem 2 yields

$$D_t \left[\mathbf{u}(t) \cdot \mathbf{v}(t) \right] = \mathbf{u}(t) \cdot \mathbf{v}'(t) + \mathbf{u}'(t) \cdot \mathbf{v}(t) = \langle 3t, -1 \rangle \cdot \langle 0, -5 \rangle + \langle 3, 0 \rangle \cdot \langle 2, -5t \rangle = 5 + 6 = 11.$$

C12S05.023: Given $\mathbf{u}(t) = \langle \cos t, \sin t \rangle$ and $\mathbf{v}(t) = \langle \sin t, -\cos t \rangle$, Theorem 2 yields

$$D_t \left[\mathbf{u}(t) \cdot \mathbf{v}(t) \right] = \mathbf{u}(t) \cdot \mathbf{v}'(t) + \mathbf{u}'(t) \cdot \mathbf{v}(t) = \langle \cos t, \sin t \rangle \cdot \langle \cos t, \sin t \rangle + \langle -\sin t, \cos t \rangle \cdot \langle \sin t, -\cos t \rangle$$

$$= \cos^2 t + \sin^2 t - \sin^2 t - \cos^2 t = 0.$$

C12S05.025: Given $\mathbf{a} = \mathbf{0} = \langle 0, 0, 0 \rangle$, it follows that $\mathbf{v}(t) = \langle c_1, c_2, c_3 \rangle$ where c_1, c_2, and c_3 are constants. Then

$$\mathbf{k} = \langle 0, 0, 1 \rangle = \mathbf{v}(0) = \mathbf{v}_0 = \langle c_1, c_2, c_3 \rangle$$

implies that $c_1 = c_2 = 0$ and $c_3 = 1$. Hence $\mathbf{v}(t) = \langle 0, 0, 1 \rangle$, and therefore

$$\mathbf{r}(t) = \langle k_1, k_2, t + k_3 \rangle$$

where k_1, k_2, and k_3 are constants. Then

$$\mathbf{i} = \langle 1, 0, 0 \rangle = \mathbf{r}(0) = \mathbf{r}_0 = \langle k_1, k_2, k_3 \rangle$$

leads to $k_1 = 1$ and $k_2 = k_3 = 0$. Therefore $\mathbf{r}(t) = \langle 1, 0, t \rangle$.

C12S05.027: Given $\mathbf{a} = \langle 2, 0, -4 \rangle$, it follows that $\mathbf{v}(t) = \langle 2t + c_1, c_2, -4t + c_3 \rangle$ where c_1, c_2, and c_3 are constants. Then

$$10\mathbf{j} = \langle 0, 10, 0 \rangle = \mathbf{v}(0) = \mathbf{v}_0 = \langle c_1, c_2, c_3 \rangle$$

implies that $c_1 = c_3 = 0$ and $c_2 = 10$. Hence $\mathbf{v}(t) = \langle 2t, 10, -4t \rangle$, and therefore

$$\mathbf{r}(t) = \langle t^2 + k_1, 10t + k_2, -2t^2 + k_3 \rangle$$

where k_1, k_2, and k_3 are constants. Then

$$\mathbf{0} = \langle 0, 0, 0 \rangle = \mathbf{r}(0) = \mathbf{r}_0 = \langle k_1, k_2, k_3 \rangle$$

leads to $k_1 = k_2 = k_3 = 0$. Therefore $\mathbf{r}(t) = \langle t^2, 10t, -2t^2 \rangle$.

C12S05.029: Given $\mathbf{a} = \langle 0, 2, -6t \rangle$, it follows that $\mathbf{v}(t) = \langle c_1, 2t + c_2, -3t^2 + c_3 \rangle$ where c_1, c_2, and c_3 are constants. Then

$$5\mathbf{k} = \langle 0, 0, 5 \rangle = \mathbf{v}(0) = \mathbf{v}_0 = \langle c_1, c_2, c_3 \rangle$$

implies that $c_1 = c_2 = 0$ and $c_3 = 5$. Hence $\mathbf{v}(t) = \langle 0, 2t, -3t^2 + 5 \rangle$, and therefore

$$\mathbf{r}(t) = \langle\, k_1,\ t^2 + k_2,\ -t^3 + 5t + k_3 \,\rangle$$

where k_1, k_2, and k_3 are constants. Then

$$2\mathbf{i} = \langle\, 2,\, 0,\, 0 \,\rangle = \mathbf{r}(0) = \mathbf{r}_0 = \langle\, k_1,\ k_2,\ k_3 \,\rangle$$

leads to $k_1 = 2$ and $k_2 = k_3 = 0$. Therefore

$$\mathbf{r}(t) = \langle\, 2,\ t^2,\ -t^3 + 5t \,\rangle.$$

C12S05.031: Given $\mathbf{a} = \langle\, t,\, t^2,\, t^3 \,\rangle$, it follows that

$$\mathbf{v}(t) = \left\langle\, \frac{1}{2}t^2 + c_1,\ \frac{1}{3}t^3 + c_2,\ \frac{1}{4}t^4 + c_3 \,\right\rangle$$

where c_1, c_2, and c_3 are constants. Then

$$10\mathbf{j} = \langle\, 0,\, 10,\, 0 \,\rangle = \mathbf{v}(0) = \mathbf{v}_0 = \langle\, c_1,\ c_2,\ c_3 \,\rangle$$

implies that $c_1 = c_3 = 0$, and $c_2 = 10$. Hence

$$\mathbf{v}(t) = \left\langle\, \frac{1}{2}t^2,\ \frac{1}{3}t^3 + 10,\ \frac{1}{4}t^4 \,\right\rangle,$$

and therefore

$$\mathbf{r}(t) = \left\langle\, \frac{1}{6}t^3 + k_1,\ \frac{1}{12}t^4 + 10t + k_2,\ \frac{1}{20}t^5 + k_3 \,\right\rangle$$

where k_1, k_2, and k_3 are constants. Then

$$10\mathbf{i} = \langle\, 10,\, 0,\, 0 \,\rangle = \mathbf{r}(0) = \mathbf{r}_0 = \langle\, k_1,\ k_2,\ k_3 \,\rangle$$

leads to $k_1 = 10$ and $k_2 = k_3 = 0$. Therefore

$$\mathbf{r}(t) = \left\langle\, \frac{1}{6}t^3 + 10,\ \frac{1}{12}t^4 + 10t,\ \frac{1}{20}t^5 \,\right\rangle.$$

C12S05.033: Given $\mathbf{a} = \langle\, \cos t,\, \sin t,\, 0 \,\rangle$, it follows that

$$\mathbf{v}(t) = \langle\, c_1 + \sin t,\ c_2 - \cos t,\ c_3 \,\rangle$$

where c_1, c_2, and c_3 are constants. Then

$$-\mathbf{i} + 5\mathbf{k} = \langle\, -1,\, 0,\, 5 \,\rangle = \mathbf{v}(0) = \mathbf{v}_0 = \langle\, c_1,\ -1 + c_2,\ c_3 \,\rangle$$

implies that $c_1 = -1$, $c_2 = 1$, and $c_3 = 5$. Hence

$$\mathbf{v}(t) = \langle\, -1 + \sin t,\ 1 - \cos t,\ 5 \,\rangle,$$

and therefore

$$\mathbf{r}(t) = \langle -t - \cos t + k_1, \ t - \sin t + k_2, \ 5t + k_3 \rangle$$

where k_1, k_2, and k_3 are constants. Then

$$\mathbf{j} = \langle 0, 1, 0 \rangle = \mathbf{r}(0) = \mathbf{r}_0 = \langle -1 + k_1, \ k_2, \ k_3 \rangle$$

leads to $k_1 = 1$, $k_2 = 1$, and $k_3 = 0$. Therefore

$$\mathbf{r}(t) = \langle 1 - t - \cos t, \ 1 + t - \sin t, \ 5t \rangle.$$

C12S05.035: The position vector of the moving point is $\mathbf{r}(t) = \langle 3\cos 2t, \ 3\sin 2t, \ 8t \rangle$. Hence its velocity, speed, and acceleration are

$$\mathbf{v}(t) = \langle -6\sin 2t, \ 6\cos 2t, \ 8 \rangle,$$

$$v(t) = |\mathbf{v}(t)| = \sqrt{36(\sin^2 2t + \cos^2 2t) + 64} \ = 10, \quad \text{and}$$

$$\mathbf{a}(t) = \langle -12\cos 2t, \ -12\sin 2t, \ 0 \rangle,$$

respectively. Therefore

$$\mathbf{v}\left(\tfrac{7}{8}\pi\right) = \langle 3\sqrt{2}, \ 3\sqrt{2}, \ 8 \rangle, \quad v\left(\tfrac{7}{8}\pi\right) = 10, \quad \text{and} \quad \mathbf{a}\left(\tfrac{7}{8}\pi\right) = \langle -6\sqrt{2}, \ 6\sqrt{2}, \ 0 \rangle.$$

C12S05.037: Given $\mathbf{u}(t) = \langle 0, 3, 4t \rangle$ and $\mathbf{v}(t) = \langle 5t, 0, -4 \rangle$, we first compute

$$\mathbf{u}(t) \times \mathbf{v}(t) = \begin{vmatrix} \mathbf{i} & \mathbf{j} & \mathbf{k} \\ 0 & 3 & 4t \\ 5t & 0 & -4 \end{vmatrix} = \langle -12, \ 20t^2, \ -15t \rangle.$$

Therefore $D_t\left[\mathbf{u}(t) \times \mathbf{v}(t)\right] = \langle 0, 40t, -15 \rangle$. Next,

$$\mathbf{u}(t) \times \mathbf{v}'(t) + \mathbf{u}'(t) \times \mathbf{v}(t) = \begin{vmatrix} \mathbf{i} & \mathbf{j} & \mathbf{k} \\ 0 & 3 & 4t \\ 5 & 0 & 0 \end{vmatrix} + \begin{vmatrix} \mathbf{i} & \mathbf{j} & \mathbf{k} \\ 0 & 0 & 4 \\ 5t & 0 & -4 \end{vmatrix}$$

$$= \langle 0, 20t, -15 \rangle + \langle 0, 20t, 0 \rangle = \langle 0, 40t, -15 \rangle = D_t\left[\mathbf{u}(t) \times \mathbf{v}(t)\right].$$

C12S05.039: Given: $|\mathbf{r}(t)| = R$, a constant. Let $\mathbf{v}(t) = \mathbf{r}'(t)$. Then $\mathbf{r}(t) \cdot \mathbf{r}(t) = R^2$, also a constant. Hence

$$0 = D_t\left[\mathbf{r}(t) \cdot \mathbf{r}(t)\right] = 2\mathbf{r}(t) \cdot \mathbf{v}(t),$$

so that $\mathbf{r}(t) \cdot \mathbf{v}(t) = 0$. Therefore $\mathbf{v}(t)$ is perpendicular to the radius of the sphere, so the velocity vector is everywhere tangent to the sphere.

C12S05.041: The ball of Example 10 has position vector $\mathbf{r}(t) = \langle t^2, \ 80t, \ 80t - 16t^2 \rangle$ and velocity vector $\mathbf{v}(t) = \langle 2t, \ 80, \ 80 - 32t \rangle$. Its maximum height occurs when the z-component of its velocity vector is zero: $80 - 32t = 0$, so that $t = \tfrac{5}{2}$. The speed of the ball at time t is

$$v(t) = \sqrt{4t^2 + 6400 + 6400 - 5120t + 1024t^2}\,,$$

so that

$$v\left(\tfrac{5}{2}\right) = \sqrt{25 + 12800 - 12800 + 6400} = \sqrt{6425} = 5\sqrt{257} \approx 80.156097709407$$

(ft/s). Its position then is

$$\mathbf{r}\left(\tfrac{5}{2}\right) = \left\langle\, \tfrac{25}{4},\; 200,\; 100 \,\right\rangle,$$

and the z-component of this vector is its maximum height, 100 ft.

C12S05.043: Because $x_0 = y_0 = 0$, the equations in (22) and (23) take the form

$$x(t) = (v_0 \cos\alpha)t, \qquad y(t) = -\frac{1}{2}gt^2 + (v_0 \sin\alpha)t.$$

To find the range, find the positive value of t for which $y(t) = 0$:

$$gt = 2v_0 \sin\alpha; \qquad t = \frac{2v_0 \sin\alpha}{g};$$

thus the range is the value of $x(t)$ then; it is

$$R = x\left(\frac{2v_0 \sin\alpha}{g}\right) = \frac{2v_0^2 \sin\alpha \cos\alpha}{g} = \frac{v_0^2 \sin 2\alpha}{g}.$$

If $\alpha = \frac{1}{4}\pi$ and $R = 5280$ (there are 5280 feet in one mile), then

$$\frac{v_0^2}{32} = 5280, \quad \text{so that} \quad v_0 = \sqrt{32 \cdot 5280} = 32\sqrt{165} \approx 411.047442517284$$

(feet per second).

C12S05.045: The formula for the range is derived in the solution of Problem 43.

C12S05.047: We saw in the solution of Problem 43 that the range of the projectile is

$$R = \frac{v_0^2 \sin 2\alpha}{g}.$$

To find its maximum height, first find when $y'(t) = 0$:

$$-gt = v_0 \sin\alpha, \quad \text{so that} \quad t = \frac{v_0 \sin\alpha}{g}.$$

Then to find the maximum height, evaluate $y(t)$ at that value of t:

$$y\left(\frac{v_0 \sin\alpha}{g}\right) = \frac{v_0^2 \sin^2\alpha}{2g}.$$

Part (a): If $v_0 = 160$ and $\alpha = \frac{1}{6}\pi$, then the range is

$$R = \frac{(160)^2 \cdot \sqrt{3}}{2 \cdot 32} = 400\sqrt{3} \approx 692.820323027551$$

636

(feet) and the maximum height is

$$\frac{v_0 \sin^2 \alpha}{2g} = \frac{(160)^2 \cdot 1}{4 \cdot 2 \cdot 32} = 100$$

(feet). Part (b): If $v_0 = 160$ and $\alpha = \frac{1}{4}\pi$, then the range is

$$R = \frac{(160)^2 \cdot 1}{32} = 800$$

(feet) and the maximum altitude is

$$\frac{v_0^2 \sin^2 \alpha}{2g} = \frac{(160)^2 \cdot 1}{2 \cdot 2 \cdot 32} = 200$$

(feet). Part (c): If $\alpha = \frac{1}{3}\pi$ and $v_0 = 160$, then the range is

$$R = \frac{(160)^2 \cdot \sqrt{3}}{2 \cdot 32} = 400\sqrt{3} \approx 692.820323027551$$

(feet) and the maximum altitude (also in feet) is

$$y_{max} = \frac{(160)^2 \cdot 3}{4 \cdot 2 \cdot 32} = 300.$$

C12S05.049: With $x_0 = 0$ and $y_0 = 100$, the equations in (22) and (23) of the text take the form

$$x(t) = (v_0 \cos \alpha)t, \qquad y(t) = -\frac{1}{2}gt^2 + (v_0 \sin \alpha)t + 100.$$

With $g = 9.8$ and $\alpha = 0$, these equations become

$$x(t) = v_0 t, \qquad y(t) = 100 - \frac{1}{2}gt^2.$$

We require $x(t) = 1000$ when $y(t) = 0$. But $y(t) = 0$ when

$$t^2 = \frac{200}{9.8} = \frac{1000}{49}, \qquad \text{so that} \qquad t = \frac{10\sqrt{10}}{7}.$$

Thus

$$1000 = x\left(\frac{10\sqrt{10}}{7}\right) = \frac{10\sqrt{10}}{7}v_0,$$

and it follows that $v_0 = 70\sqrt{10} \approx 221.359436211787$ (meters per second, approximately 726.244869461242 feet per second).

C12S05.051: First we analyze the behavior of the bomb. Suppose that it is dropped at time $t = 0$. If the projectile is fired from the origin, then the equations of motion of the bomb are

$$x(t) \equiv 800, \qquad y(t) = 800 - \frac{1}{2}gt^2$$

where $g = 9.8$ (m/s^2). Now $y(t) = 400$ when

$$800 - \frac{1}{2}gt^2 = 400; \qquad \frac{1}{2}gt^2 = 400;$$

$$t^2 = \frac{800}{9.8} = \frac{4000}{49}; \qquad t = T = \frac{20\sqrt{10}}{7}.$$

Now we turn our attention to the projectile, fired at time $t = 0$ from the origin. (We will adjust for the one-second delay later in this solution.) By Eqs. (22) and (23) of the text, its equations of motion are

$$x(t) = (v_0 \cos \alpha)t, \quad y(t) = -\frac{1}{2}gt^2 + (v_0 \sin \alpha)t.$$

We require $x(T-1) = 800$ and $y(T-1) = 400$. (This is how we take care of the one-second delay). Thus

$$T - 1 = \frac{20\sqrt{10} - 7}{7};$$

$$(v_0 \cos \alpha)(T - 1) = 800;$$

$$-\frac{1}{2}g(T-1)^2 + (v_0 \sin \alpha)(T-1) = 400;$$

$$v_0 \cos \alpha = \frac{800}{T-1}. \tag{1}$$

Also

$$-\frac{1}{2}g(T-1) + v_0 \sin \alpha = \frac{400}{T-1};$$

$$v_0 \sin \alpha = \frac{400}{T-1} + \frac{1}{2}g(T-1). \tag{2}$$

Division of Eq. (2) by Eq. (1) then yields

$$\tan \alpha = \frac{T-1}{800} \cdot \left[\frac{400}{T-1} + \frac{1}{2}g(T-1) \right] = \frac{1}{2} + \frac{9.8}{1600}(T-1)^2$$

$$= \frac{1}{2} + \frac{98}{16000} \cdot \left(\frac{20\sqrt{10} - 7}{7} \right)^2 = \frac{8049 - 280\sqrt{10}}{8000} \approx 0.895445,$$

so that $\alpha \approx 41.842705345876°$. Then, by Eq. (1),

$$v_0 = \frac{800}{(T-1)\cos\alpha} \approx 133.645951548503$$

meters per second, approximately 438.470969647319 feet per second.

C12S05.053: We give proofs for vectors with two components; these proofs generalize readily to vectors with three or more components. Let $\mathbf{u}(t) = \langle u_1(t), u_2(t) \rangle$ and $\mathbf{v}(t) = \langle v_1(t), v_2(t) \rangle$. The assumptions that

$$\lim_{t \to a} \mathbf{u}(t) \quad \text{and} \quad \lim_{t \to a} \mathbf{v}(t)$$

638

both exist means that there exist vectors $\langle p_1, p_2 \rangle$ and $\langle q_1, q_2 \rangle$ such that

$$\lim_{t \to a} u_1(t) = p_1, \quad \lim_{t \to a} u_2(t) = p_2, \quad \lim_{t \to a} v_1(t) = q_1, \quad \text{and} \quad \lim_{t \to a} v_2(t) = q_2.$$

Part (a):

$$\lim_{t \to a} [\mathbf{u}(t) + \mathbf{v}(t)] = \lim_{t \to a} \langle u_1(t) + v_1(t), u_2(t) + v_2(t) \rangle = \langle p_1 + q_1, p_2 + q_2 \rangle$$

$$= \langle p_1, p_2 \rangle + \langle q_1, q_2 \rangle = \left[\lim_{t \to a} \mathbf{u}(t) \right] + \left[\lim_{t \to a} \mathbf{v}(t) \right].$$

Part (b):

$$\lim_{t \to a} [\mathbf{u}(t) \cdot \mathbf{v}(t)] = \lim_{t \to a} [u_1(t)v_1(t) + u_2(t)v_2(t)] = p_1 q_1 + p_2 q_2$$

$$= \langle p_1, p_2 \rangle \cdot \langle q_1, q_2 \rangle = \left(\lim_{t \to a} \mathbf{u}(t) \right) \cdot \left(\lim_{t \to a} \mathbf{v}(t) \right).$$

C12S05.055: If $\mathbf{v}(t)$ is the velocity vector of the moving particle, then we are given $|\mathbf{v}(t)| = C$, a constant. Then $\mathbf{v}(t) \cdot \mathbf{v}(t) = C^2$, also a constant. Hence $0 = D_t [\mathbf{v}(t) \cdot \mathbf{v}(t)] = 2\mathbf{v}(t) \cdot \mathbf{a}(t)$ where $\mathbf{a}(t)$ is the acceleration vector of the particle. But because $\mathbf{v}(t) \cdot \mathbf{a}(t) = 0$, it follows that $\mathbf{v}$ and $\mathbf{a}$ are always perpendicular.

C12S05.057: If $\mathbf{r}(t) = \langle \cosh \omega t, \sinh \omega t \rangle$, then

$$\mathbf{v}(t) = \mathbf{r}'(t) = \langle \omega \sinh \omega t, \omega \cosh \omega t \rangle \quad \text{and}$$

$$\mathbf{a}(t) = \mathbf{v}'(t) = \langle \omega^2 \cosh \omega t, \omega^2 \sinh \omega t \rangle = \omega^2 \mathbf{r}(t) = c\mathbf{r}(t)$$

where $c = \omega^2 > 0$. An external force that would produce this sort of motion would be a central repulsive force proportional to distance from the origin.

C12S05.059: Given the acceleration vector $\mathbf{a} = \langle 0, a \rangle$, we first find the velocity and position vectors:

$$\mathbf{v}(t) = \langle c_1, at + c_2 \rangle \quad \text{and}$$

$$\mathbf{r}(t) = \langle c_1 t + k_1, \tfrac{1}{2} at^2 + c_2 t + k_2 \rangle$$

where c_1, c_2, k_1, and k_2 are constants. Thus the position $(x(t), y(t))$ of the moving point is given by

$$x(t) = c_1 t + k_1 \quad \text{and} \quad y(t) = \tfrac{1}{2} at^2 + c_2 t + k_2.$$

If $c_1 = 0$ then the point moves in a straight line. Otherwise, we solve the first of these equations for t and substitute in the second:

$$t = \frac{x - k_1}{c_1};$$

$$y = \frac{1}{2} a \left(\frac{x - k_1}{c_1} \right)^2 + c_2 \cdot \frac{x - k_1}{c_1} + k_2$$

$$= \frac{a}{2c_1^2} (x^2 - 2k_1 x + k_1^2) + \frac{c_2}{c_1} (x - k_1) + k_2$$

$$= \frac{a}{2c_1^2}x^2 + \left(\frac{c^2}{c_1} - \frac{ak_1}{c_1^2}\right)x + \left(\frac{ak_1^2}{2c_1^2} - \frac{c_2k_1}{c_1} + k_2\right)$$

$$= Ax^2 + Bx + C$$

where A, B, and C are constants. If $A \neq 0$ then the trajectory of the particle is a parabola. If $A = 0$ it is a straight line.

C12S05.061: Given: $\mathbf{r}(t) = \langle r\cos\omega t, \; r\sin\omega t \rangle$. Part (a):

$$\mathbf{v}(t) = \langle -r\omega\sin\omega t, \; r\omega\cos\omega t \rangle = r\omega\langle -\sin\omega t, \; \cos\omega t \rangle. \quad \text{So}$$

$$\mathbf{r}(t) \cdot \mathbf{v}(t) = r^2\omega(-\sin\omega t\cos\omega t + \sin\omega t\cos\omega t) = 0,$$

and therefore $\mathbf{r}$ and $\mathbf{v}$ are always perpendicular. Therefore $\mathbf{v}$ is always tangent to the circle. Moreover, the speed of motion is

$$v(t) = r\omega\sqrt{\sin^2\omega t + \cos^2\omega t} = r\omega.$$

Part (b): $\mathbf{a}(t) = r\omega^2\langle -\cos\omega t, \; -\sin\omega t \rangle = -\omega^2\mathbf{r}(t)$. Therefore $\mathbf{a}$ and $\mathbf{r}$ are always parallel and have opposite directions (because $-\omega^2 < 0$). Finally, the scalar acceleration is

$$a(t) = |\mathbf{a}(t)| = |-\omega^2| \cdot |\mathbf{r}(t)| = r\omega^2.$$

C12S05.063: With north the direction of the positive x-axis, west the direction of the positive y-axis, and upward the direction of the positive z-axis, the baseball has acceleration $\mathbf{a}(t) = \langle 0.1, 0, -32 \rangle$, initial velocity $\mathbf{v}_0 = \langle 0, 0, 160 \rangle$, and initial position $\mathbf{r}_0 = \langle 0, 0, 0 \rangle$. It follows that its position vector is

$$\mathbf{r}(t) = \left\langle \frac{1}{20}t^2, \; 0, \; 160t - 16t^2 \right\rangle.$$

The ball returns to the ground at that positive value of t for which the z-component of $\mathbf{r}$ is zero; that is, $t = 10$. At that time the x-component of $\mathbf{r}$ is 5, so the ball lands 5 feet north of the point from which it was thrown.

C12S05.065: In the "obvious" coordinate system, the acceleration of the projectile is $\mathbf{a}(t) = \langle 2, 0, -32 \rangle$; its initial velocity is $\mathbf{v}_0 = \langle 0, 200, 160 \rangle$ and its initial position is $\mathbf{r}_0 = \langle 0, 0, 384 \rangle$. Hence its velocity and position vectors are

$$\mathbf{v}(t) = \langle 2t, 200, 160 - 32t \rangle \quad \text{and} \quad \mathbf{r}(t) = \langle t^2, 200t, 384 - 160t - 16t^2 \rangle.$$

The projectile strikes the ground at that positive value of t from which the z-component of $\mathbf{r}$ is zero:

$$16t^2 - 160t - 384 = 0; \quad t^2 - 10t - 24 = 0;$$

$$(t-12)(t+2) = 0; \quad t = 12 \quad (\text{not } t = -2).$$

When $t = 12$, the position of the projectile is $\mathbf{r}(12) = \langle 144, 2400, 0 \rangle$, so it lands 2400 ft north and 144 ft east of the firing position. The projectile reaches its maximum altitude when the z-component of $\mathbf{v}(t)$ is zero; that is, when $t = 5$. Its position then is $\mathbf{r}(5) = \langle 25, 1000, 784 \rangle$, so its maximum altitude is 784 ft.

Section 12.6

C12S06.001: We first compute

$$v(t) = \sqrt{(x'(t))^2 + (y'(t))^2 + (z'(t))^2} = \sqrt{64 + 36\cos^2 2t + 36\sin^2 2t} = \sqrt{100} = 10.$$

Therefore the length of the graph is

$$s = \int_0^\pi 10 \, dt = \Big[\, 10t \,\Big]_0^\pi = 10\pi.$$

C12S06.003: First,

$$v(t) = \sqrt{[x'(t)]^2 + [y'(t)]^2 + [z'(t)]^2} = \sqrt{(6e^t \cos t - 6e^t \sin t)^2 + (6e^t \cos t + 6e^t \sin t)^2 + (17e^t)^2}$$

$$= \sqrt{289e^{2t} + 36e^{2t}\cos^2 t - 72e^{2t}\sin t \cos t + 36e^{2t}\sin t + 36e^{2t}\cos^2 t + 72e^{2t}\sin t \cos t + 36e^{2t}\sin^2 t}$$

$$= \sqrt{289e^{2t} + 36e^{2t} + 36e^{2t}} = \sqrt{361e^{2t}} = 19e^t.$$

Therefore the arc length is

$$s = \int_0^1 19e^t \, dt = \Big[\, 19e^t \,\Big]_0^1 = 19e - 19 = 19(e-1) \approx 32.647354740722.$$

C12S06.005: First,

$$v(t) = \sqrt{(3t\cos t + 3\sin t)^2 + (3\cos t - 3t\sin t)^2 + (4t)^2}$$

$$= \sqrt{16t^2 + 9\cos^2 t + 9\sin^2 t + 9t^2\cos^2 t + 9t^2\sin^2 t} = \sqrt{9 + 25t^2}.$$

Then the substitutions $u = 5t$, $a = 3$, and formula 44 of the endpapers of the text yields arc length

$$s = \int_0^{4/5} \sqrt{9 + 25t^2} \, dt = \left[\, \frac{t}{2}\sqrt{9 + 25t^2} + \frac{9}{10}\left(\ln 5t + \sqrt{9 + 25t^2}\right) \right]_0^{4/5}$$

$$= 2 + \frac{9}{10}\ln 9 - \frac{9}{10}\ln 3 = \frac{20 + 9\ln 3}{10} \approx 2.988751059801.$$

By comparison, *Mathematica 3.0* yields

$$s = \left[\, \frac{t}{2}\sqrt{9 + 25t^2} + \frac{9}{10}\sinh^{-1}\left(\frac{5t}{3}\right) \right]_0^{4/5},$$

which can be simplified to the same answer using identities found in Section 7.6.

C12S06.007: By Eq. (13) of the text,

$$\kappa(x) = \frac{|y''(x)|}{[1 + (y'(x))^2]^{3/2}} = \frac{|6x|}{(1 + 9x^4)^{3/2}},$$

and therefore $\kappa(0) = 0$.

C12S06.009: By Eq. (13) of the text,

641

$$\kappa(x) = \frac{|\cos x|}{(1 + \sin^2 x)^{3/2}},$$

and thus $\kappa(0) = 1$.

C12S06.011: By Eq. (12) of the text,

$$\kappa(t) = \frac{|x'(t)y''(t) - x''(t)y'(t)|}{[(x'(t))^2 + (y'(t))^2]^{3/2}} = \frac{|-20\sin^2 t - 20\cos^2 t|}{[25\sin^2 t + 16\cos^2 t]^{3/2}} = \frac{20}{(16 + 9\sin^2 t)^{3/2}},$$

so that

$$\kappa\left(\frac{\pi}{4}\right) = \frac{20}{\left(16 + \frac{9}{2}\right)^{3/2}} = \frac{40\sqrt{82}}{1681} \approx 0.2154761484.$$

C12S06.013: Given: $y = e^x$. By Eq. (13) of the text, the curvature at x is

$$\kappa(x) = \frac{|y''(x)|}{[1 + (y'(x))^2]^{3/2}} = \frac{e^x}{(1 + e^{2x})^{3/2}}.$$

Because $\kappa(x) > 0$ for all x, $\kappa(x) \to 0$ as $x \to \pm\infty$, and κ is continuous on the set of all real numbers, there is a maximum value. Next,

$$\kappa'(x) = \frac{e^x(1 - 2e^{2x})}{(1 + e^{2x})^{5/2}}; \quad \kappa'(x) = 0 \quad \text{when} \quad e^{2x} = \frac{1}{2}: \quad x = -\frac{1}{2}\ln 2.$$

Answer: The maximum curvature of the graph of $y = e^x$ occurs at the point

$$\left(-\frac{1}{2}\ln 2, \ \frac{1}{2}\sqrt{2}\right).$$

The curvature there is

$$\frac{\frac{1}{2}\sqrt{2}}{\left(1 + \frac{1}{2}\right)^{3/2}} = \frac{\sqrt{2}}{2} \cdot \frac{2^{3/2}}{3^{3/2}} = \frac{2\sqrt{3}}{9} \approx 0.3849001794597505.$$

C12S06.015: Given: $x = 5\cos t$, $y = 3\sin t$. By Eq. (12) of the text, the curvature at $(x(t), y(t))$ is given by

$$\kappa(t) = \frac{|x'(t)y''(t) - x''(t)y'(t)|}{[(x'(t))^2 + (y'(t))^2]^{3/2}} = \frac{|15\sin^2 t + 15\cos^2 t|}{(25\sin^2 t + 9\cos^2 t)^{3/2}} = \frac{15}{(9 + 16\sin^2 t)^{3/2}}.$$

Nothing is lost by restriction of t to the interval $[0, 2\pi]$, and $\kappa(t)$ is continuous there, so κ has both a global maximum value and a global minimum value in that interval. To find them,

$$\kappa'(t) = \frac{-15 \cdot \frac{3}{2}(9 + 16\sin^2 t)^{1/2} \cdot 32\sin t \cos t}{(9 + 16\sin^2 t)^3} = -\frac{720\sin t \cos t}{(9 + 16\sin^2 t)^{5/2}}.$$

Because $\kappa'(t) = 0$ at every integral multiple of $\pi/2$, we check these critical points (and only these):

$$\kappa(0) = \frac{5}{9} = \kappa(\pi) = \kappa(2\pi);$$

$$\kappa(\pi/2) = \frac{3}{25} = \kappa(3\pi/2).$$

Therefore the maximum curvature of the graph of the given parametric equations is $\frac{5}{9}$, which occurs at $(5, 0)$ and at $(-5, 0)$ (corresponding to $t = 0$ and $t = \pi$); the minimum curvature is $\frac{3}{25}$ and occurs at $(0, 3)$ and at $(0, -3)$.

C12S06.017: Let $\mathbf{r}(t) = \langle t, t^3 \rangle$. For the purpose of determining the direction of $\mathbf{N}$, note that the graph of $y = x^3$ is concave downward at and near the given point $(-1, -1)$. Then

$$\mathbf{v}(t) = \langle 1, 3t^2 \rangle;$$

$$\mathbf{v}(-1) = \langle 1, 3 \rangle;$$

$$\mathbf{T}(-1) = \left\langle \frac{\sqrt{10}}{10}, \frac{3\sqrt{10}}{10} \right\rangle;$$

$$\mathbf{N}(-1) = \left\langle \frac{3\sqrt{10}}{10}, -\frac{\sqrt{10}}{10} \right\rangle.$$

C12S06.019: Let $\mathbf{r}(t) = \langle 3\sin 2t, 4\cos 2t \rangle$. For the purpose of determining the direction of $\mathbf{N}$, note that the graph of the given parametric equations is concave downward at and near the given point for which $t = \pi/6$. Then

$$\mathbf{v}(t) = \langle 6\cos 2t, -8\sin 2t \rangle; \qquad \mathbf{v}(\pi/6) = \langle 3, -4\sqrt{3} \rangle;$$

$$\mathbf{T}(\pi/6) = \left\langle \frac{\sqrt{57}}{19}, -\frac{4\sqrt{19}}{19} \right\rangle; \qquad \mathbf{N}(\pi/6) = \left\langle -\frac{4\sqrt{19}}{19}, -\frac{\sqrt{57}}{19} \right\rangle.$$

C12S06.021: Let $\mathbf{r}(t) = \langle \cos^3 t, \sin^3 t \rangle$. For the purpose of determining the direction of $\mathbf{N}$, note that the graph of the given parametric equations is concave upward at and near the given point for which $t = 3\pi/4$. Then

$$\mathbf{v}(t) = \langle -3\cos^2 t \sin t, 3\sin^2 t \cos t \rangle; \qquad \mathbf{v}(3\pi/4) = \left\langle -\frac{3\sqrt{2}}{4}, -\frac{3\sqrt{2}}{4} \right\rangle; \qquad v(3\pi/4) = \frac{3}{2};$$

$$\mathbf{T}(3\pi/4) = \left\langle -\frac{\sqrt{2}}{2}, -\frac{\sqrt{2}}{2} \right\rangle; \qquad \mathbf{N}(3\pi/4) = \left\langle -\frac{\sqrt{2}}{2}, \frac{\sqrt{2}}{2} \right\rangle.$$

C12S06.023: We adjoin third component zero to form $\mathbf{r}(t) = \langle 2t+1, 3t^2-1, 0 \rangle$. Then $\mathbf{v}(t) = \langle 2, 6t, 0 \rangle$, $\mathbf{a}(t) = \langle 0, 6, 0 \rangle$, and $v(t) = (36t^2 + 4)^{1/2}$. By Eq. (26) of the text,

$$a_T = \frac{\mathbf{v} \cdot \mathbf{a}}{v} = \frac{36t}{(36t^2 + 4)^{1/2}} = \frac{18t}{\sqrt{9t^2 + 1}},$$

and by Eq. (28),

$$a_N = \frac{|\mathbf{v} \times \mathbf{a}|}{v} = \frac{1}{v} \cdot |\langle 0, 0, 12 \rangle| = \frac{6}{\sqrt{9t^2 + 1}}.$$

C12S06.025: We adjoin third component zero to form $\mathbf{r}(t) = \langle t \cos t, \, t \sin t, \, 0 \rangle$. Then

$$\mathbf{v}(t) = \langle \cos t - t \sin t, \, \sin t + t \cos t, \, 0 \rangle \qquad \mathbf{a}(t) = \langle -2 \sin t - t \cos t, \, 2 \cos t - t \sin t, \, 0 \rangle,$$

and $v(t) = \sqrt{\cos^2 t + t^2 \sin^2 t + \sin^2 t + t^2 \cos^2 t} = \sqrt{t^2 + 1}$. Next,

$$\mathbf{v}(t) \cdot \mathbf{a}(t) = -2 \sin t \cos t + 2t \sin^2 t - t \cos^2 t + t^2 \sin t \cos t + 2 \sin t \cos t - t \sin^2 t + 2t \cos^2 t - t^2 \sin t \cos t$$

$$= 2t - t = t.$$

So by Eq. (26) of the text,

$$a_T = \frac{\mathbf{v}(t) \cdot \mathbf{a}(t)}{v(t)} = \frac{t}{\sqrt{t^2 + 1}},$$

Next,

$$\mathbf{v}(t) \times \mathbf{a}(t) = \begin{vmatrix} \mathbf{i} & \mathbf{j} & \mathbf{k} \\ \cos t - t \sin t & \sin t + t \cos t & 0 \\ -2 \sin t - t \cos t & 2 \cos t - t \sin t & 0 \end{vmatrix}$$

$$= \langle 0, \, 0, \, 2 \cos^2 t - t \sin t \cos t - 2t \sin t \cos t + t^2 \sin^2 t + 2 \sin^2 t + 2t \sin t \cos t + t \sin t \cos t + t^2 \cos^2 t \rangle$$

$$= \langle 0, \, 0, \, t^2 + 2 \rangle.$$

Therefore by Eq. (28),

$$a_N = \frac{|\mathbf{v}(t) \times \mathbf{a}(t)|}{v(t)} = \frac{t^2 + 2}{\sqrt{t^2 + 1}}.$$

C12S06.027: Given $x^2 + y^2 = a^2$, implicit differentiation yields

$$2x + 2y \frac{dy}{dx} = 0, \quad \text{so that} \quad \frac{dy}{dx} = -\frac{x}{y}.$$

Differentiation of both sides of the last equation with respect to x then yields

$$\frac{d^2y}{dx^2} = \frac{x \dfrac{dy}{dx} - y}{y^2} = \frac{xy \dfrac{dy}{dx} - y^2}{y^3} = \frac{-x^2 - y^2}{y^3} = -\frac{a^2}{y^3}.$$

Next, Eq. (13) yields

$$\kappa(x) = \frac{|y''(x)|}{[1 + (y'(x))^2]^{3/2}} = \frac{a^2}{|y|^3 (1 + x^2 y^{-2})^{3/2}}.$$

If $y \neq 0$, then

644

$$\kappa(x) = \frac{a^2}{(y^2 + x^2)^{3/2}} = \frac{a^2}{a^3} = \frac{1}{a}.$$

If $y = 0$, then $x \neq 0$; interchange the roles of x and y and work with dx/dy to obtain the same result. In this case, instead of Eq. (13) use:

$$\kappa(y) = \frac{|x''(y)|}{[1 + (x'(y))^2]^{3/2}}.$$

C12S06.029: Let $x(t) = t$, $y(t) = 1 - t^2$, and $\mathbf{r}(t) = \langle x(t), y(t) \rangle$. Then $\mathbf{r}'(t) = \langle 1, -2t \rangle$, so that $\mathbf{r}'(0) = \langle 1, 0 \rangle = \mathbf{T}(1)$. The graph of the given equation is concave down everywhere, so that the unit normal vector corresponding to $t = 0$ is $\mathbf{N} = \langle 0, -1 \rangle$. Next, $\mathbf{a}(t) = \mathbf{r}''(t) = \langle 0, -2 \rangle$, and so $\mathbf{a}(0)$ is the same. By Eq. (12) the curvature is

$$\kappa(t) = \frac{|x'(t)y''(t) - x''(t)y'(t)|}{[v(t)]^3} = \frac{2}{(1 + 4t^2)^{3/2}}.$$

Because $\kappa(0) = 2$, the radius of the osculating circle at $(0, 1)$ is $\frac{1}{2}$, and by Eq. (16) the position vector of the center of that circle is

$$\mathbf{r}(0) + \frac{1}{2}\mathbf{N}(0) = \left\langle 0, \frac{1}{2} \right\rangle.$$

Therefore an equation of the osculating circle is

$$x^2 + \left(y - \frac{1}{2} \right)^2 = \frac{1}{4}.$$

C12S06.031: Let $x(t) = t$ and $y(t) = t^{-1}$; let $\mathbf{r}(t) = \langle x(t), y(t) \rangle$. Then

$$\mathbf{v}(t) = \mathbf{r}'(t) = \left\langle 1, -\frac{1}{t^2} \right\rangle \quad \text{and} \quad v(t) = |\mathbf{v}(t)| = \frac{\sqrt{t^4 + 1}}{t^2}.$$

Thus the unit tangent and unit normal vectors at $(1, 1)$ are

$$\mathbf{T}(1) = \frac{\mathbf{v}(1)}{v(1)} = \left\langle \frac{\sqrt{2}}{2}, -\frac{\sqrt{2}}{2} \right\rangle \quad \text{and} \quad \mathbf{N}(1) = \left\langle \frac{\sqrt{2}}{2}, \frac{\sqrt{2}}{2} \right\rangle,$$

respectively. By Eq. (12) of the text, the curvature at $(x(t), y(t))$ is

$$\kappa(t) = \frac{|x'(t)y''(t) - x''(t)y'(t)|}{[(x'(t))^2 + (y'(t))^2]^{3/2}} = \frac{2t^3}{(t^4 + 1)^{3/2}},$$

so the curvature at $(1, 1)$ is $\kappa(1) = \frac{1}{2}\sqrt{2}$. Therefore the osculating circle at $(1, 1)$ has radius $\rho = \sqrt{2}$. By Eq. (16) of the text, the position vector of the center of that circle is

$$\mathbf{r}(1) + \left(\sqrt{2} \right) \cdot \mathbf{N}(1) = \langle 2, 2 \rangle.$$

Therefore an equation of the osculating circle is $(x - 2)^2 + (y - 2)^2 = 2$.

C12S06.033: Given $\mathbf{r}(t) = \langle t, \sin t, \cos t \rangle$, we first compute

$$\mathbf{v}(t) = \langle 1, \, \cos t, \, -\sin t \rangle, \quad v(t) = \sqrt{1 + \cos^2 t + \sin^2 t} \equiv \sqrt{2}, \quad \text{and} \quad \mathbf{a}(t) = \langle 0, \, -\sin t, \, -\cos t \rangle.$$

Then

$$\mathbf{v}(t) \times \mathbf{a}(t) = \begin{vmatrix} \mathbf{i} & \mathbf{j} & \mathbf{k} \\ 1 & \cos t & -\sin t \\ 0 & -\sin t & -\cos t \end{vmatrix} = \langle -\cos^2 t - \sin^2 t, \, \cos t, \, -\sin t \rangle = \langle -1, \, \cos t \, -\sin t \rangle.$$

Therefore, by Eq. (27), the curvature is

$$\kappa(t) = \frac{|\mathbf{v}(t) \times \mathbf{a}(t)|}{[v(t)]^3} = \frac{\sqrt{2}}{2\sqrt{2}} \equiv \frac{1}{2}.$$

C12S06.035: Given: $\mathbf{r}(t) = \langle e^t \cos t, \, e^t \sin t, \, e^t \rangle$. Then

$$\mathbf{v}(t) = \langle e^t(\cos t - \sin t), \, e^t(\sin t + \cos t), \, e^t \rangle,$$

$$v(t) = e^t(\cos^2 t + \sin^2 t + \sin^2 t + \cos^2 t + 1)^{1/2} = e^t\sqrt{3}, \quad \text{and}$$

$$\mathbf{a}(t) = \langle e^t(\cos t - \sin t - \sin t - \cos t), \, e^t(\sin t + \cos t + \cos t - \sin t), \, e^t \rangle = e^t\langle -2\sin t, \, 2\cos t, \, 1 \rangle.$$

Therefore

$$\mathbf{v}(t) \times \mathbf{a}(t) = e^{2t} \begin{vmatrix} \mathbf{i} & \mathbf{j} & \mathbf{k} \\ \cos t - \sin t & \sin t + \cos t & 1 \\ -2\sin t & 2\cos t & 0 \end{vmatrix} = e^{2t}\langle \sin t - \cos t, \, -\sin t - \cos t, \, 2 \rangle.$$

Thus

$$|\mathbf{v}(t) \times \mathbf{a}(t)| = e^{2t}\sqrt{\sin^2 t + \cos^2 t + \sin^2 t + \cos^2 t + 4} = e^{2t}\sqrt{6},$$

and therefore

$$\kappa(t) = \frac{e^{2t}\sqrt{6}}{3e^{2t}\sqrt{3}} = \frac{\sqrt{2}}{3}e^{-t}.$$

C12S06.037: Let $x(t) = t$, $y(t) = 2t - 1$, $z(t) = 3t + 5$, and $\mathbf{r}(t) = \langle x(t), \, y(t), \, z(t) \rangle$. It follows that $\mathbf{v}(t) = \langle 1, 2, 3 \rangle$ and $\mathbf{a}(t) = 0$. Hence $\mathbf{v}(t) \cdot \mathbf{a}(t) = 0$ and $\mathbf{v}(t) \times \mathbf{a}(t) = 0$, and therefore $a_T = 0 = a_N$.

C12S06.039: Using *Mathematica* 3.0, we let

```
r[t_] := { t, t^2, t^3 }; v[t_] := r'[t]
```

so that $\mathbf{v}(t) = \langle 1, 2t, 3t^2 \rangle$, and we let

```
speed[t_] := Sqrt[v[t].v[t]]
```

(using the fact that $|\mathbf{u}| = \sqrt{\mathbf{u} \cdot \mathbf{u}}$), so that $v(t) = (1 + 4t^2 + 9t^4)^{1/2}$. We also defined

646

```
a[t_] := r''[t]
```

so that $\mathbf{a}(t) = \langle 0, 2, 6t \rangle$. Next we computed

```
v[t].a[t]
```
$$4t + 18t^3$$

and then, by Eq. (26),

```
asubT = v[t].a[t]/speed[t]
```

so that $a_T = \dfrac{4t + 18t^3}{\sqrt{1 + 4t^2 + 9t^4}}$. Next,

```
p[t_] := Cross[v[t], a[t]]
```

yielded $\mathbf{v}(t) \times \mathbf{a}(t) = \langle 6t^2, -6t, 2 \rangle$. Then

```
magp[t_] := Sqrt[p[t].p[t]]
```

revealed that $|\mathbf{v}(t) \times \mathbf{a}(t)| = (4 + 36t^2 + 36t^4)^{1/2}$. Hence, by Eq. (28),

```
asubN = magp[t]/speed[t]
```

—that is, $a_N = \dfrac{\sqrt{4 + 36t^2 + 36t^4}}{\sqrt{1 + 4t^2 + 9t^4}}$.

C12S06.041: Beginning with $\mathbf{r}(t) = \langle t \sin t, \, t \cos t, \, t \rangle$, we found:

$$\mathbf{v}(t) = \mathbf{r}'(t) = \langle t \cos t + \sin t, \, \cos t - t \sin t, \, 1 \rangle,$$

$$v(t) = |\mathbf{v}(t)| = \sqrt{1 + (t \cos t + \sin t)^2 + (\cos t - t \sin t)^2} = \sqrt{t^2 + 2},$$

$$\mathbf{a}(t) = \mathbf{v}'(t) = \langle 2 \cos t - t \sin t, \, -t \cos t - 2 \sin t, \, 0 \rangle,$$

$$\mathbf{v}(t) \cdot \mathbf{a}(t) = (-t \cos t - 2 \sin t)(\cos t - t \sin t) + (t \cos t + \sin t)(2 \cos t - t \sin t) = t,$$

$$a_T = \frac{\mathbf{v}(t) \cdot \mathbf{a}(t)}{v(t)} = \frac{t}{\sqrt{t^2 + 2}},$$

$$\mathbf{v}(t) \times \mathbf{a}(t) = \langle t \cos t + 2 \sin t, \, 2 \cos t - t \sin t, \, -(t^2 + 2) \rangle,$$

$$|\mathbf{v}(t) \times \mathbf{a}(t)| = \sqrt{t^4 + 5t^2 + 8}, \quad \text{and}$$

$$a_N = \frac{|\mathbf{v}(t) \times \mathbf{a}(t)|}{v(t)} = \frac{\sqrt{t^4 + 5t^2 + 8}}{\sqrt{t^2 + 2}}.$$

C12S06.043: Given $\mathbf{r}(t) = \langle t, \, \sin t, \, \cos t \rangle$, we will compute the unit tangent vector $\mathbf{T}$ using its definition in Eq. (17), then the principal unit normal vector $\mathbf{N}$ by means of Eq. (29). We find:

$$\mathbf{v}(t) = \mathbf{r}'(t) = \langle 1, \, \cos t, \, -\sin t \rangle,$$

$$\mathbf{v}(0) = \langle 1, 1, 0 \rangle,$$

$$v(0) = \sqrt{2},$$

$$\mathbf{T}(0) = \left\langle \frac{\sqrt{2}}{2}, \frac{\sqrt{2}}{2}, 0 \right\rangle,$$

$$\mathbf{a}(t) = \langle 0, -\sin t, -\cos t \rangle,$$

$$\mathbf{a}(0) = \langle 0, 0, -1 \rangle,$$

$$a_T(0) = \frac{\mathbf{v}(0) \cdot \mathbf{a}(0)}{v(0)} = 0,$$

$$a_N(0) = \frac{|\mathbf{v}(0) \times \mathbf{a}(0)|}{v(t)} = 1,$$

$$\kappa(0) = \frac{|\mathbf{v}(0) \times \mathbf{a}(0)|}{[v(t)]^3} = \frac{1}{2}, \quad \text{and}$$

$$\mathbf{N}(0) = \frac{\mathbf{a} - a_T \mathbf{T}}{a_N} = \langle 0, 0, -1 \rangle$$

C12S06.045: The process of computing **T** and **N** can be carried out almost automatically in *Mathematica* 3.0. Given $\mathbf{r}(t) = \langle e^t \cos t, e^t \sin t, e^t \rangle$, we entered the following commands:

```
v0 = r'[0]
        {1, 1, 1}
a0 = r''[0]
        {0, 2, 1}
speed = Sqrt[v0.v0]
```
$$\sqrt{3}$$
```
asubT = v0.a0/speed
```
$$\sqrt{3}$$
```
vcrossa = Cross[ v0, a0 ]
        {-1, -1, 2}
asubN = (Sqrt[ vcrossa.vcrossa ])/speed
```
$$\sqrt{2}$$
```
kappa = asubN/(speed*speed)
```
$$\frac{\sqrt{2}}{3}$$
```
utan = v0/speed
```
$$\left\{ \frac{1}{\sqrt{3}}, \frac{1}{\sqrt{3}}, \frac{1}{\sqrt{3}} \right\}$$
```
unorm = (a0 - asubT*utan)/asubN
```
$$\left\{ -\frac{1}{\sqrt{2}}, \frac{1}{\sqrt{2}}, 0 \right\}$$

Thus we see that

$$\mathbf{T}(0) = \left\langle \frac{\sqrt{3}}{3}, \frac{\sqrt{3}}{3}, \frac{\sqrt{3}}{3} \right\rangle \quad \text{and} \quad \mathbf{N}(0) = \left\langle -\frac{\sqrt{2}}{2}, \frac{\sqrt{2}}{2}, 0 \right\rangle.$$

C12S06.047: Because

$$\frac{ds}{dt} = \sqrt{[x'(t)]^2 + [y'(t)]^2 + [z'(t)]^2} = \sqrt{16 + 144 + 9} = 13,$$

we see that the arc length is given by $s = 13t$, and therefore the arc-length parametrization of the given curve is

$$x(s) = 2 + \frac{4s}{13}, \quad y(s) = 1 - \frac{12s}{13}, \quad z(s) = 3 + \frac{3s}{13}.$$

C12S06.049: Because

$$\frac{ds}{dt} = \sqrt{[x'(t)]^2 + [y'(t)]^2 + [z'(t)]^2} = \sqrt{9\cos^2 t + 9\sin^2 t + 16} = \sqrt{25} = 5,$$

we see that the arc length is given by $s = 5t$, and therefore the arc-length parametrization of the given curve is

$$x(s) = 3\cos \frac{s}{5}, \quad y(s) = 3\sin \frac{s}{5}, \quad z(s) = \frac{4s}{5}.$$

C12S06.051: By Newton's second law of motion, the acceleration of the particle is a scalar multiple of the force acting on the particle, so the force and acceleration vectors are parallel. Therefore the acceleration vector $\mathbf{a}$ is normal to the velocity vector $\mathbf{v}$. But then,

$$D_t(\mathbf{v} \cdot \mathbf{v}) = \mathbf{v} \cdot \mathbf{a} + \mathbf{a} \cdot \mathbf{v} = 0 + 0 = 0,$$

and therefore $\mathbf{v} \cdot \mathbf{v} = K$, a constant. Hence the speed $v(t) = \sqrt{\mathbf{v} \cdot \mathbf{v}}$ is also constant.

C12S06.053: Given $x(t) = \cos t + t\sin t$ and $y(t) = \sin t - t\cos t$, we first compute

$$[v(t)]^2 = [x'(t)]^2 + [y'(t)]^2$$

$$= (t\cos t + \sin t - \sin t)^2 + (\cos t - \cos t + t\sin t)^2 = t^2\cos^2 t + t^2\sin^2 t = t^2,$$

$$v'(t) = 1,$$

$$[x''(t)]^2 = (\cos t - t\sin t)^2,$$

$$[y''(t)]^2 = (\sin t + t\cos t)^2, \quad \text{and}$$

$$[x''(t)]^2 + [y''(t)]^2 = \cos^2 t + \sin^2 t + t^2\sin^2 t + t^2\cos^2 t = t^2 + 1.$$

Then the formula in Problem 52 yields

$$\kappa(t) = \frac{\sqrt{[x''(t)]^2 + [y''(t)]^2 - [v'(t)]^2}}{[x'(t)]^2 + [y'(t)]^2} = \frac{\sqrt{t^2 + 1 - 1}}{t^2} = \frac{1}{|t|}.$$

C12S06.055: The six conditions listed in the statement of the problem imply, in order, that

$$0 = F,$$

$$1 = A + B + C + D + E + F,$$

$$0 = E,$$

$$1 = 5A + 4B + 3C + 2D + E,$$

$$0 = 2D, \quad \text{and}$$

$$0 = 20A + 12B + 6C + 2D.$$

The last two equations were obtained by observing—via Eq. (13)—that curvature zero is equivalent to $y''(x) = 0$. Simultaneous solution of these equations yields $A = 3$, $B = -8$, $C = 6$, and $D = E = F = 0$. Answers: $y = 3x^5 - 8x^4 + 6x^3$; because $\kappa(x)$ is zero where the curved track meets the straight tracks, because their derivatives agree at the junctions, and because

$$\kappa(x) = \frac{|60x^3 - 96x^2 + 36x|}{[1 + (15x^4 - 32x^3 + 18x^2)^2]^{3/2}}$$

is continuous on $[0, 1]$, the normal forces on the train negotiating this transitional section of track will change continuously from zero to larger values and then back to zero. Thus there will be no abrupt change in the lateral forces on the train.

C12S06.057: Conversion into miles yields the semimajor axis of the orbit of Mercury to be $a = 35973972$. With eccentricity $e = 0.206$ and period $T = 87.97$ days, we use

$$b^2 = a^2(1 - e^2)$$

to find that $b \approx 35202402$. Then the formula $c^2 = a^2 - b^2$ yields $c \approx 7410638$. So at perihelion the speed of Mercury is

$$\frac{2\pi ab}{(a - c)T} \approx 3166628$$

miles per day. We divide by $24 \cdot 3600$ to convert this answer to 36.650789 miles per second. Replace $a - c$ with $a + c$ to find that its speed at aphelion is approximately 24.129956 miles per second.

C12S06.059: We are given that the semimajor axis of the Moon's orbit is $a = 238900$ (miles). With eccentricity $e = 0.055$ and period $T = 27.32$ days, we use

$$b^2 = a^2(1 - e^2)$$

to find that $b \approx 238538$. Then the formula $c^2 = a^2 - b^2$ yields $c \approx 13139$. So at perigee the speed of the Moon is

$$\frac{2\pi ab}{(a - c)T} \approx 58053$$

miles per day. We divide by $24 \cdot 3600$ to convert this answer to 0.671911 miles per second. Replace $a - c$ with $a + c$ to find that the speed at apogee is approximately 0.601854 miles per second.

C12S06.061: Equation (44), applied to the Earth-Moon system with units of miles and days, yields $(27.32)^2 = \gamma \cdot (238900)^3$. For a satellite with period $T = \frac{1}{24}$ (of a day—one hour), it yields $T^2 = \gamma \cdot r^3$ where r is the radius of the orbit of the satellite. Divide the second of these equations by the first to eliminate γ:

$$\frac{T^2}{(27.32)^2} = \frac{r^3}{(238900)^3},$$

so that

$$r^3 = \frac{(238900)^3}{(24)^2 \cdot (27.32)^2}, \quad \text{and thus} \quad r \approx 3165.35$$

miles, about 795 miles below the surface of the Earth. So it can't be done.

C12S06.063: With the usual meaning of the symbols, the data given in the problem tell us that $a + c = 4960$ and $a - c = 4060$, which we solve for $a = 4510$ (units are in days and miles). Let T be the period of the satellite in its orbit. Equation (44), applied to the Earth-Moon system, then to the Earth-satellite system, yields

$$(27.32)^2 = \gamma \cdot (238900)^3 \quad \text{and} \quad T^2 = \gamma \cdot (4510)^2,$$

which we solve for $T \approx 0.0708632854$. Multiply by 24 to convert this answer to approximately 1.7007188486 hours—about 1 h 42 min 2.588 s.

C12S06.065: We will use Eqs. (37) and (41), which are—respectively—

$$r^2 \frac{d\theta}{dt} = h \quad \text{(constant)} \quad \text{and} \quad \frac{d^2r}{dt^2} = \frac{h^2}{r^2}\left(\frac{1}{r} - \frac{1}{pe}\right).$$

But we begin with Eq. (42),

$$\mathbf{a} = \left[\frac{d^2r}{dt^2} - r\left(\frac{d\theta}{dt}\right)^2\right]\mathbf{u}_r.$$

Substitution of Eqs. (37) and (41) then yields

$$\mathbf{a} = \left[\frac{h^2}{r^2}\left(\frac{1}{r} - \frac{1}{pe}\right) - \frac{1}{r^3}\left(r^2\frac{d\theta}{dt}\right)^2\right]\mathbf{u}_r$$

$$= \left[\frac{h^2}{r^2}\left(\frac{1}{r} - \frac{1}{pe}\right) - \frac{1}{r^3}h^2\right]\mathbf{u}_r = \left[\frac{h^2}{r^3} - \frac{h^2}{per^2} - \frac{h^2}{r^3}\right]\mathbf{u}_r = -\frac{h^2}{per^2}\mathbf{u}_r.$$

C12S06.067: We begin with Eq. (33),

$$\mathbf{v} = \frac{dr}{dt}\mathbf{u}_r + r\frac{d\theta}{dt}\mathbf{u}_\theta,$$

and differentiate both sides with respect to t:

$$\mathbf{a} = \frac{d\mathbf{v}}{dt} = \left(\frac{d^2r}{dt^2}\mathbf{u}_r + \frac{dr}{dt}\cdot\frac{d\mathbf{u}_r}{dt}\right) + \left(\frac{dr}{dt}\frac{d\theta}{dt}\mathbf{u}_\theta + r\frac{d^2\theta}{dt^2}\mathbf{u}_\theta + r\frac{d\theta}{dt}\cdot\frac{d\mathbf{u}_\theta}{dt}\right)$$

$$= \frac{d^2r}{dt^2}\mathbf{u}_r + \frac{dr}{dt}\cdot\frac{d\theta}{dt}\mathbf{u}_\theta + \frac{dr}{dt}\cdot\frac{d\theta}{dt}\mathbf{u}_\theta + r\frac{d^2\theta}{dt^2}\mathbf{u}_\theta - r\left(\frac{d\theta}{dt}\right)^2\mathbf{u}_r$$

$$= \left[\frac{d^2r}{dt^2} - r\left(\frac{d\theta}{dt}\right)^2 \right]\mathbf{u}_r + \left[2\frac{dr}{dt} \cdot \frac{d\theta}{dt} + r\frac{d^2\theta}{dt^2} \right]\mathbf{u}_\theta$$

$$= \left[\frac{d^2r}{dt^2} - r\left(\frac{d\theta}{dt}\right)^2 \right]\mathbf{u}_r + \left[\frac{1}{r} \cdot \frac{d}{dt}\left(r^2\frac{d\theta}{dt}\right) \right]\mathbf{u}_\theta.$$

Section 12.7

C12S07.001: The graph of $3x + 2y + 10z = 20$ is a plane with intercepts $x = \frac{20}{3}$, $y = 10$, and $z = 2$. The graph produced by the *Mathematica* 3.0 command

```
ParametricPlot3D[ { x, y, (20 - 3*x - 2*y)/10 }, { x, -7, 7 }, { y, -7, 7 } ];
```

is shown next.

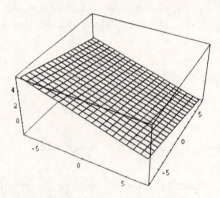

C12S07.003: The graph of $x^2 + y^2 = 9$ is a circular cylinder of radius 3 with axis the z-axis. The graph produced by the *Mathematica* 3.0 command

```
ParametricPlot3D[ { 3*Cos[t], 3*Sin[t], z }, { t, 0, 2*Pi }, { z, -4, 4 } ];
```

is shown next.

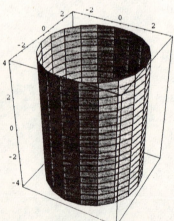

C12S07.005: The graph of $xy = 4$ is a cylinder whose rulings are parallel to the z-axis. It meets the xy-plane in the hyperbola with equation $xy = 4$. The graph produced by the *Mathematica* 3.0 command

```
ParametricPlot3D[ {{ x, 4/x, z }, { -x, -4/x, z }}, { x, 1/3, 5 }, { z, -3, 3 } ];
```

652

is shown next.

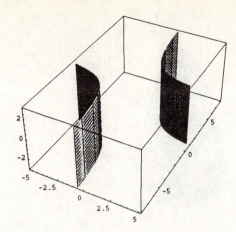

C12S07.007: The graph of the equation $z = 4x^2 + y^2$ is an elliptic paraboloid with axis the z-axis and vertex at the origin. It is elliptic because, if $z = a^2$, then the horizontal cross section there has equation $4x^2 + y^2 = a^2$. It is a paraboloid because it meets every vertical plane containing the z-axis in a parabola; for example, it meets the xz-plane in the parabola with equation $z = 4x^2$. The graph produced by the *Mathematica* 3.0 command

```
ParametricPlot3D[ { r*Cos[t], 2*r*Sin[t], 4*r*r }, { r, 0, 1 }, { t, 0, 2*Pi } ];
```

is shown next.

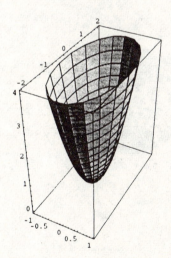

C12S07.009: The given equation $z = 4 - x^2 - y^2$ has the polar coordinates form $z = 4 - r^2$, so the graph is a surface of revolution around the z-axis. The graph meets the xz-plane in the parabola $z = 4 - x^2$, so the graph is a circular paraboloid, opening downward, with axis the z-axis, and vertex at $(0, 0, 4)$. The graph produced by the *Mathematica* 3.0 command

```
ParametricPlot3D[ { r*Cos[t], r*Sin[t], 4 - r∧2 }, { r, 0, 3 }, { t, 0, 2*Pi } ];
```

is shown next.

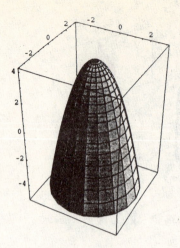

C12S07.011: The given equation $2z = x^2 + y^2$ has the polar form $z = \frac{1}{2}r^2$, so the graph is a surface of revolution around the z-axis. It meets the yz-plane in the parabola with equation $z = \frac{1}{2}x^2$, so the surface is a circular paraboloid; its axis is the z-axis and its vertex is at the origin. The graph produced by the *Mathematica* 3.0 command

```
ParametricPlot3D[ { r*Cos[t], r*Sin[t], (r*r)/2 }, { t, 0, 2*Pi }, { r, 0, 2 } ];
```

is shown next.

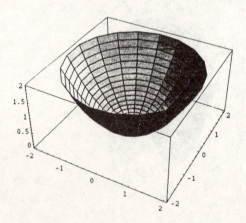

C12S07.013: The polar form of the given equation $z^2 = 4(x^2 + y^2)$ is $z = \pm 2r$, so the graph is a surface of revolution around the z-axis. Because z is proportional to r, the graph consists of both nappes of a circular cone with axis the z-axis and vertex at the origin. To produce the graph of the upper nappe, we used the *Mathematica* 3.0 command

```
ParametricPlot3D[ { r*Cos[t], r*Sin[t], 2*r }, { t, 0, 2*Pi }, { r, 0, 2 } ];
```

and the result is shown next.

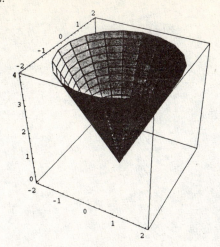

C12S07.015: The graph of the equation $x^2 = 4z + 8$ is a cylinder parallel to the y-axis. It is a parabolic cylinder because its trace in the xz-plane is the parabola with equation $z = (x^2 - 8)/4$. It opens upward and its lowest points consist of line $z = -2$, $x = 0$. The graph produced by the *Mathematica* 3.0 command

```
ParametricPlot3D[ { x, y, (x*x - 8)/4 }, { x, -4, 4 }, { y, -4, 4 } ];
```

is shown next.

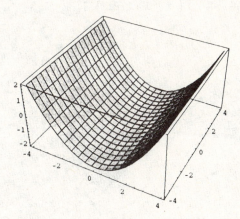

C12S07.017: The graph of the equation $4x^2 + y^2 = 4$ is a cylinder parallel to the z-axis. It is an elliptical cylinder because its trace in any horizontal plane is the ellipse with equation $4x^2 + y^2 = 4$. The graph of this surface produced by the *Mathematica* 3.0 command

```
ParametricPlot3D[ { Cos[t], 2*Sin[t], z }, { t, 0, 2*Pi }, { z, -3, 3 } ];
```

is shown next.

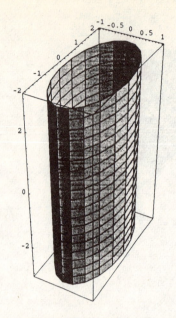

C12S07.019: The graph of the equation $x^2 = 4y^2 + 9z^2$ is an elliptical cone with axis the x-axis. The graph generated by the *Mathematica* 3.0 command

```
ParametricPlot3D[ {{ 6*r, 3*r*Cos[t], 2*r*Sin[t] }, { -6*r, 3*r*Cos[t], 2*r*Sin[t] }},

                  { r, 0, 1.4 }, { t, 0, 2*Pi } ];
```

is shown next.

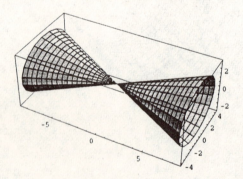

C12S07.021: The polar form of the given equation $x^2 + y^2 + 4z = 0$ is $z = -\frac{1}{4}r^2$, so its graph is a paraboloid opening downward, with axis the negative z-axis and vertex at the origin. The graph generated by the *Mathematica* 3.0 command

```
ParametricPlot3D[ { r*Cos[t], r*Sin[t], -r*r/4 }, { t, 0, 2*Pi }, { r, 0, 3.2 } ];
```

is shown next.

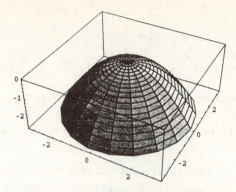

C12S07.023: There may be two versions of this problem, depending on which printing and which version (with or without matrices) of the textbook you are using. The intended version and its solution are given first.

The graph of the equation $x = 2y^2 - z^2$ is a hyperbolic paraboloid with saddle point at the origin. It meets the xz-plane in the parabola $x = -z^2$ with vertex at the origin and opening to the left; it meets the xy-plane in the parabola $x = 2y^2$ with vertex at the origin and opening to the right. The surface meets each plane parallel to the yz-plane in both branches of a hyperbola (except for the yz-plane itself, which it meets in a pair of straight lines that meet at the origin—a degenerate hyperbola). The graph generated by the *Mathematica* 3.0 command

```
ParametricPlot3D[ { 2*y*y - z*z, y, z }, { y, -1, 1 }, { z, -1, 1 } ];
```

is shown next.

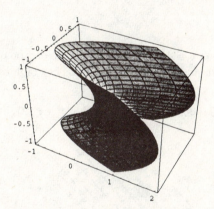

The other version of this problem is the result of a typograhical error discovered too late in the production process to be corrected in the first printing of the version of the textbook "with matrices." Its solution follows.

Given the equation $z = 2y^2 - z^2$, we complete the square in z to obtain

$$2y^2 - z^2 - z = 0;$$

$$2y^2 - z^2 - z - \frac{1}{4} = -\frac{1}{4};$$

$$2y^2 - \left(z + \frac{1}{2}\right)^2 = -\frac{1}{4}.$$

The graph of this equation is a cylinder with rulings parallel to the x-axis (because x is missing from the equation). It meets the yz-plane in the hyperbola with equation

657

$$\left(z + \frac{1}{2}\right)^2 - 2y^2 = \frac{1}{4}.$$

Thus the graph is a hyperbolic cylinder parallel to the x-axis.

C12S07.025: The graph of the equation $x^2 + y^2 - 9z^2 = 9$ is a hyperboloid of one sheet with axis the z-axis. It meets the xy-plane in a circle of radius 3 and meets parallel planes in larger circles. It meets every plane containing the z-axis in both branches of a hyperbola. The graph generated by the *Mathematica* 3.0 command

```
ParametricPlot3D[ {{ r*Cos[t], r*Sin[t], Sqrt[(r*r - 9)/9] },
                  { r*Cos[t], r*Sin[t], -Sqrt[(r*r - 9)/9] }},
                { r, 3, 4.5 }, { t, 0, 2*Pi } ];
```

is shown next.

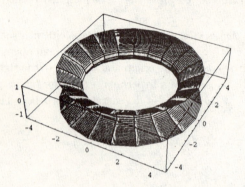

C12S07.027: The graph of the equation $y = 4x^2 + 9z^2$ is an elliptic paraboloid opening in the positive y-direction, with axis the nonnegative y-axis and vertex at the origin. The graph generated by the *Mathematica* 3.0 command

```
ParametricPlot3D[ { 3*r*Cos[t], 36*r*r, 2*r*Sin[t] }, { r, 0, 1/2 }, { t, 0, 2*Pi } ];
```

is shown next.

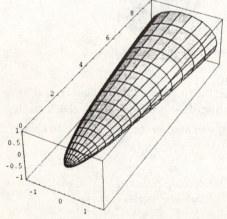

C12S07.029: The graph of the equation $y^2 - 9x^2 - 4z^2 = 36$ is a hyperboloid of two sheets with axis the y-axis, center the origin, and intercepts $(0, \pm 6, 0)$. Planes containing the y-axis meet it in both branches of a hyperbola. Planes normal to the y-axis and outside the intercepts meet it in ellipses. The graph generated by the *Mathematica* 3.0 command

```
ParametricPlot3D[ {{ 2*r*Cos[t], 6*Sqrt[r*r + 1], 3*r*Sin[t] },
                   { 2*r*Cos[t], -6*Sqrt[r*r + 1], 3*r*Sin[t] }},
                 { t, 0, 2*Pi }, { r, 0, 1 } ];
```

is shown next.

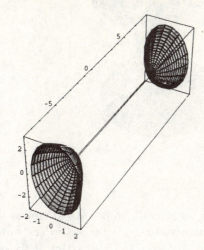

C12S07.031: The graph of the curve $x = 2z^2$ (in the xz-plane) is to be rotated around the x-axis. To obtain an equation of the resulting surface, replace z with $(y^2 + z^2)^{1/2}$ to obtain $x = 2(y^2 + z^2)$. The surface is a circular paraboloid opening along the positive x-axis. The graph generated by the *Mathematica* 3.0 command

```
ParametricPlot3D[ { 2*r*r, r*Cos[t], r*Sin[t] }, { t, 0, 2*Pi }, { r, 0, 2 } ];
```

is shown next.

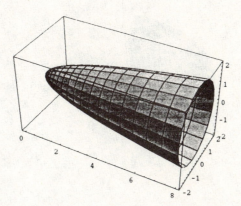

C12S07.033: The curve $y^2 - z^2 = 1$ (in the yz-plane) is to be rotated around the z-axis. To obtain an equation of the resulting surface, replace y^2 with $x^2 + y^2$ to obtain $x^2 + y^2 - z^2 = 1$. The surface is a circular hyperboloid of one sheet with axis the z-axis. The graph generated by the *Mathematica* 3.0 command

```
ParametricPlot3D[ {{ r*Cos[t], r*Sin[t], Sqrt[r*r - 1] },
                   { r*Cos[t], r*Sin[t], -Sqrt[r*r - 1] }},
                 { r, 1, 3 }, { t, 0, 2*Pi } ];
```

659

is shown next.

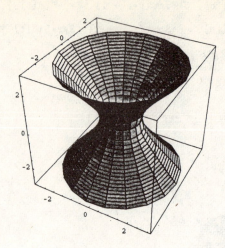

C12S07.035: The curve $y^2 = 4x$ (in the xy-plane) is to be rotated around the x-axis. Replace y^2 with $y^2 + z^2$ to obtain an equation of the resulting surface: $y^2 + z^2 = 4x$. The surface is a circular paraboloid opening along the positive x-axis, with axis that axis, and with vertex at the origin. The graph generated by the *Mathematica* 3.0 command

```
ParametricPlot3D[ { r*r/4, r*Cos[t], r*Sin[t] }, { t, 0, 2*Pi }, { r, 0, 3.5 } ];
```

is shown next.

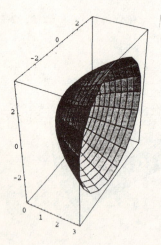

C12S07.037: Given: the curve with equation $z = \exp(-x^2)$ in the xz-plane, to be rotated around the z-axis. To obtain an equation of the resulting surface, replace x^2 with $x^2 + y^2$ to obtain $z = \exp\left(-x^2 - y^2\right)$. The graph of this surface generated by the *Mathematica* 3.0 command

```
ParametricPlot3D[ { r*Cos[t], r*Sin[t], Exp[-r*r] }, { t, 0, 2*Pi }, { r, 0, 2.4 },

          AspectRatio -> Automatic, ViewPoint -> { 1.3, -2.2, 0.6 } ];
```

is shown next.

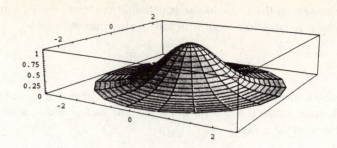

C12S07.039: The curve with equation $z = 2x$ (in the xz-plane) is to be rotated around the z-axis. To obtain an equation of the surface thereby generated, replace x with $\pm\sqrt{x^2 + y^2}$. The resulting equation is $z^2 = 4(x^2 + y^2)$ and its graph consists of both nappes of a right circular cone with vertices at the origin and axis the z-axis. The graph generated by the *Mathematica* 3.0 command

```
ParametricPlot3D[ {{ r*Cos[t], r*Sin[t], 2*r }, { r*Cos[t], r*Sin[t], -2*r }},
                  { t, 0, 2*Pi }, { r, 0, 3 } ];
```

is shown next.

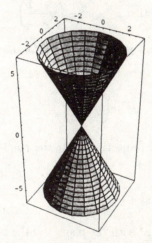

C12S07.041: The graph of $x^2 + 4y^2 = 4$ is an elliptical cylinder with centerline the z-axis. Thus its traces in horizontal planes are ellipses with semiaxes 2 and 1.

C12S07.043: The graph of $x^2 + 4y^2 + 4z^2 = 4$ is an ellipsoid. A plane parallel to the yz-plane is perpendicular to the x-axis and thus has an equation of the form $x = a$. So the trace of the ellipsoid in such a plane has equations

$$x = a, \quad 4y^2 + 4z^2 = 4 - a^2.$$

Thus the trace is a circle if $|a| < 2$, a single point if $|a| = 2$, and the empty set if $|a| > 2$.

C12S07.045: The graph of $z = 4x^2 + 9y^2$ is an elliptical paraboloid opening upward with axis the z-axis and vertex at the origin. A plane parallel to the yz-plane has equation $x = a$ and thus the trace in such a plane has equations

$$x = a, \quad z = 4a^2 + 9y^2.$$

Thus the trace is a parabola opening upward with vertex at $(a, 0, 4a^2)$.

C12S07.047: A plane containing the z-axis has an equation of the form $ax + by = 0$. So if $b \neq 0$ then the intersection of the hyperbolic paraboloid $z = xy$ with such a plane has equations

$$y = -\frac{a}{b}x, \quad z = -\frac{a}{b}x^2.$$

Hence the trace is a parabola opening downward if a and b have the same sign, opening upward if they have opposite sign, and is a horizontal line if $a = 0$ (or if $b = 0$). The surface itself resembles the one shown in Fig. 12.7.22 rotated 45° around the z-axis.

C12S07.049: The triangles OAC and OBC in Fig. 12.7.1 are congruent because the angles OCA and OCB are right angles, $|OA| = |OB|$ because they are both radii of the same sphere, and $|OC| = |OC|$. So the triangles have matching side-side-angle in the same order.

C12S07.051: The intersection I of the plane $z = y$ and the paraboloid $z = x^2 + y^2$ satisfies both equations, and thus lies on the surface with equation

$$y = x^2 + y^2; \quad \text{that is,} \quad x^2 + \left(y - \frac{1}{2}\right)^2 = \frac{1}{4}.$$

This surface is a circular cylinder perpendicular to the xy-plane. Hence the projection of I into that plane lies on the circle C with equations

$$z = 0, \quad x^2 + \left(y - \frac{1}{2}\right)^2 = \frac{1}{4}.$$

It remains to show that all of C is obtained by projecting I into the xy-plane. Suppose that $(x, y, 0)$ lies on C. Then $y = x^2 + y^2$. Let $z = y$. Then

$$x^2 + y^2 = y = z,$$

and therefore (x, y, z) satisfies the equations of both the plane and the paraboloid. Therefore (x, y, z) lies on I. This proves that the projection of I into the xy-plane is *all* of C. Therefore the projection of I into the xy-plane is a circle.

C12S07.053: The plane with equation $x + y + z = 1$ and the ellipsoid with equation $x^2 + 4y^2 + 4z^2 = 4$ have intersection that satisfies both equations, so it lies on the surface with equation

$$x^2 + 4y^2 + 4(1 - x - y)^2 = 4; \quad \text{that is,} \quad 5x^2 + 8xy + 8y^2 - 8x - 8y = 0.$$

This surface is a cylinder normal to the xy-plane, so its projection into the xy-plane—which is the projection of the intersection into the xy-plane—has the same equation. By earlier discussions of such equations, the projection must be a conic section. But it cannot be a hyperbola or a parabola because it is clearly a closed curve. (It is not any of the degenerates cases of a conic section because it contains the two points $(0, 0)$ and $(0, 1)$.) Therefore it is an ellipse. To be absolutely certain of this, the *Mathematica* 3.0 command

```
ContourPlot[ 5*x*x + 8*x*y + 8*y*y - 8*x - 8*y, { x, -1, 2 }, { y, -1.5, 1.5 },
     Axes → True, AxesOrigin → (0,0), AxesLabel → { x, y }, Contours → 3,
     ContourShading → False, PlotPoints → 47, PlotRange → { -0.01, 0.01 } ];
```

generates a plot of this curve, and it's shown next.

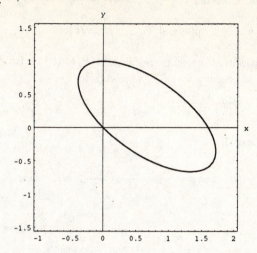

Section 12.8

C12S08.001: The formulas in (3) immediately yield

$$x = r\cos\theta = 1\cdot\cos\frac{\pi}{2} = 0, \qquad y = r\sin\theta = 1\cdot\sin\frac{\pi}{2} = 1, \qquad z = 2.$$

Answer: $(0, 1, 2)$.

C12S08.003: If $(r, \theta, z) = (2, 3\pi/4, 3)$, then $(x, y, z) = (r\cos\theta, r\sin\theta, z) = \left(-\sqrt{2}, \sqrt{2}, 3\right)$.

C12S08.005: If $(r, \theta, z) = (2, \pi/3, -5)$, then $(x, y, z) = (r\cos\theta, r\sin\theta, z) = \left(1, \sqrt{3}, -5\right)$.

C12S08.007: Given $(\rho, \phi, \theta) = (2, 0, \pi)$, the equations in (6) immediately yield

$$x = \rho\sin\phi\cos\theta = 0, \qquad y = \rho\sin\phi\sin\theta = 0, \qquad z = \rho\cos\phi = 2.$$

Answer: $(0, 0, 2)$.

C12S08.009: If $(\rho, \phi, \theta) = (3, \pi/2, \pi)$, then

$$(x, y, z) = (\rho\sin\phi\cos\theta, \rho\sin\phi\sin\theta, \rho\cos\phi) = (-3, 0, 0).$$

C12S08.011: If $(\rho, \phi, \theta) = (2, \pi/3, 3\pi/2)$, then

$$(x, y, z) = (\rho\sin\phi\cos\theta, \rho\sin\phi\sin\theta, \rho\cos\phi) = \left(0, -\sqrt{3}, 1\right).$$

C12S08.013: If $P(x, y, z) = P(0, 0, 5)$, then P has cylindrical coordinates

$$r = \sqrt{x^2 + y^2} = 0, \qquad \theta \text{ (therefore) arbitrary, and} \qquad z = 5.$$

So one correct answer is $(0, 0, 5)$. Also P has spherical coordinates

$$\rho = \sqrt{x^2 + y^2 + z^2} = \pm 5, \qquad \phi = 0 \text{ (if } \rho > 0), \qquad \theta \text{ (therefore) arbitrary.}$$

Thus one correct answer is $(5, 0, 0)$. Another is $(-5, \pi, \pi/2)$.

C12S08.015: Cylindrical $\left(\sqrt{2}, \pi/4, 0\right)$, spherical $\left(\sqrt{2}, \pi/2, \pi/4\right)$.

C12S08.017: Given: the point with Cartesian coordinates $P(1, 1, 1)$. Its cylindrical coordinates are $\left(\sqrt{2}, \pi/4, 1\right)$. To find the spherical coordinates of P, we compute

$$\rho = \sqrt{1^2 + 1^2 + 1^2} = \sqrt{3} \quad \text{and} \quad \theta = \frac{\pi}{4}.$$

To find ϕ, Fig. 12.8.10 makes it clear that

$$\cos \phi = \frac{z}{\rho} = \frac{1}{\sqrt{3}} = \frac{\sqrt{3}}{3},$$

and hence

$$\phi = \arccos\left(\frac{\sqrt{3}}{3}\right) \approx 0.9553166181245093;$$

that's approximately $54.7356103162°$, about $54°\ 44'\ 8.197''$. Thus the spherical coordinates of P are

$$\left(\sqrt{3}, \cos^{-1}\frac{\sqrt{3}}{3}, \frac{\pi}{4}\right).$$

Other ways to express ϕ are

$$\phi = \sin^{-1}\frac{\sqrt{6}}{3} \quad \text{and} \quad \phi = \tan^{-1}\left(\sqrt{2}\right).$$

C12S08.019: The given point P having the Cartesian coordinates $(2, 1, -2)$ has cylindrical coordinates $\left(\sqrt{5}, \tan^{-1}\left(\frac{1}{2}\right), -2\right)$. To find its spherical coordinates, it's clear that $\rho = 3$. Imagine the vertical plane that contains the z-axis and $P(2, 1, -2)$. If you mark the points $O(0, 0, 0)$ and $Q(2, 1, 0)$, draw the z-axis and the triangle OAP, then the angle ϕ is the obtuse angle reaching from the positive z-axis to the hypotenuse of triangle OAP. Let α be the acute angle of that triangle at O. Note that OA has length $r = \sqrt{5}$, that OP has length $\rho = 3$, and that AP has length $-z = 2$. Because

$$\phi = \alpha + \frac{\pi}{2}, \quad \sin \alpha = \frac{2}{3}, \quad \text{and} \quad \cos \alpha = \frac{\sqrt{5}}{3},$$

it now follows that

$$\cos \phi = \cos\left(\frac{\pi}{2} + \alpha\right) = -\sin \alpha = -\frac{2}{3}.$$

Therefore the spherical coordinates of P are

$$\left(3, \cos^{-1}\left(-\frac{2}{3}\right), \tan^{-1}\left(\frac{1}{2}\right)\right) \approx (3, 2.300523983822, 0.463647609001).$$

C12S08.021: The cylindrical coordinates of $P(3, 4, 12)$ are $\left(5, \arctan\left(\frac{4}{3}\right), 12\right)$ and its spherical coordinates are $\left(13, \arcsin\left(\frac{5}{13}\right), \arctan\left(\frac{4}{3}\right)\right)$.

C12S08.023: The graph of the cylindrical equation $r = 5$ is the cylinder of radius 5 with axis the z-axis.

C12S08.025: The graph of the cylindrical *or* spherical equation $\theta = \pi/4$ is the vertical plane with Cartesian equation $y = x$.

C12S08.027: The graph of the spherical equation $\phi = \pi/6$ consists of both nappes of the cone with Cartesian equation $z^2 = 3x^2 + 3y^2$ if ρ may be negative. If $\rho \geq 0$, then the graph consists of the upper nappe alone.

C12S08.029: The graph of the spherical equation $\phi = \pi/2$ is the xy-plane.

C12S08.031: The cylindrical equation $z^2 + 2r^2 = 4$ has the same graph as does the Cartesian equation $2x^2 + 2y^2 + z^2 = 4$. It is an ellipsoid (actually, a prolate spheroid) with intercepts $(\pm\sqrt{2}, 0, 0)$, $(0, \pm\sqrt{2}, 0)$, and $(0, 0, \pm 2)$.

C12S08.033: The graph of the cylindrical equation $r = 4\cos\theta$ is the same as the graph of the Cartesian equation

$$x^2 + y^2 = 4x; \qquad \text{that is,} \qquad (x - 2)^2 + y^2 = 4.$$

Therefore its graph is a circular cylinder parallel to the z-axis, of radius 2, and centerline the line with Cartesian equations $x = 2$, $y = 0$.

C12S08.035: The cylindrical equation $r^2 - 4r + 3 = 0$ can be written in the form

$$(r - 1)(r - 3) = 0,$$

so that $r = 1$ or $r = 3$. The graph consists of all points that satisfy either equation. Hence the graph consists of two concentric circular cylinders, each with axis the z-axis; their radii are 1 and 3.

C12S08.037: The cylindrical equation $z^2 = r^4$ can be written in Cartesian form as

$$z = \pm(x^2 + y^2), \quad \text{so that} \quad z = x^2 + y^2 \quad \text{or} \quad z = -(x^2 + y^2).$$

The graph consists of all points that satisfy either of the last two equations, hence it consists of two circular paraboloids, each with axis the z-axis, vertex at the origin; one opens upward and the other opens downward.

C12S08.039: The Cartesian equation $x^2 + y^2 + z^2 = 25$ has cylindrical form $r^2 + z^2 = 25$ and spherical form $\rho^2 = 25$; that is, $\rho = \pm 5$ (but the graph of $\rho = -5$ coincides with the graph of $\rho = 5$, so $\rho = 5$ is also a correct answer).

C12S08.041: The Cartesian equation $x + y + z = 1$ takes the cylindrical form

$$r\cos\theta + r\sin\theta + z = 1$$

and the spherical form

$$\rho\sin\phi\cos\theta + \rho\sin\phi\sin\theta + \rho\cos\phi = 1.$$

C12S08.043: The Cartesian equation $x^2 + y^2 + z^2 = x + y + z$ takes the cylindrical form

$$r^2 + z^2 = r\cos\theta + r\sin\theta + z$$

665

and the spherical form

$$\rho^2 = \rho \sin \phi \cos \theta + \rho \sin \phi \sin \theta + \rho \cos \phi.$$

The common factor ρ may be cancelled from both sides of the last equation without loss of any points on the graph.

C12S08.045: The surface is the part of the cylinder of radius 3 and centerline the z-axis that lies between the planes $z = -1$ and $z = 1$. The *Mathematica* 3.0 command

```
ParametricPlot3D[ { 3*Cos[t], 3*Sin[t], z }, { t, 0, 2*Pi }, { z, -1, 1 } ];
```

generates the graph, shown next.

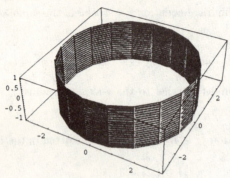

C12S08.047: The surface is the part of the sphere of radius 2 and center the origin that lies between the two horizontal planes $z = -1$ and $z = 1$. The *Mathematica* 3.0 command

```
ParametricPlot3D[ { 2*Sin[phi]*Cos[t], 2*Sin[phi]*Sin[t], 2*Cos[phi] },
            { t, 0, 2*Pi }, { phi, Pi/3, 2*Pi/3 } ];
```

generates the graph, shown next.

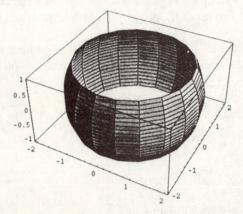

C12S08.049: The solid lies between the horizontal planes $z = -2$ and $z = 2$, is bounded on the outside by the cylinder of radius 3 with centerline the z-axis, and is bounded on the inside by the cylinder of radius 1 with centerline the z-axis. To generate an image, we will create the inner and outer bounding surfaces, the top, and display them simultaneously. For the outer cylinder, we use the *Mathematica* 3.0 command

666

```
f1 = ParametricPlot3D[ { 3*Cos[t], 3*Sin[t], z }, { t, 0, 2*Pi }, { z, -2, 2 } ];
```

For the inner cylinder, we use

```
f2 = ParametricPlot3D[ { 1*Cos[t], 1*Sin[t], z }, { t, 0, 2*Pi }, { z, -2, 2 } ];
```

For the top, we use

```
f3 = ParametricPlot3D[ { r*Cos[t], r*Sin[t], 2 }, { t, 0, 2*Pi }, { r, 1, 3 } ];
```

Finally, the command **Show[f1, f2, f3]** produces an image of the solid. The infamous "out of memory" message prohibits us from displaying it.

C12S08.051: The solid described by the spherical inequality $3 \leq \rho \leq 5$ is the region between two concentric spheres centered at the origin, one of radius 3 and the other of radius 5. We will display the half of this solid for which $y \geq 0$ (otherwise, you wouldn't be able to tell that the solid has a hollow core). We use *Mathematica* 3.0 to draw the outer hemisphere, then the inner hemisphere, then make it appear "solid" by covering the gap between the two with an annulus. Here are the commands:

```
f1 = ParametricPlot3D[ { 5*Sin[phi]*Cos[t], 5*Sin[phi]*Sin[t], 5*Cos[phi] },
                       { phi, 0, Pi }, { t, 0, Pi } ];

f2 = ParametricPlot3D[ { 3*Sin[phi]*Cos[t], 3*Sin[phi]*Sin[t], 3*Cos[phi] },
                       { phi, 0, Pi }, { t, 0, Pi } ];

f3 = ParametricPlot3D[ { r*Cos[t], 0, r*Sin[t] }, { r, 3, 5 }, { t, 0, 2*Pi } ];

Show[ f1, f2, f3 ];
```

The figure is next.

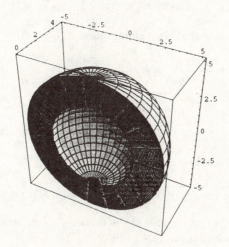

C12S08.053: Given $z = x^2$, a curve in the xz-plane to be rotated around the z-axis, replace x^2 with $x^2 + y^2$ to obtain $z = x^2 + y^2$. In cylindrical coordinates, the equation is $z = r^2$.

C12S08.055: A central cross section of the sphere-with-hole, shown after this solution, makes it clear that the figure is described in cylindrical coordinates by $1 \leq r \leq 2$ and $0 \leq \theta \leq 2\pi$; the third cylindrical coordinate z is determined by the cylindrical equation of the sphere, $r^2 + z^2 = 4$, so that

$$-\sqrt{4-r^2} \le z \le \sqrt{4-r^2}.$$

The figure also makes it clear that the spherical coordinates of the figure satisfy the inequalities $0 \le \theta \le 2\pi$ and $\pi/3 \le \phi \le 2\pi/3$. The value of the third spherical coordinate ρ is $\sec\phi$ on the surface of the cylinder and 2 on the surface of the sphere, and hence $\sec\phi \le \rho \le 2$.

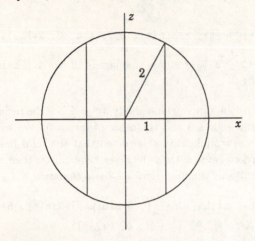

C12S08.057: Take $\rho = 3960$ for the radius of the earth throughout this solution. The spherical coordinates of Fairbanks are then $(\rho,\ \phi,\ \theta)$ where

$$\phi = \frac{7\pi}{50} \quad \text{and} \quad \theta = \frac{4243\pi}{3600}.$$

Then the formulas in (6) yield its Cartesian coordinates:

$$x_1 \approx -1427.537970311664, \quad y_1 \approx -897.229805845151, \quad z_1 \approx 3583.115127765437.$$

Similarly, the angular spherical coordinates of St. Petersburg are

$$\phi = \frac{1003\pi}{6000} \quad \text{and} \quad \theta = \frac{3043\pi}{18000},$$

and hence its Cartesian coordinates are

$$x_2 \approx 1711.895024130100, \quad y_2 \approx 1005.568095522826, \quad z_2 \approx 3426.346192611774.$$

Thus the straight-line distance between the two cities is

$$d = \sqrt{(x_1-x_2)^2 + (y_1-y_2)^2 + (z_1-z_2)^2} \approx 3674.405513694586.$$

Now we use the triangle with vertices at O (the center of the earth), F (Fairbanks), and S (St. Petersburg) and the law of cosines to find the angle α between OF and OS:

$$d^2 = \rho^2 + \rho^2 - 2\rho^2 \cos\alpha;$$

$$\cos\alpha = \frac{2\rho^2 - d^2}{2\rho^2} \approx 0.569519185572;$$

$$\alpha \approx 0.964875538222.$$

Therefore the great circle distance from Fairbanks to St. Petersburg is $\rho\alpha \approx 3820.907131357640$ miles, approximately 6149.153966407630 kilometers.

C12S08.059: Take $r = 3960$ as the radius of the earth. Recall from the solution of Problem 57 that the Cartesian coordinates of Fairbanks are $F(x_1, y_1, z_1)$ where

$$x_1 \approx -1427.537970311664, \quad y_1 \approx -897.229805845151, \quad z_1 \approx 3583.115127765437$$

and that those of St. Petersburg are $S(x_2, y_2, z_2)$ where

$$x_2 \approx 1711.895024130100, \quad y_2 \approx 1005.568095522826, \quad z_2 \approx 3426.346192611774.$$

Thus a normal to the plane $\mathcal{P}$ containing the center of the earth O and the two points F and P is

$$\mathbf{n} = \langle n_1, n_2, n_3 \rangle = \overrightarrow{OF} \times \overrightarrow{OS} = \langle -6677286.184221, \ 11025156.247493, \ 100476.602035 \rangle.$$

The intersection of this plane with the surface of the earth is the great-circle route between F and S. The point on this route closest to the north pole will be the point where the plane $\mathcal{Q}$ containing the z-axis and normal to $\mathcal{P}$ intersects the route. (We are in effect constructing the geodesics from F and P to the north pole and bisecting the angle between them with $\mathcal{Q}$.) Observe that $\mathcal{Q}$ has an equation of the form $Ax + By = 0$, thus a normal of the form $\mathbf{m} = \langle A, B, 0 \rangle$ having the property that $\mathbf{m} \cdot \mathbf{n} = 0$. Indeed, we choose $B = 1$ and solve for A to find that $A \approx -6.056409573098$. Then we asked *Mathematica* to solve for the point $C(x, y, z)$ on the plane's great circle route closest to the north pole:

```
Solve[ { A*x + B*y == 0, n1*x + n2*y + n3*z == 0, x*x + y*y + z*z == r*r },
    { x, y, z } ]
```

The response was $x \approx -6.620560438727$, $y \approx -40.096825620356$, and $z \approx 3959.791460765913$ (and their negatives, which we rejected). Let $N(0, 0, r)$ denote the north pole. The straight-line distance d between C and N is given by

$$d = \sqrt{x^2 + y^2 + (z - r)^2} \approx 40.640260013511.$$

Then the law of cosines will tell us the angle α between OC and ON:

$$d^2 = r^2 + r^2 - 2r^2 \cos\alpha;$$

$$\cos\alpha = \frac{2r^2 - d^2}{2r^2} \approx 0.999947338577;$$

$$\alpha \approx 0.010262736960.$$

Thus the great-circle distance from C to N is $r\alpha \approx 40.640438363451$ (miles).

C12S08.061: The cone has the following description in spherical coordinates:

$$0 \leqq \rho \leqq \sqrt{R^2 + H^2}, \qquad 0 \leqq \theta \leqq 2\pi, \qquad \phi = \arctan\left(\frac{R}{H}\right).$$

C12S08.063: Begin with the equation of the given circle,

$$(y - a)^2 + z^2 = b^2 \qquad (x = 0),$$

669

and replace y with r to get the cylindrical-coordinates equation

$$(r - a)^2 + z^2 = b^2$$

of the torus. Then substitute $\sqrt{x^2 + y^2}$ for r and rationalize to get the rectangular-coordinates equation

$$4a^2(x^2 + y^2) = (x^2 + y^2 + z^2 + a^2 - b^2)^2$$

of the torus. We recognize the combination $x^2 + y^2 = r^2$ and $x^2 + y^2 + z^2 = \rho^2$ and substitute to obtain

$$4a^2 r^2 = (\rho^2 + a^2 - b^2)^2.$$

Finally, we take positive square roots and substitute $\rho \sin \phi$ for r to obtain the spherical-coordinates equation

$$2a\rho \sin \phi = \rho^2 + a^2 - b^2$$

of the torus.

(—C.H.E.)

Chapter 12 Miscellaneous Problems

C12S0M.001: Note that $\overrightarrow{PM} = \overrightarrow{MQ}$. Hence

$$\frac{1}{2}\left(\overrightarrow{AP} + \overrightarrow{AQ}\right) = \frac{1}{2}\left(\overrightarrow{AM} - \overrightarrow{PM} + \overrightarrow{AM} + \overrightarrow{MQ}\right) = \overrightarrow{AM}.$$

C12S0M.003: Given the distinct points P and Q in space, let L be the straight line through both. Then a vector equation of L is $\mathbf{r}(t) = \overrightarrow{OP} + t\overrightarrow{PQ}$. First suppose that R is a point on L. Then for some scalar t,

$$\overrightarrow{OR} = \overrightarrow{OP} + t\overrightarrow{PQ} = \overrightarrow{OP} + t\left(\overrightarrow{OQ} - \overrightarrow{OP}\right) = (1 - t)\overrightarrow{OP} + t\overrightarrow{OQ} = a\overrightarrow{OP} + b\overrightarrow{OQ}$$

where $a = 1 - t$ and $b = t$; moreover, $a + b = 1$, as desired. Next suppose that there exist scalars a and b such that $a + b = 1$ and

$$\overrightarrow{OR} = a\overrightarrow{OP} + b\overrightarrow{OQ}.$$

Let $t = b$, so that $1 - t = a$. Then

$$\overrightarrow{OR} = (1 - t)\overrightarrow{OP} + t\overrightarrow{OQ} = \overrightarrow{OP} + t\left(\overrightarrow{OQ} - \overrightarrow{OP}\right) = \overrightarrow{OP} + t\overrightarrow{PQ}.$$

Therefore R lies on the line L. Consequently an alternative form of the vector equation of L is

$$\mathbf{r}(t) = t\overrightarrow{OP} + (1 - t)\overrightarrow{OQ}.$$

C12S0M.005: Given $P(x_0, y_0)$, $Q(x_1, y_1)$, and $R(x_2, y_2)$, we think of $\overrightarrow{PQ}$ and $\overrightarrow{PR}$ as vectors in space, so that

$$\overrightarrow{PQ} = \langle x_1 - x_0, y_1 - y_0, 0 \rangle \quad \text{and} \quad \overrightarrow{PR} = \langle x_2 - x_0, y_2 - y_0, 0 \rangle.$$

By Eq. (10) in Section 12.3, the area A of the triangle PQR is half the magnitude of $\overrightarrow{PQ} \times \overrightarrow{OR}$. But

$$\overrightarrow{PQ} \times \overrightarrow{PR} = \begin{vmatrix} \mathbf{i} & \mathbf{j} & \mathbf{k} \\ x_1 - x_0 & y_1 - y_0 & 0 \\ x_2 - x_0 & y_2 - y_0 & 0 \end{vmatrix} = \left[(x_1 - x_0)(y_2 - y_0) - (x_2 - x_0)(y_1 - y_0) \right] \mathbf{k}.$$

Therefore

$$A = \frac{1}{2} \left| (x_1 - x_0)(y_2 - y_0) - (x_2 - x_0)(y_1 - y_0) \right|.$$

C12S0M.007: A vector equation of the line is

$$\mathbf{r}(t) = \overrightarrow{OP_1} + t\overrightarrow{P_1P_2} = \langle 1, -1, 2 \rangle + t \langle 2, 3, -3 \rangle.$$

Therefore the line has parametric equations

$$x = 1 + 2t, \quad y = -1 + 3t, \quad z = 2 - 3t, \quad -\infty < t < +\infty$$

and symmetric equations

$$\frac{x - 1}{2} = \frac{y + 1}{3} = \frac{z - 2}{-3}.$$

C12S0M.009: The line L_1 with symmetric equations $x - 1 = 2(y + 1) = 3(z - 2)$ contains the points $P_1(1, -1, 2)$ and $Q_1(7, 2, 4)$, and thus L_1 is parallel to the vector $\mathbf{u}_1 = \langle 6, 3, 2 \rangle$. The line L_2 with symmetric equations $x - 3 = 2(y - 1) = 3(z + 1)$ contains the points $P_2(3, 1, -1)$ and $Q_2(9, 4, 1)$, and thus L_2 is parallel to the vector $\mathbf{u}_2 = \langle 6, 3, 2 \rangle = \mathbf{u}_1$. Therefore L_1 and L_2 are parallel. The vector $\mathbf{v} = \overrightarrow{P_1P_2} = \langle 2, 2, -3 \rangle$ lies in the plane $\mathcal{P}$ containing L_1 and L_2, as does $\mathbf{u}_1$, and therefore a normal to $\mathcal{P}$ is

$$\mathbf{n} = \mathbf{u}_1 \times \mathbf{v} = \begin{vmatrix} \mathbf{i} & \mathbf{j} & \mathbf{k} \\ 6 & 3 & 2 \\ 2 & 2 & -3 \end{vmatrix} = \langle -13, 22, 6 \rangle.$$

Therefore an equation of $\mathcal{P}$ has the form $13x - 22y - 6z = D$. The point $P_1(1, -1, 2)$ lies in the plane $\mathcal{P}$, and therefore $D = 13 \cdot 1 - 22 \cdot (-1) - 6 \cdot 2 = 23$. Therefore an equation of $\mathcal{P}$ is $13x - 22y - 6z = 23$.

C12S0M.011: Given the four points $A(2, 3, 2)$, $B(4, 1, 0)$, $C(-1, 2, 0)$, and $D(5, 4, -2)$, we are to find an equation of the plane containing A and B and parallel to $\overrightarrow{CD} = \langle 6, 2, -2 \rangle$. Because $\overrightarrow{AB} = \langle 2, -2, -2 \rangle$ is not parallel to $\overrightarrow{CD}$, a vector normal to the plane is

$$\mathbf{m} = \overrightarrow{AB} \times \overrightarrow{CD} = \begin{vmatrix} \mathbf{i} & \mathbf{j} & \mathbf{k} \\ 2 & -2 & -2 \\ 6 & 2 & -2 \end{vmatrix} = \langle 8, -8, 16 \rangle;$$

to simplify matters slightly we will use the parallel vector $\mathbf{n} = \langle 1, -1, 2 \rangle$. Therefore the plane has a Cartesian equation of the form $x - y + 2z = K$. Because $A(2, 3, 2)$ lies in the plane, $K = 2 - 3 + 2 \cdot 2 = 3$. Answer: $x - y + 2z = 3$.

C12S0M.013: A vector normal to the plane $\mathcal{P}$ is $\mathbf{n} = \langle a, b, c \rangle$ and $\overrightarrow{OP_0} = \langle x_0, y_0, z_0 \rangle$ connects the origin to $\mathcal{P}$. Hence the distance D from the origin to $\mathcal{P}$ is the length of the projection of $\overrightarrow{OP_0}$ in the direction of $\mathbf{n}$. But the projection of $\overrightarrow{OP_0}$ in the direction of $\mathbf{n}$ is

$$\frac{\overrightarrow{OP_0} \cdot \mathbf{n}}{|\mathbf{n}|} = \frac{ax_0 + by_0 + cz_0}{\sqrt{a^2 + b^2 + c^2}} = \frac{-d}{\sqrt{a^2 + b^2 + c^2}}.$$

Therefore $D = \dfrac{|d|}{\sqrt{a^2 + b^2 + c^2}}$.

C12S0M.015: The planes have the common normal $\mathbf{n} = \langle 2, -1, 2 \rangle$, a point on the first plane $\mathcal{P}_1$ is $P_1(2, 0, 0)$, and a point on the second plane $\mathcal{P}_2$ is $P_2(7, 1, 0)$. So a vector that connects the two planes is $\mathbf{c} = \overrightarrow{P_1 P_2} = \langle 5, 1, 0 \rangle$. So the distance D between $\mathcal{P}_1$ and $\mathcal{P}_2$ is the absolute value of the projection of $\mathbf{c}$ in the direction of $\mathbf{n}$; that is,

$$D = \frac{|\mathbf{n} \cdot \mathbf{c}|}{|\mathbf{n}|} = \frac{9}{3} = 3.$$

C12S0M.017: Given the isosceles triangle with sides AB and AC of equal length and M the midpoint of the third side BC, let $\mathbf{u} = \overrightarrow{BM} = \overrightarrow{MC}$ and let $\mathbf{v} = \overrightarrow{AM}$. Then $\overrightarrow{AB} = \mathbf{v} - \mathbf{u}$ and $\overrightarrow{AC} = \mathbf{v} + \mathbf{u}$. Now

$$|\mathbf{v} - \mathbf{u}| = |\mathbf{v} + \mathbf{u}|, \quad \text{so}$$

$$(\mathbf{v} - \mathbf{u}) \cdot (\mathbf{v} - \mathbf{u}) = (\mathbf{v} + \mathbf{u}) \cdot (\mathbf{v} + \mathbf{u});$$

$$\mathbf{v} \cdot \mathbf{v} - 2\mathbf{v} \cdot \mathbf{u} + \mathbf{u} \cdot \mathbf{u} = \mathbf{v} \cdot \mathbf{v} + 2\mathbf{v} \cdot \mathbf{u} + \mathbf{u} \cdot \mathbf{u}.$$

Therefore $\mathbf{v} \cdot 2\mathbf{u} = 0$, and therefore the segment AM is perpendicular to the segment BC.

C12S0M.019: Given $\mathbf{a}(t) = \langle \sin t, -\cos t \rangle$, the velocity vector is

$$\mathbf{v}(t) = \langle c_1 - \cos t, \ c_2 - \sin t \rangle.$$

But $\mathbf{v}(0) = \langle -1, 0 \rangle = \langle c_1 - 1, c_2 \rangle$, and it follows that $c_1 = c_2 = 0$. So $\mathbf{v}(t) = \langle -\cos t, -\sin t \rangle$. Hence the position vector of the particle is

$$\mathbf{r}(t) = \langle k_1 - \sin t, \ k_2 + \cos t \rangle.$$

But $\langle 0, 1 \rangle = \mathbf{r}(0) = \langle k_1, k_2 + 1 \rangle$, so that $k_1 = k_2 = 0$. Thus $\mathbf{r}(t) = \langle -\sin t, \cos t \rangle$. The parametric equations of the motion of the particle are $x(t) = -\sin t$, $y(t) = \cos t$. Because $x^2 + y^2 = 1$ for all t, the particle moves in a circle of radius 1 centered at the origin.

C12S0M.021: The trajectory of the projectile fired by the gun is given by

$$x_1(t) = (320 \cos \alpha)t, \quad y_1(t) = -16t^2 + (320 \sin \alpha)t.$$

The trajectory of the moving target is given by

$$x_2(t) = 160 + 80t, \quad y_2(t) \equiv 0.$$

We require the angle of elevation α so that, at some time $T > 0$,

$$y_1(T) = 0 \quad \text{and} \quad x_1(T) = x_2(T); \quad \text{that is,}$$

$$-16T^2 + (320\sin\alpha)T = 0 \quad \text{and} \quad (320\cos\alpha)T = 160 + 80T.$$

It follows that $T = 20\sin\alpha$ and, consequently, that

$$6400\sin\alpha\cos\alpha = 160 + 1600\sin\alpha;$$

$$40\sin\alpha\cos\alpha = 1 + 10\sin\alpha.$$

Let $u = \sin\alpha$. Then the last equation yields

$$40u\sqrt{1-u^2} = 1 + 10u;$$

$$1600u^2(1-u^2) = 1 + 20u + 100u^2;$$

$$1600u^4 - 1500u^2 + 20u + 1 = 0.$$

The graph of the last equation shows solutions between 0 and $\pi/2$ near 0.03 and 0.96. Five iterations of Newton's method with these starting values yields the two solutions

$$u_1 \approx 0.0333580866275847 \quad \text{and} \quad u_2 \approx 0.9611546539773131.$$

These correspond to the two angles

$$\alpha_1 \approx 0.033364276320 \quad \text{(about } 1^\circ\ 54'\ 41.876'') \quad \text{and}$$

$$\alpha_2 \approx 1.291155565633 \quad \text{(about } 73^\circ\ 58'\ 39.953'').$$

C12S0M.023: Let $\mathbf{r}(t) = \langle\, t,\ t^2,\ \frac{4}{3}t^{3/2}\,\rangle$. Then

$$\mathbf{v}(t) = \mathbf{r}'(t) = \langle\, 1,\ 2t,\ 2t^{1/2}\,\rangle,$$

so that $\mathbf{v}(1) = \langle\, 1,\ 2,\ 2\,\rangle$. Also

$$\mathbf{a}(t) = \mathbf{r}''(t) = \langle\, 0,\ 2,\ t^{-1/2}\,\rangle,$$

and thus $\mathbf{a}(1) = \langle\, 0,\ 2,\ 1\,\rangle$. Moreover,

$$\mathbf{v}(1) \times \mathbf{a}(1) = \begin{vmatrix} \mathbf{i} & \mathbf{j} & \mathbf{k} \\ 1 & 2 & 2 \\ 0 & 2 & 1 \end{vmatrix} = \langle\, -2,\ -1,\ 2\,\rangle,$$

and thus $|\mathbf{v}(1) \times \mathbf{a}(1)| = 3$. Thus

$$a_T(1) = \frac{\mathbf{v}(1) \cdot \mathbf{a}(1)}{v(1)} = \frac{6}{3} = 2,$$

$$\kappa(1) = \frac{|\mathbf{v}(1) \times \mathbf{a}(1)|}{[v(1)]^3} = \frac{3}{27} = \frac{1}{9}, \quad \text{and}$$

$$a_N(1) = \frac{|\mathbf{v}(1) \times \mathbf{a}(1)|}{v(1)} = \frac{3}{3} = 1.$$

C12S0M.025: A vector normal to the plane is

$$\mathbf{n} = \mathbf{v}_1 \times \mathbf{v}_2 = \begin{vmatrix} \mathbf{i} & \mathbf{j} & \mathbf{k} \\ a_1 & b_1 & c_1 \\ a_2 & b_2 & c_2 \end{vmatrix} = \langle b_1 c_2 - b_2 c_1, \; a_2 c_1 - a_1 c_2, \; a_1 b_2 - a_2 b_1 \rangle.$$

Hence the plane has equation

$$(b_1 c_2 - b_2 c_1)(x - x_0) + (a_2 c_1 - a_1 c_2)(y - y_0) + (a_1 b_2 - a_2 b_1)(z - z_0) = 0.$$

But this is exactly the equation you obtain when the matrix in the equation

$$\begin{vmatrix} x - x_0 & y - y_0 & z - z_0 \\ a_1 & b_1 & c_1 \\ a_2 & b_2 & c_2 \end{vmatrix} = 0$$

given in Problem 25 is expanded along its first row.

C12S0M.027: Let $\mathbf{r}(t) = \langle t, t^2, t^3 \rangle$. Then

$$\mathbf{r}'(t) = \langle 1, 2t, 3t^2 \rangle, \quad \text{so that} \quad \mathbf{r}'(1) = \langle 1, 2, 3 \rangle.$$

Also

$$\mathbf{r}''(t) = \langle 0, 2, 6t \rangle, \quad \text{and so} \quad \mathbf{r}''(1) = \langle 0, 2, 6 \rangle.$$

Thus by the results in Problems 25 and 26, an equation of the osculating plane to the twisted cubic $\mathbf{r}(t)$ at the point $\mathbf{r}(1)$ is

$$\begin{vmatrix} x - 1 & y - 1 & z - 1 \\ 1 & 2 & 3 \\ 0 & 2 & 6 \end{vmatrix} = 0.$$

Expansion of this determinant along its first row yields the equation of the osculating plane in more conventional form:

$$6(x - 1) - 6(y - 1) + 2(z - 1) = 0; \quad \text{that is,} \quad 3x - 3y + z = 1.$$

C12S0M.029: Equation (6) in Section 12.8, with $\rho = 1$, yields

$$x = \sin \phi \cos \theta, \quad y = \sin \phi \sin \theta, \quad z = \cos \phi.$$

Mathematica 3.0 can solve this problem for us.

```
x[t_] := Sin[phi[t]]*Cos[theta[t]]

y[t_] := Sin[phi[t]]*Sin[theta[t]]
```

```
z[t_] := Cos[phi[t]]
term1 = (x'[t])^2
  (Cos[phi[t]] Cos[theta[t]] phi'[t] - Sin[phi[t]] Sin[theta[t]] theta'[t])²
term2 = (y'[t])^2
  (Cos[phi[t]] Sin[theta[t]] phi'[t] + Cos[theta[t]] Sin[phi[t]] theta'[t])²
term3 = (z'[t])^2
  Sin[phi[t]]² phi't]²
Simplify[ term1 + term2 + term3 ]
  phi'[t]² + Sin[phi[t]]² theta'[t]²
```

That is,

$$\left(\frac{dx}{dt}\right)^2 + \left(\frac{dy}{dt}\right)^2 + \left(\frac{dz}{dt}\right)^2 = \left(\frac{d\phi}{dt}\right)^2 + (\sin^2\phi)\left(\frac{d\theta}{dt}\right)^2.$$

Then the result in Problem 13 follows from Eq. (2) in Section 12.6.

C12S0M.031: The helix of Example 7 of Section 12.1 has position vector

$$\mathbf{r}(t) = \langle\, a\cos\omega t,\ a\sin\omega t,\ bt\,\rangle.$$

(Assume that a, b, and ω are all positive.) In that example we found that the velocity and acceleration vectors are

$$\mathbf{v}(t) = \langle\, -a\omega\sin\omega t,\ a\omega\cos\omega t,\ b\,\rangle \quad \text{and} \quad \mathbf{a}(t) = \langle\, -a\omega^2\cos\omega t,\ -a\omega^2\sin\omega t,\ 0\,\rangle.$$

We also found that the speed is given by $v(t) = \sqrt{a^2\omega^2 + b^2}$. Note that $\mathbf{v}(t)\cdot\mathbf{a}(t) = 0$ and that

$$\mathbf{v}(t)\times\mathbf{a}(t) = \begin{vmatrix} \mathbf{i} & \mathbf{j} & \mathbf{k} \\ -a\omega\sin\omega t & a\omega\cos\omega t & b \\ -a\omega^2\cos\omega t & -a\omega^2\sin\omega t & 0 \end{vmatrix} = \langle\, ab\omega^2\sin\omega t,\ -ab\omega^2\cos\omega t,\ a^2\omega^3\,\rangle.$$

By Eq. 26 of Section 12.6, the tangential component of acceleration is zero: $a_T = 0$. Next,

$$|\mathbf{v}(t)\times\mathbf{a}(t)| = \sqrt{a^2b^2\omega^4 + a^4\omega^6} = a\omega^2\sqrt{b^2 + a^2\omega^2}.$$

So by Eq. (28) in Section 12.6, the normal component of acceleration is

$$a_N = \frac{|\mathbf{v}(t)\times\mathbf{a}(t)|}{v(t)} = a\omega^2.$$

Then the unit tangent vector is

$$\mathbf{T}(t) = \frac{\mathbf{v}(t)}{v(t)} = \frac{1}{\sqrt{a^2\omega^2 + b^2}}\,\langle\, -a\omega\sin\omega t,\ a\omega\cos\omega t,\ b\,\rangle.$$

Thus by Eq. (29) of Section 12.6, the principal unit normal vector is

675

$$N(t) = \frac{\mathbf{a} - a_T \mathbf{T}}{a_N} = \frac{1}{a\omega^2} \left(\langle -a\omega^2 \cos \omega t, \ -a\omega^2 \sin \omega t, \ 0 \rangle - 0 \cdot \mathbf{T} \right) = \langle -\cos \omega t, \ -\sin \omega t, \ 0 \rangle.$$

Therefore the unit binormal vector is

$$\mathbf{B} = \mathbf{T} \times \mathbf{N} = \frac{1}{\sqrt{a^2\omega^2 + b^2}} \begin{vmatrix} \mathbf{i} & \mathbf{j} & \mathbf{k} \\ -a\omega \sin \omega t & a\omega \cos \omega t & b \\ -\cos \omega t & -\sin \omega t & 0 \end{vmatrix} = \frac{1}{\sqrt{a^2\omega^2 + b^2}} \langle b \sin \omega t, \ -b \cos \omega t, \ a\omega \rangle.$$

Then

$$\frac{d\mathbf{B}}{ds} = \frac{d\mathbf{B}}{dt} \cdot \frac{dt}{ds} = \frac{1}{v(t)} \cdot \frac{d\mathbf{B}}{dt} = \frac{b\omega}{a^2\omega^2 + b^2} \langle \cos \omega t, \ \sin \omega t, \ 0 \rangle = -\frac{b\omega}{a^2\omega^2 + b^2} \mathbf{N}.$$

Therefore, by definition, the torsion is

$$\tau = \frac{b\omega}{a^2\omega^2 + b^2}.$$

C12S0M.033: The spherical surface with center $(0, 0, 1)$ (in Cartesian coordinates) and radius 1 has Cartesian equation $x^2 + y^2 + (z-1)^2 = 1$; that is,

$$x^2 + y^2 + z^2 = 2z.$$

In spherical coordinates, this equation takes the form $\rho^2 = 2\rho \cos \phi$. No points on the surface are lost by cancellation of ρ from the last equation (take $\phi = \pi/2$), so a slightly simpler spherical equation is $\rho = 2 \cos \phi$.

C12S0M.035: Replace y^2 with $x^2 + y^2$ to obtain a Cartesian equation of the surface of revolution:

$$(x^2 + y^2 + z^2)^2 = 2(z^2 - x^2 - y^2).$$

In cylindrical coordinates this equation takes the form

$$(r^2 + z^2)^2 = 2(z^2 - r^2),$$

and thus its spherical form is

$$\rho^4 = 2\rho^2 (\cos^2 \phi - \sin^2 \phi); \quad \text{that is,} \quad \rho^2 = 2 \cos 2\phi.$$

(No points on the graph are lost by cancellation of ρ^2: Take $\phi = \pi/2$.) The graph of this equation is next.

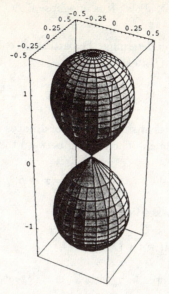

C12S0M.037: If $\mathbf{a} = \langle a_1, a_2, a_3 \rangle$ and $\mathbf{b} = \langle b_1, b_2, b_3 \rangle$, then by Eq. (17) in Section 12.3,

$$\mathbf{i} \cdot (\mathbf{a} \times \mathbf{b}) = \begin{vmatrix} 1 & 0 & 0 \\ a_1 & a_2 & a_3 \\ b_1 & b_2 & b_3 \end{vmatrix} = \begin{vmatrix} a_2 & a_3 \\ b_2 & b_3 \end{vmatrix};$$

similar results hold for $\mathbf{j} \cdot (\mathbf{a} \times \mathbf{b})$ and $\mathbf{k} \cdot (\mathbf{a} \times \mathbf{b})$. Hence by the result in Problem 36,

$$A^2 = \begin{vmatrix} a_2 & a_3 \\ b_2 & b_3 \end{vmatrix}^2 + \begin{vmatrix} a_3 & a_1 \\ b_3 & b_1 \end{vmatrix}^2 + \begin{vmatrix} a_1 & a_2 \\ b_1 & b_2 \end{vmatrix}^2.$$

Rotation of Axes and Second-Degree Curves

In Section 10.6 we studied the second-degree equation

$$Ax^2 + Cy^2 + Dx + Ey + F = 0, \tag{1}$$

which contains no xy-term. We found that the graph is always a conic section, apart from exceptional cases of the following types:

$$2x^2 + 3y^2 = -1 \qquad \text{(no locus),}$$
$$2x^2 + 3y^2 = 0 \qquad \text{(a single point),}$$
$$(2x - 1)^2 = 0 \qquad \text{(a straight line),}$$
$$(2x - 1)^2 = 1 \qquad \text{(two parallel lines),}$$
$$x^2 - y^2 = 0 \qquad \text{(two intersecting lines).}$$

677

We may therefore say that the graph of Eq. (1) is a conic section, possibly **degenerate**. If either A or C is zero (but not both), then the graph is a parabola. It is an ellipse if $AC > 0$, a hyperbola if $AC < 0$ (by results in Section 10.6).

Let us assume that $AC \neq 0$. Then we can determine the particular conic section represented by Eq. (1) by completing squares; that is, we write Eq. (1) in the form

$$A(x - h)^2 + C(y - k)^2 = G. \tag{2}$$

This equation can be simplified further by a **translation of coordinates** to a new $x'y'$-coordinate system centered at the point (h, k) in the old xy-system. The relation between the old and new coordinates is

$$x' = x - h, \quad y' = y - k; \quad \text{that is,} \quad x = x' + h, y = y' + k. \tag{3}$$

In the new $x'y'$-coordinate system, Eq. (2) takes the simpler form

$$A(x')^2 + C(y')^2 = G, \tag{2'}$$

from which it is clear whether we have an ellipse, a hyperbola, or a degenerate case.

We now turn to the general second-degree equation

$$Ax^2 + Bxy + Cy^2 + Dx + Ey + F = 0. \tag{4}$$

Note the presence of the "cross-product," or xy-, term. In order to recognize its graph, we need to change to a new $x'y'$-coordinate system obtained by a **rotation of axes**.

We obtain the $x'y'$-axes from the xy-axes by a rotation through an angle α in the counterclockwise direction. The next figure shows that

$$x = OQ = OP \cos(\phi + \alpha) \quad \text{and} \quad y = PQ = OP \sin(\phi + \alpha). \tag{5}$$

Similarly,

$$x' = OR = OP \cos \phi \quad \text{and} \quad y' = PR = OP \sin \phi. \tag{6}$$

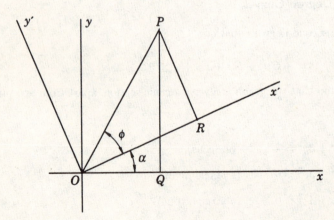

Recall the addition formulas

$$\cos(\phi + \alpha) = \cos\phi\cos\alpha - \sin\phi\sin\alpha,$$

$$\sin(\phi + \alpha) = \sin\phi\cos\alpha + \cos\phi\sin\alpha.$$

With the aid of these identities and the substitution of the equations in (6) into those in (5), we obtain this result:

Equations for Rotation of Axes:

$$x = x'\cos\alpha - y'\sin\alpha,$$

$$y = x'\sin\alpha + y'\cos\alpha. \tag{7}$$

These equations express the old xy-coordinates of the point P in terms of its new $x'y'$-coordinates and the rotation angle α.

Example 1: The xy-axes are rotated through an angle of $\alpha = 45°$. Find the equation of the curve $2xy = 1$ in the new coordinates x' and y'.

Solution: Because $\cos 45° = \sin 45° = \frac{1}{2}\sqrt{2}$, the equations in (7) yield

$$x = \frac{x' - y'}{\sqrt{2}} \quad \text{and} \quad y = \frac{x' + y'}{\sqrt{2}}.$$

The original equation $2xy = 1$ then becomes

$$(x')^2 + (y')^2 = 1.$$

So, in the $x'y'$-coordinate system, we have a hyperbola with $a = b = 1$ (in the notation of Section 10.8), $c = \sqrt{2}$, and foci $(\pm\sqrt{2}, 0)$. In the original xy-coordinate system, its foci are $(1, 1)$ and $(-1, -1)$ and its asymptotes are the x- and y-axes. A hyperbola of this form, one which has equation $xy = k$, is called a **rectangular** hyperbola (because its asymptotes are perpendicular). ◀

Example 1 suggests that the cross-product term Bxy in Eq. (4) may disappear upon rotation of the coordinate axes. One can, indeed, always choose an appropriate angle α of rotation so that, in the new coordinate system, there is no xy-term.

To determine the appropriate rotation angle, we substitute the equations in (7) for x and y into the general second-degree equation in (4). We obtain the following new second-degree equation:

$$A'(x')^2 + B'x'y' + C'(y')^2 + D'x' + E'y' + F' = 0. \tag{8}$$

The new coefficients are given in terms of the old ones and the angle α by the following equations:

$$A' = A\cos^2\alpha + B\cos\alpha\sin\alpha + C\sin^2\alpha,$$

$$B' = B(\cos^2\alpha - \sin^2\alpha) + 2(C - A)\sin\alpha\cos\alpha,$$

$$C' = A\sin^2\alpha - B\sin\alpha\cos\alpha + C\cos^2\alpha,$$

$$D' = D\cos\alpha + E\sin\alpha, \tag{9}$$

$$E' = -D\sin\alpha + E\cos\alpha, \quad \text{and}$$

$$F' = F.$$

Now suppose that an equation of the form in (4) is given, with $B \neq 0$. We simply choose α so that $B' = 0$ in the list of new coefficients in (9). Then Eq. (8) will have no cross-product term, and we can identify and sketch the curve with little trouble in the $x'y'$-coordinate system. But is it really easy to choose such an angle α?

It is. Recall that

$$\cos 2\alpha = \cos^2 \alpha - \sin^2 \alpha \quad \text{and} \quad \sin 2\alpha = 2 \sin \alpha \cos \alpha.$$

So the equation for B' in (9) may be written

$$B' = B \cos 2\alpha + (C - A) \sin 2\alpha.$$

Thus we can cause B' to be zero by choosing α to be that (unique) acute angle such that

$$\cot 2\alpha = \frac{A - C}{B}. \tag{10}$$

If we plan to use the equations in (9) to calculate the coefficients in the transformed Eq. (8), we shall need the values of $\sin \alpha$ and $\cos \alpha$ that follow from Eq. (10). It is sometimes convenient to calculate these values directly from $\cot 2\alpha$, as follows. Draw a right triangle containing an acute angle 2α with opposite side B and adjacent side $A - C$, so that Eq. (10) is satisfied. Then the numerical value of $\cos 2\alpha$ can be read directly from this triangle. Because the cosine and cotangent are both positive in the first quadrant and both negative in the second quadrant, we give $\cos 2\alpha$ the same sign as $\cot 2\alpha$. Then we use the half-angle formulas to find $\sin \alpha$ and $\cos \alpha$:

$$\sin \alpha = \left(\frac{1 - \cos 2\alpha}{2} \right)^{1/2}, \qquad \cos \alpha = \left(\frac{1 + \cos 2\alpha}{2} \right)^{1/2}. \tag{11}$$

Once we have the values of $\sin \alpha$ and $\cos \alpha$, we can compute the coefficients in the resulting Eq. (8) by means of the equations in (9). Alternatively, it's frequently simpler to get Eq. (8) directly by substitution of the equations in (9), with the numerical values of $\sin \alpha$ and $\cos \alpha$ obtained from Eq. (11), into Eq. (4).

Example 2: Determine the graph of the equation

$$73x^2 - 72xy + 52y^2 - 30x - 40y - 75 = 0.$$

Solution: We begin with Eq. (10) and find that $\cot 2\alpha = -\frac{7}{24}$, so that $\cos 2\alpha = -\frac{7}{25}$. Thus

$$\sin \alpha = \left(\frac{1 - \left(-\frac{7}{25}\right)}{2} \right)^{1/2} = \frac{4}{5}, \qquad \cos \alpha = \left(\frac{1 + \left(-\frac{7}{25}\right)}{2} \right)^{1/2} = \frac{3}{5}.$$

Then, with $A = 73$, $B = -72$, $C = 52$, $D = -30$, $E = -40$, and $F = -75$, the equations in (9) yield

$$A' = 25, \qquad\qquad D' = -50,$$

$$B' = 0 \quad \text{(this was the point)}, \qquad E' = 0,$$

$$C' = 100, \qquad\qquad F' = -75.$$

Consequently the equation in the new $x'y'$-coordinate system, obtained by rotation through an angle of $\alpha = \arcsin\left(\frac{4}{5}\right) \approx 53.13°$, is

680

$$25(x')^2 + 100(y')^2 - 50x' = 75.$$

If you prefer, you could obtain this equation by substitution of

$$x = \frac{3}{5}x' - \frac{4}{5}y', \quad y = \frac{4}{5}x' + \frac{3}{5}y'$$

in the original equation.

By completing the square in x' we finally obtain

$$25(x' - 1)^2 + 100(y')^2 = 100,$$

which we put into the standard form

$$\frac{(x' - 1)^2}{4} + \frac{(y')^2}{1} = 1.$$

Thus the original curve is an ellipse with major semiaxis 2, minor semiaxis 1, and center $(1, 0)$ in the $x'y'$-coordinate system. ◄

Example 2 illustrates the general procedure for finding the graph of a second-degree equation. First, if there is a cross-product (xy) term, rotate axes to eliminate it. Then translate axes as necessary to reduce the equation to the standard form of a parabola, ellipse, or hyperbola (or one of the degenerate cases of a conic section).

There *is* a test by which the nature of the curve may be discovered without actually carrying out the transformations described here. This test derives from the fact that, whatever the angle α of rotation, the equations in (9) imply that

$$(B')^2 - 4A'C' = B^2 - 4AC \tag{12}$$

(it is easy to verify this for yourself). Thus the **discriminant** $B^2 - 4AC$ is an *invariant* under any rotation of axes. If α is so chosen that $B' = 0$, then the left-hand side of Eq. (12) is simply $-4A'C'$. Because A' and C' are the coefficients of the squared terms, our earlier discussion of Eq. (1) now applies. It follows that the graph will be:

- a *parabola* if $B^2 - 4AC = 0$,
- an *ellipse* if $B^2 - 4AC < 0$, and
- a *hyperbola* if $B^2 - 4AC > 0$.

Of course, degenerate cases may occur.

Here are some examples:

- $x^2 + 2xy + y^2 = 1$ is a (degenerate) parabola,
- $x^2 + xy + y^2 = 1$ is an ellipse, and
- $x^2 + 3xy + y^2 = 1$ is a hyperbola.

C12S0M.039: First Solution   Without loss of generality, we may assume that the ellipse in the xy-plane is centered at the origin, and by rotation (if necessary) that the plane containing the curve K has equation $z = by + c$ (that is, parallel to the x-axis). Then the ellipse has an equation of the form

$$Ax^2 + Bxy + Cy^2 + D = 0$$

681

where $B^2 - 4AC < 0$ (by the preceding discussion of rotation of axes). Parametrize the plane with a uv-coordinate system that projects vertically onto the xy-coordinate system in the plane:

$$u = x, \qquad v = y\sqrt{1 + b^2} \ .$$

This parametrization preserves distance; that is,

$$\sqrt{x^2 + y^2 + (z - c)^2}$$

on the plane is

$$\sqrt{u^2 + v^2} = \sqrt{x^2 + (1 + b^2)y^2} = \sqrt{x^2 + b^2y^2 + y^2} = \sqrt{x^2 + (z - c)^2 + y^2} \ .$$

Now suppose that the point (x, y, z) is on the intersection of the plane and the cylinder. The equation

$$Ax^2 + Bxy + Cy^2 + D = 0$$

takes the form

$$Au^2 + Bu\frac{v}{\sqrt{1 + b^2}} + C\frac{v^2}{1 + b^2} + D = 0;$$

that is,

$$Au^2 + \frac{B}{\sqrt{1 + b^2}}uv + \frac{C}{1 + b^2}v^2 + D = 0.$$

Is this the equation of an ellipse? We check the discriminant:

$$\left(\frac{B}{\sqrt{1 + b^2}}\right)^2 - \frac{4AC}{1 + b^2} = \frac{B^2 - 4AC}{1 + b^2} < 0.$$

Because the descriminant is negative, the curve K is an ellipse.

Note: If you have access to the version of the sixth edition of the textbook *with early transcendentals functions and matrices*, you can use instead the following much simpler solution of this problem.

C12S0M.039: Second Solution   Without loss of generality, we may suppose that the elliptical cylinder E with vertical sides has equation $z = ax^2 + by^2$ where a and b are both positive. We may also suppose that the nonvertical plane P has equation $z = px + qy + r$ where the coefficients have the property that the intersection C of P and E is a closed curve. Substitution of the equation of P for z in the equation of E shows that C lies in the vertical cylinder with equation

$$ax^2 - px + by^2 - qy = r, \tag{1}$$

and the equation of the vertical projection D of C into the xy-plane has the same equation (together with $z = 0$). The matrix associated with the form in Eq. (1) is

$$\mathbf{A} = \begin{bmatrix} a & 0 \\ 0 & b \end{bmatrix}$$

and has eigenvalues $\lambda_1 = a$ and $\lambda_2 = b$. Hence, in an appropriately rotated and translated uv-coordinate system, D has equation

$$au^2 + bv^2 = s.$$

Because D is a closed curve, $s > 0$; thus D is an ellipse. Therefore, by the result in Problem 38, C itself is also an ellipse.

C12S0M.041: The intersection of $z = Ax + By$ with the ellipsoid

$$\left(\frac{x}{a}\right)^2 + \left(\frac{y}{b}\right)^2 + \left(\frac{z}{c}\right)^2 = 1$$

has the simultaneous equations

$$\left(\frac{x}{a}\right)^2 + \left(\frac{y}{b}\right)^2 + \frac{A^2 x^2 + 2ABxy + B^2 y^2}{c^2} = 1, \quad z = Ax + By.$$

The first of these two equations is the equation of the projection of the intersection into the xy-plane; write it in the form

$$Px^2 + Qxy + Ry^2 = 1$$

and show that its discriminant is negative—this shows that the projection is an ellipse (see the discussion of rotation of axes immediately preceding the solution of Problem 39). It then follows from the result in Problem 38 that the intersection itself must be an ellipse. (Of course, if the plane is tangent to the ellipse or misses it altogether, then the intersection is not an ellipse—it is either empty or else consists of a single point.)

C12S0M.043: By Eq. (13) in Section 12.6, the curvature of $y = \sin x$ at the point $(x, \sin x)$ is

$$\kappa(x) = \frac{|\sin x|}{(1 + \cos^2 x)^{3/2}}.$$

The curvature is minimal when it is zero, and this occurs at every integral multiple of π. To maximize the curvature, note that the numbers that maximize the numerator in the curvature formula—the odd integral multiples of $\pi/2$—also minimize the denominator and thereby maximize the curvature itself.

C12S0M.045: With $\mathbf{r}(t) = \langle t\cos t,\ t\sin t \rangle$, we have

$$\mathbf{v}(t) = \mathbf{r}'(t) = \langle \cos t - t\sin t,\ t\cos t + \sin t \rangle;$$

$$\mathbf{v}(\pi/2) = \langle -\pi/2,\ 1 \rangle;$$

$$v(\pi/2) = \frac{1}{2}\sqrt{\pi^2 + 4}\ ;$$

$$\mathbf{T}(\pi/2) = \frac{1}{\sqrt{\pi^2 + 4}}\langle -\pi,\ 2 \rangle.$$

Because the curve turns left as t increases,

$$\mathbf{N}\left(\frac{\pi}{2}\right) = \frac{1}{\sqrt{\pi^2 + 4}}\langle -2,\ -\pi \rangle.$$

C12S0M.047: We will ask *Mathematica* 3.0 to solve this problem (but we will include intermediate computations so you can check your work).

```
x[t_] := r[t]*Cos[t]
```

```
y[t_] := r[t]*Sin[t]

x'[t]

    -r[t] Sin[t] + Cos[t] r'[t]

y'[t]

    Cos[t] r[t] + Sin[t] r'[t]

x''[t]

    -Cos[t] r[t] - 2 Sin[t] r'[t] + Cos[t] r''[t]

y''[t]

    -r[t] Sin[t] + 2 Cos[t] r'[t] + Sin[t] r''[t]

num = Simplify[ x'[t]*y''[t] - x''[t]*y'[t] ]

    r[t]² + 2 r'[t]² - r[t] r''[t]

den1 = Expand[ (x'[t])² ]

    r[t]² r[t]² - 2 Cos[t] r[t] Sin[t] r'[t] + Cos[t]² r'[t]²

den2 = Expand[ (y'[t])² ]

    Cos[t]² r[t]² + 2 Cos[t] r[t] Sin[t] r'[t] + Sin[t]² r'[t]²

den = Simplify[ den1 + den2 ]

    r[t]² + r'[t]²
```

Finally, when we asked for `Abs[num]/den∧(3/2)`, the response was

$$\frac{|(r(t))^2 + 2(r'(t))^2 - r(t)r''(t)|}{[(r(t))^2 + (r'(t))^2]^{3/2}}.$$

C12S0M.049: The function f is said to be *odd* if $f(-x) = -f(x)$ for all x; f is said to be *even* if $f(-x) = f(x)$ for all x. Because $y(x) = Ax + Bx^3 + Cx^5$ is an odd function, the condition $y(1) = 1$ will imply that $y(-1) = -1$. Because $y'(x) = A + 3Bx^2 + 5Cx^4$ is an even function, the condition $y'(1) = 0$ will imply that $y'(-1) = 0$ as well. Because the graph of y is symmetric around the origin (every odd function has this property), the condition that the curvature is zero at $(1, 1)$ will imply that it is also zero at $(-1, -1)$. By Eq. (13) of Section 12.6, the curvature at x is

$$\kappa(x) = \frac{|6Bx + 20Cx^3|}{\left[1 + (A + 3Bx^2 + 5Cx^4)^2\right]^{3/2}},$$

so the curvature at $(1, 1)$ will be zero when $6B + 20C = 0$. Thus we obtain the simultaneous equations

$$A + B + C = 1,$$

$$A + 3B + 5C = 0,$$

$$3B + 10C = 0.$$

These equations are easy to solve for $A = \frac{15}{8}$, $B = -\frac{5}{4}$, and $C = \frac{3}{8}$. Thus an equation of the connecting curve is

$$y(x) = \frac{15}{8}x - \frac{5}{4}x^3 + \frac{3}{8}x^5.$$

C12S0M.051: Without loss of generality we may suppose that the tetrahedron lies in the first octant with the solid right angle at the origin. Let the coordinates of its other three vertices be $(a, 0, 0)$, $(0, b, 0)$, and $(0, 0, c)$. Then the vectors $\mathbf{a} = \langle a, 0, 0 \rangle$, $\mathbf{b} = \langle 0, b, 0 \rangle$, and $\mathbf{c} = \langle 0, 0, c \rangle$ form three of the edges of the tetrahedron and the other three edges are $\mathbf{b} - \mathbf{a}$, $\mathbf{c} - \mathbf{b}$, and $\mathbf{a} - \mathbf{c}$. The area A of the triangle with these three edges is half the magnitude of the cross product of $\mathbf{b} - \mathbf{a}$ and $\mathbf{c} - \mathbf{a}$, and

$$(\mathbf{b} - \mathbf{a}) \times (\mathbf{c} - \mathbf{a}) = \begin{vmatrix} \mathbf{i} & \mathbf{j} & \mathbf{k} \\ -a & b & 0 \\ -a & 0 & c \end{vmatrix} = \langle bc, ac, ab \rangle.$$

Therefore

$$A^2 = \frac{1}{4}(a^2b^2 + a^2c^2 + b^2c^2).$$

Part (a): The triangular face of the tetrahedron that lies in the xz plane has area $B = \frac{1}{2}ac$, the triangle in the xy plane has area $C = \frac{1}{2}ab$, and the triangle in the yz-plane has area $D = \frac{1}{2}bc$. It follows immediately that

$$A^2 = B^2 + C^2 + D^2.$$

Part (b): This is a generalization of the Pythagorean theorem to three dimensions. How would you generalize it to higher dimensions?

Section 13.2

C13S02.001: Because $f(x, y) = 4 - 3x - 2y$ is defined for all x and y, the domain of f is the entire two-dimensional plane.

C13S02.003: If either x or y is nonzero, then $x^2 + y^2 > 0$, and so $f(x, y)$ is defined—but not if $x = y = 0$. Hence the domain of f consists of all points (x, y) in the plane other than the origin.

C13S02.005: The real number z has a unique cube root $z^{1/3}$ regardless of the value of z. Hence the domain of $f(x, y) = (y - x^2)^{1/3}$ consists of all points in the xy-plane.

C13S02.007: Because $\arcsin z$ is a real number if and only if $-1 \leq z \leq 1$, the domain of the given function $f(x, y) = \sin^{-1}(x^2 + y^2)$ consists of those points (x, y) in the xy-plane for which $x^2 + y^2 \leq 1$; that is, the set of all points on and within the unit circle.

C13S02.009: For every real number z, $\exp(z)$ is defined and unique. Therefore the domain of the given function $f(x, y) = \exp(-x^2 - y^2)$ consists of all points (x, y) in the entire xy-plane.

C13S02.0011: Because $\ln z$ is a unique real number if and only if $z > 0$, then domain of $f(x, y) = \ln(y - x)$ consists of those points (x, y) for which $y > x$. This is the region *above* the graph of the straight line with equation $y = x$ (the line itself is *not* part of the domain of f).

C13S02.013: If x and y are real numbers, then so are xy, $\sin xy$, and $1 + \sin xy$. Hence the only obstruction to computation of

$$f(x, y) = \frac{1 + \sin xy}{xy}$$

is the possibility of division by zero. So the domain of f consists of those points (x, y) for which $xy \neq 0$; that is, all points in the xy-plane other than those on the coordinate axes.

C13S02.015: If x and y are real numbers, then so are xy and $x^2 - y^2$. So the only obstruction to the computation of

$$f(x, y) = \frac{xy}{x^2 - y^2}$$

is the possibility that $x^2 - y^2 = 0$. If so, then $f(x, y)$ is undefined, and therefore the domain of f consists of those points (x, y) for which $x^2 \neq y^2$. That is, the domain of f consists of those points in the xy-plane other than the two straight lines with equations $y = x$ and $y = -x$.

C13S02.017: If w is any real number, then $\exp(w)$ is a unique real number. So the only obstruction to the computation of

$$f(x, y, z) = \exp\left(\frac{1}{x^2 + y^2 + z^2}\right)$$

is the possibility that $x^2 + y^2 + z^2 = 0$. If so, then f is undefined, and therefore its domain consists of all those point (x, y, z) in space other than the origin $(0, 0, 0)$

C13S02.019: If w is a positive real number, then $\ln w$ is a unique real number, but not otherwise. So the domain of $f(x, y, z) = \ln(z - x^2 - y^2)$ consists of those points (x, y, z) in space for which $x^2 + y^2 < z$.

686

These are the points strictly above the circular paraboloid with equation $z = x^2 + y^2$; such a paraboloid is shown in Fig. 12.3.15.

C13S02.021: The graph is the horizontal (parallel to the xy-plane) plane passing through the point $(0, 0, 10)$.

C13S02.023: The graph of $f(x, y) = x + y$ is the plane with equation $z = x + y$. The trace of this plane in the xy-plane is the straight line with equations $z = 0$, $y = -x$. Its trace in the xz-plane is the straight line with equations $y = 0$, $z = x$ and its trace in the yz-plane is the straight line with equations $x = 0$, $z = y$.

C13S02.025: The graph of $z = x^2 + y^2$ is a circular paraboloid with axis the nonnegative z-axis, opening upward, with its vertex at the origin. See Fig. 13.2.4.

C13S02.027: If $z = \sqrt{4 - x^2 - y^2}$, then $z^2 = 4 - x^2 - y^2$ and $z \geq 0$; that is, $x^2 + y^2 + z^2 = 4$ and $z \geq 0$. Consequently the graph of $f(x, y) = \sqrt{4 - x^2 - y^2}$ is a hemisphere—the upper half of the sphere with radius 2 and center $(0, 0, 0)$.

C13S02.029: In cylindrical coordinates, the graph of $f(x, y) = 10 - \sqrt{x^2 + y^2}$ has the equation $z = 10 - r$ where $r \geq 0$. Therefore the graph of f is the lower nappe of a circular cone with axis the z-axis and vertex at $(0, 0, 10)$.

C13S02.031: The level curves of $f(x, y) = x - y$ are the straight lines $x - y = c$ where c is a constant. Some of these are shown next.

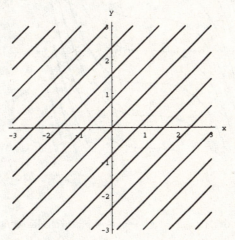

C13S02.033: The level curves of $f(x, y) = 4x^2 + y^2$ are ellipses centered at the origin, with major axes on the x-axis and minor axes on the y-axis. The *Mathematica* 3.0 command

```
ContourPlot[ x*x + 4*y*y, { x, -3, 3 }, { y, -3, 3 }, Axes -> True,

        AxesLabel -> { x, y }, AxesOrigin -> { 0, 0 }, Contours -> 15,

        ContourShading -> False, Frame -> False, PlotPoints -> 47 ];
```

generates some of these level curves; the output resulting from the preceding command is shown next.

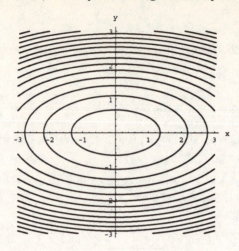

C13S02.035: The level curves of $f(x, y) = y - x^3$ are curves with the equation $y = x^3 + C$ for various values of C. Their inflection points all lie on the y-axis. A few of these level curves are shown next.

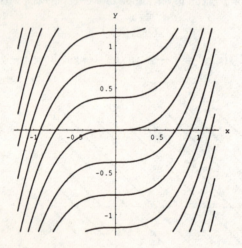

C13S02.037: The level curves of $f(x, y) = x^2 + y^2 - 4x$ are circles centered at the point $(2, 0)$. The *Mathematica* 3.0 command

```
ContourPlot[ x*x + y*y - 4*x, { x, -5, 5 }, { y, -5, 5 }, Axes -> True,

        AxesLabel -> { x, y }, AxesOrigin -> { 0, 0 }, Contours -> 15,

        ContourShading -> False, Frame -> False, PlotPoints -> 47 ];
```

produces several of these level curves, shown next.

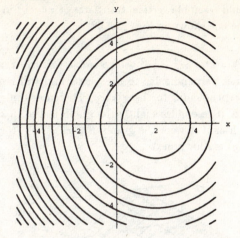

C13S02.039: Because $\exp(-x^2 - y^2)$ is constant exactly when $x^2 + y^2$ is constant, the level curves of $f(x, y) = \exp(-x^2 - y^2)$ are circles centered at the origin. When they are close the graph is steep; when they are far apart the slope of the graph is more moderate. Some level curves of f are shown next.

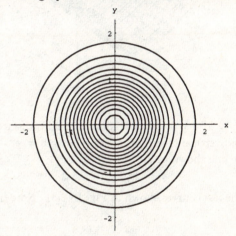

C13S02.041: Because $f(x, y, z) = x^2 + y^2 - z$ is constant when $z = x^2 + y^2 + C$ (where C is a constant), the level surfaces of f are circular paraboloids with axis the z-axis, all opening upward and all having the same shape.

C13S02.043: The function $f(x, y, z) = x^2 + y^2 + z^2 - 4x - 2y - 6z$ is constant when

$$(x - 2)^2 + (y - 1)^2 + (z - 3)^2 = C$$

for some nonnegative constant C. Therefore the level surfaces of f are spherical surfaces all centered at the point $(2, 1, 3)$.

C13S02.045: The function $f(x, y, z) = x^2 + 4y^2 - 4x - 8y + 17$ is constant exactly when $(x - 2)^2 + 4(y - 1)^2$ is a nonnegative constant. Hence the level surfaces of f are elliptical cylinders parallel to the z-axis and centered on the vertical line that meets the xy-plane at the point $(2, 1, 0)$. The ellipse in which each such cylinder meets the xy-plane has major axis parallel to the x-axis, minor axis parallel to the y-axis, and the major axis is twice the length of the minor axis.

C13S02.047: The graph of $f(x, y) = x^3 + y^2$ should resemble vertical translates of $z = x^3$ in planes perpendicular to the y-axis and should resemble vertical translates of $z = y^2$ in planes perpendicular to the x-axis. Hence the graph must be the one shown in Fig. 13.2.32.

C13S02.049: The graph of $f(x, y) = x^3 - 3x^2 + \frac{1}{2}y^2$ will resemble vertical translates of the cubic equation $z = x^3 - 3x^2 = x^2(x - 3)$ in planes perpendicular to the y-axis and will resemble vertical translates of the parabola $z = \frac{1}{2}y^2$ in planes perpendicular to the x-axis. Therefore the graph of f does not appear among Figs. 13.3.27 through 13.3.32. It resembles slightly the graph in Fig.13.2.30, but that graph appears to be linear in the y-direction, much as if it were the graph of $g(x, y) = \frac{1}{4}(5x^3 - 15x + 3y)$ instead. For comparison, the graph of $z = g(x, y)$ is shown next.

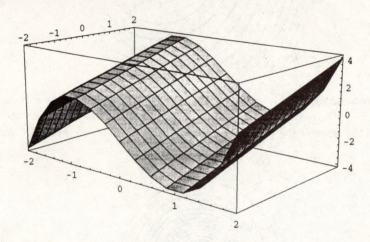

C13S02.051: The graph of $f(x, y) = x^2 + y^4 - 4y^2$ will show vertical translates of the parabola $z = x^2$ in planes perpendicular to the y-axis and vertical translates of the quartic $z = y^4 - 4y^2$ in planes perpendicular to the x-axis. Thus this graph is the one shown in Fig. 13.2.28.

C13S02.053: Figure 13.2.33 has the level curves shown in Fig. 13.2.41. Note how the level curves are close together where the graph is steep.

C13S02.055: Figure 13.2.35 has the level curves shown in Fig. 13.2.42. Moving outward from the center, Fig. 13.2.34 is alternately steep and almost flat; this is reflected in the behavior of the level curves, alternately close and far apart.

C13S02.057: Figure 13.2.57 has the level curves shown in Fig. 13.2.44. The latter indicates three peaks in a row, or two peaks with a deep hole between them, or some variation of this idea.

C13S02.059: The *Mathematica* 3.0 commands

```
a = 2; b = 1;

Plot3D[ (a*x + b*y)*Exp[ -x^2 - y^2 ], { x, -3, 3 }, { y, -3, 3 },
    PlotRange → { -1, 1 }, PlotPoints → { 25, 25 },
    AxesLabel → { "x", "y", "z" } ];
```

produce the graph shown next.

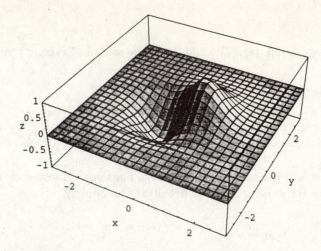

Plots with various values of a and b (not both zero) indicate that the surface always has one pit and one peak, both lying on the same straight line in the xy-plane. The values of a and b determine the orientation of this line and the distances of the pit and the peak from the origin.　　　　—C.H.E.

C13S02.061: The *Mathematica* 3.0 commands

```
n = 3;
x = r*Cos[t]; y = r*Sin[t];
z = (r^2)*(Sin[ n*t ])*Exp[ -r^2 ];
ParametricPlot3D[ { x, y, z }, { r, 0, 3 }, { t, 0, 2*Pi },
   PlotPoints → { 30, 30 }, PlotRange → { -1, 1 },
   BoxRatios → { 1, 1, 1 } ];
```

produce the following graph.

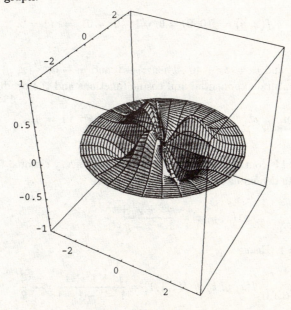

Apparently n peaks and n pits alternately surround the origin. —C.H.E.

Section 13.3

C13S03.001: The limit laws in Eqs. (5), (6), and (7) and the result in Example 4 yield

$$\lim_{(x,y)\to(0,0)} \left(7 - x^2 + 5xy\right)$$

$$= \left(\lim_{(x,y)\to(0,0)} 7\right) - \left(\lim_{(x,y)\to(0,0)} x\right)^2 + 5\left(\lim_{(x,y)\to(0,0)} x\right) \cdot \left(\lim_{(x,y)\to(0,0)} y\right) = 7 - 0^2 + 5\cdot 0\cdot 0 = 7.$$

C13S03.003: The product and composition of continuous functions is continuous where defined, hence $f(x, y) = e^{-xy}$ is continuous everywhere. Then, by definition of continuity,

$$\lim_{(x,y)\to(1,-1)} f(x, y) = f(1, -1) = e^1 = e.$$

C13S03.005: The sum, product, and quotient of continuous functions is continuous where defined, so

$$f(x, y) = \frac{5 - x^2}{3 + x + y}$$

is continuous where $x + y \neq -3$. Therefore

$$\lim_{(x,y)\to(0,0)} f(x, y) = \frac{5 - 0^2}{3 + 0 + 0} = \frac{5}{3}.$$

C13S03.007: The sum, product, and composition of continuous functions is continuous where defined, so

$$f(x, y) = \ln\sqrt{1 - x^2 - y^2}$$

is continuous if $x^2 + y^2 < 1$. Therefore

$$\lim_{(x,y)\to(0,0)} f(x, y) = f(0, 0) = \ln\sqrt{1 - 0^2 - 0^2} = \ln 1 = 0.$$

C13S03.009: Let $z = x + 2y$ and $w = 3x + 4y$. Then $z \to 0$ and $w \to 0$ as $(x, y) \to (0, 0)$ by Example 4. Hence, by continuity of the natural exponential and cosine functions and the product law for limits,

$$\lim_{(x,y)\to(0,0)} e^{x+2y}\cos(3x + 4y) = e^0\cos 0 = 1\cdot 1 = 1.$$

C13S03.011: Every polynomial is continuous everywhere, and hence every rational function (even of three variables) is continuous where its denominator is not zero. Therefore

$$f(x, y, z) = \frac{x^2 + y^2 + z^2}{1 - x - y - z}$$

is continuous where $x + y + z \neq 1$. Hence

$$\lim_{(x,y,z)\to(1,1,1)} f(x, y, z) = f(1,1,1) = \frac{1^2 + 1^2 + 1^2}{1 - 1 - 1 - 1} = -\frac{3}{2}.$$

692

C13S03.013: The sum, product, quotient, and composition of continuous functions is continuous where defined. Hence

$$f(x,\,y,\,z) = \frac{xy - z}{\cos xyz}$$

is continuous provided that xyz is not an odd integral multiple of $\pi/2$. Therefore

$$\lim_{(x,y,z)\to(1,1,0)} f(x,\,y,\,z) = f(1,1,0) = \frac{1\cdot 1 - 0}{\cos 0} = \frac{1}{1} = 1.$$

C13S03.015: First,

$$f(z) = \tan\frac{3\pi z}{4}$$

is continuous provided that $3\pi z/4$ is not an odd integral multiple of $\pi/2$; that is, provided that z is not two-thirds of an odd integer. Hence $f(z)$ is continuous at $z = 1$. Next,

$$g(x,\,y) = \sqrt{xy}$$

is the composition of continuous functions, so $g(x,\,y)$ is continuous provided that $xy > 0$. So g is continuous at $(2, 8)$. Finally, the product of continuous functions is continuous, so $h(x,\,y,\,z) = g(x,\,y)\cdot f(z)$ is continuous at $(2, 8, 1)$. Therefore

$$\lim_{(x,y,z)\to(2,8,1)} h(x,\,y,\,z) = h(2,8,1) = 16^{1/2}\cdot\tan\frac{3\pi}{4} = 4\cdot(-1) = -4.$$

C13S03.017: If $f(x,\,y) = xy$, then

$$\lim_{h\to 0}\frac{f(x+h,\,y) - f(x,\,y)}{h} = \lim_{h\to 0}\frac{xy + hy - xy}{h} = \lim_{h\to 0}\frac{hy}{h} = y \quad\text{and}$$

$$\lim_{k\to 0}\frac{f(x,\,y+k) - f(x,\,y)}{k} = \lim_{k\to 0}\frac{xy + kx - xy}{k} = \lim_{k\to 0}\frac{kx}{k} = x.$$

C13S03.019: If $f(x,\,y) = xy^2 - 2$, then

$$\lim_{h\to 0}\frac{f(x+h,\,y) - f(x,\,y)}{h} = \lim_{h\to 0}\frac{(x+h)y^2 - 2 - xy^2 + 2}{h} = \lim_{h\to 0}\frac{hy^2}{h} = y^2 \quad\text{and}$$

$$\lim_{k\to 0}\frac{f(x,\,y+k) - f(x,\,y)}{k} = \lim_{k\to 0}\frac{x(y+k)^2 - 2 - xy^2 + 2}{k} = \lim_{k\to 0}\frac{2kxy + k^2 x}{k} = 2xy.$$

C13S03.021: $\displaystyle\lim_{(x,y)\to(1,1)}\frac{1 - xy}{1 + xy} = \frac{1 - 1\cdot 1}{1 + 1\cdot 1} = \frac{0}{2} = 0.$

C13S03.023: $\displaystyle\lim_{(x,y,z)\to(1,1,1)}\frac{xyz}{yz + xz + xy} = \frac{1\cdot 1\cdot 1}{1\cdot 1 + 1\cdot 1 + 1\cdot 1} = \frac{1}{3}.$

C13S03.025: $\displaystyle\lim_{(x,y)\to(0,0)}\ln(1 + x^2 + y^2) = \ln(1 + 0 + 0) = \ln 1 = 0.$

C13S03.027: Convert to polar coordinates, then replace r^2 with z. Thus

$$\lim_{(x,y)\to(0,0)} \frac{\cot(x^2+y^2)}{x^2+y^2} = \lim_{r\to0} \frac{\cot r^2}{r^2} = \lim_{z\to0^+} \frac{\cos z}{z\sin z},$$

which does not exist because $\cos z \to 1$ as $z \to 0^+$, but $z\sin z \to 0$. Because the last numerator is approaching $+\infty$ as $z \to 0^+$ but the last denominator is approaching zero through *positive* values, it is also correct to write

$$\lim_{(x,y)\to(0,0)} \frac{\cot(x^2+y^2)}{x^2+y^2} = +\infty.$$

C13S03.029: Convert to polar coordinates. Then

$$\lim_{(x,y)\to(0,0)} \exp\left(-\frac{1}{x^2+y^2}\right) = \lim_{r\to0} \exp\left(-\frac{1}{r^2}\right) = \lim_{z\to\infty} e^{-z} = 0.$$

C13S03.031: For continuity of $f(x,y) = \sqrt{x+y}$, we require $x+y > 0$. Thus f is continuous on the set of all points (x,y) that lie strictly above the graph of $y = -x$.

C13S03.033: For continuity of $f(x,y) = \ln(x^2+y^2-1)$, we require that $x^2+y^2-1 > 0$; that is, that $x^2+y^2 > 1$. So f will be continuous on the points strictly outside the unit circle centered at $(0,0)$; that is, strictly outside the circle with equation $x^2+y^2 = 1$.

C13S03.035: Because the inverse tangent function is continuous on the set of all real numbers, the only discontinuity of

$$f(x,y) = \arctan\left(\frac{1}{x^2+y^2}\right)$$

will occur when the denominator in the fraction is zero. Hence f is continuous on the set of all points in the xy-plane other than $(0,0)$. This discontinuity is removable because conversion to polar coordinates shows that

$$\lim_{(x,y)\to(0,0)} \arctan\left(\frac{1}{x^2+y^2}\right) = \lim_{r\to0^+} \arctan\left(\frac{1}{r}\right) = \frac{\pi}{2}.$$

C13S03.037: Using polar coordinates yields

$$\lim_{(x,y)\to(0,0)} \frac{x^2-y^2}{\sqrt{x^2+y^2}} = \lim_{r\to0} \frac{r^2\cos^2\theta - r^2\sin^2\theta}{r} = \lim_{r\to0} \frac{r^2\cos2\theta}{r} = \lim_{r\to0} r\cos2\theta = 0.$$

C13S03.039: Use of polar coordinates yields

$$\lim_{(x,y)\to(0,0)} \frac{x^4+y^4}{(x^2+y^2)^{3/2}} = \lim_{r\to0} \frac{r^4\cos^4\theta + r^4\sin^4\theta}{r^3} = \lim_{r\to0} r(\cos^4\theta + \sin^4\theta) = 0.$$

C13S03.041: Using spherical coordinates yields

$$\lim_{(x,y,z)\to(0,0,0)} \frac{xyz}{x^2+y^2+z^2} = \lim_{\rho\to0} \frac{\rho^3(\sin^2\phi\cos\phi\sin\theta\cos\theta)}{\rho^2} = \lim_{\rho\to0} \rho(\sin^2\phi\cos\phi\sin\theta\cos\theta) = 0.$$

C13S03.043: The substitution $y = mx$ yields

$$\lim_{(x,y)\to(0,0)} \frac{x^2 - y^2}{x^2 + y^2} = \lim_{x\to 0} \frac{x^2(1 - m^2)}{x^2(1 + m^2)} = \frac{1 - m^2}{1 + m^2}.$$

This is the limit as $(x, y) \to (0, 0)$ along the straight line of slope m. Different values of m (such as 0 and 1) give different values for the limit (such as 1 and 0), and therefore the original limit does not exist (see the Remark following Example 9).

C13S03.045: If $(x, y, z) \to (0, 0, 0)$ along the positive x-axis—where $y = z = 0$—we obtain

$$\lim_{x\to 0^+} \frac{x + y + z}{x^2 + y^2 + z^2} = \lim_{x\to 0^+} \frac{1}{x} = +\infty.$$

Therefore the original limit does not exist.

C13S03.047: The graph that follows this solution suggests that as $(x, y) \to (0, 0)$ along straight lines of various slopes, the values of

$$\lim_{(x,y)\to(0,0)} \frac{x^2 - 2y^2}{x^2 + y^2}$$

range from -2 to 1, so that the limit in question does not exist. To be certain that it does not, let (x, y) approach $(0, 0)$ along the x-axis to get limit 1, then along the y-axis to get limit -2. The figure was generated by executing the *Mathematica* 3.0 command

```
ParametricPlot3D[ { r*Cos[t], r*Sin[t], (Cos[t])^2 - 2*(Sin[t])^2 },
        { t, 0, 2*Pi }, { r, 0.01, 2 } ];
```

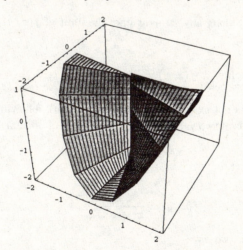

C13S03.049: The graph that follows this solution suggests that as $(x, y) \to (0, 0)$ along straight lines of various slopes, the values of

$$\lim_{(x,y)\to(0,0)} \frac{xy}{2x^2 + 3y^2}$$

695

range from -0.2 to 0.2, so that the limit in question does not exist. To be sure that it does not, let (x, y) approach $(0, 0)$ along the line $y = x$ to get limit 0.2, then long the line $y = -x$ to get limit -0.2.

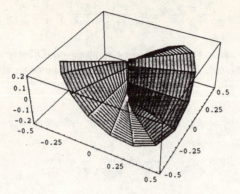

C13S03.051: Given:

$$f(x, y) = \frac{2x^2 y}{x^4 + y^2}.$$

Suppose that (x, y) approaches $(0, 0)$ along the nonvertical straight line with equation $y = mx$. If $m \neq 0$, then—on that line—

$$\lim_{(x,y)\to(0,0)} f(x, y) = \lim_{x\to 0} \frac{2mx^3}{x^4 + m^2 x^2} = \lim_{x\to 0} \frac{2mx}{x^2 + m^2} = \frac{0}{m^2} = 0.$$

Clearly if $m = 0$ the result is the same. And if (x, y) approaches $(0, 0)$ along the y-axis, then—on that line—

$$\lim_{(x,y)\to(0,0)} f(x, y) = \lim_{y\to 0} \frac{2 \cdot 0 \cdot y}{0 + y^2} = 0.$$

Therefore as (x, y) approaches $(0, 0)$ along any straight line, the limit of $f(x, y)$ is zero. But on the curve $y = x^2$ we have

$$\lim_{(x,y)\to(0,0)} f(x, y) = \lim_{x\to 0} \frac{2x^4}{x^4 + x^4} = 1,$$

and therefore the limit of $f(x, y)$ does not exist at $(0, 0)$ (see the Remark following Example 9). For related paradoxical results involving functions of two variables, see Problem 60 of Section 13.5 and the miscellaneous problems of Chapter 13.

C13S03.053: Given:

$$f(x, y) = \frac{xy}{x^2 + y^2}.$$

After we convert to polar coordinates, we have

$$f(r, \theta) = \frac{r^2 \cos \theta \sin \theta}{r^2} = \cos \theta \sin \theta = \frac{1}{2} \sin 2\theta.$$

On the hyperbolic spiral $r\theta = 1$, we have $\theta \to +\infty$ as r approaches zero through positive values. Hence $f(r, \theta)$ takes on all values between $-\frac{1}{2}$ and $\frac{1}{2}$ infinitely often as $r \to 0^+$. Therefore, as we discovered in Example 9, $f(x, y)$ has no limit as $(x, y) \to (0, 0)$.

696

C13S03.055: Let $z = x^2 - y^2$. By the discussions following Examples 4 and 5 and the "basic trigonometric limit" (Theorem 1 in Section 2.3),

$$\lim_{(x,y)\to(0,0)} \frac{\sin(x^2 - y^2)}{x^2 - y^2} = \lim_{z\to 0} \frac{\sin z}{z} = 1 = f(0, 0).$$

Therefore f is continuous at $(0, 0)$. By the discussion following Example 5, f is continuous at every other point of the xy-plane. Therefore f is continuous everywhere.

Section 13.4

C13S04.001: If $f(x, y) = x^4 - x^3y + x^2y^2 - xy^3 + y^4$, then

$$\frac{\partial f}{\partial x} = 4x^3 - 3x^2y + 2xy^2 - y^3 \quad \text{and}$$

$$\frac{\partial f}{\partial y} = -x^3 + 2x^2y - 3xy^2 + 4y^3.$$

C13S04.003: If $f(x, y) = e^x(\cos y - \sin y)$, then

$$\frac{\partial f}{\partial x} = e^x(\cos y - \sin y) \quad \text{and} \quad \frac{\partial f}{\partial y} = -e^x(\cos y + \sin y).$$

C13S04.005: If $f(x, y) = \dfrac{x + y}{x - y}$, then

$$\frac{\partial f}{\partial x} = -\frac{2y}{(x - y)^2} \quad \text{and} \quad \frac{\partial f}{\partial y} = \frac{2x}{(x - y)^2}.$$

C13S04.007: If $f(x, y) = \ln(x^2 + y^2)$, then

$$\frac{\partial f}{\partial x} = \frac{2x}{x^2 + y^2} \quad \text{and} \quad \frac{\partial f}{\partial y} = \frac{2y}{x^2 + y^2}.$$

C13S04.009: If $f(x, y) = x^y$, then

$$\frac{\partial f}{\partial x} = yx^{y-1} \quad \text{and} \quad \frac{\partial f}{\partial y} = x^y \ln x.$$

C13S04.011: If $f(x, y, z) = x^2y^3z^4$, then

$$\frac{\partial f}{\partial x} = 2xy^3z^4, \quad \frac{\partial f}{\partial y} = 3x^2y^2z^4, \quad \text{and} \quad \frac{\partial f}{\partial z} = 4x^2y^3z^3.$$

C13S04.013: If $f(x, y, z) = e^{xyz}$, then

$$\frac{\partial f}{\partial x} = yze^{xyz}, \quad \frac{\partial f}{\partial y} = xze^{xyz}, \quad \text{and} \quad \frac{\partial f}{\partial z} = xye^{xyz}.$$

C13S04.015: If $f(x, y, z) = x^2e^y \ln z$, then

$$\frac{\partial f}{\partial x} = 2xe^y \ln z, \qquad \frac{\partial f}{\partial y} = x^2 e^y \ln z, \qquad \text{and} \qquad \frac{\partial f}{\partial z} = \frac{x^2 e^y}{z}.$$

C13S04.017: If $f(r,\, s) = \dfrac{r^2 - s^2}{r^2 + s^2}$. then

$$\frac{\partial f}{\partial r} = \frac{4rs^2}{(r^2 + s^2)^2} \qquad \text{and} \qquad \frac{\partial f}{\partial s} = -\frac{4r^2 s}{(r^2 + s^2)^2}.$$

C13S04.019: If $f(u,\, v,\, w) = ue^v + ve^w + we^u$, then

$$\frac{\partial f}{\partial u} = we^u + e^v, \qquad \frac{\partial f}{\partial v} = ue^v + e^w, \qquad \text{and} \qquad \frac{\partial f}{\partial w} = e^u + ve^w.$$

C13S04.021: If $z(x,\, y) = x^2 - 4xy + 3y^2$, then

$$z_x(x,\, y) = 2x - 4y, \qquad z_y(x,\, y) = -4x + 6y,$$

$$z_{xy}(x,\, y) = -4, \qquad z_{yx}(x,\, y) = -4.$$

C13S04.023: If $z(x,\, y) = x^2 \exp(-y^2)$, then

$$z_x(x,\, y) = 2x \exp(-y^2), \qquad z_y(x,\, y) = -2x^2 y \exp(-y^2),$$

$$z_{xy}(x,\, y) = -4xy \exp(-y^2), \qquad z_{yx}(x,\, y) = -4xy \exp(-y^2).$$

C13S04.025: If $z(x,\, y) = \ln(x + y)$, then

$$z_x(x,\, y) = \frac{1}{x + y} = z_y(x,\, y) \qquad \text{and} \qquad z_{xy}(x,\, y) = -\frac{1}{(x + y)^2} = z_{yx}(x,\, y).$$

C13S04.027: If $z(x,\, y) = e^{-3x} \cos y$, then

$$z_x(x,\, y) = -3e^{-3x} \cos y, \qquad z_y(x,\, y) = -e^{-3x} \sin y,$$

$$z_{xy}(x,\, y) = 3e^{-3x} \sin y, \qquad z_{yx}(x,\, y) = 3e^{-3x} \sin y.$$

C13S04.029: If $z(x,\, y) = x^2 \cosh\left(\dfrac{1}{y^2}\right)$, then

$$z_x(x,\, y) = 2x \cosh\left(\frac{1}{y^2}\right), \qquad z_y(x,\, y) = -\frac{2x^2}{y^3} \sinh\left(\frac{1}{y^2}\right),$$

$$z_{xy}(x,\, y) = -\frac{4x}{y^3} \sinh\left(\frac{1}{y^2}\right), \qquad z_{yx}(x,\, y) = -\frac{4x}{y^3} \sinh\left(\frac{1}{y^2}\right).$$

C13S04.031: Given: $f(x,\, y) = x^2 + y^2$ and the point $P(3,\, 4,\, 25)$ on its graph. Then

$$f_x(x, y) = 2x; \qquad f_x(3, 4) = 6;$$

$$f_y(x, y) = 2y; \qquad f_y(3, 4) = 8.$$

Hence by Eq. (11) of Section 13.4, an equation of the plane tangent to the graph of $z = f(x, y)$ at the point P is

$$z - 25 = 6(x - 3) + 8(y - 4); \qquad \text{that is,} \quad 6x + 8y - z = 25.$$

C13S04.033: Given: $f(x, y) = \sin \dfrac{\pi xy}{2}$ and the point $P(3, 5, -1)$ on its graph. Then

$$f_x(x, y) = \frac{\pi y}{2} \cos \frac{\pi xy}{2}; \qquad f_y(x, y) = \frac{\pi x}{2} \cos \frac{\pi xy}{2};$$

$$f_x(3, 5) = 0; \qquad f_y(3, 5) = 0.$$

The plane tangent to the graph of $z = f(x, y)$ at the point P is horizontal, so its has equation $z = -1$.

C13S04.035: Given: $f(x, y) = x^3 - y^3$ and the point $P(3, 2, 19)$ on its graph. Then $f_x(x, y) = 3x^2$ and $f_y(x\,y) = -3y^2$, so $f_x(3, 2) = 27$ and $f_y(3, 2) = -12$. So by Eq. (11) an equation of the plane tangent to the graph of $z = f(x, y)$ at the point P is

$$z - 19 = 27(x - 3) - 12(y - 2); \qquad \text{that is,} \quad 27x - 12y - z = 38.$$

C13S04.037: Given: $f(x, y) = xy$ and the point $P(1, -1, -1)$ on its graph. Then $f_x(x, y) = y$ and $f_y(x, y) = x$, so $f_x(1, -1) = -1$ and $f_y(1, -1) = 1$. By Eq. (11) an equation of the plane tangent to the graph of $z = f(x, y)$ at the point P is $z + 1 = -(x - 1) + (y - 1)$; that is, $x - y + z = 1$.

C13S04.039: Given: $f(x, y) = x^2 - 4y^2$ and the point $P(5, 2, 9)$ on its graph. Then $f_x(x, y) = 2x$ and $f_y(x, y) = -8y$, so $f_x(5, 2) = 10$ and $f_y(5, 2) = -16$. By Eq. (11) an equation of the plane tangent to the graph of $z = f(x, y)$ at the point P is

$$z - 9 = 10(x - 5) - 16(y - 2); \qquad \text{that is,} \quad 10x - 16y - z = 9.$$

C13S04.041: If $f_x(x, y) = 2xy^3$ and $f_y(x, y) = 3x^2y^2$, then

$$f_{xy}(x, y) = 6xy^2 = f_{yx}(x, y).$$

In Section 15.3 we will find that there must exist a function f having the given partial derivatives. Here we can find one by inspection; it is $f(x, y) = x^2y^3$.

C13S04.043: If $f_x(x, y) = \cos^2 xy$ and $f_y(x, y) = \sin^2 xy$, then

$$f_{xy}(x, y) = -2x \sin xy \, \cos xy \neq -2y \sin xy \, \cos xy = f_{yx}(x, y).$$

By the *Note* preceding and following Eq. (14), there can be no function $f(x, y)$ having the given first-order partial derivatives.

C13S04.045: The graph of $z = f(x, y)$ is shown in Fig. 13.4.14. The key to solving this group of six problems is first to locate f. Do this by sketching cross sections of each graph parallel to the xz- and

yz-planes. This will make it easy to eliminate all but one candidate for the graph of f. Then the same sketches—the ones you created for the graph that turned out to be the graph of f—will quickly identify the other five graphs.

C13S04.047: The graph of $z = f_y(x, y)$ is shown in Fig. 13.4.13.

C13S04.049: The graph of $z = f_{xy}(x, y)$ is shown in Fig. 13.4.15.

C13S04.051: If m and n are positive integers and $f(x, y) = x^m y^n$, then $f_x(x, y) = mx^{m-1}y^n$ and $f_y(x, y) = nx^m y^{n-1}$. Hence $f_{xy}(x, y) = mnx^{m-1}y^{n-1} = f_{yx}(x, y)$.

C13S04.053: If $f(x, y, z) = e^{xyz}$, then

$$f_x(x, y, z) = yze^{xyz}, \quad f_y(x, y, z) = xze^{xyz}, \quad \text{and} \quad f_z(x, y, z) = xye^{xyz}.$$

Therefore

$$f_{xx}(x, y, z) = y^2 z^2 e^{xyz}, \quad f_{xy}(x, y, z) = f_{yx}(x, y, z) = (xyz^2 + z)e^{xyz},$$

$$f_{xz}(x, y, z) = f_{zx}(x, y, z) = (y + xy^2 z)e^{xyz}, \quad f_{yz}(x, y, z) = f_{zy}(x, y, z) = (x + x^2 yz)e^{xyz},$$

$$f_{yy}(x, y, z) = x^2 z^2 e^{xyz}, \quad f_{zz}(x, y, z) = x^2 y^2 e^{xyz}, \quad \text{and}$$

$$f_{xyz}(x, y, z) = (1 + 3xyz + x^2 y^2 z^2)e^{xyz}.$$

C13S04.055: If $u(x, t) = \exp(-n^2 kt)\sin nx$ where n and k are constants, then

$$u_t(x, t) = -n^2 k \exp(-n^2 kt)\sin nx, \quad u_x(x, t) = n\exp(-n^2 kt)\cos nx,$$

$$\text{and} \quad u_{xx}(x, t) = -n^2 \exp(-n^2 kt)\sin nx.$$

Therefore $u_t = ku_{xx}$ for any choice of the constants k and n.

C13S04.057: Part (a): If $y(x, t) = \sin(x + at)$ (where a is a constant), then

$$y_t(x, t) = a\cos(x + at), \qquad y_x(x, t) = \cos(x + at),$$

$$y_{tt}(x, t) = -a^2 \sin(x + at), \qquad y_{xx}(x, t) = -\sin(x + at).$$

Therefore $y_{tt} = a^2 y_{xx}$.

Part (b): If $y(x, t) = \cosh\big(3(x - at)\big)$, then

$$y_t(x, t) = -3a\sinh\big(3(x - at)\big), \qquad y_x(x, t) = 3\sinh\big(3(x - at)\big),$$

$$y_{tt}(x, t) = 9a^2 \cosh\big(3(x - at)\big), \qquad y_{xx}(x, t) = 9\cosh\big(3(x - at)\big).$$

Therefore $y_{tt} = a^2 y_{xx}$.

Part (c): If $y(x, t) = \sin kx \cos kat$ (where k is a constant), then

$$y_t(x,\, t) = -ka \sin kx \sin kat, \qquad y_x(x,\, t) = k \cos kx \cos kat,$$

$$y_{tt}(x,\, t) = -k^2 a^2 \sin kx \cos kat, \qquad y_{xx}(x,\, t) = -k^2 \sin kx \cos kat.$$

Therefore $y_{tt} = a^2 y_{xx}$.

C13S04.059: Given: f and g are twice-differentiable functions of a single variable, a is a constant, and $y(x,\, t) = f(x + at) + g(x - at)$. Then

$$y_t(x,\, t) = af'(x + at) - ag'(x - at), \qquad y_{tt}(x,\, t) = a^2 f''(x + at) + a^2 g''(x - at),$$

$$y_x(x,\, t) = f'(x + at) + g'(x - at), \qquad y_{xx}(x,\, t) = f''(x + at) + g''(x - at).$$

It's now clear that $y(x,\, t)$ satisfies the wave equation $y_{tt} = a^2 y_{xx}$.

C13S04.061: Given:

$$u(x,\, t) = T_0 + a_0 \exp\left(-x\sqrt{\omega/2k}\,\right) \cos\left(\omega t - x\sqrt{\omega/2k}\,\right)$$

where T_0, a_0, ω, and k are constants. First note that

$$u(0,\, t) = T_0 + a_0 e^0 \cos(\omega t - 0) = T_0 + a_0 \cos \omega t.$$

Next,

$$u_t(x,\, t) = -a_0 \omega \exp\left(-x\sqrt{\omega/2k}\,\right) \sin\left(\omega t - x\sqrt{\omega/2k}\,\right),$$

$$u_x(x,\, t) = -a_0 \left(\sqrt{\omega/2k}\,\right) \exp\left(-x\sqrt{\omega/2k}\,\right) \left[\cos\left(\omega t - x\sqrt{\omega/2k}\,\right) - \sin\left(\omega t - x\sqrt{\omega/2k}\,\right)\right], \quad \text{and}$$

$$u_{xx}(x,\, t) = -\frac{a_0 \omega}{k} \exp\left(-x\sqrt{\omega/2k}\,\right) \sin\left(\omega t - x\sqrt{\omega/2k}\,\right).$$

Therefore $y(x,\, t)$ satisfies the one-dimensional heat equation $u_t = k u_{xx}$.

C13S04.063: Given $pV = nRT$ where n and R are constants, solve for p, V, and T in turn; then compute

$$\frac{\partial p}{\partial V} = -\frac{nRT}{V^2},$$

$$\frac{\partial V}{\partial T} = \frac{nR}{p}, \quad \text{and}$$

$$\frac{\partial T}{\partial p} = \frac{V}{nR}.$$

Then

$$\frac{\partial p}{\partial V} \cdot \frac{\partial V}{\partial T} \cdot \frac{\partial T}{\partial p} = -\frac{nRT}{pV} = -1.$$

C13S04.065: If $f(x,\, y) = x^2 + 2xy + 2y^2 - 6x + 8y$, then the equations

701

$$f_x(x, y) = 0, \quad f_y(x, y) = 0$$

are

$$2x + 2y - 6 = 0, \quad 2x + 4y = -8,$$

which have the unique solution $x = 10$, $y = -7$. So the surface that is the graph of $z = f(x, y)$ contains exactly one point at which the tangent plane is horizontal, and that point is $(10, -7, -58)$.

C13S04.067: Given: van der Waals' equation

$$\left(p + \frac{a}{V^2}\right)(V - b) = (82.06)T$$

where p denotes pressure (in atm), V volume (in cm^3), and T temperature (in K). For CO_2, the empirical constants are $a = 3.59 \times 10^6$ and $b = 42.7$. We are also given $V = 25600$ when $p = 1$ and $T = 313$. Part (a): We differentiate with respect to p while holding T constant. We obtain

$$\left(1 - \frac{2a}{V^3}V_p\right)(V - b) + \left(p + \frac{a}{V^2}\right) \cdot V_p = 0;$$

$$V - b - \frac{2a}{V^2}V_p + \frac{2ab}{V^3}V_p + pV_p + \frac{a}{V^2}V_p = 0;$$

$$\left(\frac{a}{V^2} - \frac{2ab}{V^3} - p\right)V_p = V - b;$$

$$\frac{aV - 2ab - pV^3}{V^3}V_p = V - b;$$

$$V_p = \frac{V^3(V - b)}{aV - 2ab - pV^3}.$$

If p is changed from 1 to 1.1, then the resulting change in V will be predicted by

$$\Delta V \approx V_p \, \Delta p = \frac{1}{10}V_p,$$

and substitution of the data $V = 25600$, $p = 1$ yields $\Delta V \approx -2569.76$ (cm^3).

Part (b): We hold p constant and differentiate both sides of van der Waals' equation with respect to T. The result:

$$\left(p + \frac{a}{V^2}\right)V_T - \frac{2a}{V^3}(V - b)V_T = 82.06;$$

$$(pV^3 + aV)\,V_T - 2a(V - b)V_T = (82.06)V^3;$$

$$V_T = \frac{(82.06)V^3}{pV^3 - aV + 2ab}.$$

If T is changed from 313 to 314 while the pressure is maintained at the constant value $p = 1$, the resulting change in V is predicted by

$$\Delta V \approx V_T \, \Delta T = 1 \cdot V_T.$$

Substitution of the data $V = 25600$, $p = 1$, $T = 313$ yields $\Delta V \approx 82.5105$.

The exact values of the two answers are closer to -2347.60 and 69.9202. Part of the discrepancy is caused by the fact that $V = 25600$ is itself only an approximation; the true value is closer to 25669.920232. Reworking parts (a) and (b) with this value of V yields the different approximations -2557.17 and 82.5095.

C13S04.069: Part (a): If $f(x, y) = \sin x \sinh(\pi - y)$, then

$$f_x(x, y) = \cos x \sinh(\pi - y), \qquad f_{xx}(x, y) = -\sin x \sinh(\pi - y),$$

$$f_y(x, y) = -\sin x \cosh(\pi - y), \qquad f_{yy}(x, y) = \sin x \sinh(\pi - y).$$

Clearly f is harmonic.

Part (b): If $f(x, y) = \sinh 2x \sin 2y$, then

$$f_x(x, y) = 2\cosh 2x \sin 2y, \qquad f_{xx}(x, y) = 4\sinh 2x \sin 2y,$$

$$f_y(x, y) = 2\sinh 2x \cos 2y, \qquad f_{yy}(x, y) = -4\sinh 2x \sin 2y.$$

Clearly f is harmonic.

Part (c): If $f(x, y) = \sin 3x \sinh 3y$, then

$$f_x(x, y) = 3\cos 3x \sinh 3y, \qquad f_{xx}(x, y) = -9\sin 3x \sinh 3y,$$

$$f_y(x, y) = 3\sin 3x \cosh 3y, \qquad f_{yy}(x, y) = 9\sin 3x \sinh 3y.$$

Again it is clear that f is harmonic.

Part (d): If $f(x, y) = \sinh 4(\pi - x) \sin 4y$, then

$$f_x(x, y) = -4\cosh 4(\pi - x) \sin 4y, \qquad f_{xx}(x, y) = 16\sinh 4(\pi - x) \sin 4y,$$

$$f_y(x, y) = 4\sinh 4(\pi - x) \cos 4y, \qquad f_{yy}(x, y) = -16\sinh 4(\pi - y) \sin 4y.$$

Therefore f is harmonic.

C13S04.071: If $f(x, y) = 100 + \dfrac{1}{100}(x^2 - 3xy + 2y^2)$, then

$$f_x(x, y) = \frac{1}{100}(2x - 3y), \qquad f_y(x, y) = \frac{1}{100}(4y - 3x),$$

$$f_x(100, 100) = -\frac{100}{100} = -1, \qquad f_y(100, 100) = \frac{100}{100} = 1.$$

Part (a): You will initially be descending at a $45°$ angle. Part (b): You will initially be ascending at a $45°$ angle.

C13S04.073: Given:

$$f(x, y) = \begin{cases} \dfrac{xy}{x^2 + y^2} & \text{if } (x, y) \neq (0, 0), \\ 0 & \text{if } (x, y) = (0, 0). \end{cases}$$

Part (a): To find the first-order partial derivatives of f at a point other than $(0, 0)$, we proceed normally:

$$f_x(x, y) = \frac{y^3 - x^2 y}{(x^2 + y^2)^2} \quad \text{and} \quad f_y(x, y) = \frac{x^3 - xy^2}{(x^2 + y^2)^2}.$$

Clearly both are defined and continuous everywhere except possibly at the origin. Next,

$$f_x(0, 0) = \lim_{h \to 0} \frac{f(0 + h, 0) - f(0, 0)}{h} = \lim_{h \to 0} \frac{0}{h} = 0$$

and

$$f_y(0, 0) = \lim_{k \to 0} \frac{f(0, 0 + k) - f(0, 0)}{k} = \lim_{k \to 0} \frac{0}{k} = 0.$$

Therefore both f_x and f_y are defined everywhere.

Part (b): At points of the line $y = mx$ other than $(0, 0)$, we have

$$f_x(x, y) = f_x(x, mx) = \frac{m^3 - m}{(m^2 + 1)^2 x},$$

and hence (taking $m = 0$) we have $f_x(x, y) = f_x(x, 0) \equiv 0$, whereas (taking $m = 2$) we have

$$f_x(x, y) = f_x(x, 2x) = \frac{6}{25x},$$

so that f_x is not continuous at $(0, 0)$. Similarly, at points of the line $y = mx$ other than $(0, 0)$, we have

$$f_y(x, y) = f_y(x, mx) = \frac{(1 - m^2)}{(1 + m^2)^2 x},$$

and hence (taking $m = 1$) we have $f_y(x, y) = f_y(x, x) \equiv 0$, whereas (taking $m = 0$) we have

$$f_y(x, y) = f_y(x, 0) = \frac{1}{x}.$$

Hence f_y is also not continuous at $(0, 0)$.

Part (c): If $(x, y) \neq (0, 0)$, then differentiation of f_x with respect to x yields

$$f_{xx}(x, y) = \frac{2xy(x^2 - 3y^2)}{(x^2 + y^2)^3}.$$

Similarly,

$$f_{xy}(x, y) = f_{yx}(x, y) = \frac{6x^2 y^2 - x^4 - y^4}{(x^2 + y^2)^3} \quad \text{and} \quad f_{yy}(x, y) = \frac{2xy(y^2 - 3x^2)}{(x^2 + y^2)^3}.$$

Therefore the second-order partial derivatives of f are all defined and continuous except possibly at the origin.

Part (d): Here we have

$$f_{xx}(0, 0) = \lim_{h \to 0} \frac{f_x(0 + h, 0) - f_x(0, 0)}{h} = \lim_{h \to 0} \frac{0}{h} = 0$$

and, similarly, $f_{yy}(0, 0) = 0$. Hence both second-order partial derivatives f_{xx} and f_{yy} exist at the origin (but you can use polar coordinates to show that neither is continuous there). On the other hand,

$$f_{xy}(0,\,0) = \lim_{k \to 0} \frac{f_x(0,\,0+k) - f_x(0,\,0)}{k} = \lim_{k \to 0} \frac{1}{k^2},$$

which does not exist; $f_{yx}(0,\,0)$ also does not exist by a similar computation.

Section 13.5

C13S05.001: If $z = x - 3y + 5$, then $z_x \equiv 1$ and $z_y \equiv -3$. Hence there is no point on the graph at which the tangent plane is horizontal. Indeed, the graph of $z = x - 3y + 5$ is itself a plane with nonvertical normal vector $\langle 1,\,-3,\,-1 \rangle$, and that's another reason why no tangent plane is horizontal.

C13S05.003: If $z = xy + 5$, then $z_x = y$ and $z_y = x$. Both vanish at $(0,\,0)$, so there is exactly one point on the graph of $z = xy + 5$ at which the tangent plane is horizontal—$(0,\,0,\,5)$.

C13S05.005: If $z = f(x,\,y) = x^2 + y^2 - 6x + 2y + 5$, then $z_x = 2x - 6$ and $z_y = 2y + 2$. Both are zero only at the point $(3,\,-1)$, so the graph of $z = f(x,\,y)$ has a horizontal tangent plane at the point $(3,\,-1,\,-5)$.

C13S05.007: If $z = f(x,\,y) = x^2 + 4x + y^3$, then $z_x = 2x + 4$ and $z_y = 3y^2$. Both are zero at the point $(-2,\,0)$, so the graph of $z = f(x,\,y)$ has exactly one horizontal tangent plane—the one that is tangent at the point $(-2,\,0,\,-4)$.

C13S05.009: If $z = f(x,\,y) = 3x^2 + 12x + 4y^3 - 6y^2 + 5$, then $z_x = 6x + 12$ and $z_y = 12y(y - 1)$. Both partial derivatives are zero at $(-2,\,0)$ and $(-2,\,1)$, so the graph of $z = f(x,\,y)$ has two horizontal tangent planes. One is tangent at the point $(-2,\,0,\,-7)$ and the other is tangent at the point $(-2,\,1,\,-9)$.

C13S05.011: If $f(x,\,y) = (2x^2 + 3y^2)\exp(-x^2 - y^2)$, then

$$\frac{\partial f}{\partial x} = 4x\exp(-x^2 - y^2) - 2x(2x^2 + 3y^2)\exp(-x^2 - y^2) = -2x(2x^2 + 3y^2 - 2)\exp(-x^2 - y^2) \quad \text{and}$$

$$\frac{\partial f}{\partial y} = 6y\exp(-x^2 - y^2) - 2y(2x^2 + 3y^2)\exp(-x^2 - y^2) = -2y(2x^2 + 3y^2 - 3)\exp(-x^2 - y^2).$$

We note that $\exp(-x^2 - y^2)$ is never zero, so to find when both partial derivatives are zero, it is enough to solve simultaneously

$$x(2x^2 + 3y^2 - 2) = 0,$$

$$y(2x^2 + 3y^2 - 3) = 0.$$

One solution is obvious: $x = y = 0$. Next, if $x \neq 0$ but $y = 0$, then the first of these equations implies that $2x^2 = 2$, so we get the two critical points $(1,\,0)$ and $(-1,\,0)$. If $x = 0$ but $y \neq 0$, then the second of these equations implies that $3y^2 = 3$, so we obtain two more critical points, $(0,\,-1)$ and $(0,\,1)$. There is no solution if both x and y are nonzero, for that would imply that $2x^2 + 3y^2 = 2$ and $2x^2 + 3y^2 = 3$. But are there five horizontal tangent planes? No, because two of them are tangent at two critical points. One plane is tangent to the graph of $z = f(x,\,y)$ at $(-1,\,0,\,2e^{-1})$ and $(1,\,0,\,2e^{-1})$, a second is tangent at $(0,\,-1,\,3e^{-1})$ and $(0,\,1,\,3e^{-1})$, and the third is tangent at $(0,\,0,\,0)$.

C13S05.013: Given: $z = f(x,\,y) = x^2 - 2x + y^2 - 2y + 3$. Then

$$f_x(x,\,y) = 2x - 2 \quad \text{and} \quad f_y(x,\,y) = 2y - 2,$$

705

so both partials are zero at only one point: $(1, 1)$. So the graph of $z = f(x, y)$ has only one horizontal tangent plane; it is tangent at the point $(1, 1, 1)$. This is clearly the lowest point on the graph of f.

C13S05.015: If $z = f(x, y) = 2x - x^2 + 2y^2 - y^4$, then

$$f_x(x, y) = -2x + 2 \quad \text{and} \quad f_y(x, y) = -4y^3 + 4y = -4y(y + 1)(y - 1).$$

Thus there are three critical points: $(1, -1)$, $(1, 0)$, and $(1, 1)$. But there are only two horizontal planes tangent to the graph of f because one is tangent at two points—namely, at $(1, -1, 2)$ and at $(1, 1, 2)$; the other horizontal tangent plane is tangent to the graph at $(1, 0, 1)$. The first two of these points are the equally high highest points on the graph of f.

C13S05.017: Given: $z = f(x, y) = 3x^4 - 4x^3 - 12x^2 + 2y^2 - 12y$, the following sequence of *Mathematica* 3.0 commands will find the points of tangency of all horizontal tangent planes. (Recall that % refers to the "last output.")

```
f[x_, y_ ] := 3*x^4 - 4*x^3 - 12*x^2 + 2*y^2 - 12*y

d1 = D[ f[x,y], x ]
```
$$-24x - 12x^2 + 12x^3$$
```
d2 = D[ f[x,y], y ]
```
$$-12 + 4y$$
```
Solve[ { d1 == 0, d2 == 0 }, { x, y } ];
```
$$\{\{x \to -1, y \to 3\}, \{x \to 0, y \to 3\}, \{x \to 2, y \to 3\}\}$$
```
f[x,y] /. %
```
$$\{-23, -18, -50\}$$

Thus there are three horizontal planes tangent to the graph of $z = f(x, y)$; the points of tangency are $(-1, 3, -23)$, $(0, 3, -18)$, and $(2, 3, -50)$. The last of these is the lowest point on the graph of f.

C13S05.019: If $f(x, y) = 2x^2 + 8xy + y^4$, then

$$f_x(x, y) = 4x + 8y = 4(x + 2y) \quad \text{and} \quad f_y(x, y) = 8x + 4y^3 = 4(2x + y^3).$$

Thus both partial derivatives are zero at the three points $(-4, 2)$, $(0, 0)$, and $(4, -2)$. But there are only two horizontal tangent planes because one is tangent to the graph of $z = f(x, y)$ at two points: $(-4, 2, -16)$ and $(4, -2, -16)$. The other plane is tangent to the graph at the origin. The two equally low lowest points on the graph of f are $(-4, 2, -16)$ and $(4, -2, -16)$.

Detail: Solving simultaneous nonlinear equations is an *ad hoc* procedure. One method that frequently works is to solve one equation for one of the variables, then substitute in the others. Here we begin with

$$4(x + 2y) = 0 \quad \text{and} \quad 4(2x + y^3) = 0.$$

We solve the first for $x = -2y$ and substitute in the second to obtain

$$-4y + y^3 = 0; \quad \text{that is,} \quad y(y + 2)(y - 2) = 0.$$

This yields the three solutions $y = -2$, $y = 0$, and $y = 2$, and the corresponding values of x are 4, 0, and -4.

706

C13S05.021: If $z = f(x, y) = \exp(2x - 4y - x^2 - y^2)$, then

$$f_x(x, y) = (2 - 2x)\exp(2x - 4y - x^2 - y^2) \quad \text{and} \quad f_y(x, y) = -(2y + 4)\exp(2x - 4y - x^2 - y^2).$$

Hence both partial derivatives are zero when

$$2x - 2 = 0 \quad \text{and} \quad 2y + 4 = 0;$$

that is, at the single point $(1, -2)$. Hence there is exactly one horizontal plane tangent to the graph of $z = f(x, y)$; it is tangent at the point $(1, -2, e^5)$. This is the highest point on the graph of f.

C13S05.023: The graph of $f(x, y) = x + 2y$ is a plane, so its maximum and minimum values on a polygonal region must occur at the vertices of the polygon. Here we have

$$f(-1, -1) = -3, \quad f(-1, 1) = 1, \quad f(1, -1) = -1, \quad \text{and} \quad f(1, 1) = 3.$$

At this point it is clear what are the maximum and minimum values of $f(x, y)$ on R.

C13S05.025: Given: $f(x, y) = x^2 + y^2 - 2x$ on the triangular region R with vertices at $(0, 0)$, $(2, 0)$, and $(0, 2)$. Then

$$f_x(x, y) = 2x - 2 \quad \text{and} \quad f_y(x, y) = 2y,$$

so the only critical point of f is $(1, 0)$, which is a point of R. Because

$$f(x, y) = (x - 1)^2 + y^2 - 1,$$

the global minimum value of f on R is $f(1, 0) = -1$. The maximum value of f must occur on the boundary of R, which we explore next.

- On the lower edge of R, we have $f(x, 0) = x^2 - 2x$, which must attain its maximum at one endpoint of that edge. Hence the maximum value of f there is $f(0, 0) = f(2, 0) = 0$.

- On the left-hand edge of R, we have $f(0, y) = y^2$, with maximum value $f(0, 2) = 4$.

- On the diagonal edge of R, which has equation $y = 2 - x$, we have

$$f(x, 2 - x) = g(x) = x^2 + (2 - x)^2 - 2x = 2x^2 - 6x + 4,$$

and $g'(x) = 4x - 6$, so the minimum value of f there is $g\left(\frac{3}{2}\right) = -\frac{1}{2}$ and the maximum value there (because it must occur at an endpoint of the diagonal edge) is $f(0, 2) = 4$.

In summary, the minimum value of f on R is $f(1, 0) = -1$ and its maximum value is $f(0, 2) = 4$.

C13S05.027: Given: $f(x, y) = 2xy$ on the circular disk R described by the inequality $x^2 + y^2 \leq 1$. Then $f_x(x, y) = 2y$ and $f_x(x, y) = 2x$, so the only critical point of f is $(0, 0)$. On the line $y = x$ we have $f(x, x) = 2x^2$, but on the line $y = -x$ we have $f(x, -x) = -2x^2$. Therefore f does not have an extremum at $(0, 0)$. (The graph of $z = 2xy$ is a hyperbolic paraboloid with a *saddle point* at $(0, 0)$; to see its graph, rotate the graph shown in Fig. 13.10.1 $45°$ around the z-axis.) Because f must have a global maximum and a global minimum on R, both must occur on its boundary.

- We describe the boundary $x^2 + y^2 = 1$ of R in polar coordinates: $r = 1$, $0 \leq \theta \leq 2\pi$. Thus on the boundary, we have

$$f(x, y) = 2xy = g(\theta) = 2 \sin \theta \cos \theta.$$

Then

$$g'(\theta) = 2 \cos^2 \theta - 2 \sin^2 \theta = 2 \cos 2\theta,$$

and $g'(\theta) = 0$ when 2θ is an odd integral multiple of $\pi/2$; that is, when θ is an odd integral multiple of $\pi/4$. Now

$$g(\pi/4) = g(5\pi/4) = 1 \quad \text{and} \quad g(3\pi/4) = g(7\pi/4) = -1,$$

so we have discovered the global extrema of f on R.

Summary: The global maximum value of f is 1 and occurs at each of the two points $(\frac{1}{2}\sqrt{2}, \frac{1}{2}\sqrt{2})$ and $(-\frac{1}{2}\sqrt{2}, -\frac{1}{2}\sqrt{2})$. The global minimum value of f is -1 and occurs at each of the two points $(-\frac{1}{2}\sqrt{2}, \frac{1}{2}\sqrt{2})$ and $(\frac{1}{2}\sqrt{2}, -\frac{1}{2}\sqrt{2})$.

C13S05.029: The square of the distance between the plane and the origin is

$$f(x, y) = x^2 + y^2 + \left(\frac{169 - 12x - 4y}{3} \right)^2.$$

Setting both partials of $f(x, y)$ equal to zero yields the equations

$$2x - \frac{8}{3}(169 - 12x - 4y) = 0, \quad 2y - \frac{8}{9}(169 - 12x - 4y) = 0.$$

These equations are easiest to solve if you note first that they imply immediately that $6x = 18y$, because each is equal to $8(169 - 12x - 4y)$. It follows that their solution is $x = 12$, $y = 4$. The corresponding value of z on the given plane is 3, so the point of the plane closest to the origin is $(12, 4, 3)$. (The formula for f makes it plain that f has a global minimum and no maximum.) The distance between the plane and the origin is therefore $\sqrt{f(12, 4)} = 13$.

C13S05.031: The square of the distance between the plane and the point Q is

$$f(x, y) = (x - 7)^2 + (y + 7)^2 + (49 - 2x - 3y)^2.$$

When we set both partials of f equal to zero, we get the simultaneous equations

$$2(x - 7) - 4(49 - 2x - 3y) = 0, \quad 2(y + 7) - 6(49 - 2x - 3y) = 0,$$

with solution $x = 15$, $y = 5$; the corresponding z-coordinate on the plane is 4. So the point on the plane closest to Q is $(15, 5, 4)$. The distance between the two is $4\sqrt{14}$.

C13S05.033: The square of the distance from the origin to the point (x, y, z) of the surface is

$$f(x, y) = x^2 + y^2 + \frac{16}{x^4 y^4}.$$

The equations $f_x(x, y) = 0 = f_y(x, y)$ are

$$2x - \frac{64}{x^5 y^4} = 0, \quad 2y - \frac{64}{x^4 y^5} = 0.$$

708

They are easiest to solve if you begin with the observation that they imply that $2x^2 = 2y^2$. There are four solutions—all possible combinations of $x = \pm\sqrt{2}$, $y = \pm\sqrt{2}$. It follows that the point on the surface in the first octant closest to the origin is $\left(\sqrt{2},\ \sqrt{2},\ 1\right)$; its distance from the origin is $\sqrt{5}$.

C13S05.035: We will find the maximum possible product of three *nonnegative* real numbers with sum 120—the reason in a moment. If x, y, and z are the three numbers, then we are to maximize xyz given $x + y + z = 120$. So we solve for z, substitute, and maximize

$$f(x,\ y) = xy(120 - x - y), \quad 0 \le x, \quad 0 \le y, \quad x + y \le 120.$$

Thus by allowing one or two of the numbers to be zero, the domain of f is now a closed and bounded subset of the plane—the triangle with two sides on the nonnegative coordinate axes and the third side part of the graph of $y = 120 - x$. Write $f(x,\ y) = 120xy - x^2 y - xy^2$ and set both partial derivatives equal to zero to obtain

$$120y - 2xy - y^2 = 0, \quad 120x - 2xy - x^2 = 0.$$

Because neither x nor y is zero—that would minimize the product, not maximize it—we may cancel to obtain

$$120 - 2x - y = 0, \quad 120 - 2y - x = 0,$$

and it follows that $2x + y = x + 2y$, so that $y = x$, and then the equation $3x = 120$ yields $x = 40$ and $y = 40$. It follows that $z = 40$ as well. The maximum of $f(x,\ y)$ does not occur on the boundary of its domain, for $f(x,\ y) = 0$ there. Hence this lone interior critical point must yield the global maximum, which is $40 \cdot 40 \cdot 40 = 64000$.

C13S05.037: Let the dimensions of the box be x by y by z. We are to minimize total surface area $A = 2xy + 2xz + 2yz$ given $xyz = 1000$. Solve the latter equation for z and substitute in A to obtain the function to be minimized:

$$A(x,\ y) = \frac{2000}{x} + \frac{2000}{y} + 2xy, \quad 0 < x, \quad 0 < y.$$

Although A is continuous on its domain, the domain is neither closed nor bounded. But an argument similar to the one given in Example 7 makes it clear that A has a global minimum, so it must occur at a critical point of the domain. When we set both partial derivatives of A equal to zero, we obtain the equations

$$2y - \frac{2000}{x^2} = 0, \quad 2x - \frac{2000}{y^2} = 0.$$

These equations imply that $2x^2 y = 2xy^2$, so that $y = x$. Then either of the two preceding equations implies that $x = 10 = y$. Finally, $xyz = 1000$ implies that $z = 10$ as well. This is the only critical point, so we have found the global minimum of A. To minimize the total surface area, make a cube of edge length 10.

C13S05.039: Suppose that the dimensions of the base of the box are x by y, the front and back have dimensions x by z, and the sides have dimensions y by z. The cost of the base is then $6xy$ and the total cost of the other four sides is $2 \cdot 5xz + 2 \cdot 5yz$. So we are to minimize total cost $C = 6xy + 10xz + 10yz$ given $xyz = 600$. Solve the last equation for z and substitute in the cost expression to obtain the function to be minimized:

$$C(x,\ y) = \frac{6000}{x} + \frac{6000}{y} + 6xy, \quad 0 < x, \quad 0 < y.$$

By an argument similar to the one used in the solution of Example 7, $C(x, y)$ has a global minimum and it occurs at a critical point. When we set both partial derivatives of C equal to zero, we get the equations

$$6y - \frac{6000}{x^2} = 0, \quad 6x - \frac{6000}{y^2} = 0.$$

It follows immediately that $x^2 y = xy^2$, so that $x = y$. Then either of the displayed equations yields $x = y = 10$. The corresponding value of z is 6, so the dimensions of the least expensive such box are these: base 10 inches by 10 inches, height 6 inches. It will cost \$18.00.

C13S05.041: Let the base and top of the box have dimensions x by y, the front and back dimensions x by z, and the sides dimensions y by z. Then the top and base cost $3xy$ cents each, the front and back cost $6xz$ cents each, and the two sides cost $9yz$ cents each. Hence the total cost of the box will be $C = 6xy + 12xz + 18yz$. But $xyz = 750$, so that $z = 750/(xy)$. To substitute this into the formula for C and simplify, we use the *Mathematica* 3.0 command

```
6*x*y + 12*x*z + 18*y*z /. z -> 750/(x*y)
```

(Recall that `/.` translates roughly as "evaluate subject to.") The response is

$$\frac{13500}{x} + \frac{9000}{y} + 6xy.$$

Next we construct the total surface area function, the quantity to be minimized:

```
f[x_, y_] := 13500/x + 9000/y + 6*x*y
```

Then we compute both partial derivatives:

```
d1 = D[ f[x,y], x]
```

$$6y - \frac{13500}{x^2}$$

```
d2 = D[ f[x,y], y]
```

$$6x - \frac{9000}{y^2}$$

Then we set both partial derivatives equal to zero and solve simultaneously:

```
Solve[ { d1 == 0, d2 == 0 }, { x, y } ]
```

The response is

$$\{\{x \to 15, \ y \to 10\}, \ \{x \to -15\,(-1)^{1/3}, \ y \to -10\,(-1)^{1/3}\}, \ \{x \to 15\,(-1)^{2/3}, \ y \to 10\,(-1)^{2/3}\}\}$$

We ignore the two pairs of non-real roots and evaluate z:

```
750/(x*y) /. {x -> 15, y -> 10}
```

5

Finally, we evaluate f at $(15, 10)$ to find the minimum cost:

710

f[15,10]

2700

The domain of f is not a closed and bounded set, but instead the interior of the entire first quadrant. Nevertheless, $f(x, y)$ has a global minimum at a critical point by an argument similar to the one used in Example 7. Because we have found only one critical point, we have found the global minimum as well. The box should have dimensions $x = 15$, $y = 10$, and $z = 5$ inches. It will cost \$27.00.

C13S05.043: Suppose that the base of the building measures x feet by y feet, that its front and back measure x by z, and that its two sides measure y by z. Then the total heating and cooling costs will be $C = 2xy + 4xz + 8yz$. Solve $xyz = 8000$ for z and substitute in the expression for cost to obtain the quantity to be minimized:

$$C(x, y) = \frac{64000}{x} + \frac{32000}{y} + 2xy, \quad 0 < x, \quad 0 < y.$$

Although the domain of C is not a closed and bounded subset of the plane, nevertheless $C(x, y)$ has a global minimum at a critical point by an argument similar to the one used in the solution of Example 7. When we set both partial derivatives of $C(x, y)$ equal to zero, we obtain

$$2y - \frac{64000}{x^2} = 0, \quad 2x - \frac{32000}{y^2} = 0,$$

having the only real solutions $x = 40$, $y = 20$. The corresponding value of z is 10, so the building should be 40 feet wide (in front), 20 feet deep, and 10 feet high. The annual heating and cooling costs will thereby have their minimum possible value, $C(40, 20) = 4800$ dollars per year.

C13S05.045: Suppose that (x, y, z) is the vertex of the box that lies on the given plane with equation $x + 3y + 7z = 11$. We are to maximize the volume $V = xyz$ of the box. Solve the equation of the plane for z and subsitute to obtain

$$V(x, y) = \frac{xy(11 - x - 3y)}{7}, \quad 0 \le x, \quad 0 \le y, \quad x + 3y \le 11.$$

The domain of V is a closed and bounded subset of the xy-plane—it consists of the sides and interior of the triangle with vertices at $(0, 0)$, $(11, 0)$, and $\left(0, \frac{11}{3}\right)$. Therefore the continuous function (it's a polynomial) $V(x, y)$ has a global maximum on its domain. The maximum does not occur on the boundary because $V(x, y)$ is identically zero there. Hence the maximum occurs at an interior critical point. When we set the partial derivatives of V equal to zero, we get the simultaneous equations

$$\frac{(11 - x - 3y)y - xy}{7} = 0, \quad \frac{(11 - x - 3y)x - 3xy}{7} = 0.$$

To solve these equations, multiply through by 7 and factor to obtain

$$(11 - 2x - 3y)y = 0, \quad (11 - x - 6y)x = 0.$$

- If $x = 0$ and $y = 0$, we have one solution.
- If $x = 0$ and $y \ne 0$, then $y = \frac{11}{3}$.
- If $x \ne 0$ and $y = 0$, then $x = 11$.
- If $x \ne 0$ and $y \ne 0$, then $2x + 3y = 11$ and $x + 6y = 11$. It follows that $x = \frac{11}{3}$ and $y = \frac{11}{9}$.

711

Only the last of these solutions will produce a box of positive volume. The corresponding value of z is $\frac{11}{21}$. Thus we have found the maximizing values of x, y, and z. The maximum possible volume is

$$\frac{11}{3} \cdot \frac{11}{9} \cdot \frac{11}{21} = \frac{1331}{567} \approx 2.347442680776.$$

C13S05.047: Suppose that the dimensions of the rectangular box are x by y by z (in inches). Without loss of generality we may suppose that $x \leqq y \leqq z$. Then the length of the box is z, so its girth is $2x + 2y$. We are to maximize box volume $V = xyz$ given the side condition $2x + 2y + z \leq 108$; of course, the maximum occurs when $2x + 2y + z = 108$, so that is the side condition we use. Solve for z in the last equation and substitute in the expression for volume to obtain the function to be maximized:

$$V(x, y) = xy(108 - 2x - 2y), \quad 0 \leqq x, \quad 0 \leqq y, \quad x + y \leqq 54.$$

Now V is continuous ($V(x, y)$ is a polynomial) on its domain, a closed and bounded region in the xy-plane, so V has a global maximum there. The maximum does not occur on the boundary because $V(x, y)$ is identically zero on the boundary of its domain. So the global maximum occurs at an interior critical point where both partial derivatives are zero; that is,

$$(108 - 2x - 2y)y - 2xy = 0, \quad (108 - 2x - 2y)x - 2xy = 0.$$

We may cancel y from the first of these equations and x from the second because neither is zero at the maximum. Thus we are to solve

$$108 - 4x - 2y = 0, \quad 108 - 2x - 4y = 0.$$

It follows that $4x + 2y = 2x + 4y$, and thus that $x = y$. This implies in turn that $x = y = 18$ and $z = 36$. So the maximum volume of such a box is $18 \cdot 18 \cdot 36 = 11664$ cubic inches, exactly 6.25 cubic feet. If it were filled with osmium (the heaviest element known) it would weigh over 4338 kg, about 4.78 tons.

C13S05.049: Suppose that the upper corner of the box in the first octant meets the paraboloid at the point (x, y, z), so that $z = 1 - x^2 - y^2$. We are to maximize box volume $V = 2x \cdot 2y \cdot z$; that is,

$$V(x, y) = 4xy(1 - x^2 - y^2) = 4xy - 4x^3y - 4xy^2, \quad 0 \leqq x, \quad 0 \leqq y, \quad x^2 + y^2 \leqq 1.$$

Because V is continuous ($V(x, y)$ is a polynomial) and its domain is a closed and bounded subset of the xy-plane, V has a global maximum—which does not occur on the boundary of its domain because $V(x, y)$ is identically zero there. Hence the maximum we seek occurs at an interior critical point. When we set the partial derivatives of V simultaneously equal to zero, we obtain the equations

$$4y(1 - x^2 - y^2) - 8x^2y = 0, \quad 4x(1 - x^2 - y^2) - 8xy^2 = 0;$$

that is, because neither x nor y is zero at maximum box volume,

$$1 - 3x^2 - y^2 = 0 \quad \text{and } 1 - x^2 - 3y^2 = 0.$$

It follows in the usual way that $y = x$, and thus that $x = y = \frac{1}{2}$. The corresponding value of z is also $\frac{1}{2}$, so the dimensions of the box of maximum volume are 1 by 1 by $\frac{1}{2}$ and its volume is $\frac{1}{2}$.

C13S05.051: Let r be the common radius of the two cones and the cylinder, h the height of the cylinder, and z the height of each cone. Note that the slant height of each cone is $(r^2 + z^2)^{1/2}$, so each has curved surface area

$$2\pi \cdot \frac{r}{2} \cdot (r^2 + z^2)^{1/2} = \pi r(r^2 + z^2)^{1/2}.$$

We are to minimize the total surface area

$$A = 2\pi r(r^2 + z^2)^{1/2} + 2\pi rh \tag{1}$$

of the buoy given fixed volume $V = \frac{2}{3}\pi r^2 z + \pi r^2 h$. We first solve this last equation for

$$h = \frac{3V - 2\pi r^2 z}{3\pi r^2},$$

then substitute in (1) to express A as a function of r and z:

$$A(r, z) = \frac{2V}{r} - \frac{4\pi rz}{3} + 2\pi r(r^2 + z^2)^{1/2}.$$

The domain of A is described by the inequalities

$$0 < r, \quad 0 \leqq z, \quad r^2 z \leqq \frac{3V}{2\pi},$$

and though it is neither closed nor bounded, it can be shown by an argument similar to the one in Example 7 that $A(r, z)$ has a global minimum that does not occur on the boundary of its domain (unless it occurs where $h = 0$ or where $z = 0$; we will attend to those possibilities later). Moreover, intuition and experience suggest that the minimal surface area will occur when the figure can be inscribed in a nearly spherical ellipsoid.

Next,

$$A_z(r, z) = \frac{2[3\pi rz - 2\pi r(r^2 + z^2)^{1/2}]}{3(r^2 + z^2)^{1/2}},$$

so $A_z(r, z) = 0$ when $3z = 2(r^2 + z^2)^{1/2}$. Therefore, when $A_z(r, z) = 0$, we have both $2r = z\sqrt{5}$ and $(r^2 + z^2)^{1/2} = \frac{3}{2}z$.

Also,

$$A_r(r, z) = \frac{2[6\pi r^4 + 3\pi r^2 z^2 - 3V(r^2 + z^2)^{1/2} - 2\pi r^2 z(r^2 + z^2)^{1/2}]}{3r^2(r^2 + z^2)^{1/2}},$$

so $A_r(r, z) = 0$ when

$$6\pi r^4 + 3\pi r^2 z^2 - 3V(r^2 + z^2)^{1/2} - 2\pi r^2 z(r^2 + z^2)^{1/2} = 0.$$

We substitute $\frac{3}{2}z$ for $(r^2 + z^2)^{1/2}$ in this last equation to find that when both partials vanish, also

$$\frac{9Vz}{2} = 6\pi r^4,$$

then we replace r with $\frac{1}{2}z\sqrt{5}$ to find that when both partials vanish,

$$25\pi z^3 = 12V.$$

Thus the minimum surface area seems to occur when

$$z = \left(\frac{12V}{25\pi}\right)^{1/3} \approx (0.534601847029)V^{1/3},$$

713

for which

$$r = \left(\frac{9V^2}{20\pi^2}\right)^{1/6} \approx (0.597703035427)V^{1/3}$$

and

$$h = \left(\frac{12V}{25\pi}\right)^{1/3} \approx (0.534601847029)V^{1/3}.$$

At these values of the variables, the surface area is

$$A = 5^{1/6}(18\pi V^2)^{1/3} \approx (5.019214931473)V^{2/3}.$$

The symmetry of the solution—that $h = z$—suggests that we have found the minimum, but we have yet to check the cases $h = 0$ and $z = 0$.

• If $z = 0$, then the buoy is a cylinder with radius r, height h, total surface area $A = 2\pi r^2 + 2\pi rh$, and fixed volume $V = \pi r^2 h$; we are to minimize its total surface area. We substitute

$$h = \frac{V}{\pi r^2}$$

in the surface area formula to obtain the function to be minimized:

$$A(r) = 2\pi r^2 + \frac{2V}{r}, \quad 0 < r < \infty.$$

Then

$$A'(r) = 4\pi r - \frac{2V}{r^2} = \frac{4\pi r^3 - 2V}{r^2}.$$

Thus $A'(r) = 0$ when

$$r = r_0 = \left(\frac{V}{2\pi}\right)^{1/3} \approx (0.541926070139)V^{1/3}.$$

The corresponding value of $A(r)$ is a global minimum by the first derivative test, and it is

$$A(r_0) = (54\pi V^2)^{1/3} \approx (5.535810445932)V^{2/3}.$$

This is somewhat larger than the minimum we found in the case $z > 0$.

• If $h = 0$, then the buoy consists of two congruent right circular cones with their bases, circles of radii r, coinciding, and each cone of height z. The total volume of the two cones is

$$V = \frac{2}{3}\pi r^2 z,$$

which we solve for z and substitute in the surface area formula $A = 2\pi r(r^2 + z^2)^{1/2}$ to obtain the function to be minimized,

$$A(r) = 2\pi r\left(r^2 + \frac{9V^2}{4\pi^2 r^4}\right)^{1/2} = \frac{(4\pi^2 r^6 + 9V^2)^{1/2}}{r}, \quad 0 < r.$$

Now

714

$$A'(r) = \frac{8\pi^2 r^6 - 9V^2}{r^2(4\pi^2 r^6 + 9V^2)^{1/2}};$$

$A'(r) = 0$ when $8\pi^2 r^6 = 9V^2$, so that

$$r = r_0 = \frac{\sqrt{2}}{2}\left(\frac{3V}{\pi}\right)^{1/3} \approx (0.696319882685)V^{1/3}.$$

Then

$$A(r_0) = 3^{7/6} \cdot \pi^{1/3} \cdot V^{2/3} \approx (5.276647566071)V^{2/3},$$

which is larger than the minimum found in the first part of this solution. Therefore the minimum possible surface area of the buoy is $5^{1/6}(18\pi V^2)^{1/3} \approx (5.019214931473)V^{2/3}$.

A final observation: The buoy of minimal possible surface area cannot be inscribed in a sphere. If $V = 1$ (say), then it can be inscribed in an ellipsoid (actually, a prolate spheroid) with approximate equation

$$\frac{x^2}{(0.69016802)^2} + \frac{y^2}{(0.69016802)^2} + \frac{z^2}{(1.06923700)^2} = 1$$

with the axis of symmetry of the buoy on the z-axis. A figure showing the elliptical cross-section of this ellipsoid in the xz-plane and the cross-section of the buoy in the xz plane (for the case $V = 1$) is next.

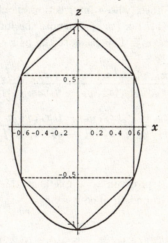

C13S05.053: We want to minimize

$$f(x, y) = x^2 + (y-1)^2 + x^2 + y^2 + (x-2)^2 + y^2 = (x-2)^2 + 2x^2 + (y-1)^2 + 2y^2$$

with domain the entire xy-plane. An argument similar to the one used in Example 7 establishes that $f(x, y)$ has a global minimum value, which must be at a point where both partial derivatives are zero. These equations are $6x - 4 = 0$, $6y - 2 = 0$, with solution $x = \frac{2}{3}$, $y = \frac{1}{3}$. Because this is the only critical point, we have located the point that minimizes $f(x, y)$. It is $\left(\frac{2}{3}, \frac{1}{3}\right)$, and the value of $f(x, y)$ there is $\frac{10}{3}$.

It would seem somewhat more practical to find the point (x, y) such that the sum of the distances (not their squares) from (x, y) to the three points $(0, 1)$, $(0, 0)$, and $(2, 0)$ is a minimum. For example, where should a power company be located to minimize the total length of its cables to industries located at the three points $(0, 1)$, $(0, 0)$, and $(2, 0)$? This is a much more difficult problem. The function to be minimized is

$$h(x, y) = \sqrt{x^2 + (y-1)^2} + \sqrt{x^2 + y^2} + \sqrt{(x-2)^2 + y^2},$$

and its partial derivatives are

$$f(x, y) = h_x(x, y) = \frac{x}{\sqrt{x^2 + (y-1)^2}} + \frac{x-2}{\sqrt{(x-2)^2 + y^2}} + \frac{x}{\sqrt{x^2 + y^2}} \quad \text{and}$$

$$g(x, y) = h_y(x, y) = \frac{y-1}{\sqrt{x^2 + (y-1)^2}} + \frac{y}{\sqrt{(x-2)^2 + y^2}} + \frac{y}{\sqrt{x^2 + y^2}}.$$

It is doubtful that any present-day computer algebra system could solve the simultaneous equations

$$f(x, y) = 0, \qquad g(x, y) = 0 \tag{1}$$

exactly with a reasonable amount of memory and in a reasonable time. But recall Newton's method from the last section of Chapter 3. Techniques of this chapter can be used to extend Newton's method to two simultaneous equations in two unknowns, as in Eq. (1). Here is the extension.

The graph of $f(x, y) = 0$ is, generally, a curve in the xy-plane, as is the graph of $g(x, y) = 0$. The point where these curves meet is the simultaneous solution that we seek. If (x_0, y_0) is an initial "guess" for the simultaneous solution, then $(x_0, y_0, f(x_0, y_0))$ is a point on the surface $z = f(x, y)$ and $(x_0, y_0, g(x_0, y_0))$ is a point on the surface $z = g(x, y)$. The tangent planes to these respective surfaces at these respective points generally meet the xy-plane in a pair of lines, whose intersection should be close to the desired solution (x_*, y_*) of the simultaneous equations in (1). Let (x_1, y_1) denote the intersection of these lines, and repeat the process with (x_0, y_0) replaced with (x_1, y_1). This leads to the pair of iterative formulas

$$x_{k+1} = x_k - \frac{f(x_k, y_k)g_y(x_k, y_k) - g(x_k, y_k)f_y(x_k, y_k)}{f_x(x_k, y_k)g_y(x_k, y_k) - g_x(x_k, y_k)f_y(x_k, y_k)},$$

$$y_{k+1} = y_k - \frac{g(x_k, y_k)f_x(x_k, y_k) - f(x_k, y_k)g_x(x_k, y_k)}{f_x(x_k, y_k)g_y(x_k, y_k) - g_x(x_k, y_k)f_y(x_k, y_k)}$$

for $k = 0, 1, 2, \ldots$.

We wrote a simple *Mathematica* 3.0 program to implement this procedure with the initial guess $(x_0, y_0) = (0.5, 0.5)$. Here are the results (rounded, of course):

$(x_1, y_1) = (0.201722209269, 0.369098300563),$ $(x_2, y_2) = (0.250232659645, 0.283132873843),$

$(x_3, y_3) = (0.254890752792, 0.304117462143),$ $(x_4, y_4) = (0.254568996440, 0.304503089648),$

$(x_5, y_5) = (0.254569313597, 0.304503701206),$ $(x_6, y_6) = (0.254569313597, 0.304503701206),$

and—to the number of digits shown—$(x_7, y_7) = (x_6, y_6)$. The sum of the distances from (x_6, y_6) to the three points $(0, 1)$, $(0, 0)$, and $(2, 0)$ is approximately 2.909312911180.

Warning: A much better "initial guess" is required than in the case of a function of a single variable. For example, with the initial guess $(x_0, y_0) = (2/3, 1/3)$ (the solution to the original version of Problem 53, which one would think would be a fairly good initial guess), $(x_6, y_6) \approx (1422.779, 732.866)$. Thus it is clear that Newton's method is not converging to the correct solution.

C13S05.055: Let $2x$ denote the length of the base of each isosceles triangle, z the height of each triangle, and y the distance between the triangles. We are to minimize the area $A = 2xz + 2y\sqrt{x^2 + z^2}$ given the house has fixed volume $V = xyz$. Solve the latter equation for y and subsitute in the area formula to obtain the function

$$A(x, z) = 2xz + \frac{2V\sqrt{x^2 + z^2}}{xz}, \quad 0 < x, \quad 0 < z.$$

An argument similar to the one used in the solution of Example 7 shows that A must have a global minimum value even though its domain is neither closed nor bounded. Therefore, if there is a unique critical point of A in its domain, that will be the location of its global minimum.

To find the critical point or points, we use *Mathematica* 3.0 and first define

```
a[x_, z_] := 2*x*z + (2*v*Sqrt[x*x + z*z])/(x*z)
```

Then we compute and simplify both partial derivatives.

```
d1 = D[ a[x,z], x ]
```

$$2z + \frac{2v}{z\sqrt{x^2 + z^2}} - \frac{2v\sqrt{x^2 + z^2}}{x^2 z}$$

```
d1 = Together[d1]
```

$$\frac{2z\left(-v + x^2\sqrt{x^2 + z^2}\right)}{x^2\sqrt{x^2 + z^2}}$$

```
n1 = Numerator[ d1 ]
```

$$2z\left(-v + x^2\sqrt{x^2 + z^2}\right)$$

```
d2 = D[ a[x,z], z ]
```

$$2x + \frac{2v}{x\sqrt{x^2 + z^2}} - \frac{2v\sqrt{x^2 + z^2}}{x z^2}$$

```
d2 = Together[ d2 ]
```

$$\frac{2x\left(-v + z^2\sqrt{x^2 + z^2}\right)}{z^2\sqrt{x^2 + z^2}}$$

```
n2 = Numerator[ d2 ]
```

$$2x\left(-v + z^2\sqrt{x^2 + z^2}\right)$$

Because neither $x = 0$ nor $z = 0$ (because $V > 0$), we may cancel:

```
n1 = n1/(2*z) // Cancel
```

$$-v + x^2\sqrt{x^2 + z^2}$$

```
n2 = n2/(2*x) // Cancel
```

$$-v + z^2\sqrt{x^2 + z^2}$$

Both of the last expressions must be zero when both partial derivatives are set equal to zero, and it follows that $z = x$. We substitute this information into the numerator **n1** of $A_x(x, z)$ and set the result equal to zero.

```
n1 /.  z -> x
```

$$-v + \sqrt{2}\ (x^2)^{3/2}$$

```
Solve[ % == 0, x ]
```

And *Mathematica* returns six solutions, only two of which are real, and only one of the two is positive: We find that

$$x = z = \frac{V^{1/3}}{2^{1/6}}, \quad \text{and} \quad y = 2^{1/3} \cdot V^{1/3},$$

so that the minimum possible surface area is $3 \cdot 2^{2/3} \cdot V^{2/3}$.

C13S05.057: Let x, y, and z be the lengths of the edges of the three squares. We are to maximize and minimize their total area $A = x^2 + y^2 + z^2$ given the condition $4x + 4y + 4z = 120$; that is, $x + y + z = 30$. Using this side condition to eliminate z in the expression for A, we obtain the function

$$A(x, y) = x^2 + y^2 + (30 - x - y)^2, \quad 0 \leqq x, \quad 0 \leqq y, \quad x + y \leqq 30.$$

The domain D of A is the triangular region in the xy-plane with vertices at $(0, 0)$, $(30, 0)$, and $(0, 30)$, including the boundary segments. Hence D is a closed and bounded subset of the plane and A is continuous there (because $A(x, y)$, when expanded, is a polynomial). Therefore there is both a global maximum and a global minimum value of $A(x, y)$. We proceed in the usual way, first setting both partial derivatives equal to zero:

$$2x - 2(30 - x - y) = 0 \quad \text{and} \quad 2y - 2(30 - x - y) = 0.$$

It follows that $y = x$. By symmetry (rework the problem eliminating y rather than z) $z = x$, so that $x = y = z$. Therefore $x = y = z = 10$ may yield an extremum of A.

- On the boundary segment of D on which $y = 0$,

$$A(x, 0) = f(x) = x^2 + (30 - x)^2, \quad 0 \leqq x \leqq 30.$$

Methods of single-variable calculus yield the critical point $x = 15$; we have also the two endpoints of the domain of f to check.

- On the boundary segment of D on which $x = 0$,

$$A(0, y) = g(y) = y^2 + (30 - y)^2, \quad 0 \leqq y \leqq 30.$$

Methods of single-variable calculus yield the critical point $y = 15$; we have also the two endpoints of the domain of g to check.

- On the boundary segment of D with equation $x + y = 30$, we have $y = 30 - x$, so that

$$A(x, 30 - x) = h(x) = x^2 + (30 - x)^2, \quad 0 \leqq x \leqq 30.$$

Again, methods of single-variable calculus yield the critical point $x = 15$; the endpoints of this boundary segment will be checked in the other two cases.

Results:

$$A(10, 10) = 300, \qquad A(15, 0) = 450,$$

$$A(0, 15) = 450, \qquad A(0, 0) = 900,$$

$$A(30, 0) = 900, \qquad A(0, 30) = 900, \quad \text{and}$$

$$A(15, 15) = 450.$$

Answer: For maximum total area, make only one square, measuring 30 cm on each side, with area 900 cm^2. For minimum total area, make three equal squares, each measuring 10 cm on each side, with total area 300 cm^2.

C13S05.059: Using the notation in Fig. 13.5.16, we have cross-sectional area

$$A(x, \theta) = (L - 2x)x \sin \theta + x^2 \sin \theta \cos \theta, \quad 0 \leqq \theta \leqq \frac{\pi}{2}, \quad 0 \leqq x \leqq \frac{L}{2}.$$

to be maximized. When we set the partial derivatives of A equal to zero, we obtain the equations

$$(L - 2x)\sin \theta - 2x \sin \theta + 2x \sin \theta \cos \theta = 0, \quad (L - 2x)x \cos \theta + x^2 \cos^2 \theta - x^2 \sin^2 \theta = 0. \tag{1}$$

The first equation in (1) yields

$$(L - 2x) - 2x + 2x \cos \theta = 0 \quad \text{or} \quad \sin \theta = 0,$$

but the second of these can be rejected as $\theta = 0$ minimizes the cross-sectional area. The second equation in (1) yields

$$(L - 2x)\cos \theta + x \cos^2 \theta - x \sin^2 \theta = 0 \quad \text{or} \quad x = 0,$$

and the second of these can also be rejected because $x = 0$ minimizes the area. Thus we have the simultaneous equations

$$L - 2x - 2x + 2x \cos \theta = 0 \quad \text{and} \quad (L - 2x)\cos \theta + x \cos^2 \theta - x \sin^2 \theta = 0. \tag{2}$$

We solve the first of these for $L - 2x = 2x - 2x \cos \theta$ and substitute in the second to obtain

$$2x(1 - \cos \theta)\cos \theta + x \cos^2 \theta - x \sin^2 \theta = 0;$$

$$2(1 - \cos \theta)\cos \theta + \cos^2 \theta - \sin^2 \theta = 0;$$

$$2 \cos \theta - 2 \cos^2 \theta + \cos^2 \theta - 1 + \cos^2 \theta = 0;$$

$$\cos \theta = \frac{1}{2};$$

$$\theta = \frac{\pi}{3}.$$

719

In the second step we used the fact that $x = 0$ minimizes A to cancel x with impunity; in the last step we used the domain of A to determine θ. In any case, the first equation in (2) lets us determine the maximizing value of x as well:

$$L - 4x + x = 0, \quad \text{and thus} \quad x = \frac{L}{3}.$$

We have found only one critical point, and the only endpoint that might produce a larger cross-sectional area occurs when $\theta = \pi/2$, in which case the first equation in (2) implies that

$$L - 4x = 0, \quad \text{so that} \quad x = \frac{L}{4}.$$

In the latter case the cross-sectional area of the gutter is $L^2/8$, but in the previous case, when $\theta = \pi/3$, we find that $A = \left(L^2\sqrt{3}\,\right)/12 \approx (0.1443)L^2$, so this is the maximum possible cross-sectional area of the gutter.

C13S05.061: Given:

$$P(x) = -2x^2 + 12x + xy - y - 10,$$

$$Q(y) = -3y^2 + 18y + 2xy - 2x - 15.$$

Part (a): This is known as a *game of perfect information*—each player (manager) knows every strategy available to his opponent. Each manager computes

$$P'(x) = 12 - 4x + y, \quad Q'(y) = 18 + 2x - 6y,$$

sets both equal to zero (knowing that the other manager is doing the same), and solves for $x = 45/11$, $y = 48/11$. Thus each maximizes his profit knowing that the other manager is doing the same; indeed, if either player (manager) deviates from his *optimal strategy*, his profit will decrease and that of his opponent (the other manager) is likely to increase. With these values of x and y, the profits will be

$$P = \frac{2312}{121} \approx 19.107 \quad \text{and} \quad Q = \frac{4107}{121} \approx 33.942.$$

Part (b): After the merger and the agreement to maximize total profit, the new partners plan to maximize

$$R(x, y) = P(x) + Q(y) = -2x^2 - 3y^2 + 3xy + 10x + 17y - 25.$$

The junior partner computes both partial derivatives and sets both equal to zero:

$$10 - 4x + 3y = 0, \quad 17 + 3x - 6y = 0.$$

The simultaneous solution is

$$x = \frac{37}{5}, \quad y = \frac{98}{15}$$

for a combined profit of

$$\frac{1013}{15} \approx 67.533 > 53.050 \approx \frac{6419}{121} = \frac{2312}{121} + \frac{4107}{121}.$$

Thus the merger increases total profit.

C13S05.063: Let x be the number of sheep, y the number of hogs, and z the number of head of cattle. Suppose that 60 cattle use 1 unit of land. Then each head of cattle uses $\frac{1}{60}$ units of land. By similar reasoning, each hog uses $\frac{1}{120}$ units of land and each sheep uses $\frac{1}{80}$ units of land. This leads to the side condition

$$\frac{x}{80} + \frac{y}{120} + \frac{z}{60} = 1$$

for each unit of land available. Let's write this in the simpler form $3x + 2y + 4z = 240$. An additional condition in the problem is that $y \geq x + z$. We are now to maximize the profit P per unit of land, given by $P(x, y, z) = 10x + 8y + 20z$.

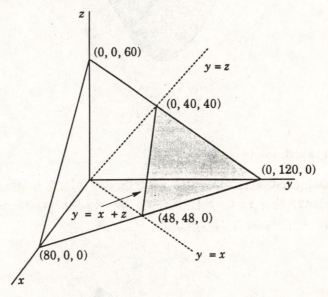

The domain of P is the triangular region shown shaded in the preceding figure. It is obtained as follows: Draw the part of the plane $3x + 2y + 4z = 240$ that lies in the first octant (because none of x, y, or z can be negative). The intersections of that plane with the positive coordinate planes are shown as solid lines. The intersection of the plane $y = x + z$ with the positive coordinate planes is shown as a pair of dashed lines. The condition $y \geq x + z$ implies that only the part of the first plane to the *right* of the second may be used as the domain of P. Therefore we arrive at the shaded triangle as the domain of the profit function. Finally, because P is a linear function of x, y, and z, its maximum and minimum values occur at the vertices of the shaded triangle. Here, then, are the results:

Vertex: $(48, 48, 0)$; Profit: $864 per unit of land.

Vertex: $(0, 120, 0)$; Profit: $960 per unit of land.

Vertex: $(0, 40, 40)$; Profit: $1120 per unit of land.

Were it not for the restriction of the state law mentioned in the problem, the farmer could maximize her profit per unit of land by raising only cattle: $P(0, 0, 60) = 1200$. *Answer:* Raise 40 hogs and 40 cattle per unit of land, but no sheep.

C13S05.065: Here is the case $a = 3$, $b = 1$, $c = 2$ (via *Mathematica 3.0*):

```
ParametricPlot3D[ { r*Cos[t], r*Sin[t], r*r*(a*(Cos[t])^2 + 2*b*Cos[t]*Sin[t] +
```

```
    c*(Sin[t])∧2) }, { r, 0, 1 }, { t, 0, 2*Pi }, BoxRatios → { 1, 1, 1 },
    ViewPoint → { −1.2, 2.1, 1.6 } ];
```

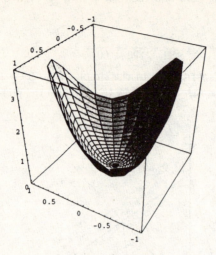

Here is the case $a = 1$, $b = 3$, $c = 1$:

```
ParametricPlot3D[ { r*Cos[t], r*Sin[t], r*r*(a*(Cos[t])∧2 + 2*b*Cos[t]*Sin[t] +
    c*(Sin[t])∧2) }, { r, 0, 2 }, { t, 0, 2*Pi }, BoxRatios → { 1, 1, 1 },
    ViewPoint → { −1.2, 2.1, 1.6 } ];
```

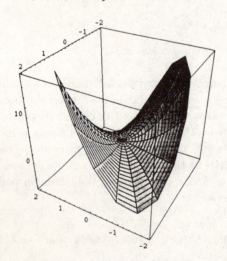

Space prohibits our further experimentation along these lines, but these two examples certainly support the conclusions given in the statement of Problem 65.

C13S05.067: Upon changing to polar coordinates we get $f(x, y) = x^4 + 2bx^2y^2 + y^4 = r^4 g(\theta)$ where

$$g(\theta) = \cos^4 \theta + 2b \cos^2 \theta \sin^2 \theta + \sin^4 \theta.$$

Upon differentiating and simplifying we find that

$$g'(\theta) = 4(b - 1)(\cos^2 \theta - \sin^2 \theta) \cos \theta \sin \theta.$$

722

Hence the critical points of g are

- multiples of $\pi/2$, where $g(\theta) = 1$, and
- odd multiples of $\pi/4$, where $\cos^2\theta = \sin^2\theta = \frac{1}{2}$ so $g(\theta) = \frac{1}{4} + \frac{1}{2}b + \frac{1}{4} = \frac{1}{2}(b+1)$.

If $b > -1$ it follows that $g(\theta)$ is always positive, so $f(x, y) = r^4 g(\theta)$ is positive except at the origin. But if $b < -1$ it follows that $g(\theta)$ attains both positive and negative values, so $z = f(x, y) = r^4 g(\theta)$ exhibits a saddle point at the origin. —C.H.E.

C13S05.069: When we set all first-order partial derivative of g equal to zero, we get the simultaneous equations

$$g_x(x,\, y,\, z) = 4x^3 - 16xy^2 = 0,$$
$$g_y(x,\, y,\, z) = 4y^3 - 16x^2 y = 0,$$
$$g_z(x,\, y,\, z) = z = 0.$$

The only solution of these equations is $x = y = z = 0$, at which $g(0, 0, 0) = 12$. But

$$g(1,\, 1,\, 0) = 6 < 12 \quad \text{and} \quad g(0,\, 0,\, 2) = 28 > 12.$$

Therefore g has no global maximum or minimum at the origin. Examination of the behavior of $g(x,\, y,\, z)$ on the two lines

$$y = x, \quad z = 0 \qquad \text{and} \qquad x = 0, \quad y = 0$$

is enough to establish that g has no extrema, global or local. Thus one cannot conclude that g has an extremum at a point where all its partial derivatives are zero.

Section 13.6

C13S06.001: If $w = 3x^2 + 4xy - 2y^3$, then $dw = (6x + 4y)\, dx + (4x - 6y^2)\, dy$.

C13S06.003: If $w = \sqrt{1 + x^2 + y^2}$, then

$$dw = \frac{x}{\sqrt{1 + x^2 + y^2}}\, dx + \frac{y}{\sqrt{1 + x^2 + y^2}}\, dy = \frac{x\, dx + y\, dy}{\sqrt{1 + x^2 + y^2}}.$$

C13S06.005: If $w(x,\, y) = \arctan\left(\dfrac{x}{y}\right)$, then $dw = \dfrac{y\, dx - x\, dy}{x^2 + y^2}$.

C13S06.007: If $w = \ln(x^2 + y^2 + z^2)$, then

$$dw = \frac{2x\, dx}{x^2 + y^2 + z^2} + \frac{2y\, dy}{x^2 + y^2 + z^2} + \frac{2z\, dz}{x^2 + y^2 + z^2} = \frac{2x\, dx + 2y\, dy + 2z\, dz}{x^2 + y^2 + z^2}.$$

C13S06.009: If $w = x \tan yz$, then $dw = \tan yz\, dx + xz \sec^2 yz\, dy + xy \sec^2 yz\, dz$.

C13S06.011: If $w = e^{-xyz}$, then $dw = -yz e^{-xyz}\, dx - xz e^{-xyz}\, dy - xy e^{-xyz}\, dz$.

C13S06.013: If $w = u^2 \exp(-v^2)$, then $dw = 2u \exp(-v^2)\, du - 2u^2 v \exp(-v^2)\, dv$.

C13S06.015: If $w = \sqrt{x^2 + y^2 + z^2}$, then

$$dw = \frac{x\,dx}{\sqrt{x^2 + y^2 + z^2}} + \frac{y\,dy}{\sqrt{x^2 + y^2 + z^2}} + \frac{z\,dz}{\sqrt{x^2 + y^2 + z^2}} = \frac{x\,dx + y\,dy + z\,dz}{\sqrt{x^2 + y^2 + z^2}}.$$

C13S06.017: If $w = f(x, y) = \sqrt{x^2 + y^2}$, then

$$dw = \frac{x\,dx + y\,dy}{\sqrt{x^2 + y^2}}.$$

Choose $x = 3$, $y = 4$, $dx = -0.03$, and $dy = 0.04$. Then

$$f(2.97, 4.04) \approx f(3, 4) + \frac{3 \cdot (-0.03) + 4 \cdot (0.04)}{\sqrt{3^2 + 4^2}} = \frac{2507}{500} = 5.014.$$

Compare with the true value of

$$f(2.97, 4.04) = \frac{\sqrt{10057}}{20} \approx 5.014229751417.$$

C13S06.019: If $w = f(x, y) = \dfrac{1}{1 + x + y}$, then

$$dw = -\frac{dx + dy}{(1 + x + y)^2}.$$

Choose $x = 3$, $y = 6$, $dx = 0.02$, and $dy = 0.05$. Then

$$f(3.02, 6.05) \approx f(3, 6) - \frac{0.02 + 0.05}{(1 + 3 + 6)^2} = \frac{1}{10} - \frac{7}{10000} = \frac{993}{10000} = 0.0993.$$

Compare with the true value of

$$f(3.02, 6.05) = \frac{100}{1007} \approx 0.0993048659384310.$$

C13S06.021: If $w = f(x, y, z) = \sqrt{x^2 + y^2 + z^2}$, then

$$dw = \frac{x\,dx + y\,dy + z\,dz}{\sqrt{x^2 + y^2 + z^2}}.$$

Choose $x = 3$, $y = 4$, $z = 12$, $dx = 0.03$, $dy = -0.04$, and $dz = 0.05$. Then

$$f(3.03, 3.96, 12.05) \approx f(3, 4, 12) + \frac{3 \cdot (0.03) - 4 \cdot (0.04) + 12 \cdot (0.05)}{\sqrt{3^2 + 4^2 + 12^2}}$$

$$= 13 + \frac{53}{1300} = \frac{16953}{1300} \approx 13.040769230769.$$

Compare with the true value of

$$f(3.03, 3.96, 12.05) = \frac{\sqrt{68026}}{20} \approx 13.040897208398.$$

C13S06.023: If $w = f(x, y, z) = e^{-xyz}$, then

$$dw = -e^{-xyz}(yz\,dx + xz\,dy + xy\,dz).$$

Take $x = 1$, $y = 0$, $z = -2$, $dx = 0.02$, $dy = 0.03$, and $dz = -0.02$. Then

$$f(1.02,\,0.03,\,-2.02) \approx f(1,\,0,\,-2) - e^0(0 - 2\cdot(0.03) + 0) = 1 + \frac{3}{50} = 1.06.$$

Compare with the exact value, which is

$$f(1.02,\,0.03,\,-2.02) = \exp\left(\frac{15453}{250000}\right) \approx 1.0637623386083891.$$

C13S06.025: If $w = f(x,\,y) = \left(\sqrt{x} + \sqrt{y}\,\right)^2$, then

$$dw = \frac{\sqrt{x} + \sqrt{y}}{\sqrt{x}}\,dx + \frac{\sqrt{x} + \sqrt{y}}{\sqrt{y}}\,dy.$$

Take $x = 16$, $y = 100$, $dx = -1$, and $dy = -1$. Then

$$f(15,\,99) \approx f(16,\,100) + \frac{4 + 10}{4}\cdot(-1) + \frac{4 + 10}{10}\cdot(-1) = 196 - \frac{49}{10} = \frac{1911}{10} = 191.1.$$

By comparison, the exact value is

$$f(15,\,99) = \left(\sqrt{15} + \sqrt{99}\,\right)^2 \approx 191.0713954719907741.$$

C13S06.027: If $w = f(x,\,y) = \exp(x^2 - y^2)$, then

$$dw = 2x\exp(x^2 - y^2)\,dx - 2y\exp(x^2 - y^2)\,dy.$$

Take $x = 1$, $y = 1$, $dx = 0.1$, and $dy = -0.1$. Then

$$f(1.1,\,0.9) \approx f(1,\,1) + 2\cdot(0.1) - 2\cdot(-0.1) = 1 + \frac{2}{5} = \frac{7}{5} = 1.4.$$

Compare with the true value, which is

$$f(1.1,\,0.9) = e^{2/5} \approx 1.4918246976412703.$$

C13S06.029: If $w = f(x,\,y,\,z) = \sqrt{x^2 + y^2 + z^2}$, then

$$dw = \frac{x\,dx + y\,dy + z\,dz}{\sqrt{x^2 + y^2 + z^2}}.$$

Take $x = 3$, $y = 4$, $z = 12$, $dx = 0.1$, $dy = 0.2$, and $dz = -0.3$. Then

$$f(3.1,\,4.2,\,11.7) \approx f(3,\,4,\,12) + \frac{0.3 + 0.8 - 3.6}{13} = 13 - \frac{5}{26} = \frac{333}{26} \approx 12.8076923076923077.$$

The true value is

$$f(3.1,\,4.2,\,11.7) = \frac{\sqrt{16414}}{10} \approx 12.8117133904876283.$$

C13S06.031: Given: The point $Q(1, 2)$ on the curve $f(x, y) = 0$, where $f(x, y) = 2x^3 + 2y^3 - 9xy$. Then

$$df = (6x^2 - 9y) \, dx + (6y^2 - 9x) \, dy = 0.$$

Choose $x = 1$, $y = 2$, and $dx = 0.1$. Then

$$(6 - 18) \cdot \frac{1}{10} + (24 - 9) \, dy = 0;$$

$$dy = \frac{1}{15} \cdot \frac{12}{10} = \frac{12}{150} = \frac{2}{25} = 0.08.$$

So the point P on the curve $f(x, y) = 0$ near Q and with x-coordinate 1.1 has y-coordinate

$$y \approx 2 + \frac{2}{25} = \frac{52}{25} = 2.08.$$

The true value of the y-coordinate is approximately 2.0757642703016864.

C13S06.033: Suppose that the base of the rectangle has length x and that its height is y. Then its area is $w = f(x, y) = xy$, and $dw = y \, dx + x \, dy$. Choose $x = 10$, $y = 15$, $dx = 0.1$, and $dy = 0.1$. Then $dw = 2.5$; this is the estimate of the maximum error in computing the area of the rectangle. The actual maximum error possible is $f(10.1, 15.1) - f(10, 15) = 2.51$.

C13S06.035: The volume of the cone is given by

$$w = f(r, h) = \frac{\pi}{3} r^2 h, \quad \text{so that} \quad dw = \frac{2\pi}{3} rh \, dr + \frac{\pi}{3} r^2 \, dh.$$

Choose $r = 5$, $h = 10$, and $dr = dh = 0.1$. Then

$$dw = \frac{2\pi}{3} \cdot 5 \cdot 10 \cdot (0.1) + \frac{\pi}{3} \cdot 5^2 \cdot (0.1) = \frac{25\pi}{6} \approx 13.0899693899574718$$

is an estimate of the maximum error in measuring the volume of the cylinder. The true value of the maximum error is

$$f(5.1, 10.1) - f(5, 10) = \frac{12701\pi}{3000} \approx 13.3004560977479880.$$

C13S06.037: If the sides of the field are x and y and the angle between them is θ, then the area of the field is given by

$$w = f(x, y, \theta) = \frac{1}{2} xy \sin \theta,$$

so that

$$dw = \frac{1}{2} y \sin \theta \, dx + \frac{1}{2} x \sin \theta \, dy + \frac{1}{2} xy \cos \theta \, d\theta.$$

If $x = 500$, $y = 700$, $\theta = \pi/6$, $dx = dy = 1$, and $d\theta = \pi/720$, then

$$dw = 350 \cdot \frac{1}{2} \cdot 1 + 250 \cdot \frac{1}{2} \cdot 1 + \frac{1}{2} \cdot 500 \cdot 700 \cdot \frac{\sqrt{3}}{2} \cdot \frac{\pi}{720} = 300 + \frac{4375\pi\sqrt{3}}{36} \approx 961.2810182103919247$$

(in square feet) is an estimate of the maximum error in computing the area of the field. The true value of the maximum error is

$$f(501, 701, (\pi/6) + (\pi/720)) - f(500, 700, \pi/6) \approx 962.9622561829376760$$

(in square feet). The former amounts to approximately 0.0220679756246646 acres (there are 43560 square feet in one acre).

C13S06.039: The period T of a pendulum of length L is given (approximately) by

$$T = 2\pi \left(\frac{L}{g} \right)^{1/2}, \quad \text{for which} \quad dT = \left(\frac{g}{L} \right)^{1/2} \cdot \frac{\pi g \, dL - \pi L \, dg}{g^2}.$$

If $L = 2$, $dL = 1/12$, $g = 32$, and $dg = 0.2$, then

$$dT = \frac{17\pi}{1920} \approx 0.0278161849536596.$$

The true value of the increase in the period is

$$T(2 + 1/12, \ 32.2) - T(2, \ 32) \approx 0.0274043631738259.$$

C13S06.041: Given: $R(v_0, \alpha) = \dfrac{1}{32} (v_0)^2 \sin 2\alpha$, we first compute

$$dR = \frac{1}{16} \left(v_0 \sin 2\alpha \, dv_0 + (v_0)^2 \cos 2\alpha \, d\alpha \right).$$

Substitution of $v_0 = 400$, $dv_0 = 10$, $\alpha = \pi/6$, and $d\alpha = \pi/180$ yields

$$dR = 125\sqrt{3} + \frac{250\pi}{9} \approx 303.7728135458261405$$

as an estimate of the increase in the range. The true value of the increase is

$$R(410, (\pi/6) + (\pi/180)) - R(400, \pi/6) = -2500\sqrt{3} + \frac{42025}{8} \sin \left(\frac{31\pi}{90} \right) \approx 308.1070548148573585.$$

C13S06.043: Part (a): If $(x, y) \to (0, 0)$ along the line $y = x$, then

$$\lim_{(x,y) \to (0,0)} f(x, y) = \lim_{x \to 0} f(x, x) = \lim_{x \to 0} 1 = 1.$$

But if $(x, y) \to (0, 0)$ along the line $y = 0$, then

$$\lim_{(x,y) \to (0,0)} f(x, y) = \lim_{x \to 0} f(x, 0) = \lim_{x \to 0} 0 = 0.$$

Therefore f is not continuous at $(0, 0)$.

Part (b): We compute the partial derivatives of f at $(0, 0)$ by the definition:

$$f_x(0, 0) = \lim_{h \to 0} \frac{f(0 + h, 0) - f(0, 0)}{h} = \lim_{h \to 0} \frac{f(h, 0)}{h} = \lim_{h \to 0} \frac{0}{h} = 0;$$

$$f_y(0, 0) = \lim_{k \to 0} \frac{f(0, 0 + k) - f(0, 0)}{k} = \lim_{k \to 0} \frac{f(0, k)}{k} = \lim_{k \to 0} \frac{0}{k} = 0.$$

Therefore both f_x and f_y exist at $(0, 0)$ but f is not continuous at $(0, 0)$.

C13S06.045: Given:

$$f(x, y) = y^2 + x^2 \sin \frac{1}{x}$$

if $x \neq 0$; $f(0, y) = y^2$. Then

$$f_x(0, 0) = \lim_{h \to 0} \frac{f(h, 0) - f(0, 0)}{h} = \lim_{h \to 0} \frac{1}{h} \cdot h^2 \sin \frac{1}{h} = \lim_{h \to 0} h \sin \frac{1}{h} = 0$$

and

$$f_y(0, 0) = \lim_{k \to 0} \frac{f(0, k) - f(0, 0)}{k} = \lim_{k \to 0} \frac{k^2}{k} = \lim_{k \to 0} k = 0.$$

Therefore the linear approximation to f at $(0, 0)$ can only be $z = 0$. Moreover,

$$0 \leq f(x, y) \leq x^2 + y^2 = g(x, y)$$

for all (x, y), and $z = g(x, y)$ has the tangent plane $z = 0$ at $(0, 0)$. Therefore $z = 0$ does approximate $f(x, y)$ accurately at and near $(0, 0)$. That is, f is differentiable at $(0, 0)$. But if $x \neq 0$, then

$$f_x(x, y) = 2x \sin \frac{1}{x} - \cos \frac{1}{x},$$

so $f_x(x, y)$ has no limit as $(x, y) \to (0, 0)$ along the x-axis. Therefore $f_x(x, y)$ is not continuous at $(0, 0)$, and thus f is not continuously differentiable at $(0, 0)$ even though it is differentiable there.

C13S06.047: Suppose that the function f of $n \geq 2$ variables is differentiable at $\mathbf{a}$. Then there exists a constant vector $\mathbf{c} = \langle c_1, c_2, \ldots, c_n \rangle$ such that

$$\lim_{\mathbf{h} \to 0} \frac{f(\mathbf{a} + \mathbf{h}) - f(\mathbf{a}) - \mathbf{c} \cdot \mathbf{h}}{|\mathbf{h}|} = 0.$$

Therefore

$$\left(\lim_{\mathbf{h} \to 0} |\mathbf{h}| \right) \cdot \left(\lim_{\mathbf{h} \to 0} \frac{f(\mathbf{a} + \mathbf{h}) - f(\mathbf{a}) - \mathbf{c} \cdot \mathbf{h}}{|\mathbf{h}|} \right) = 0 \cdot 0 = 0,$$

and therefore

$$\lim_{\mathbf{h} \to 0} \left[f(\mathbf{a} + \mathbf{h}) - f(\mathbf{a}) - \mathbf{c} \cdot \mathbf{h} \right] = 0.$$

But $\mathbf{c} \cdot \mathbf{h} \to 0$ as $\mathbf{h} \to 0$ because $\mathbf{c}$ is a constant vector. Therefore

$$\left(\lim_{\mathbf{h} \to 0} \left[f(\mathbf{a} + \mathbf{h}) - f(\mathbf{a}) - \mathbf{c} \cdot \mathbf{h} \right] \right) + \left(\lim_{\mathbf{h} \to 0} \left[\mathbf{c} \cdot \mathbf{h} + f(\mathbf{a}) \right] \right) = 0 + 0 + f(\mathbf{a}) = f(\mathbf{a}),$$

and thus we see that

$$\lim_{\mathbf{h} \to 0} f(\mathbf{a} + \mathbf{h}) = f(\mathbf{a}).$$

Therefore f is continuous at $\mathbf{x} = \mathbf{a}$. That is, the function f is continuous wherever it is differentiable.

Section 13.7

C13S07.001: If $w = \exp(-x^2 - y^2)$, $x = t$, and $y = t^{1/2}$, then

$$\frac{dw}{dt} = \frac{\partial w}{\partial x} \cdot \frac{dx}{dt} + \frac{\partial w}{\partial y} \cdot \frac{dy}{dt}$$

$$= -2x \exp(-x^2 - y^2) - yt^{-1/2} \exp(-x^2 - y^2) = -2t \exp(-t^2 - t) - \exp(-t^2 - t).$$

Alternatively, $w = \exp(-t^2 - t)$, and hence

$$\frac{dw}{dt} = -(2t + 1) \exp(-t^2 - t).$$

C13S07.003: If $w(x, y, z) = \sin xyz$, $x = t$, $y = t^2$, and $z = t^3$, then

$$\frac{dw}{dt} = \frac{\partial w}{\partial x} \cdot \frac{dx}{dt} + \frac{\partial w}{\partial y} \cdot \frac{dy}{dt} + \frac{\partial w}{\partial z} \cdot \frac{dz}{dt}$$

$$= yz \cos xyz + 2txz \cos xyz + 3t^2 xy \cos xyz = t^5 \cos t^6 + 2t^5 \cos t^6 + 3t^5 \cos t^6 = 6t^5 \cos t^6.$$

Alternatively, $w = \sin t^6$, and thus $\dfrac{dw}{dt} = 6t^5 \cos t^6$.

C13S07.005: If $w(x, y, z) = \ln(x^2 + y^2 + z^2)$, $x = s - t$, $y = s + t$, and $z = 2(st)^{1/2}$, then

$$\frac{\partial w}{\partial s} = \frac{\partial w}{\partial x} \cdot \frac{\partial x}{\partial s} + \frac{\partial w}{\partial y} \cdot \frac{\partial y}{\partial s} + \frac{\partial w}{\partial z} \cdot \frac{\partial z}{\partial s}$$

$$= \frac{2x}{x^2 + y^2 + z^2} + \frac{2y}{x^2 + y^2 + z^2} + \frac{2tz}{(st)^{1/2}(x^2 + y^2 + z^2)} = \frac{2(st)^{1/2}x + 2(st)^{1/2}y + 2tz}{(st)^{1/2}(x^2 + y^2 + z^2)}$$

$$= \frac{2(s - t)(st)^{1/2} + 4t(st)^{1/2} + 2(s + t)(st)^{1/2}}{(st)^{1/2}[(s - t)^2 + 4st + (s + t)^2]} = \frac{2(2s + 2t)}{2s^2 + 4st + 2t^2} = \frac{2}{s + t}$$

and

$$\frac{\partial w}{\partial t} = \frac{\partial w}{\partial x} \cdot \frac{\partial x}{\partial t} + \frac{\partial w}{\partial y} \cdot \frac{\partial y}{\partial t} + \frac{\partial w}{\partial z} \cdot \frac{\partial z}{\partial t}$$

$$= -\frac{2x}{x^2 + y^2 + z^2} + \frac{2y}{x^2 + y^2 + z^2} + \frac{2sz}{(st)^{1/2}(x^2 + y^2 + z^2)} = \frac{-2(st)^{1/2}x + 2(st)^{1/2}y + 2sz}{(st)^{1/2}(x^2 + y^2 + z^2)}$$

$$= \frac{4s(st)^{1/2} - 2(s - t)(st)^{1/2} + 2(s + t)(st)^{1/2}}{(st)^{1/2}[(s - t)^2 + 4st + (s + t)^2]} = \frac{2(2s + 2t)}{2s^2 + 4st + 2t^2} = \frac{2}{s + t}.$$

Alternatively,

$$w(s, t) = \ln\left((s - t)^2 + 4st + (s + t)^2\right) = \ln(2s^2 + 4st + 2t^2),$$

and therefore

$$\frac{\partial w}{\partial s} = \frac{2}{s + t} \quad \text{and} \quad \frac{\partial w}{\partial t} = \frac{2}{s + t}.$$

729

C13S07.007: If $w(u, v, z) = \sqrt{u^2 + v^2 + z^2}$, $u = 3e^t \sin s$, $v = 3e^t \cos s$, and $z = 4e^t$, then

$$\frac{\partial w}{\partial s} = \frac{3ue^t \cos s}{\sqrt{u^2 + v^2 + z^2}} - \frac{3ve^t \sin s}{\sqrt{u^2 + v^2 + z^2}} + 0 = \frac{3e^t(u \cos s - v \sin s)}{\sqrt{u^2 + v^2 + z^2}} = 0$$

because $u \cos s - v \sin s = 3e^t \sin s \cos s - 3e^t \cos s \sin s = 0$. But

$$\frac{\partial w}{\partial t} = \frac{3ue^t \sin s}{\sqrt{u^2 + v^2 + z^2}} + \frac{3ve^t \cos s}{\sqrt{u^2 + v^2 + z^2}} + \frac{4ze^t}{\sqrt{u^2 + v^2 + z^2}}$$

$$= \frac{e^t(3u \sin s + 3v \cos s + 4z)}{\sqrt{u^2 + v^2 + z^2}} = \frac{e^t(16e^t + 9e^t \cos^2 s + 9e^t \sin^2 s)}{\sqrt{16e^{2t} + 9e^{2t} \cos^2 s + 9e^{2t} \sin^2 s}} = 5e^t.$$

Alternatively,

$$w(s, t) = \sqrt{16e^{2t} + 9e^{2t} \cos^2 s + 9e^{2t} \sin^2 s} = 5e^t,$$

and therefore

$$\frac{\partial w}{\partial s} = 0 \quad \text{and} \quad \frac{\partial w}{\partial t} = 5e^t.$$

C13S07.009: Because $r(x, y, z) = \exp(yz + xz + xy)$, we find that

$$\frac{\partial r}{\partial x} = (y + z) \exp(yz + xz + xy), \qquad \frac{\partial r}{\partial y} = (x + z) \exp(yz + xz + xy),$$

$$\text{and} \qquad \frac{\partial r}{\partial z} = (x + y) \exp(yz + xz + xy).$$

C13S07.011: Because

$$r(x, y, z) = \sin\left(\frac{\sqrt{xy^2z^3}}{\sqrt{x + 2y + 3z}}\right),$$

we find that

$$\frac{\partial r}{\partial x} = \frac{(2y + 3z)\sqrt{xy^2z^3}}{2x(x + 2y + 3z)^{3/2}} \cos\left(\frac{\sqrt{xy^2z^3}}{\sqrt{x + 2y + 3z}}\right),$$

$$\frac{\partial r}{\partial y} = \frac{(x + y + 3z)\sqrt{xy^2z^3}}{y(x + 2y + 3z)^{3/2}} \cos\left(\frac{\sqrt{xy^2z^3}}{\sqrt{x + 2y + 3z}}\right), \quad \text{and}$$

$$\frac{\partial r}{\partial z} = \frac{3(x + 2y + 2z)\sqrt{xy^2z^3}}{2z(x + 2y + 3z)^{3/2}} \cos\left(\frac{\sqrt{xy^2z^3}}{\sqrt{x + 2y + 3z}}\right).$$

C13S07.013: If $p = f(x, y)$, $x = x(u, v, w)$, and $y = y(u, v, w)$, then

$$\frac{\partial p}{\partial u} = \frac{\partial f}{\partial x} \cdot \frac{\partial x}{\partial u} + \frac{\partial f}{\partial y} \cdot \frac{\partial y}{\partial u},$$

$$\frac{\partial p}{\partial v} = \frac{\partial f}{\partial x} \cdot \frac{\partial x}{\partial v} + \frac{\partial f}{\partial y} \cdot \frac{\partial y}{\partial v}, \quad \text{and}$$

$$\frac{\partial p}{\partial w} = \frac{\partial f}{\partial x} \cdot \frac{\partial x}{\partial w} + \frac{\partial f}{\partial y} \cdot \frac{\partial y}{\partial w}.$$

C13S07.015: If $p = f(u, v, w)$, $u = u(x, y, z)$, $v = v(x, y, z)$, and $w = w(x, y, z)$, then

$$\frac{\partial p}{\partial x} = \frac{\partial f}{\partial u} \cdot \frac{\partial u}{\partial x} + \frac{\partial f}{\partial v} \cdot \frac{\partial v}{\partial x} + \frac{\partial f}{\partial w} \cdot \frac{\partial w}{\partial x},$$

$$\frac{\partial p}{\partial y} = \frac{\partial f}{\partial u} \cdot \frac{\partial u}{\partial y} + \frac{\partial f}{\partial v} \cdot \frac{\partial v}{\partial y} + \frac{\partial f}{\partial w} \cdot \frac{\partial w}{\partial y}, \quad \text{and}$$

$$\frac{\partial p}{\partial z} = \frac{\partial f}{\partial u} \cdot \frac{\partial u}{\partial z} + \frac{\partial f}{\partial v} \cdot \frac{\partial v}{\partial z} + \frac{\partial f}{\partial w} \cdot \frac{\partial w}{\partial z}.$$

C13S07.017: If $p = f(w)$ and $w = w(x, y, z, u, v)$, then

$$\frac{\partial p}{\partial x} = f'(w) \cdot \frac{\partial w}{\partial x}, \quad \frac{\partial p}{\partial y} = f'(w) \cdot \frac{\partial w}{\partial y}, \quad \frac{\partial p}{\partial z} = f'(w) \cdot \frac{\partial w}{\partial z}, \quad \frac{\partial p}{\partial u} = f'(w) \cdot \frac{\partial w}{\partial u}, \quad \text{and} \quad \frac{\partial p}{\partial v} = f'(w) \cdot \frac{\partial w}{\partial v}.$$

C13S07.019: Let $F(x, y, z) = x^{2/3} + y^{2/3} + z^{2/3} - 1$. Then

$$\frac{\partial z}{\partial x} = -\frac{F_x}{F_z} = -\frac{\frac{2}{3}x^{-1/3}}{\frac{2}{3}z^{-1/3}} = -\frac{z^{1/3}}{x^{1/3}} \quad \text{and}$$

$$\frac{\partial z}{\partial y} = -\frac{F_y}{F_z} = -\frac{\frac{2}{3}y^{-1/3}}{\frac{2}{3}z^{-1/3}} = -\frac{z^{1/3}}{y^{1/3}}.$$

C13S07.021: Let $F(x, y, z) = xe^{xy} + ye^{zx} + ze^{xy} - 3$. Then

$$\frac{\partial z}{\partial x} = -\frac{F_x}{F_z} = -\frac{e^{xy} + xye^{xy} + yze^{zx} + yze^{xy}}{xye^{zx} + e^{xy}} \quad \text{and} \quad \frac{\partial z}{\partial y} = -\frac{F_y}{F_z} = -\frac{x^2e^{xy} + e^{zx} + xze^{xy}}{xye^{zx} + e^{xy}}.$$

C13S07.023: Let

$$F(x, y, z) = \frac{x^2}{a^2} + \frac{y^2}{b^2} + \frac{z^2}{c^2} - 1.$$

Then

$$\frac{\partial z}{\partial x} = -\frac{F_x}{F_z} = -\frac{c^2 x}{a^2 z} \quad \text{and} \quad \frac{\partial z}{\partial y} = -\frac{F_y}{F_z} = -\frac{c^2 y}{b^2 z}.$$

C13S07.025: If $w = u^2 + v^2 + x^2 + y^2$, $u = x - y$, and $v = x + y$, then

731

$$\frac{\partial w}{\partial x} = 2u \cdot u_x + 2v \cdot v_x + 2x = 2(x-y) + 2(x+y) + 2x = 6x \quad \text{and}$$

$$\frac{\partial w}{\partial y} = 2u \cdot u_y + 2v \cdot v_y + 2y = -2(x-y) + 2(x+y) + 2y = 6y.$$

C13S07.027: If $w(u, v, x, y) = xy \ln(u+v)$, $u = (x^2 + y^2)^{1/3}$, and $v = (x^3 + y^3)^{1/2}$, then

$$w_x = w_u u_x + w_v v_x + w_x \cdot 1 + w_y \cdot 0$$

$$= \frac{2x^2 y}{3(u+v)(x^2+y^2)^{2/3}} + \frac{3x^3 y}{2(u+v)(x^3+y^3)^{1/2}} + y \ln(u+v).$$

Then substitution of the formulas for u and v yields

$$\frac{\partial w}{\partial x} = \frac{2x^2 y}{3(x^2+y^2)^{2/3} \left[(x^2+y^2)^{1/3} + (x^3+y^3)^{1/2} \right]}$$

$$+ \frac{3x^3 y}{2(x^3+y^3)^{1/2} \left[(x^2+y^2)^{1/3} + (x^3+y^3)^{1/2} \right]} + y \ln \left((x^2+y^2)^{1/3} + (x^3+y^3)^{1/2} \right).$$

Similarly,

$$\frac{\partial w}{\partial y} = \frac{2xy^2}{3(x^2+y^2)^{2/3} \left[(x^2+y^2)^{1/3} + (x^3+y^3)^{1/2} \right]}$$

$$+ \frac{3xy^3}{2(x^3+y^3)^{1/2} \left[(x^2+y^2)^{1/3} + (x^3+y^3)^{1/2} \right]} + x \ln \left((x^2+y^2)^{1/3} + (x^3+y^3)^{1/2} \right).$$

C13S07.029: We differentiate the equation $x^2 + y^2 + z^2 = 9$ implicitly, first with respect to x, then with respect to y, and obtain

$$2x + 2z \cdot z_x = 0, \qquad 2y + 2z \cdot z_y = 0.$$

We substitute the coordinates of the point of tangency $P(1, 2, 2)$ and find that

$$2 + 4z_x = 0 \quad \text{and} \quad 4 + 4z_y = 0,$$

and it follows that at P, $z_x = -\frac{1}{2}$ and $z_y = -1$. Hence an equation of the plane tangent to the given surface at the point P is

$$z - 2 = -\frac{1}{2}(x-1) - (y-2);$$

that is, $x + 2y + 2z = 9$.

C13S07.031: Given: The surface with equation $x^3 + y^3 + z^3 = 5xyz$ and the point $P(2, 1, 1)$ on it. We differentiate the equation implicitly, first with respect to x, then with respect to y, and thereby obtain

$$3x^2 + 3z^2 \cdot z_x = 5yz + 5xy \cdot z_x, \qquad 3y^2 + 3z^2 \cdot z_y = 5xz + 5xy \cdot z_y.$$

Then we substitute the coordinates of P and find that

$$12 + 3z_x = 5 + 10z_x \quad \text{and} \quad 3 + 3z_y = 10 + 10z_y,$$

and it follows that $z_x = 1$ and $z_y = -1$ at the point P. Hence an equation of the plane tangent to the surface at P is

$$z - 1 = (x - 2) - (y - 1);$$

that is, $x - y - z = 0$.

C13S07.033: Suppose that the square base of the box measures x (inches) on each side and that its height is z. Suppose also that time t is measured in hours. Then the volume of the box is $V = x^2 z$, and by the chain rule

$$\frac{dV}{dt} = \frac{\partial V}{\partial x} \cdot \frac{dx}{dt} + \frac{\partial V}{\partial z} \cdot \frac{dz}{dt} = 2xz \cdot (-3) + x^2 \cdot (-2).$$

Thus when $x = 24$ and $z = 12$, we have

$$\frac{dV}{dt} = 2 \cdot 24 \cdot 12 \cdot (-3) + 24^2 \cdot (-2) = -2880$$

cubic inches per hour; that is, $-\frac{5}{3}$ cubic feet per hour.

C13S07.035: Let r denote the radius of the conical sandpile and h its height. Units: feet and minutes. We are given that, at the time t when when $h = 5$ and $r = 2$,

$$\frac{dh}{dt} = 0.4 \quad \text{and} \quad \frac{dr}{dt} = 0.7.$$

The volume of the sandpile is given by $V = \dfrac{1}{3}\pi r^2 h$, and thus

$$\frac{dV}{dt} = \frac{\partial V}{\partial r} \cdot \frac{dr}{dt} + \frac{\partial V}{\partial h} \cdot \frac{dh}{dt} = \left(\frac{2}{3}\pi r h\right) \cdot \frac{7}{10} + \left(\frac{1}{3}\pi r^2\right) \cdot \frac{2}{5}.$$

Thus when $h = 5$ and $r = 2$,

$$\frac{dV}{dt} = \frac{20\pi}{3} \cdot \frac{7}{10} + \frac{4\pi}{3} \cdot \frac{2}{5} = \frac{26\pi}{5} \approx 16.336282$$

(cubic feet per minute).

C13S07.037: For this gas sample, we have $V = 10$ when $p = 2$ and $T = 300$. Substitution in the equation $pV = nRT$ yields $nR = \frac{1}{15}$. Moreover, with time t measured in minutes, we have

$$V = \frac{nRT}{p}, \quad \text{so that} \quad \frac{dV}{dt} = nR\left(\frac{1}{p} \cdot \frac{dT}{dt} - \frac{T}{p^2} \cdot \frac{dp}{dt}\right).$$

Finally, substitution of the data $nR = \frac{1}{15}$, $V = 10$, $p = 2$, $T = 300$, $dT/dt = 10$, and $dp/dt = 1$ yields the conclusion that the volume of the gas sample is decreasing at the rate of $\frac{13}{3}$ liters per minute at the time in question.

C13S07.039: Given: $x = h(y, z)$ satisfies the equation $F(x, y, z) = 0$. Thus $F(h(y, z), y, z) \equiv 0$, and so implicit differentiation with respect to y yields

$$\frac{\partial F}{\partial x} \cdot \frac{\partial x}{\partial y} + \frac{\partial F}{\partial y} \cdot \frac{\partial y}{\partial y} + \frac{\partial F}{\partial z} \cdot 0 = 0.$$

Thus if $F_x \neq 0$, we find that $\dfrac{\partial x}{\partial y} = -\dfrac{F_y}{F_x}$.

C13S07.041: If $w = f(u)$ and $u = x + y$, then

$$\frac{\partial w}{\partial x} = f'(u) \cdot \frac{\partial u}{\partial x} = f'(u) = f'(u) \cdot \frac{\partial u}{\partial y} = \frac{\partial w}{\partial y}.$$

C13S07.043: If $w = f(x, y)$, $x = u + v$, and $y = u - v$, then

$$\frac{\partial w}{\partial v} = \frac{\partial w}{\partial x} \cdot \frac{\partial x}{\partial v} + \frac{\partial w}{\partial y} \cdot \frac{\partial y}{\partial v} = \frac{\partial w}{\partial x} - \frac{\partial w}{\partial y}.$$

Therefore

$$\frac{\partial^2 w}{\partial u \, \partial v} = \frac{\partial}{\partial u} \left(\frac{\partial w}{\partial x} - \frac{\partial w}{\partial y} \right)$$

$$= \frac{\partial^2 w}{\partial x^2} \cdot \frac{\partial x}{\partial u} + \frac{\partial^2 w}{\partial y \, \partial x} \cdot \frac{\partial y}{\partial u} - \frac{\partial^2 w}{\partial x \, \partial y} \cdot \frac{\partial x}{\partial u} - \frac{\partial^2 w}{\partial y^2} \cdot \frac{\partial y}{\partial u}$$

$$= \frac{\partial^2 w}{\partial x^2} + \frac{\partial^2 w}{\partial y \, \partial x} - \frac{\partial^2 w}{\partial x \, \partial y} - \frac{\partial^2 w}{\partial y^2} = \frac{\partial^2 w}{\partial x^2} - \frac{\partial^2 w}{\partial y^2}.$$

C13S07.045: If $w = f(x, y)$, $x = r \cos \theta$, and $y = r \sin \theta$, then

$$\frac{\partial w}{\partial r} = \frac{\partial w}{\partial x} \cos \theta + \frac{\partial w}{\partial y} \sin \theta;$$

$$\frac{\partial^2 w}{\partial r^2} = \frac{\partial}{\partial r} \left(\frac{\partial w}{\partial x} \cos \theta + \frac{\partial w}{\partial y} \sin \theta \right)$$

$$= \frac{\partial^2 w}{\partial x^2} \cdot \frac{\partial x}{\partial r} \cos \theta + \frac{\partial^2 w}{\partial y \, \partial x} \cdot \frac{\partial y}{\partial r} \cos \theta + \frac{\partial^2 w}{\partial x \, \partial y} \cdot \frac{\partial x}{\partial r} \sin \theta + \frac{\partial^2 w}{\partial y^2} \cdot \frac{\partial y}{\partial r} \sin \theta$$

$$= \frac{\partial^2 w}{\partial x^2} \cos^2 \theta + \frac{\partial^2 w}{\partial y \, \partial x} \sin \theta \cos \theta + \frac{\partial^2 w}{\partial x \, \partial y} \sin \theta \cos \theta + \frac{\partial^2 w}{\partial y^2} \sin^2 \theta;$$

$$\frac{1}{r} \cdot \frac{\partial w}{\partial r} = \frac{1}{r} \cdot \frac{\partial w}{\partial x} \cos \theta + \frac{1}{r} \cdot \frac{\partial w}{\partial y} \sin \theta;$$

$$\frac{\partial w}{\partial \theta} = \frac{\partial w}{\partial x} \cdot \frac{\partial x}{\partial \theta} + \frac{\partial w}{\partial y} \cdot \frac{\partial y}{\partial \theta} = \frac{\partial w}{\partial x} (-r \sin \theta) + \frac{\partial w}{\partial y} (r \cos \theta);$$

$$\frac{\partial^2 w}{\partial \theta^2} = \frac{\partial}{\partial \theta} \left(-r \cdot \frac{\partial w}{\partial x} \sin \theta + r \cdot \frac{\partial w}{\partial y} \cos \theta \right)$$

$$= -r \left(\frac{\partial w}{\partial x} \cos \theta + \frac{\partial^2 w}{\partial x^2} \cdot \frac{\partial x}{\partial \theta} \sin \theta + \frac{\partial^2 w}{\partial y \, \partial x} \cdot \frac{\partial y}{\partial \theta} \sin \theta \right)$$

$$+r\left(-\frac{\partial w}{\partial y}\sin\theta+\frac{\partial^2 w}{\partial x\,\partial y}\cdot\frac{\partial x}{\partial\theta}\cos\theta+\frac{\partial^2 w}{\partial y^2}\cdot\frac{\partial y}{\partial\theta}\cos\theta\right)$$

$$=-r\cdot\frac{\partial w}{\partial x}\cos\theta+r^2\cdot\frac{\partial^2 w}{\partial x^2}\sin^2\theta-r^2\cdot\frac{\partial^2 w}{\partial y\,\partial x}\cos\theta\sin\theta$$

$$-r\cdot\frac{\partial w}{\partial y}\sin\theta-r^2\cdot\frac{\partial^2 w}{\partial x\,\partial y}\sin\theta\cos\theta+r^2\cdot\frac{\partial^2 w}{\partial y^2}\cos^2\theta.$$

Therefore

$$\frac{\partial^2 w}{\partial r^2}+\frac{1}{r}\cdot\frac{\partial w}{\partial r}+\frac{1}{r^2}\cdot\frac{\partial^2 w}{\partial\theta^2}=\frac{\partial^2 w}{\partial x^2}\cos^2\theta+2\frac{\partial^2 w}{\partial y\,\partial x}\sin\theta\cos\theta+\frac{\partial^2 w}{\partial y^2}\sin^2\theta$$

$$+\frac{1}{r}\cdot\frac{\partial w}{\partial x}\cos\theta+\frac{1}{r}\cdot\frac{\partial w}{\partial y}\sin\theta-\frac{1}{r}\cdot\frac{\partial w}{\partial x}\cos\theta-\frac{1}{r}\cdot\frac{\partial w}{\partial y}\sin\theta$$

$$+\frac{\partial^2 w}{\partial x^2}\sin^2\theta+\frac{\partial^2 w}{\partial y^2}\cos^2\theta-2\frac{\partial^2 w}{\partial y\,\partial x}\sin\theta\cos\theta=\frac{\partial^2 w}{\partial x^2}+\frac{\partial^2 w}{\partial y^2}.$$

C13S07.047: Suppose that $w=f(r)$ where $r=(x^2+y^2+z^2)^{1/2}$. Then

$$\frac{\partial w}{\partial x}=f'(r)\frac{\partial r}{\partial x}=\frac{x}{(x^2+y^2+z^2)^{1/2}}f'(r),$$

and thus

$$\frac{\partial^2 w}{\partial x^2}=\frac{\partial}{\partial x}\left(\frac{x}{(x^2+y^2+z^2)^{1/2}}f'(r)\right)$$

$$=\frac{x}{(x^2+y^2+z^2)^{1/2}}\cdot\frac{\partial}{\partial x}\big(f'(r)\big)+f'(r)\cdot\frac{\partial}{\partial x}\left(\frac{x}{(x^2+y^2+z^2)^{1/2}}\right)$$

$$=-\frac{x^2 f'(r)}{(x^2+y^2+z^2)^{3/2}}+\frac{f'(r)}{(x^2+y^2+z^2)^{1/2}}+\frac{x^2 f''(r)}{x^2+y^2+z^2}$$

$$=-\frac{x^2 f'(r)}{r^3}+\frac{f'(r)}{r}+\frac{x^2 f''(r)}{r^2}.$$

Similarly,

$$\frac{\partial^2 w}{\partial y^2}=-\frac{y^2 f'(r)}{r^3}+\frac{f'(r)}{r}+\frac{y^2 f''(r)}{r^2}\quad\text{and}\quad\frac{\partial^2 w}{\partial z^2}=-\frac{z^2 f'(r)}{r^3}+\frac{f'(r)}{r}+\frac{z^2 f''(r)}{r^2}.$$

Hence

$$\frac{\partial^2 w}{\partial x^2}+\frac{\partial^2 w}{\partial y^2}+\frac{\partial^2 w}{\partial z^2}=-\frac{r^2 f'(r)}{r^3}+\frac{3f'(r)}{r}+\frac{r^2 f''(r)}{r^2}$$

$$=-\frac{f'(r)}{r}+\frac{3f'(r)}{r}+f''(r)=\frac{d^2 w}{dr^2}+\frac{2}{r}\cdot\frac{dw}{dr}.$$

C13S07.049: If $w=f(u,v)$, $u=x+y$, and $v=x-y$, then

$$\frac{\partial w}{\partial x} = \frac{\partial w}{\partial u} \cdot \frac{\partial u}{\partial x} + \frac{\partial w}{\partial v} \cdot \frac{\partial v}{\partial x} = \frac{\partial w}{\partial u} + \frac{\partial w}{\partial v} \quad \text{and}$$

$$\frac{\partial w}{\partial y} = \frac{\partial w}{\partial u} \cdot \frac{\partial u}{\partial y} + \frac{\partial w}{\partial v} \cdot \frac{\partial v}{\partial y} = \frac{\partial w}{\partial u} - \frac{\partial w}{\partial v}.$$

Therefore

$$\frac{\partial w}{\partial x} \cdot \frac{\partial w}{\partial y} = \left(\frac{\partial w}{\partial u}\right)^2 - \left(\frac{\partial w}{\partial v}\right)^2.$$

C13S07.051: We are given $w = f(x, y)$ and the existence of a constant α such that

$$x = u \cos \alpha - v \sin \alpha \quad \text{and} \quad y = u \sin \alpha + v \cos \alpha.$$

Then

$$w_u = w_x x_u + w_y y_u = w_x \cos \alpha + w_y \sin \alpha;$$

$$w_v = w_x x_v + w_y y_v = -w_x \sin \alpha + w_y \cos \alpha;$$

$$(w_u)^2 + (w_v)^2 = (w_x)^2 \cos^2 \alpha + 2 w_x w_y \sin \alpha \cos \alpha + (w_y)^2 \sin^2 \alpha$$

$$+ (w_x)^2 \sin^2 \alpha - 2 w_x w_y \sin \alpha \cos \alpha + (w_y)^2 \cos^2 \alpha$$

$$= (w_x)^2 + (w_y)^2.$$

C13S07.053: Using the *Suggestion* and the notation in the equations in (17), we have

$$\left(\frac{\partial x}{\partial y}\right)_z \cdot \left(\frac{\partial y}{\partial z}\right)_x \cdot \left(\frac{\partial z}{\partial x}\right)_y = \left(-\frac{F_y}{F_x}\right) \cdot \left(-\frac{F_z}{F_y}\right) \cdot \left(-\frac{F_x}{F_z}\right) = -1.$$

C13S07.055: If the equation $pV - nRT = 0$ implicitly defines the functions $T = f(p, V)$, $V = g(p, T)$, and $p = h(V, T)$, then

$$p + V \frac{\partial h}{\partial V} = 0, \quad \text{so} \quad -\frac{p}{V} = \frac{\partial h}{\partial V} = \left(\frac{\partial p}{\partial V}\right)_T,$$

$$p \frac{\partial g}{\partial T} - nR = 0, \quad \text{so} \quad \frac{nR}{p} = \frac{\partial g}{\partial T} = \left(\frac{\partial V}{\partial T}\right)_p, \quad \text{and}$$

$$V - nR \frac{\partial f}{\partial p} = 0, \quad \text{so} \quad \frac{V}{nR} = \frac{\partial f}{\partial p} = \left(\frac{\partial T}{\partial p}\right)_V.$$

Therefore

$$\left(\frac{\partial p}{\partial V}\right)_T \left(\frac{\partial V}{\partial T}\right)_p \left(\frac{\partial T}{\partial p}\right)_V = \left(-\frac{p}{V}\right)\left(\frac{nR}{p}\right)\left(\frac{V}{nR}\right) = -1.$$

C13S07.057: First note that

$$\frac{\partial p}{\partial T} = \frac{\alpha}{\beta} = \frac{1.8 \times 10^6}{3.9 \times 10^4} = \frac{600}{13}.$$

Hence an increase of 5° in the Celsius temperature multiplies the initial pressure (1 atm) by $\frac{3000}{13} \approx 230.77$, so the bulb will burst.

C13S07.059: Here we have

$$T'(\rho, \phi, \theta) = \begin{bmatrix} \sin\phi\cos\theta & \rho\cos\phi\cos\theta & -\rho\sin\phi\sin\theta \\ \sin\phi\sin\theta & \rho\cos\phi\sin\theta & \rho\sin\phi\cos\theta \\ \cos\phi & -\rho\sin\phi & 0 \end{bmatrix}.$$

Therefore

$$|T'(\rho, \phi, \theta)| = \rho^2\sin^3\phi\sin^2\theta + \rho^2\sin\phi\cos^2\phi\cos^2\theta + \rho^2\cos^2\phi\sin\phi\sin^2\theta + \rho^2\sin^3\phi\cos^2\theta$$

$$= \rho^2(\sin^3\phi + \sin\phi\cos^2\phi) = \rho^2\sin\phi.$$

C13S07.061: Here we compute

$$\begin{bmatrix} F_x & F_y & F_z \end{bmatrix} \begin{bmatrix} \sin\phi\cos\theta & \rho\cos\phi\cos\theta & -\rho\sin\phi\sin\theta \\ \sin\phi\sin\theta & \rho\cos\phi\sin\theta & \rho\sin\phi\cos\theta \\ \cos\phi & -\rho\sin\phi & 0 \end{bmatrix} = \begin{bmatrix} \dfrac{\partial w}{\partial\rho} & \dfrac{\partial w}{\partial\phi} & \dfrac{\partial w}{\partial\theta} \end{bmatrix},$$

which has first component

$$\frac{\partial w}{\partial\rho} = F_x\sin\phi\cos\theta + F_y\sin\phi\sin\theta + F_z\cos\phi,$$

second component

$$\frac{\partial w}{\partial\phi} = F_x\,\rho\cos\phi\cos\theta + F_y\,\rho\cos\phi\sin\theta - F_z\,\rho\sin\phi,$$

and third component

$$\frac{\partial w}{\partial\theta} = -F_x\,\rho\sin\phi\sin\theta + F_y\,\rho\sin\phi\cos\theta.$$

Section 13.8

C13S08.001: If $f(x, y) = 3x - 7y$ and $P(17, 39)$ are given, then

$$\nabla f(x, y) = \langle 3, -7 \rangle, \quad \text{and thus} \quad \nabla f(17, 39) = \langle 3, -7 \rangle.$$

C13S08.003: If $f(x, y) = \exp(-x^2 - y^2)$ and $P(0, 0)$ are given, then

$$\nabla f(x, y) = \langle -2x\exp(-x^2 - y^2), -2y\exp(-x^2 - y^2) \rangle, \quad \text{and thus} \quad \nabla f(0, 0) = \langle 0, 0 \rangle = \mathbf{0}.$$

C13S08.005: Given $f(x, y, z) = y^2 - z^2$ and $P(17, 3, 2)$, then

$$\nabla f(x, y, z) = \langle 0, 2y, -2z \rangle, \quad \text{and therefore} \quad \nabla f(17, 3, 2) = \langle 0, 6, -4 \rangle.$$

C13S08.007: Given $f(x, y, z) = e^x \sin y + e^y \sin z + e^z \sin x$ and $P(0, 0, 0)$, then

$$\nabla f(x, y, z) = \langle e^z \cos x + e^x \sin y, \; e^x \cos y + e^y \sin z, \; e^y \cos z + e^z \sin x \rangle,$$

and therefore $\nabla f(0, 0, 0) = \langle 1, 1, 1 \rangle$.

C13S08.009: Given $f(x, y, z) = 2\sqrt{xyz}$ and $P(3, -4, -3)$, then

$$\nabla f(x, y, z) = \left\langle \frac{yz}{\sqrt{xyz}}, \frac{xz}{\sqrt{xyz}}, \frac{xy}{\sqrt{xyz}} \right\rangle, \quad \text{and so} \quad \nabla f(3, -4, -3) = \left\langle 2, -\frac{3}{2}, -2 \right\rangle.$$

C13S08.011: Given $f(x, y) = x^2 + 2xy + 3y^2$, $P(2, 1)$, and $\mathbf{v} = \langle 1, 1 \rangle$, we first compute a unit vector with the same direction as $\mathbf{v}$:

$$\mathbf{u} = \frac{\mathbf{v}}{|\mathbf{v}|} = \left\langle \tfrac{1}{2}\sqrt{2}, \tfrac{1}{2}\sqrt{2} \right\rangle.$$

Also $\nabla f(x, y) = \langle 2x + 2y, 2x + 6y \rangle$, so $\nabla f(P) = \langle 6, 10 \rangle$. Therefore

$$D_{\mathbf{u}}f(P) = (\nabla f(P)) \cdot \mathbf{u} = \langle 6, 10 \rangle \cdot \left\langle \tfrac{1}{2}\sqrt{2}, \tfrac{1}{2}\sqrt{2} \right\rangle = 8\sqrt{2}.$$

C13S08.013: Given $f(x, y) = x^3 - x^2 y + xy^2 + y^3$, $P(1, -1)$, and $\mathbf{v} = \langle 2, 3 \rangle$, we first compute a unit vector with the same direction as $\mathbf{v}$:

$$\mathbf{u} = \frac{\mathbf{v}}{|\mathbf{v}|} = \left\langle \tfrac{2}{13}\sqrt{13}, \tfrac{3}{13}\sqrt{13} \right\rangle.$$

Also $\nabla f(x, y) = \langle 3x^2 - 2xy + y^2, 3y^2 + 2xy - x^2 \rangle$, so $\nabla f(P) = \langle 6, 0 \rangle$. Therefore

$$D_{\mathbf{u}}f(P) = (\nabla f(P)) \cdot \mathbf{u} = \langle 6, 0 \rangle \cdot \left\langle \tfrac{2}{13}\sqrt{13}, \tfrac{3}{13}\sqrt{13} \right\rangle = \tfrac{12}{13}\sqrt{13}.$$

C13S08.015: Given: $f(x, y) = \sin x \cos y$, the point $P\left(\tfrac{1}{3}\pi, -\tfrac{2}{3}\pi\right)$, and the vector $\mathbf{v} = \langle 4, -3 \rangle$, we first construct a unit vector with the same direction as $\mathbf{v}$:

$$\mathbf{u} = \frac{\mathbf{v}}{|\mathbf{v}|} = \left\langle \frac{4}{5}, -\frac{3}{5} \right\rangle.$$

Next, $\nabla f(x, y) = \langle \cos x \cos y, -\sin x \sin y \rangle$, and hence $\nabla f(P) = \langle -\tfrac{1}{4}, \tfrac{3}{4} \rangle$. Therefore

$$D_{\mathbf{u}}f(P) = (\nabla f(P)) \cdot \mathbf{u} = -\frac{13}{20}.$$

C13S08.017: Given $f(x, y, z) = \sqrt{xyz}$, the point $P(2, -1, -2)$, and the vector $\mathbf{v} = \langle 1, 2, -2 \rangle$, we first construct a unit vector $\mathbf{u}$ with the same direction as $\mathbf{v}$:

$$\mathbf{u} = \frac{\mathbf{v}}{|\mathbf{v}|} = \left\langle \frac{1}{3}, \frac{2}{3}, -\frac{2}{3} \right\rangle.$$

738

Next,

$$\nabla f(x, y, z) = \left\langle \frac{yz}{2\sqrt{xyz}}, \frac{xz}{2\sqrt{xyz}}, \frac{xy}{2\sqrt{xyz}} \right\rangle,$$

and hence $D_{\mathbf{u}}f(P) = \big(\nabla f(P)\big) \cdot \mathbf{u} = \left\langle \frac{1}{2}, -1, -\frac{1}{2} \right\rangle \cdot \left\langle \frac{1}{3}, \frac{2}{3}, -\frac{2}{3} \right\rangle = -\frac{1}{6}.$

C13S08.019: Given $f(x, y, z) = \exp(xyz)$, the point $P(4, 0, -3)$, and the vector $\mathbf{v} = \langle 0, 1, -1 \rangle$, we first construct a unit vector with the same direction as $\mathbf{v}$:

$$\mathbf{u} = \frac{\mathbf{v}}{|\mathbf{v}|} = \left\langle 0, \frac{\sqrt{2}}{2}, -\frac{\sqrt{2}}{2} \right\rangle.$$

Next, $\nabla f(x, y, z) = \exp(xyz)\langle yz, xz, xy \rangle$, and so $\nabla f(P) = \langle 0, -12, 0 \rangle$. Therefore

$$D_{\mathbf{u}}f(P) = \big(\nabla f(P)\big) \cdot \mathbf{u} = -6\sqrt{2}.$$

C13S08.021: Given $f(x, y) = 2x^2 + 3xy + 4y^2$ and the point $P(1, 1)$, we first compute

$$\nabla f(x, y) = \langle 4x + 3y, 3x + 8y \rangle.$$

So the direction in which f is increasing the most rapidly at P is $\nabla f(P) = \langle 7, 11 \rangle$ and its rate of increase in that direction is $|\langle 7, 11 \rangle| = \sqrt{170}$.

C13S08.023: Given $f(x, y) = \ln(x^2 + y^2)$ and the point $P(3, 4)$, we first compute

$$\nabla f(x, y) = \left\langle \frac{2x}{x^2 + y^2}, \frac{2y}{x^2 + y^2} \right\rangle.$$

Therefore the direction in which f is increasing the most rapidly at P is

$$\mathbf{v} = \nabla f(P) = \left\langle \frac{6}{25}, \frac{8}{25} \right\rangle,$$

and its rate of increase in that direction is $|\mathbf{v}| = \dfrac{2}{5}$.

C13S08.025: Given $f(x, y, z) = 3x^2 + y^2 + 4z^2$ and the point $P(1, 5, -2)$, we first compute

$$\nabla f(x, y, z) = \langle 6x, 2y, 8z \rangle.$$

Therefore the direction in which f is increasing the most rapidly at P is $\mathbf{v} = \nabla f(P) = \langle 6, 10, -16 \rangle$ and its rate of increase in that direction is $|\mathbf{v}| = 14\sqrt{2}$.

C13S08.027: We are given $f(x, y, z) = \sqrt{xy^2z^3}$ and the point $P(2, 2, 2)$. We first compute the gradient of f:

$$\nabla f(x, y, z) = \left\langle \frac{y^2z^3}{2\sqrt{xy^2z^3}}, \frac{xyz^3}{\sqrt{xy^2z^3}}, \frac{3xy^2z^2}{2\sqrt{xy^2z^3}} \right\rangle.$$

Thus the direction in which f is increasing the most rapidly at P is $\nabla f(P) = \langle 2, 4, 6 \rangle$ and its rate of increase in that direction is $|\nabla f(P)| = 2\sqrt{14}$.

C13S08.029: Let $f(x, y) = \exp(25 - x^2 - y^2) - 1$. Then

$$\nabla f(x, y) = \langle -2x \exp(25 - x^2 - y^2), \; -2y \exp(25 - x^2 - y^2) \rangle,$$

so at $P(3, 4)$ we have $\nabla f(P) = \langle -6, -8 \rangle$, a vector normal to the graph of $f(x, y) = 0$ at the point P. Hence, as in Example 7, an equation of the line tangent to the graph at P is $-6(x - 3) - 8(y - 4) = 0$; simplified, this is $3x + 4y = 25$.

C13S08.031: Let $f(x, y) = x^4 + xy + y^2 - 19$. Then $\nabla f(x, y) = \langle 4x^3 + y, \; x + 2y \rangle$, so a vector normal to the graph of $f(x, y) = 0$ at the point $P(2, -3)$ is $\nabla f(P) = \langle 29, -4 \rangle$. So the tangent line at P has equation $29(x - 2) - 4(y + 3) = 0$; that is, $29x - 4y = 70$.

C13S08.033: Let $f(x, y, z) = x^{1/3} + y^{1/3} + z^{1/3} - 1$. Then

$$\nabla f(x, y, z) = \left\langle \frac{1}{3x^{2/3}}, \; \frac{1}{3y^{2/3}}, \; \frac{1}{3z^{2/3}} \right\rangle,$$

and thus a vector normal to the surface $f(x, y, z) = 0$ at the point $P(1, -1, 1)$ is

$$\nabla f(P) = \left\langle \frac{1}{3}, \; \frac{1}{3}, \; \frac{1}{3} \right\rangle.$$

Therefore an equation of the plane tangent to the surface at P is

$$\frac{1}{3}(x - 1) + \frac{1}{3}(y + 1) + \frac{1}{3}(z - 1) = 0; \quad \text{that is,} \quad x + y + z = 1.$$

C13S08.035: If u and v are differentiable functions of x and y and a and b are constants, then

$$\nabla\big(au(x, y) + bv(x, y)\big) = \left\langle \frac{\partial}{\partial x}\big(au(x, y) + bv(x, y)\big), \; \frac{\partial}{\partial y}\big(au(x, y) + bv(x, y)\big) \right\rangle$$

$$= \langle au_x + bv_x, \; au_y + bv_y \rangle = \langle au_x, \; au_y \rangle + \langle bv_x, \; bv_y \rangle$$

$$= a\langle u_x, \; u_y \rangle + b\langle v_x, \; v_y \rangle = a\nabla u(x, y) + b\nabla v(x, y).$$

C13S08.037: If u and v are differentiable functions of x and y and $v(x, y) \neq 0$, then

$$\nabla\left(\frac{u(x, y)}{v(x, y)} \right) = \left\langle \frac{\partial}{\partial x}\left(\frac{u}{v}\right), \; \frac{\partial}{\partial y}\left(\frac{u}{v}\right) \right\rangle = \left\langle \frac{vu_x - uv_x}{v^2}, \; \frac{vu_y - uv_y}{v^2} \right\rangle$$

$$= \left\langle \frac{vu_x}{v^2}, \; \frac{vu_y}{v^2} \right\rangle - \left\langle \frac{uv_x}{v^2}, \; \frac{uv_y}{v^2} \right\rangle = \frac{v\nabla u}{v^2} - \frac{u\nabla v}{v^2} = \frac{v\nabla u - u\nabla v}{v^2}.$$

C13S08.039: We know that $\mathbf{v} = \nabla f(P)$ gives the direction in which f is increasing the most rapidly at P. Then $\mathbf{v}$ is the direction in which $-f$ is decreasing the most rapidly at P. But $\nabla\big(-f(P)\big) = -\nabla f(P) = -\mathbf{v}$, so that $-\mathbf{v}$ is the direction in which $-f$ is increasing the most rapidly at P and, therefore, is the direction in which f is decreasing the most rapidly at P.

C13S08.041: Let $f(x, y) = Ax^2 + Bxy + Cy^2 - D$. Then

$$\nabla f(x, y) = \langle 2Ax + By, \; 2Cy + Bx \rangle,$$

so a vector normal to the graph of $f(x, y) = 0$ at the point $P(x_0, y_0)$ is

$$\nabla f(P) = \langle 2Ax_0 + By_0, \, 2Cy_0 + Bx_0 \rangle.$$

Hence, as in Example 7, an equation of the line tangent to the graph at P is

$$(2Ax_0 + By_0)(x - x_0) + (2Cy_0 + Bx_0)(y - y_0) = 0;$$

$$2Ax_0 x + By_0 x - 2A(x_0)^2 - Bx_0 y_0 + 2Cy_0 y + Bx_0 y - 2C(y_0)^2 - Bx_0 y_0 = 0;$$

$$2Ax_0 x + By_0 x + Bx_0 y + 2Cy_0 y = 2A(x_0)^2 + 2Bx_0 y_0 + 2C(y_0)^2;$$

$$2(Ax_0)x + B(y_0 x + x_0 y) + 2(Cy_0)y = 2D;$$

$$(Ax_0)x + \frac{1}{2}B(y_0 x + x_0 y) + (Cy_0)y = D.$$

C13S08.043: The equation of the paraboloid can be written in the form

$$H(x, y, z) = Ax^2 + By^2 - z = 0, \quad \text{and} \quad \nabla H(x, y, z) = \langle 2Ax, 2By, -1 \rangle.$$

A vector normal to the paraboloid at the point $P(x_0, y_0, z_0)$ is $\mathbf{n} = \langle 2Ax_0, 2By_0, -1 \rangle$, and hence the plane tangent to the paraboloid at P has an equation of the form

$$2Ax_0 x + 2By_0 y - z = d.$$

But the point P also lies on the plane, and hence

$$d = 2A(x_0)^2 + 2B(y_0)^2 - z_0 = 2(Ax_0^2 + By_0^2) - z_0 = 2z_0 - z_0 = z_0.$$

Hence an equation of the tangent plane is $2Ax_0 x + 2By_0 y - z = z_0$, and the result in Problem 43 follows immediately.

C13S08.045: In the solution of Problem 44 we calculated $\nabla f(P) = \langle 5, 4, 3 \rangle$, and the unit vector in the direction from P to Q is

$$\mathbf{u} = \frac{\overrightarrow{PQ}}{|\overrightarrow{PQ}|} = \left\langle \frac{2}{3}, \frac{2}{3}, \frac{1}{3} \right\rangle.$$

Then

$$D_{\mathbf{u}} f(P) = \nabla f(P) \cdot \mathbf{u} = \langle 5, 4, 3 \rangle \cdot \langle \tfrac{2}{3}, \tfrac{2}{3}, \tfrac{1}{3} \rangle = 7$$

(degrees per kilometer). Hence

$$\frac{dw}{dt} = \frac{dw}{ds} \cdot \frac{ds}{dt} = \left(7 \, \frac{\text{deg}}{\text{km}} \right) \left(2 \, \frac{\text{km}}{\text{min}} \right) = 14 \, \frac{\text{deg}}{\text{min}}$$

as the hawk's rate of change of temperature at P.

C13S08.047: Part (a): If $W(x, y, z) = 50 + xyz$, then $\nabla W = \langle yz, xz, xy \rangle$, so at the point $P(3, 4, 1)$ we have $\nabla W(P) = \langle 4, 3, 12 \rangle$. The unit vector with the same direction as $\mathbf{v} = \langle 1, 2, 2 \rangle$ is

$$\mathbf{u} = \frac{\mathbf{v}}{|\mathbf{v}|} = \left\langle \frac{1}{3}, \frac{2}{3}, \frac{2}{3} \right\rangle,$$

and so the rate of change of temperature at P in the direction of $\mathbf{v}$ is

$$\left(\nabla W(P)\right) \cdot \mathbf{u} = \langle 4,\, 3,\, 12 \rangle \cdot \left\langle \frac{1}{3}, \frac{2}{3}, \frac{2}{3} \right\rangle = \frac{34}{3}.$$

Because distance is measured in feet, the units for this rate of change are degrees Celsius per foot.

Part (b): The maximal directional derivative of W at P is $|\nabla W(P)| = |\langle 4,\, 3,\, 12 \rangle| = 13$ and the direction in which it occurs is $\nabla W(P) = \langle 4,\, 3,\, 12 \rangle$.

C13S08.049: Part (a): Given $f(x,\, y) = \frac{1}{10}(x^2 - xy + 2y^2)$, let

$$g(x,\, y,\, z) = z - f(x,\, y); \quad \text{then} \quad \nabla g(x,\, y,\, z) = \frac{1}{10}\langle y - 2x,\, x - 4y,\, 10 \rangle.$$

Thus a normal to the surface $z = f(x,\, y)$ at the point $P\left(2,\, 1,\, \frac{2}{5}\right)$ is $\frac{1}{10}\langle -3,\, -2,\, 10 \rangle$. Hence an equation of the plane tangent to this surface at P is

$$3x + 2y - 10z = 4; \quad \text{that is,} \quad z = \frac{3}{10}x + \frac{1}{5}y - \frac{2}{5}.$$

Part (b): Let Q denote the point $(2,\, 1)$ and R the point $(2.2,\, 0.9)$. Then $\mathbf{v} = \overrightarrow{QR} = \langle 0.2,\, -0.1 \rangle$. Thus an approximation to $f(2.2,\, 0.9)$ is

$$f(2,\, 1) + \left(\nabla f(Q)\right) \cdot \mathbf{v} = 0.4 + \langle 0.3,\, 0.2 \rangle \cdot \langle 0.2,\, -0.1 \rangle = \frac{11}{25} = 0.44.$$

The true value is $f(2.2,\, 0.9) = \frac{56}{125} = 0.448$.

C13S08.051: Let $F(x,\, y,\, z) = z^2 - x^2 - y^2$ and $G(x,\, y,\, z) = 2x + 3y + 4z + 2$. Then the cone is the graph of $F(x,\, y,\, z) = 0$ and the plane is the graph of $G(x,\, y,\, z) = 0$. At the given point $P(3,\, 4,\, -5)$ we have

$$\nabla F(3,\, 4,\, -5) = \langle -6,\, -8,\, -10 \rangle \quad \text{and} \quad \nabla G(3,\, 4,\, -5) = \langle 2,\, 3,\, 4 \rangle.$$

Let $\mathcal{P}$ denote the plane normal to the ellipse (the intersection of the cone and the first plane) at the point P. Then a normal to $\mathcal{P}$ is

$$\mathbf{n} = \langle -6,\, -8,\, -10 \rangle \times \langle 2,\, 3,\, 4 \rangle = \begin{vmatrix} \mathbf{i} & \mathbf{j} & \mathbf{k} \\ -6 & -8 & -10 \\ 2 & 3 & 4 \end{vmatrix} = \langle -2,\, 4,\, -2 \rangle.$$

We will use instead the parallel vector $\langle 1,\, -2,\, 1 \rangle$. In the usual way we find that $\mathcal{P}$ has Cartesian equation $x - 2y + z + 10 = 0$.

C13S08.053: Let $F(x,\, y,\, z) = x^2 + y^2 + z^2 - r^2$ and $G(x,\, y,\, z) = z^2 - a^2 x^2 - b^2 y^2$. Then the sphere is the graph of $F(x,\, y,\, z) = 0$ and the cone is the graph of $G(x,\, y,\, z) = 0$. At a point where the sphere and the cone meet, these vectors are the normals to their tangent planes. To show that the tangent planes are perpendicular, it is sufficient to show that their normals are perpendicular. But

$$(\nabla F) \cdot (\nabla G) = \langle 2x,\, 2y,\, 2z \rangle \cdot \langle -2a^2 x,\, -2b^2 y,\, 2z \rangle = -4a^2 x^2 - 4b^2 y^2 + 4z^2 = 4(z^2 - a^2 x^2 - b^2 y^2) = 0$$

because (x, y, z) lies on the cone. Therefore the tangent planes are perpendicular at every point of the intersection of the sphere and the cone.

C13S08.055: The surface is the graph of the equation $G(x, y, z) = 0$ where

$$G(x, y, z) = xyz - 1, \quad \text{so that} \quad \nabla G(x, y, z) = \langle yz, xz, xy \rangle.$$

Suppose that $P(a, b, c)$ is a point strictly within the first octant (so that a, b, and c are all positive). Note that $abc = 1$. A vector normal to the surface at P is $\mathbf{n} = \langle bc, ac, ab \rangle$, and hence the plane tangent to the surface at P has equation

$$bcx + acy + abz = d$$

for some constant d. Moreover, because P is a point of the surface,

$$bca + acb + abc = d; \quad \text{that is,} \quad d = 3abc.$$

Hence an equation of the tangent plane is $bcx + acy + abz = 3abc$. The intercepts of the pyramid therefore occur at $(3a, 0, 0)$, $(0, 3b, 0)$, and $(0, 0, 3c)$. Therefore, because of the right angle at the origin, the pyramid has volume

$$V = \frac{1}{6}(3a)(3b)(3c) = \frac{27}{6}abc = \frac{9}{2},$$

independent of the choice of P, as we were to show.

C13S08.057: The hill is steepest in the direction of $\nabla z(-100, -100) = \langle 0.6, 0.8 \rangle$. The slope of the hill in that direction is $|\langle 0.6, 0.8 \rangle| = 1$, so that your initial angle of climb would be $45°$. The compass heading in the direction you are climbing is

$$\frac{\pi}{2} - \arctan \frac{4}{3}$$

radians, approximately $36° \ 52' \ 11.632''$.

C13S08.059: Given

$$z = f(x, y) = \frac{1000}{1 + (0.00003)x^2 + (0.00007)y^2},$$

we first compute

$$\nabla f(x, y) = \left\langle -\frac{600\,000\,000x}{(100000 + 3x^2 + 7y^2)^2}, \ -\frac{1\,400\,000\,000y}{(100000 + 3x^2 + 7y^2)^2} \right\rangle.$$

Hence to climb the most steeply, the initial direction should be

$$\nabla f(100, 100) = \left\langle -\frac{3}{2}, -\frac{7}{2} \right\rangle,$$

and the initial rate of ascent will be $|\nabla f(100, 100)| = \frac{2}{2}\sqrt{58} \approx 3.807886553$ feet per foot. The initial angle of climb will be $\arctan\left(\frac{1}{2}\sqrt{58}\right) \approx 1.313982409$ radians, approximately $75° \ 17' \ 8.327''$. The compass heading is

$$270° - \arctan\left(\frac{7}{3}\right)° \approx 203° \ 11' \ 54.926''.$$

C13S08.061: Let

$$f(x, y) = \frac{1}{1000}(3x^2 - 5xy + y^2).$$

Then

$$\nabla f(x, y) = \frac{1}{1000}\langle 6x - 5y, \ 2y - 5x \rangle,$$

and therefore $\mathbf{v} = \nabla f(100, 100) = \langle \frac{1}{10}, -\frac{3}{10} \rangle$.

Part (a): A unit vector in the northeast direction is $\mathbf{u} = \frac{1}{2}\langle \sqrt{2}, \sqrt{2} \rangle$, so the directional derivative of f at $(100, 100)$ in the northeast direction is

$$\mathbf{v} \cdot \mathbf{u} = \langle \tfrac{1}{10}, -\tfrac{3}{10} \rangle \cdot \tfrac{1}{2}\langle \sqrt{2}, \sqrt{2} \rangle = -\tfrac{1}{10}\sqrt{2}.$$

Hence you will initially be descending the hill, and at an angle of $\arctan\left(\frac{1}{10}\sqrt{2}\right)$ below the horizontal, approximately $8° \ 2' \ 58.081''$.

Part (b): A unit vector in the direction $30°$ north of east is $\mathbf{u} = \frac{1}{2}\langle \sqrt{3}, 1 \rangle$, so the directional derivative of f at $(100, 100)$ in the direction of $\mathbf{u}$ is

$$\langle \tfrac{1}{10}, -\tfrac{3}{10} \rangle \cdot \tfrac{1}{2}\langle \sqrt{3}, 1 \rangle = -\frac{3 - \sqrt{3}}{20} \approx -0.06339746.$$

Hence you will initially be descending the hill, and at an angle of approximately $3° \ 37' \ 56.665''$.

C13S08.063: Because $\mathbf{u} = \langle a, b \rangle$ and $\mathbf{v} = \langle c, d \rangle$ are not collinear, neither is zero and neither is a scalar multiple of the other. Hence, as vectors, they are linearly independent, and this implies that the simultaneous equations

$$af_x(P) + bf_y(P) = D_\mathbf{u}f(P),$$

$$cf_x(P) + df_y(P) = D_\mathbf{v}f(P)$$

have a unique solution for the values of $f_x(P)$ and $f_y(P)$. Thus $\nabla f(P) = \langle f_x(P), f_y(P) \rangle$ is uniquely determined, and therefore so is the directional derivative

$$D_\mathbf{w}f(P) = \nabla f(P) \cdot \mathbf{w}$$

in the direction of the arbitrary unit vector $\mathbf{w}$.   —C.H.E.

Section 13.9

C13S09.001: Given $f(x, y) = 2x + y$ and the constraint $g(x, y) = x^2 + y^2 - 1 = 0$, the equation $\nabla f = \lambda \nabla g$ yields

$$2 = 2\lambda x \quad \text{and} \quad 1 = 2\lambda y,$$

so that $\lambda \neq 0$. Hence

$$\frac{1}{\lambda} = x = 2y, \quad \text{and thus} \quad 4y^2 + y^2 = 1.$$

Thus we have two solutions:

$$(x,\, y) = \left(\frac{2}{5}\sqrt{5},\ \frac{1}{5}\sqrt{5}\right) \quad \text{and} \quad (x,\, y) = \left(-\frac{2}{5}\sqrt{5},\ -\frac{1}{5}\sqrt{5}\right).$$

Clearly the first maximizes $f(x,\, y)$ and the second minimizes $f(x,\, y)$. So the global maximum value of $f(x,\, y)$ is $\sqrt{5}$ and its global minimum value is $-\sqrt{5}$.

Note: The function f is continuous on the circle $x^2 + y^2 = 1$ and a continuous function defined on a closed and bounded subset of euclidean space (such as that circle) must have both a global maximum value and a global minimum value. We will use this argument to identify extrema when possible and without stating the argument explicitly in various solutions in this section.

C13S09.003: The Lagrange multiplier equation $\langle 2x,\, -2y \rangle = \lambda \langle 2x,\, 2y \rangle$ yields the scalar equations $x = \lambda x$ and $-y = \lambda y$. To solve them with a minimum number of cases, note that multiplication of the first by y and the second by x yields

$$xy = \lambda xy = -xy, \quad \text{so that} \quad xy = 0.$$

If $x = 0$ then $y = \pm 2$; if $y = 0$ then $x = \pm 2$. So there are four critical points:

$$f(2,\, 0) = 4: \quad \text{global maximum;} \qquad f(-2,\, 0) = 4: \quad \text{global maximum;}$$

$$f(0,\, 2) = -4: \quad \text{global minimum;} \qquad f(0,\, -2) = -4: \quad \text{global minimum.}$$

C13S09.005: The Lagrange multiplier equation $\langle y,\, x \rangle = \lambda \langle 8x,\, 18y \rangle$ yields the scalar equations $y = 8\lambda x$ and $x = 18\lambda y$. Multiply the first by $9y$ and the second by $4x$ to obtain

$$9y^2 = 72\lambda xy = 4x^2, \quad \text{so that} \quad 3y = \pm 2x.$$

The constraint equation takes the form $18y^2 = 36$, so that $y^2 = 2$. Thus there are four critical points. But $f(x,\, y)$ must have both a global maximum and a global minimum value on the ellipse $4x^2 + 9y^2 = 36$, so we may identify all of the critical points:

$$f\left(\tfrac{3}{2}\sqrt{2},\ \sqrt{2}\right) = 3: \quad \text{global maximum;} \qquad f\left(-\tfrac{3}{2}\sqrt{2},\ \sqrt{2}\right) = -3: \quad \text{global minimum;}$$

$$f\left(\tfrac{3}{2}\sqrt{2},\ -\sqrt{2}\right) = -3: \quad \text{global minimum;} \qquad f\left(-\tfrac{3}{2}\sqrt{2},\ -\sqrt{2}\right) = 3: \quad \text{global maximum.}$$

C13S09.007: It is clear that $f(x,\, y,\, z) = x^2 + y^2 + z^2$ can have no global maximum value on the plane $g(x,\, y,\, z) = 3x + 2y + z - 6 = 0$, and almost as clear that there is a unique global minimum. The Lagrange multiplier equation

$$\langle 2x,\, 2y,\, 2z \rangle = \lambda \langle 3,\, 2,\, 1 \rangle$$

then yields $4x = 6y = 12z = 6\lambda$, so that $x = 3z$ and $y = 2z$. Then the constraint takes the form $9z + 4z + z = 6$, and therefore the global minimum value of f on the plane $g(x,\, y,\, z) = 0$ is $f\left(\tfrac{9}{7},\ \tfrac{6}{7},\ \tfrac{3}{7}\right) = \tfrac{18}{7}$.

We have also discovered three distinct positive rational numbers whose sum is equal to the sum of their squares.

C13S09.009: The continuous function $f(x, y, z) = x + y + z$ must have both a global maximum value and a global minimum value on the ellipsoidal surface with equation $g(x, y, z) = x^2 + 4y^2 + 9z^2 - 36 = 0$. The Lagrange multiplier method yields $\langle 1, 1, 1 \rangle = \lambda \langle 2x, 8y, 18z \rangle$, and thus $\lambda \neq 0$. Therefore

$$\frac{1}{\lambda} = 2x = 8y = 18z, \quad \text{so that} \quad x = 4y = 9z.$$

Substitution of $x = 9z$ and $y = \frac{9}{4}z$ in the equation $g(x, y, z) = 0$ then yields $z = \pm\frac{4}{7}$. Results:

$$\text{Global maximum:} \quad f\left(\tfrac{36}{7}, \tfrac{9}{7}, \tfrac{4}{7}\right) = 7;$$

$$\text{Global minimum:} \quad f\left(-\tfrac{36}{7}, -\tfrac{9}{7}, -\tfrac{4}{7}\right) = -7.$$

C13S09.011: Clearly $f(x, y, z) = xy + 2z$ is continuous on the closed and bounded spherical surface $g(x, y, z) = x^2 + y^2 + z^2 - 36 = 0$, and hence f has both a global maximum and a global minimum there. The Lagrange multiplier equation is

$$\langle y, x, 2 \rangle = \lambda \langle 2x, 2y, 2z \rangle, \quad \text{and hence} \quad y = 2\lambda x, \quad x = 2\lambda y, \quad \text{and} \quad 2 = 2\lambda z.$$

Now $\lambda z = 1$, so $\lambda \neq 0$. If neither x nor y is zero, then

$$\frac{1}{\lambda} = \frac{2x}{y} = \frac{2y}{x} = z.$$

So $y^2 = x^2$ in this case. If $y = x$, then $z = 2$, so $2x^2 + 4 = 36$ and we obtain two critical points and the values $f(4, 4, 2) = 20$ and $f(-4, -4, 2) = 20$. If $y = -x$, then $z = -2$, and we obtain two critical points and the values $f(4, -4, -2) = -20$ and $f(-4, 4, -2) = -20$. Finally, if either of x and y is zero, then so is the other; $z^2 = 36$, and we obtain two more critical points and the values $f(0, 0, 6) = 12$ and $f(0, 0, -6) = -12$. Hence the global maximum value of $f(x, y, z)$ is 20 and its global minimum value is -20.

C13S09.013: The continuous function $f(x, y, z) = x^2 y^2 z^2$ clearly has a global maximum value and a global minimum value on the ellipsoidal surface $g(x, y, z) = x^2 + 4y^2 + 9z^2 - 27 = 0$. The Lagrange multiplier method yields the equations

$$xy^2 z^2 = \lambda x, \quad x^2 yz^2 = 4\lambda y, \quad \text{and} \quad x^2 y^2 z = 9\lambda z.$$

If any one of x, y, or z is zero, then $f(x, y, z) = 0$, and this is clearly the global minimum value of $f(x, y, z)$. Otherwise, $\lambda \neq 0$, and hence

$$x^2 y^2 z^2 = \lambda x^2 = 4\lambda y^2 = 9\lambda z^2,$$

and thus $x^2 = 4y^2 = 9z^2 = 9$; that is, $x^2 = 9$, $y^2 = \frac{9}{4}$, and $z^2 = 1$. Thus the global maximum value of $f(x, y, z)$ occurs at eight different critical points, and that maximum value is $9 \cdot \frac{9}{4} \cdot 1 = \frac{81}{4}$.

C13S09.015: We are to find the extrema of $f(x, y, z) = x^2 + y^2 + z^2$ subject to the two constraints $g(x, y, z) = x + y + z - 1 = 0$ and $h(x, y, z) = x + 2y + 3z - 6 = 0$. The Lagrange multiplier equation is

$$\langle 2x, 2y, 2z \rangle = \lambda \langle 1, 1, 1 \rangle + \mu \langle 1, 2, 3 \rangle,$$

746

from which we obtain the scalar equations $\lambda = 2x - \mu = 2y - 2\mu = 2z - 3\mu$, so that

$$2x + \mu = 2y \quad \text{and} \quad 2y + \mu = 2z, \quad \text{and thus}$$

$$\mu = 2(y - x) = 2(z - y): \quad 2y = x + z.$$

Thus we are to solve the following three *linear* equations in three unknowns:

$$x + y + z = 1,$$

$$x - 2y + z = 0,$$

$$x + 2y + 3z = 6.$$

We find that $x = -\frac{5}{3}$, $y = \frac{1}{3}$, and $z = \frac{7}{3}$. A geometric interpretation of this problem is to find the point on the intersection of two planes—a line—closest to and farthest from the origin. Hence there is no maximum but surely a minimum, and we have found it: $f\left(-\frac{5}{3}, \frac{1}{3}, \frac{7}{3}\right) = \frac{25}{3}$.

C13S09.017: We are to find the extrema of the function $f(x, y, z) = z$ subject to the *two* constraints $g(x, y, z) = x + y + z - 1 = 0$ and $h(x, y, z) = x^2 + y^2 - 1 = 0$. Geometrically, this is the problem of finding the highest (maximum z-coordinate) and lowest points on the ellipse formed by the intersection of a plane and a vertical cylinder. Hence there will be a unique global maximum and a unique global minimum unless the plane is horizontal or vertical—which it is not. The Lagrange multiplier equation is

$$\langle 0, 0, 1 \rangle = \lambda \langle 1, 1, 1 \rangle + \mu \langle 2x, 2y, 0 \rangle.$$

We thus obtain the three (simultaneous) scalar equations

$$2\mu x + \lambda = 0, \quad 2\mu y + \lambda = 0, \quad \lambda = 1.$$

Thus $2\mu x = 2\mu y = -1$ and so, because $\mu \neq 0$, we find that $y = x$. Then the second constraint implies that $2x^2 = 1$, and because $y = x$ there are only two cases:

Case 1: $x = \frac{1}{2}\sqrt{2} = y$, $z = 1 - \sqrt{2}$. This is the lowest point on the ellipse.

Case 2: $x = -\frac{1}{2}\sqrt{2} = y$, $z = 1 + \sqrt{2}$. This is the highest point on the ellipse.

C13S09.019: We should find only one possible extremum because there is no point on the line farthest from the origin. We minimize $f(x, y) = x^2 + y^2$ given the constraint $g(x, y) = 3x + 4y - 100 = 0$. The Lagrange multiplier method yields

$$\langle 2x, 2y \rangle = \lambda \langle 3, 4 \rangle, \quad \text{so that} \quad 2x = 3\lambda \quad \text{and} \quad 2y = 4\lambda.$$

Thus $8x = 12\lambda = 6y$, so that $y = \frac{4}{3}x$. Substitution in the constraint yields $x = 12$ and $y = 16$. Answer: The point on the line $g(x, y) = 0$ closest to the origin is $(12, 16)$.

C13S09.021: Please refer to Problem 29 of Section 13.5. We are to minimize

$$f(x, y, z) = x^2 + y^2 + z^2 \quad \text{given} \quad g(x, y, z) = 12x + 4y + 3z - 169 = 0.$$

Geometrically, we are to find the point (x, y, z) on the plane with equation $g(x, y, z) = 0$ closest to the origin, hence we can be sure that a unique solution exists. The Lagrange multiplier method yields the scalar equations

747

$$2x = 12\lambda, \quad 2y = 4\lambda, \quad \text{and} \quad 2z = 3\lambda,$$

so that $2x = 6y = 8z = 12\lambda$, and thus $y = \frac{1}{3}x$ and $z = \frac{1}{4}x$. Then the constraint yields

$$12x + \frac{4}{3}x + \frac{3}{4}x = 169;$$

$$144x + 16x + 9x = 169 \cdot 12;$$

$$x = 12, \quad y = 4, \quad z = 3.$$

Answer: The point on the plane $g(x, y, z) = 0$ closest to the origin is $(12, 4, 3)$.

C13S09.023: Please refer to Problem 31 of Section 13.5. We are to minimize

$$f(x, y, z) = (x - 7)^2 + (y + 7)^2 + z^2 \quad \text{given} \quad g(x, y, z) = 2x + 3y + z - 49 = 0.$$

This is the geometric problem of finding the point (x, y, z) on the plane $g(x, y, z) = 0$ closest to the point $Q(7, -7, 0)$. Hence we can be sure that a unique minimum exists and, because there can be no maximum of $f(x, y\ z)$, we anticipate a single critical point. The Lagrange multiplier method yields the scalar equations

$$2(x - 7) = 2\lambda, \quad 2(y + 7) = 3\lambda, \quad 2z = \lambda.$$

It follows that $6\lambda = 6x - 42 = 4y + 28 = 12z$, and thus

$$x = \frac{12z + 42}{6} = 2z + 7 \quad \text{and}$$

$$y = \frac{12z - 28}{4} = 3z - 7.$$

Substitution in the constraint gives the equation

$$4z + 14 + 9z - 21 + z = 49, \quad \text{so that} \quad 14z = 56.$$

Therefore $z = 4$, $y = 5$, and $x = 15$. The point on the plane $g(x, y, z) = 0$ closest to Q is thus $(15, 5, 4)$.

C13S09.025: Please refer to Problem 33 of Section 13.5. We are to find the point (x, y, z) in the first octant and on the surface $g(x, y, z) = x^2 y^2 z - 4 = 0$ closest to the origin, so we minimize

$$f(x, y, z) = x^2 + y^2 + z^2 \quad \text{subject to the constraint} \quad g(x, y, z) = 0.$$

The surface is unbounded, so we anticipate no maximum and only one minimum in the first octant. The Lagrange multiplier method yields the scalar equations

$$2x = 2\lambda x y^2 z, \quad 2y = 2\lambda x^2 y z, \quad 2z = \lambda x^2 y^2.$$

It follows that $2\lambda x^2 y^2 z = 2x^2 = 2y^2 = 4z^2$, and hence that $x^2 = y^2 = 2z^2$. The restriction to the first octant then implies that $x = y = z\sqrt{2}$, and then the constraint $x^2 y^2 z = 4$ implies that

$$(2z^2)(2z^2)(z) = 4; \quad 4z^5 = 4; \quad z = 1.$$

So $x = y = \sqrt{2}$. The point on $g(x, y, z) = 0$ in the first octant closest to the origin is $(\sqrt{2}, \sqrt{2}, 1)$. Its distance from the origin is $\sqrt{5}$.

C13S09.027: Please see Problem 35 of Section 13.5. We are to maximize $f(x, y, z) = xyz$ given $g(x, y, z) = x + y + z - 120 = 0$ and the additional condition that x, y, and z are all positive. The Lagrange multiplier equations are

$$\lambda = yz = xz = xy, \quad \text{and hence} \quad x = y = z = 40.$$

Therefore the maximum value of f is $f(40, 40, 40) = 40^3 = 64000$.

To establish that $f(x, y, z)$ actually has a global maximum value on its domain (the set of all triples of positive real numbers (x, y, z) such that $x+y+z = 120$), extend the domain slightly to include all triples of *nonnegative* triples of real numbers (x, y, z) such that $x+y+z = 120$. Then f is continuous on this domain, which is a closed and bounded subset of three-dimensional space (it is the part of the plane $x + y + z = 120$ that lies in the first octant or on the adjacent coordinate planes). Hence f has a global maximum there, and the maximum does not occur on the boundary because $xyz = 0$ if any of x, y, or z is zero. Hence this maximum must occur at an interior critical point, and in the previous paragraph we found only one such critical point. This is enough to establish that 64000 is the correct answer.

C13S09.029: Please see Problem 37 of Section 13.5. Suppose that the box has dimensions x by y by z (units are in inches). Then we are to minimize total surface area $A(x, y, z) = 2xy + 2xz + 2yz$ given $V(x, y, z) = xyz - 1000 = 0$. The Lagrange multipliers equations are

$$2(y + z) = \lambda yz, \quad 2(x + z) = \lambda xz, \quad \text{and} \quad 2(x + y) = \lambda xy;$$

note also that in the solution, all three of x, y, and z must be positive. Hence we can eliminate λ from all three of the previous equations:

$$\frac{\lambda}{2} = \frac{y + z}{yz} = \frac{x + z}{xz} = \frac{x + y}{xy};$$

$$\frac{1}{z} + \frac{1}{y} = \frac{1}{z} + \frac{1}{x} = \frac{1}{y} + \frac{1}{x};$$

$$x = y = z \quad \text{and} \quad xyz = 1000;$$

$$x = y = z = 10.$$

Answer: The minimum possible surface area occurs when the box is a cube measuring 10 in. along each edge. The minimum possible surface area is 600 in.^{2}

C13S09.031: Please see Problem 39 of Section 13.5. Units in this problem will be cents and inches. Suppose that the bottom of the box has dimensions x by y and that its height is z. We are to minimize total cost $C(x, y, z) = 6xy + 10xz + 10yz$ given the constraint $V(x, y, z) = xyz - 600 = 0$. The Lagrange multiplier equations are

$$6y + 10z = \lambda yz, \quad 6x + 10z = \lambda xz, \quad \text{and} \quad 10x + 10y = \lambda xy;$$

also note that x, y, and y are all positive. Multiply each of the three previous equations by the "missing" variable to find that

$$\lambda xyz = 6xy + 10xz = 6xy + 10yz = 10xz + 10yz,$$

and thus $x = y$ and $6y = 10z$; that is, $x = y = \frac{5}{3}z$. Substitution in the constraint equation yields

$$600 = xyz = \frac{25}{9}z^3: \quad z^3 = \frac{5400}{25} = 216.$$

Therefore $z = 6$ and $x = y = 10$. Answer: The dimensions that minimize the total cost are these: base 10 in. by 10 in., height 6 in.; its total cost will be $18.00.

C13S09.033: Please refer to Problem 41 of Section 13.5. Suppose that the front of the box has width x and height z (units are in inches and cents) and that the bottom of the box measures x by y. We are to minimize total cost $C(x, y, z) = 6xy + 12xz + 18yz$ given the constraint $V(x, y, z) = xyz - 750 = 0$. The Lagrange multiplier equations are

$$6y + 12z = \lambda yz, \quad 6x + 18z = \lambda xz, \quad \text{and} \quad 12x + 18y = \lambda xy.$$

Note that x, y, and z are positive. Multiply each of the three preceding equations by the "missing" variable to obtain

$$\lambda xyz = 6xy + 12xz = 6xy + 18yz = 12xz + 18yz.$$

It follows that $12xz = 18yz$, so that $2x = 3y$; also, $6xy = 12xz$, so that $y = 2z$. Thus $x = \frac{3}{2}y = 3z$. Then substitution in the constraint equation yields

$$750 = xyz = 6z^3: \quad z^3 = 125,$$

and so $z = 5$, $y = 10$, and $x = 15$. The front of the cheapest box should be 15 in. wide and 5 in. high; its depth (from front to back) should be 10 in. (This box will cost $27.00.)

C13S09.035: We are to minimize $f(x, y, z) = x^2 + y^2 + z^2$ given (x, y, z) lies on the surface with equation $g(x, y, z) = xy + 5 - z = 0$. The Lagrange multiplier equations are

$$2x = \lambda y, \quad 2y = \lambda x, \quad \text{and} \quad 2z = -\lambda.$$

Therefore $\lambda = -2z$, and thus $2x = -2yz$ and $2y = -2xz$. So

$$2x^2 = 2y^2 = -2xyz: \quad x^2 = y^2 = -xyz.$$

Case 1: $y = x$. Then $x^2 = -x^2z$, so $x = 0$ or $z = -1$. In the latter case, the constraint equation yields $x^2 = -6$, so that case is rejected. We obtain only the critical point $(0, 0, 5)$.

Case 2: $y = -x$. Then $x^2 = x^2z$, so $x = 0$ or $z = 1$. If $x = 0$, then we obtain only the critical point of Case 1. If $z = 1$ then the constraint equation yields $x^2 = 4$, and we obtain two more critical points: $(2, -2, 1)$ and $(-2, 2, 1)$.

Now $f(0, 0, 5) = 25$, $f(2, -2, 1) = 9$, and $f(-2, 2, 1) = 9$. Hence there are two points of the surface $g(x, y, z) = 0$ closest to the origin. They are $(2, -2, 1)$ and $(-2, 2, 1)$; each is at distance $\sqrt{9} = 3$ from the origin.

C13S09.037: See Fig. 13.9.9 of the text. There we see three small isosceles triangles, each with two equal sides of length 1 meeting at the center of the circle. Their total area

$$A = \frac{1}{2}\sin\alpha + \frac{1}{2}\sin\beta + \frac{1}{2}\sin\gamma$$

is the area of the large triangle, the quantity to be maximized given the constraint

$$g(\alpha, \beta, \gamma) = \alpha + \beta + \gamma - 2\pi = 0.$$

It is clear that, at maximum area, the center of the circle is *within* the large triangle or, at worst, on its boundary. Thus we have the additional restrictions

$$0 \leqq \alpha \leqq \pi, \quad 0 \leqq \beta \leqq \pi, \quad \text{and} \quad 0 \leqq \gamma \leqq \pi. \qquad (1)$$

The Lagrange multiplier equations are

$$\lambda = \frac{1}{2} \cos \alpha = \frac{1}{2} \cos \beta = \frac{1}{2} \cos \gamma,$$

and the restrictions in (1) imply that $\alpha = \beta = \gamma = \frac{2}{3}\pi$. Hence the triangle of maximal area is equilateral (because it is equiangular). If x denotes the length of each side, then by the law of cosines

$$x^2 = 1^2 + 1^2 - 2 \cdot 1 \cdot 1 \cdot \cos \frac{2\pi}{3} = 3,$$

and so each side of the maximal-area triangle has length $\sqrt{3}$. The area of the maximal-area triangle is

$$A = \frac{3}{2} \sin \frac{2\pi}{3} = \frac{3}{2} \cdot \frac{\sqrt{3}}{2} = \frac{3\sqrt{3}}{4} \approx 1.299038106,$$

and the ratio of this area to that of the circumscribed circle is

$$\frac{A}{\pi} \approx 0.413496672.$$

So the inscribed triangle of maximal area occupies about 41% of the area of the circle; this result meets the test of plausibility.

C13S09.039: We are to use the *Suggestion* in Problem 38 to find the point or points on the hyperbola $g(x, y) = x^2 + 12xy + 6y^2 - 130 = 0$ that are closest to the origin. To do so we minimize $f(x, y) = x^2 + y^2$ subject to the constraint $g(x, y) = 0$. The Lagrange multiplier equations are

$$2x = \lambda(2x + 12y) \quad \text{and} \quad 2y = \lambda(12y + 12x). \qquad (1)$$

When put into the form suggested in Problem 38, they become

$$(1 - \lambda)x - 6\lambda y = 0,$$

$$6\lambda x + (6\lambda - 1)y = 0.$$

Because $(x, y) = (0, 0)$ is not a solution of this system and because it must therefore have a nontrivial solution, the determinant of this system must be zero:

$$(1 - \lambda)(6\lambda - 1) + 36\lambda^2 = 0;$$

$$30\lambda^2 + 7\lambda - 1 = 0;$$

$$(3\lambda + 1)(10\lambda - 1) = 0.$$

Case 1: $\lambda = -\frac{1}{3}$. Then (1) implies that

751

$$\frac{4}{3}x + 2y = 0 : \quad y = -\frac{2}{3}x.$$

Substitution of the last equation for y into the constraint equation $g(x, y) = 0$ yields $x^2 + 30 = 0$, so there is no solution in Case 1.

Case 2: $\lambda = \frac{1}{10}$. Then (1) implies that

$$\frac{9}{10}x - \frac{6}{10}y = 0 : \quad y = \frac{3}{2}x.$$

Substitution of the last equation for y into the constraint equation yields $x^2 = 4$, so we obtain two solutions in this case: $(-2, -3)$ and $(2, 3)$. These two points of the hyperbola are its points closest to $(0, 0)$, each at distance $\sqrt{13}$.

C13S09.041: The highest point has the largest z-coordinate; the lowest point, the smallest. Consequently we are to maximize and minimize

$$f(x, y\ z) = z \quad \text{given} \quad g(x, y, z) = x^2 + y^2 - 1 = 0 \quad \text{and} \quad h(x, y, z) = 2x + y - z - 4 = 0.$$

Because the ellipse is formed by the intersection of a plane with a vertical cylinder, there will be a unique highest point and a unique lowest point unless the plane is horizontal or vertical—and it is not; one of its normal vectors is $\mathbf{n} = \langle 2, 1, -1 \rangle$. So we expect to find exactly two critical points. The Lagrange multiplier equations are

$$2\lambda x + 2\mu = 0, \qquad 2\lambda y + \mu = 0, \qquad -\mu = 1,$$

and when the third is substituted into the first two, we find that

$$2\lambda x = -2 = 4\lambda y.$$

But then $\lambda \neq 0$, and therefore $x = 2y$. Substitution of this information into the first constraint yields $5y^2 = 1$ and thus leads to the following two cases.

Case 1: $y = \frac{1}{5}\sqrt{5}$. Then $x = \frac{2}{5}\sqrt{5}$, and the second constraint implies that $z = \sqrt{5} - 4$.

Case 2: $y = -\frac{1}{5}\sqrt{5}$. Then $x = -\frac{2}{5}\sqrt{5}$, and the second constraint implies that $z = -\sqrt{5} - 4$.

Therefore the lowest point on the ellipse is $\left(-\frac{2}{5}\sqrt{5}, -\frac{1}{5}\sqrt{5}, -\sqrt{5} - 4\right)$ and the highest point on the ellipse is $\left(\frac{2}{5}\sqrt{5}, \frac{1}{5}\sqrt{5}, \sqrt{5} - 4\right)$. See Problem 48 of Section 13.8 for an alternative method of solving such problems.

C13S09.043: We are to maximize and minimize $f(x, y, z) = x^2 + y^2 + z^2$ given the two constraints $g(x, y, z) = x^2 + y^2 - z^2 = 0$ and $h(x, y, z) = x + 2y + 3z - 3 = 0$. From the geometry of the problem we see that there will be a unique maximum distance and a unique minimum distance, but there may be more than two critical points. The Lagrange multiplier vector equation is

$$\langle 2x, 2y, 2z \rangle = \lambda \langle 2x, 2y, -2z \rangle + \mu \langle 1, 2, 3 \rangle,$$

and the corresponding scalar equations are

$$2x = 2\lambda x + \mu, \qquad 2y = 2\lambda y + 2\mu, \qquad \text{and} \qquad 2z = -2\lambda z + 3\mu.$$

It follows that

$$6\mu = 12x - 12\lambda x = 6y - 6\lambda y = 4z + 2\lambda z,$$

and hence $12x(1 - \lambda) = 6y(1 - \lambda)$.

Case 1: $\lambda = 1$. Then $6z = 0$, and so $z = 0$. Then the first constraint implies that $x = y = 0$, which contradicts the second constraint. This leaves only the second case.

Case 2: $\lambda \neq 1$. Then $y = 2x$. Then the first constraint takes the form $5x^2 = z^2$, and it is convenient to consider separately two subcases.

Case 2a: $z = x\sqrt{5}$. Then the second constraint yields

$$x = \frac{3}{20}\left(-5 + 3\sqrt{5}\right), \quad y = \frac{3}{10}\left(-5 + 3\sqrt{5}\right), \quad \text{and} \quad z = \frac{3}{4}\left(3 - \sqrt{5}\right).$$

Case 2b: $z = -x\sqrt{5}$. Then the second constraint yields

$$x = -\frac{3}{20}\left(5 + 3\sqrt{5}\right), \quad y = -\frac{3}{10}\left(5 + 3\sqrt{5}\right), \quad \text{and} \quad z = \frac{3}{4}\left(3 + \sqrt{5}\right).$$

The coordinates in Case 2a are those of the point closest to the origin; its distance from the origin is approximately 0.81027227. The coordinates of the point farthest from the origin are those given in Case 2b; the distance from this point to the origin is approximately 5.55368876.

C13S09.045: Suppose that $f(x, y, z)$ and $g(x, y, z)$ have continuous first-order partial derivatives. Suppose also that the maximum (or minimum) of $f(x, y, z)$ subject to the constraint

$$g(x, y, z) = 0$$

occurs at a point P at which $\nabla g(P) \neq \mathbf{0}$. Prove that

$$\nabla f(P) = \lambda \nabla g(P)$$

for some number λ.

Proof: Parametrize the surface $g(x, y, z) = 0$ at and near the point P with a smooth function $\mathbf{r}(u, v)$ and in such a way that $\mathbf{r}_u(u, v)$ and $\mathbf{r}_v(u, v)$ are nonzero at and near P. Suppose that $\mathbf{r}(u_0, v_0) = \overrightarrow{OP}$. Then $f(\mathbf{r}(u, v))$ has a maximum (or minimum) at (u_0, v_0). Hence

$$D_u(f(\mathbf{r}(u, v))) = 0 \quad \text{and} \quad D_v(f(\mathbf{r}(u, v))) = 0.$$

Hence

$$\nabla f(P) \cdot \mathbf{r}_u(u_0, v_0) = 0 \quad \text{and} \quad \nabla f(P) \cdot \mathbf{r}_v(u_0, v_0) = 0.$$

But $g(\mathbf{r}(u, v)) \equiv 0$, and hence

$$\nabla g(P) \cdot \mathbf{r}_u(u_0, v_0) = 0 \quad \text{and} \quad \nabla g(P) \cdot \mathbf{r}_v(u_0, v_0) = 0.$$

Therefore $\nabla f(P)$ and $\nabla g(P)$ are parallel to $\mathbf{r}_u(u_0, v_0) \times \mathbf{r}_v(u_0, v_0)$. Consequently, because $\nabla g(P) \neq \mathbf{0}$, $\nabla f(P) = \lambda \nabla g(P)$ for some scalar λ. ◀

C13S09.047: Given the constant P, we are to maximize

753

$$A(x, y, z) = \frac{1}{2}xy$$

given the constraints $g(x, y, z) = x + y + z - P = 0$ and $h(x, y, z) = x^2 + y^2 - z^2 = 0$. The Lagrange multiplier equations are

$$\frac{1}{2}y = \lambda + 2\mu x, \quad \frac{1}{2}x = \lambda + 2\mu y, \quad \text{and} \quad 0 = \lambda - 2\mu z.$$

We solve the last equation for $\lambda = 2\mu z$ and substitute in the other two to obtain

$$y = 4\mu z + 4\mu x,$$

$$x = 4\mu z + 4\mu y.$$

We subtract the second of these from the first to find that $y - x = -4\mu(y - x)$. There are two cases to consider.

Case 1: $y = x$. Then the constraint equations become $2x + z = P$ and $z^2 = 2x^2$. We solve the first for z and substitute in the second to obtain

$$(P - 2x)^2 = 2x^2;$$

$$P - 2x = \pm x\sqrt{2};$$

$$2x \pm x\sqrt{2} = P;$$

$$x = \frac{P}{2 \pm \sqrt{2}}.$$

We must take the plus sign in the last denominator because $x \leq P$. Thus we obtain our first critical point:

$$x = \frac{P}{2 + \sqrt{2}} = \frac{2 - \sqrt{2}}{2}P, \quad y = x, \quad z = P - 2x = \left(\sqrt{2} - 1\right)P.$$

In this case the area of the triangle is

$$A = \frac{1}{2}xy = \frac{\left(2 - \sqrt{2}\right)^2}{8}P^2 = \frac{3 - 2\sqrt{2}}{4}P^2 \approx (0.042893218813)P^2.$$

Case 2: $y \neq x$. Then $4\mu = -1$, so that $\mu = -\frac{1}{4}$. Our earlier equations

$$y = 4\mu(z + x) \quad \text{and} \quad x = 4\mu(z + y)$$

now become $y = -z - x$ and $x = -z - y$, each of which is impossible as, at maximum, x, y, and z are all positive.

Thus Case 1 is the only case that produces a critical point. We may conclude that $y = x$ and that the right triangle with fixed perimeter and maximum area is isosceles.

C13S09.049: The hexagon in Fig. 13.9.13 is the union of four congruent trapezoids, one in each quadrant. The area of the trapezoid in the first quadrant is $\frac{1}{2}(1 + y)x$, so the total area of the hexagon is

$$A(x, y) = 2x(1 + y),$$

which we are to maximize subject to the constraint $x^2 + y^2 - 1 = 0$. Note also that $0 \leq x \leq 1$ and $0 \leq y \leq 1$. The Lagrange multiplier equations are

$$2 + 2y = 2\lambda x \qquad \text{and} \qquad 2x = 2\lambda y,$$

which yield (multiply the first by y, the second by x)

$$2y + 2y^2 = 2\lambda xy = 2x^2, \qquad \text{so that} \qquad y^2 + y = x^2.$$

Substitute for x^2 in the constraint equation to obtain $2y^2 + y - 1 = 0$, so that $(2y - 1)(y + 1) = 0$. Thus $y = \frac{1}{2}$ because $y \neq -1$. So the only critical point is $\left(\frac{1}{2}\sqrt{3}, \frac{1}{2}\right)$. To verify that the resulting hexagon is regular, it is sufficient (by the various symmetries in the figure) to verify that the distances from $\left(\frac{1}{2}\sqrt{3}, \frac{1}{2}\right)$ to $(0, 1)$ and from $\left(\frac{1}{2}\sqrt{3}, \frac{1}{2}\right)$ to $\left(\frac{1}{2}\sqrt{3}, -\frac{1}{2}\right)$ are equal. They are; each distance is 1.

C13S09.051: We are to minimize $f(x, y) = x^2 + y^2$ given the constraint $(x - 1)^2 - y = 0$. The Lagrange multiplier equations are

$$2x = 2\lambda(x - 1) \qquad \text{and} \qquad 2y = -\lambda,$$

and elimination of λ leads to the equation $x = -2y(x - 1)$. We combine this with the constraint equation and ask *Mathematica* 3.0 to solve the resulting two equations:

```
Solve[ { x == -2*y*(x - 1), y == (x - 1)^2 }, { x, y } ]
```

The only real solution is

$$x = 1 + \frac{\left(-9 + \sqrt{87}\,\right)^{1/3}}{6^{2/3}} - \frac{1}{\left[6 \cdot \left(-9 + \sqrt{87}\,\right)\right]^{1/3}} \approx 0.4102454876985416,$$

$$y = \frac{1}{36}\left\{ -12 + \frac{6^{4/3}}{\left(-9 + \sqrt{87}\,\right)^{2/3}} + \left[6 \cdot \left(-9 + \sqrt{87}\,\right)\right]^{2/3} \right\} \approx 0.3478103847799310.$$

C13S09.053: We are to find the first-quadrant point of the hyperbola $xy = 24$ that is closest to the point $P(1, 4)$, so we minimize $f(x, y) = (x - 1)^2 + (y - 4)^2$ subject to the constraint $xy = 24$ and the observation that $x > 0$ and $y > 0$. The Lagrange multiplier equations are

$$2(x - 1) = \lambda y \qquad \text{and} \qquad 2(y - 4) = \lambda x.$$

To eliminate λ, multiply the first equation by x and the second by y to obtain

$$2x(x - 1) = \lambda xy = 2y(y - 4).$$

We simplified this equation and asked *Mathematica* 3.0 to solve it simultaneously with the constraint equation:

```
Solve[ { x*x - x == y*y - y, x*y == 24 }, { x, y } ]
```

The resulting output is too long to reproduce here, but to evaluate the results numerically to 20 places we entered

755

and obtained four pairs of solutions, only two of which were real:

$$x_1 = 4, \ y_1 = 6 \quad \text{and} \quad x_2 \approx -5.5338281384822297, \ y_2 \approx -4.3369615751352589.$$

Hence the point in the first quadrant on the hyperbola $xy = 24$ closest to P is $(4, 6)$. It seems very likely that (x_2, y_2) is the point on the hyperbola in the third quadrant closest to P.

C13S09.055: To find the point on the spherical surface with equation

$$(x - 1)^2 + (y - 2)^2 + (z - 3)^2 - 36 = 0 \tag{1}$$

that are closest to and farthest from the origin, note that there do exist such points, and probably only one of each type, so we expect to find only two critical points. We maximize and minimize $f(x, y, z) = x^2 + y^2 + z^2$ subject to the constraint in Eq. (1). The Lagrange multiplier equations are

$$2x = 2\lambda(x - 1), \quad 2y = 2\lambda(y - 2), \quad \text{and} \quad 2z = 2\lambda(z - 3).$$

To solve these, we entered the *Mathematica* 3.0 command

```
Solve[ { x == lambda*(x - 1), y == lambda*(y - 2), z == lambda*(z - 3),
    (x - 1)^2 + (y - 2)^2 + (z - 3)^2 == 36 }, { x, y, z, lambda } ]
```

As expected, there are two solutions. Omitting the values of λ, they are

$$x_1 = \frac{7 - 3\sqrt{14}}{7}, \qquad y_1 = \frac{14 - 6\sqrt{14}}{7}, \qquad z_1 = \frac{21 - 9\sqrt{14}}{7} \quad \text{and}$$

$$x_2 = \frac{7 + 3\sqrt{14}}{7}, \qquad y_2 = \frac{14 + 6\sqrt{14}}{7}, \qquad z_2 = \frac{21 + 9\sqrt{14}}{7}.$$

Their numerical values are

$$x_1 \approx -0.6035674514745463, \qquad y_1 \approx -1.2071349029490926, \qquad z_1 \approx -1.8107023544236389 \quad \text{and}$$

$$x_2 \approx 2.6035674514745463, \qquad y_2 \approx 5.2071349029490926, \qquad z_2 \approx 7.8107023544236389.$$

The first of these is obviously much closer to the origin than the second, so (x_1, y_1, z_1) is the point of the spherical surface closest to the origin and (x_2, y_2, z_2) is the point of the spherical surface farthest from the origin.

C13S09.057: We are to find the points of the ellipse with equation $4x^2 + 9y^2 = 36$ closest to, and farthest from, the line with equation $x + y = 10$. What if the ellipse and the line intersect? If they do, then the equation $4x^2 + 9(10 - x)^2 = 36$ will have one or two real solutions. But this equation reduces to $13x^2 - 180x + 864 = 0$, which has discriminant

$$\Delta = 180^2 - 4 \cdot 13 \cdot 864 = -12528 < 0.$$

Because the quadratic has no real solutions, the ellipse and the line do not meet. To find the answers, we could assume that (x, y) is a point of the ellipse, that (u, v) is a point of the line, and maximize and minimize

$$f(x, y, u, v) = (x - u)^2 + (y - v)^2$$

subject to the constraints $4x^2 + 9y^2 - 36 = 0$ and $u + v - 10 = 0$. The Lagrange multiplier equations are

$$2(x - u) = 8\lambda x, \qquad 2(y - v) = 18\lambda y,$$

$$-2(x - u) = \mu, \qquad -2(y - v) = \mu.$$

Mathematica 3.0 can solve the system of six simultaneous equations (the four Lagrange multiplier equations and the two constraint equations) exactly in a few tenths of a second, but this problem can be solved by hand. Observe that if the line is moved without rotation toward the ellipse, when it first touches the ellipse it will be touching the point of the ellipse closest to the original line and the moving line will be tangent to the ellipse at that point. As the line continues to move across the ellipse, it will last touch the ellipse at the point of the ellipse farthest from the original line and the moving line will be tangent to the ellipse at that point. Because the line has slope -1, all we need do is find the points of the ellipse where the tangent line has slope -1. Implicit differentiation of the equation of the ellipse yields

$$8x + 18y \frac{dy}{dx} = 0, \qquad \text{so that} \qquad \frac{dy}{dx} = -\frac{8x}{18y} = -\frac{4x}{9y}.$$

The extrema occur when

$$\frac{dy}{dx} = -1; \qquad 4x = 9y; \qquad y = \frac{4}{9}x.$$

Substitution of the last equation for y in the equation of the ellipse yields the two solutions $x = \pm \frac{9}{13}\sqrt{13}$. So the points of the ellipse closest to, and farthest from, the line are (respectively)

$$\left(\frac{9}{13}\sqrt{13}, \frac{4}{13}\sqrt{13} \right) \qquad \text{and} \qquad \left(-\frac{9}{13}\sqrt{13}, -\frac{4}{13}\sqrt{13} \right).$$

(You can tell the maximum from the minimum by observing that the line is to the "northeast" of the ellipse.)

C13S09.059: The sides of the box must be parallel to the coordinate planes. Let (x, y, z) be the upper vertex of the box that lies in the first octant, so that at maximum volume we have x, y, and z all positive. The box has width $2x$, depth $2y$, and height $z = 9 - x^2 - 2y^2$, so we are to maximize box volume $V(x, y, z) = 4xyz$ given the constraint $x^2 + 2y^2 + z - 9 = 0$. The Lagrange multiplier equations are

$$4yz = 2\lambda x, \quad 4xz = 4\lambda y, \quad \text{and} \quad 4xy = \lambda.$$

These equations (together with the constraint equation) are relatively easy to solve by hand. To solve them using *Mathematica* 3.0, we entered the command

```
Solve[ { 4*y*z == 2*lambda*x, 4*x*z == 4*lambda*y, 4*x*y == lambda,
     x*x + 2*y*y + z == 9 }, { x, y, z, lambda } ]
```

The computer returned nine solutions, but eight involve negative or zero values of x, y, or z; the only viable solution is

$$x = \frac{3}{2}, \quad y = \frac{3}{4}\sqrt{2}, \quad z = \frac{9}{2},$$

and therefore the box of maximum volume has volume $V = \dfrac{81}{4}\sqrt{2}$.

C13S09.061: With $n = 3$ and $k = 2$ (for instance), the equations in (15) are

$$g_1(x_1, x_2, x_3, x_4) = 0,$$
$$g_2(x_1, x_2, x_3, x_4) = 0,$$
$$g_3(x_1, x_2, x_3, x_4) = 0$$

and the scalar component equations of the vector equation

$$\nabla f(x_1, x_2, x_3, x_4) = \lambda_1 \nabla g_1(x_1, x_2, x_3, x_4) + \lambda_2 \nabla g_2(x_1, x_2, x_3, x_4) + \lambda_3 \nabla g_3(x_1, x_2, x_3, x_4)$$

are

$$D_1 f(x_1, x_2, x_3, x_4) = \lambda_1 D_1 g_1(x_1, x_2, x_3, x_4) + \lambda_2 D_1 g_2(x_1, x_2, x_3, x_4) + \lambda_3 D_1 g_3(x_1, x_2, x_3, x_4),$$
$$D_2 f(x_1, x_2, x_3, x_4) = \lambda_1 D_2 g_1(x_1, x_2, x_3, x_4) + \lambda_2 D_2 g_2(x_1, x_2, x_3, x_4) + \lambda_3 D_2 g_3(x_1, x_2, x_3, x_4),$$
$$D_3 f(x_1, x_2, x_3, x_4) = \lambda_1 D_3 g_1(x_1, x_2, x_3, x_4) + \lambda_2 D_3 g_2(x_1, x_2, x_3, x_4) + \lambda_3 D_3 g_3(x_1, x_2, x_3, x_4),$$
$$D_4 f(x_1, x_2, x_3, x_4) = \lambda_1 D_4 g_1(x_1, x_2, x_3, x_4) + \lambda_2 D_4 g_2(x_1, x_2, x_3, x_4) + \lambda_3 D_4 g_3(x_1, x_2, x_3, x_4).$$

Consequently we have altogether $3 + 4 = 7$ equations in the seven unknowns λ_1, λ_2, λ_3, x_1, x_2, x_3, and x_4. —C.H.E.

C13S09.063: With

$$f(x, y) = x^2 + y^2 \qquad \text{and} \qquad g(x, y) = \frac{a}{x} + \frac{b}{y} - 1,$$

the Lagrange multiplier equations are

$$2x = -\frac{\lambda}{x^2} \qquad \text{and} \qquad 2y = -\frac{\lambda}{y^2},$$

and it follows that $2x^2 = -\lambda a$ and $2y^2 = -\lambda b$. Then division of the last equation by the one before it yields $y = xb^{1/3}a^{-1/3}$, and then substitution of this value in the constraint equation readily gives $x = a^{1/3}(a^{2/3} + b^{2/3})$. Thus $y = b^{1/3}(a^{2/3} + b^{2/3})$, and substitution of these values of x and y gives

$$L_{\min} = \sqrt{x^2 + y^2} = \left(a^{2/3} + b^{2/3}\right)^{3/2}. \qquad \text{—C.H.E.}$$

C13S09.065: We write $P_1(x, y, z)$ and $P_2(u, v, w)$ because the two points are independent. *Mathematica* then yields the solution, as follows.

```
f = (x − u)∧2 + (y − v)∧2 + (z − w)∧2;

eq1 = 2*x + y + 2*z == 15;

eq2 = x + 2*y + 3*z == 30;

eq3 = u − v − 2 *w == 15;

eq4 = 3*u − 2*v − 3*w == 20;

eq5 = 2*(x −u)  == 2*λ₁ + λ₂;
```

```
eq6 = 2*(y −v) == λ₁ + 2*λ₂;
eq7 = 2*(z −w) == 2*λ₁ + 3*λ₂;
eq8 = −2*(x − u) == λ₃ + 3*λ₄;
eq9 = −2*(y −v) == −λ₃ − 2*λ₄;
eq10 = −2*(z − w) == −2*λ₃ − 3*λ₄;

eqs = { eq1. eq2. eq3. eq4. eq5. eq6. eq7. eq8. eq9. eq10 };

vars = { x, y, z, u, v, w, λ₁, λ₂, λ₃, λ₄ };

soln = Solve[ eqs, vars ]
```

$$x = 7, \qquad y = 43, \qquad z = -21, \qquad u = 12, \qquad v = 41, \qquad w = -22,$$
$$\lambda_1 = -8, \qquad \lambda_2 = 6, \qquad \lambda_3 = -8, \qquad \lambda_4 = 6.$$

Thus the closest points are $P_1(7, 43, -21)$ on line L_1 and $P_2(12, 41, -22)$ on line L_2.   —C.H.E.

Section 13.10

C13S10.001: Given $f(x, y) = 2x^2 + y^2 + 4x - 4y + 5$, in the notation of Section 13.10 we have $A = 4$, $B = 0$, and $C = 2$. Hence $\Delta = 8 > 0$ and $A > 0$ for all points (x, y), and so—by Theorem 1—any critical point where both partials vanish is a local minimum. Here we have

$$f_x(x, y) = 4x + 4 \quad \text{and} \quad f_y(x, y) = 2y - 4,$$

so the only such critical point is $(-1, 2)$. The *Mathematica* 3.0 command

```
ContourPlot[ 2*x*x + y*y + 4*x - 4*y + 5, {x, -4, 2}, {y, -2, 4},
        AspectRatio → 1.0, Axes → True, Contours → 15, ContourShading → False,
        Frame → False, PlotPoints → 31, AxesOrigin → {0,0}];
```

generates level curves for f in the vicinity of the critical point $(-1, 2)$, shown next. Because the **graph of** f is an elliptic paraboloid opening upward, the local minimum at $(-1, 2)$ is actually a global minimum.

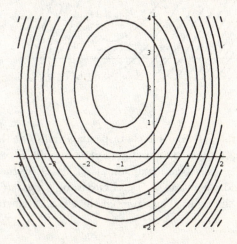

C13S10.003: Given $f(x, y) = 2x^2 - 3y^2 + 2x - 3y + 7$, both partial derivatives are zero at the single point $P\left(-\frac{1}{2}, -\frac{1}{2}\right)$. In the notation of Section 13.10 we have $A = 4$, $B = 0$, and $C = -6$. Hence $\Delta = -24 < 0$, and so—by Theorem 2—there is a saddle point at P. The *Mathematica* 3.0 command `ContourPlot` on the rectangle $-3 \leqq x \leqq 2$, $-3 \leqq y \leqq 2$ produced the level curve diagram shown next.

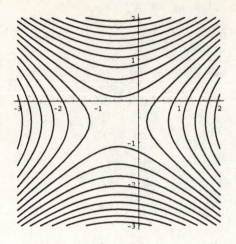

C13S10.005: Given $f(x, y) = 2x^2 + 2xy + y^2 + 4x - 2y + 1$, both partial derivatives are zero at the single point $P(-3, 4)$. In the notation of Section 13.10, $A = 4$, $B = 2$, and $C = 2$, so that $\Delta = 4 > 0$. Because $A > 0$, there is a local minimum at P. A diagram of some level curves of f near P generated by *Mathematica* 3.0 is next.

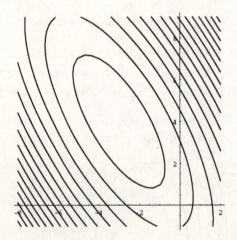

C13S10.007: Given $f(x, y) = x^3 + y^3 + 3xy + 3$, both partials vanish at the two points $P(-1, -1)$ and $Q(0, 0)$. Also $f_{xx}(x, y) = 6x$, $f_{xy}(x, y) = 3$, and $f_{yy}(x, y) = 6y$. At P we have $A = -6$, $B = 3$, and $C = -6$, so $\Delta = 27 > 0$. Thus there is a local maximum at P. At Q we have $A = 0$, $B = 3$, and $C = 0$, so $\Delta = -9 < 0$. So there is a saddle point at Q. Level curves of f near P are shown next, on the left; level curves near Q are shown on the right. The figure on the left was generated by the *Mathematica* 3.0 command

```
ContourPlot[ f[x,y], {x, -1.5, -0.5}, {y, -1.5, -0.5}, AspectRatio → 1.0
```

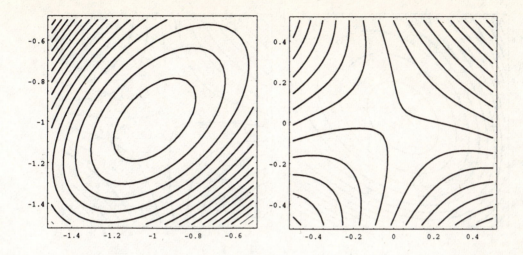

C13S10.009: Given $f(x, y) = 6x - x^3 - y^3$, both partial derivatives are zero at the two points $P\left(-\sqrt{2}, 0\right)$ and $Q\left(\sqrt{2}, 0\right)$. Because $f_{xx}(x, y) = -6x$, $f_{xy}(x, y) = 0$, and $f_{yy}(x, y) = -6y$, at P we have $A = 6\sqrt{2}$ and $B = C = 0$, so $\Delta = 0$; $\Delta = 0$ at Q as well. Hence Theorem 2 provides us with no information. But on the vertical line $x = -\sqrt{2}$ through P, $f(x, y)$ takes the form $-4\sqrt{2} - y^3$, so there is no extremum at P; similarly, there is none at Q. Hence f has no extrema. We generated level curves of f near P and Q using *Mathematica* 3.0 commands similar to those in the solution of Problem 7. The results are shown next.

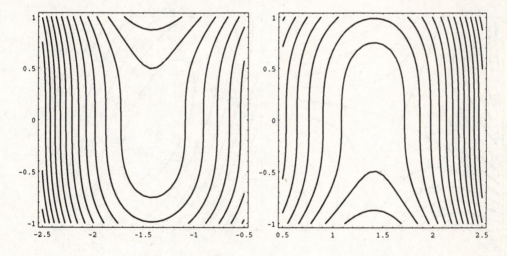

C13S10.011: Given $f(x, y) = x^4 + y^4 - 4xy$, both partial derivatives are zero at $P(-1, -1)$, $Q(0, 0)$, and $R(1, 1)$. Also $f_{xx}(x, y) = 12x^2$, $f_{xy}(x, y) = -4$, and $f_{yy}(x, y) = 12y^2$. Thus at both P and R, we have $A = 12$, $B = -4$, and $C = 12$, so that $\Delta = 128 > 0$; thus there are local minima at P and R. At Q we have $A = C = 0$ and $B = -4$, so that $\Delta = -16 < 0$: There is a saddle point at Q. The minima at P and R are global because $x^4 + y^4$ greatly exceeds $4xy$ if x and y are large in magnitude. Level curves of f near

761

the two critical points P and Q, generated by *Mathematica* 3.0, are shown next.

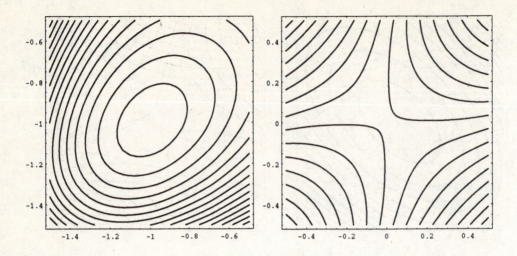

C13S10.013: Given $f(x, y) = x^3 + 6xy + 3y^2 - 9x$, both partial derivatives are zero at the two points $P(-1, 1)$ and $Q(3, -3)$. Next, $f_{xx}(x, y) = 6x$ and $f_{xy}(x, y) = f_{yy}(x, y) = 6$. Hence at P we have $A = -6$ and $B = C = 6$, so that $\Delta = -72 < 0$: There is a saddle point at P. At Q we have $A = 18$ and $B = C = 6$, so $\Delta = 72 > 0$. Because $A > 0$ there is a local minimum at Q. It cannot be global because the dominant term in $f(x, y)$ is x^3. We used *Mathematica* 3.0 with commands similar to those in previous solutions to show level curves of f near these two critical points; the diagrams are next.

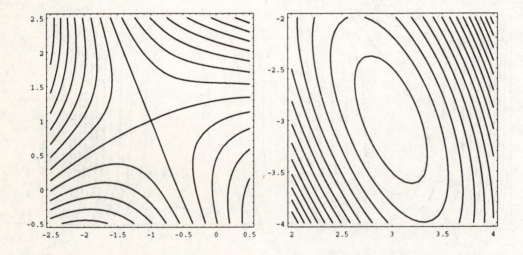

C13S10.015: If $f(x, y) = 3x^2 + 6xy + 2y^3 + 12x - 24y$, then both partial derivatives of f are zero at the two points $P(-5, 3)$ and $Q(0, -2)$. Next, $f_{xx}(x, y) = f_{xy}(x, y) = 6$ and $f_{yy}(x, y) = 12y$. Thus at P we have $A = B = 6$ and $C = 36$, so $\Delta = 180 > 0$. Because $A > 0$ there is a local minimum at P (it cannot be global because the dominant term in $f(x, y)$ is $2y^3$.) At Q we have $A = B = 6$ and $C = -24$, so $\Delta = -180 < 0$; there is a saddle point at Q. We used *Mathematica* 3.0 commands similar to those in previous solutions to generate level curves of f near its two critical points; these diagrams are next.

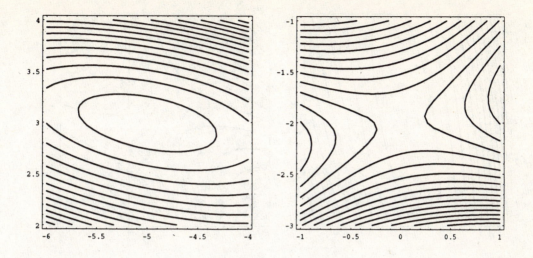

C13S10.017: If $f(x, y) = 4xy - 2x^4 - y^2$, then both partial derivatives are zero when

$$8x^3 = 4y \quad \text{and} \quad 4x = 2y;$$

$$y = 2x^3 \quad \text{and} \quad y = 2x;$$

$$x^3 - x = 0; \quad \text{that is,} \quad x(x-1)(x+1) = 0.$$

Thus we obtain three critical points: $P(-1, -2)$, $Q(0, 0)$, and $R(1, 2)$. For the next step, we compute $f_{xx}(x, y) = -24x^2$, $f_{xy}(x, y) = 4$, and $f_{yy}(x, y) = -2$. Thus an application of Theorem 2 yields this information.

At P: $A = -24$, $B = 4$, $C = -2$, $\Delta = 32$:   Local maximum.

At Q: $A = 0$, $B = 4$, $C = -2$, $\Delta = -16$:   Saddle point.

At R: $A = -24$, $B = 4$, $C = -2$, $\Delta = 32$:   Local maximum.

The extrema are global maxima with the same value, $f(-1, -2) = 2 = f(1, 2)$. To see that they are global, examine the behavior of $f(x, y)$ on straight lines through the origin. Diagrams of level curves near P and

763

Q were generated using *Mathematica* 3.0 and appear next.

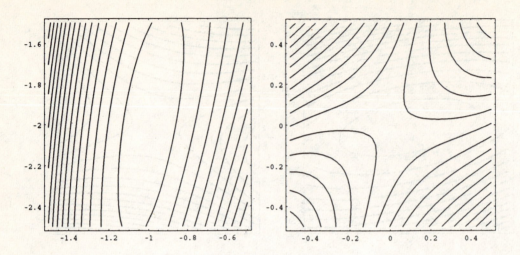

C13S10.019: Given $f(x, y) = 2x^3 - 3x^2 + y^2 - 12x + 10$, we set both partial derivatives equal to zero and solve to find the only two critical points: $P(-1, 0)$ and $Q(2, 0)$. At P we find that $A = -18$, $B = 0$, and $C = 2$, so there is a saddle point at P. At Q we find that $A = 18$, $B = 0$, and $C = 2$, so there is a local minimum at Q. (There are no global extrema because the dominant term in $f(x, y)$ is $2x^3$.) We used *Mathematica* 3.0 in the usual way to generate level curves of f near these two critical points; the resulting diagram is next.

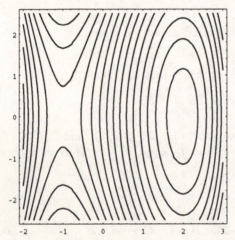

C13S10.021: If $f(x, y) = xy \exp(-x^2 - y^2)$, then

$$f_x(x, y) = (1 - 2x^2)y \exp(-x^2 - y^2) \quad \text{and} \quad f_y(x, y) = x(1 - 2y^2) \exp(-x^2 - y^2).$$

Thus there are a total of five critical points. They are $P(0, 0)$, $Q\left(-\frac{1}{2}\sqrt{2}, -\frac{1}{2}\sqrt{2}\right)$, $R\left(-\frac{1}{2}\sqrt{2}, \frac{1}{2}\sqrt{2}\right)$, $S\left(\frac{1}{2}\sqrt{2}, -\frac{1}{2}\sqrt{2}\right)$, and $T\left(\frac{1}{2}\sqrt{2}, \frac{1}{2}\sqrt{2}\right)$. The values of f at these points are (in the same order) 0, $1/(2e)$, $-1/(2e)$, $-1/(2e)$, and $1/(2e)$. The second-order partial derivatives of f are

$$f_{xx}(x, y) = (4x^3y - 6xy)\exp(-x^2 - y^2) = 2xy(2x^2 - 3)\exp(-x^2 - y^2),$$

$$f_{xy}(x, y) = (4x^2y^2 - 2x^2 - 2y^2 + 1)\exp(-x^2 - y^2), \quad \text{and}$$

$$f_{yy}(x, y) = (4xy^3 - 6xy)\exp(-x^2 - y^2) = 2xy(2y^2 - 3)\exp(-x^2 - y^2).$$

Application of Theorem 2 yields these results:

$$\text{At } P: \quad A = 0, \quad B = 1, \quad C = 0, \quad \Delta = -1: \quad \text{Saddle point.}$$

$$\text{At } Q: \quad A = -\frac{2}{e}, \quad B = 0, \quad C = -\frac{2}{e}, \quad \Delta = \frac{4}{e^2}: \quad \text{Local maximum.}$$

$$\text{At } R: \quad A = \frac{2}{e}, \quad B = 0, \quad C = \frac{2}{e}, \quad \Delta = \frac{4}{e^2}: \quad \text{Local minimum.}$$

$$\text{At } S: \quad A = \frac{2}{e}, \quad B = 0, \quad C = \frac{2}{e}, \quad \Delta = \frac{4}{e^2}: \quad \text{Local minimum.}$$

$$\text{At } T: \quad A = -\frac{2}{e}, \quad B = 0, \quad C = -\frac{2}{e}, \quad \Delta = \frac{4}{e^2}: \quad \text{Local maximum.}$$

The extrema are all global because $f(x, y) \to 0$ as $|x|$ and/or $|y|$ increase without bound. To generate a diagram of these critical points and nearby level curves of f, we executed the *Mathematica* 3.0 command

```
ContourPlot[ f[x,y], {x, -1.5, 1.5}, {y, -1.5, 1.5}, AspectRatio → 1.0,
        Contours → 15, ContourShading → False, PlotPoints → 48 ];
```

and the result is shown next.

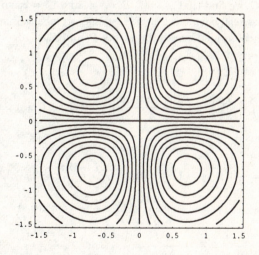

C13S10.023: If $f(x, y) = x^4 + y^4$, then

$$\Delta = AC - B^2 = (12x^2)(12y^2) - 0^2 = 144x^2y^2,$$

and so $\Delta(0, 0) = 0$. But $f(x, y)$ has the global minimum value 0 at $(0, 0)$ because $f(x, y) > 0$ for all $(x, y) \neq (0, 0)$.

C13S10.025: If $f(x, y) = \exp(-x^4 - y^4)$, then

$$\Delta = AC - B^2$$

$$= \left[4x^2(4x^4 - 3)\exp(-x^4 - y^4)\right] \cdot \left[4y^2(4y^4 - 3)\exp(-x^4 - y^4)\right] - \left[16x^3y^3\exp(-x^4 - y^4)\right]^2,$$

and so $\Delta(0, 0) = 0$. To maximize $f(x, y)$, make $x^4 + y^4$ as small as possible, and hence $1 = f(0, 0)$ is the global maximum value of $f(x, y)$.

C13S10.027: Let $f(x, y) = x^2 + y^2 + (xy - 2)^2$. Then

$$f_x(x, y) = 2x + 2y(xy - 2) = 2x + 2xy^2 - 4y \quad \text{and}$$

$$f_y(x, y) = 2y + 2x(xy - 2) = 2y + 2x^2y - 4x.$$

Both partial derivatives are zero when

$$x + xy^2 - 2y = 0 \quad \text{and} \quad y + x^2y - 2x = 0;$$

$$x^2 + x^2y^2 - 2xy = 0 \quad \text{and} \quad y^2 + x^2y^2 - 2xy = 0;$$

$$x^2 = y^2, \quad \text{so that} \quad y = x \quad \text{or} \quad y = -x.$$

If $y = x$, then $x^3 - x = 0$, and we obtain the three critical points $P(-1, -1)$, $Q(0, 0)$, and $R(1, 1)$. If $y = -x$ then $x^3 + 3x = 0$ and we obtain the critical point Q again but no others. Next

$$f_{xx}(x, y) = 2(y^2 + 1), \quad f_{xy}(x, y) = 4(xy - 1), \quad \text{and} \quad f_{yy}(x, y) = 2(x^2 + 1).$$

At P and at R, we find that $A = 4$, $B = 0$, and $C = 4$, so $f(x, y)$ has a local minimum at those two points. Its value at each is 3. But at Q we find that $A = 2$, $B = -4$, and $C = 2$, so $\Delta = -12$ and hence there is no extremum at Q. It is geometrically evident that $f(x, y)$ has a global minimum, so we have found it: Its global minimum value is 3, which occurs at both the points P and R.

C13S10.029: The sum of the areas of the squares is

$$f(x, y) = \left(\frac{x}{4}\right)^2 + \left(\frac{y}{4}\right)^2 + \left(\frac{120 - x - y}{4}\right)^2, \quad 0 \le x, \ 0 \le y, \ x + y \le 120.$$

Thus

$$\frac{\partial f}{\partial x} = \frac{1}{8}(2x + y - 120) \quad \text{and} \quad \frac{\partial f}{\partial y} = \frac{1}{8}(x + 2y - 120),$$

and both partial derivatives are zero at the single point $(x, y) = (40, 40)$. Because

$$\frac{\partial^2 f}{\partial x^2} = \frac{1}{4}, \quad \frac{\partial^2 f}{\partial y \, \partial x} = \frac{1}{8}, \quad \text{and} \quad \frac{\partial^2 f}{\partial y^2} = \frac{1}{4},$$

we see that $\Delta = \frac{3}{64} > 0$, so that $f(x, y)$ has a local minimum at $(40, 40)$. Because there must be a global maximum, it must occur on the boundary of the domain of f. If $y = 0$ then it is easy to show that $x = 60$ is a critical point of $f(x, 0)$, and $f(60, 0) = 450$. Similar results hold if $x = 0$ and if $x + y = 120$. At the corner points of the domain we find the global maximum value of $f(x, y)$ to be $f(0, 120) = 900$. It also follows that the local minimum at $(40, 40)$ is global.

C13S10.031: Given

766

$$f(x, y) = \sin\left(\frac{\pi x}{2}\right) \sin\left(\frac{\pi y}{2}\right),$$

we first compute

$$f_x(x, y) = \frac{\pi}{2} \cos\left(\frac{\pi x}{2}\right) \sin\left(\frac{\pi y}{2}\right) \quad \text{and} \quad f_y(x, y) = \frac{\pi}{2} \sin\left(\frac{\pi x}{2}\right) \cos\left(\frac{\pi y}{2}\right).$$

Now $\cos(\pi x/2) = 0$ when x is an odd integer and $\sin(\pi x/2) = 0$ when x is an even integer. In order that both partial derivatives are zero, it is necessary and sufficient that x and y are both odd integers or both even integers. Next,

$$f_{xx}(x, y) = -\frac{\pi^2}{4} \sin\left(\frac{\pi x}{2}\right) \sin\left(\frac{\pi y}{2}\right),$$

$$f_{xy}(x, y) = \frac{\pi^2}{4} \cos\left(\frac{\pi x}{2}\right) \cos\left(\frac{\pi y}{2}\right), \quad \text{and}$$

$$f_{yy}(x, y) = -\frac{\pi^2}{4} \sin\left(\frac{\pi x}{2}\right) \sin\left(\frac{\pi y}{2}\right).$$

Note that every integer has exactly one of the following forms: $4k$, $4k + 1$, $4k + 2$, $4k + 3$ (where k is an integer). The reason is that division of an integer by 4 produces eactly one of the remainders 0, 1, 2, or 3.

Case 1: x and y are both of the form $4k$ or both of the form $4k + 2$. Then (in the notation of Theorem 1) $A = 0$, $B = \frac{1}{4}\pi^2$, and $C = 0$. Hence $\Delta < 0$, and so (x, y) is a saddle point.

Case 2: One of x and y is of the form $4k$ and the other is of the form $4k + 2$. Then $A = 0$, $B = -\frac{1}{4}\pi^2$, and $C = 0$. Hence $\Delta < 0$, and so (x, y) is a saddle point.

Case 3: x and y are both of the form $4k + 1$ or both of the form $4k + 3$. Then $A = C = -\frac{1}{4}\pi^2$ and $B = 0$. Thus $\Delta > 0$ and $A < 0$, so (x, y) is a local maximum (actually, a global maximum; the value of f there is 1).

Case 4: One of x and y is of the form $4k + 1$ and the other is of the form $4k + 3$. Then $A = C = \frac{1}{4}\pi^2$ and $B = 0$. Thus $\Delta > 0$ and $A > 0$, so (x, y) is a local minimum (actually, a global minimum; the value of f there is -1).

Execute the *Mathematica* 3.0 command

```
ContourPlot[ f[x,y], {x, -3.5, 3.5}, {y, -3.5, 3.5}, AspectRatio → 1.0,
          Contours → 21, ContourShading → False, PlotPoints → 61 ];
```

to see the maxima and minima in a checkerboard pattern with saddle points arranged neatly among them.

C13S10.033: If $f(x, y) = 4xy(x^2 - y^2)$, then

$$f_x(x, y) = 4y(3x^2 - y^2) \quad \text{and} \quad f_y(x, y) = 4x(x^2 - 3y^2).$$

Hence $(0, 0)$ is the only critical point of f. On the line $y = mx$ we have

$$f(x, y) = f(x, mx) = -4m(m + 1)(m - 1)x^4.$$

In the first-quadrant region between the lines $y = 0$ and $y = x$, where $0 < m < 1$, $f(x, y) > 0$. In the first-quadrant region between the lines $y = x$ and $x = 0$, where $1 < m$, $f(x, y) < 0$. Continuing in this way counterclockwise around the origin, we find eight regions on which $f(x, y)$ is alternately positive and negative. Hence the origin qualifies as a "dog saddle."

C13S10.035: Given $f(x, y) = 2x^4 - 12x^2 + y^2 + 8x$, note first that $f(x, y)$ is dominated by the term $2x^4$. Therefore it can have no global maximum and must have a global minimum. We continue with the aid of *Mathematica* 3.0, but as usual, we will rewrite its responses for a little more clarity.

```
{D[f[x,y],x], D[f[x,y],y]}
```

$$f_x(x, y) = 8x^3 - 24x + 8, \qquad f_y(x, y) = 2y.$$

```
N[Solve[ % == 0, {x,y} ], 20]
```

(We let *Mathematica* find the exact solutions, then approximate them to 20 places.)

$$x_1 \approx 1.532088886238, \quad y_1 = 0;$$
$$x_2 \approx 0.347296245334, \quad y_2 = 0;$$
$$x_3 \approx -1.879385241572, \quad y_3 = 0.$$

```
{D[f[x,y], {x,2}], D[f[x,y], x, y], D[f[x,y], {y,2}]}
```

$$f_{xx}(x, y) = 24x^2 - 24, \quad f_{xy}(x, y) = 0, \quad f_{yy}(x, y) = 2.$$

Then we asked *Mathematica* to evaluate the second-order partial derivatives and $f(x, y)$ itself at the three critical points. The response was

$$(x_1, y_1): \quad A \approx 32.335113, \quad B = 0, \quad C = 2, \quad f(x_1, y_1) \approx -4.891245,$$
$$(x_2, y_2): \quad A \approx -21.105246, \quad B = 0, \quad C = 2, \quad f(x_2, y_2) \approx 1.360090,$$
$$(x_3, y_3): \quad A \approx 60.770133, \quad B = 0, \quad C = 2, \quad f(x_3, y_3) \approx -32.468845.$$

Hence there is a local minimum at (x_1, y_1), a saddle point at (x_2, y_2), and the global minimum at (x_3, y_3).

C13S10.037: Given $f(x, y) = x^4 + 12xy + 6y^2 + 4x + 10$, note first that $f(x, y)$ is dominated by the term x^4. Therefore it can have no global maximum. We continue with the aid of *Mathematica* 3.0, but as usual, we will rewrite its responses for a little more clarity.

```
{D[f[x,y],x], D[f[x,y],y]}
```

$$f_x(x, y) = 4x^3 + 12y + 4, \qquad f_y(x, y) = 12x + 12y.$$

```
N[Solve[ % == 0, {x,y} ], 20]
```

(We let *Mathematica* find the exact solutions, then approximate them to 20 places.)

$$x_1 \approx -1.879385241572, \quad y_1 \approx 1.879385241572;$$
$$x_2 \approx 0.347296355334, \quad y_2 \approx -0.347296355334;$$
$$x_3 \approx 1.532088886238, \quad y_3 \approx -1.532088886238.$$

```
{D[f[x,y], {x,2}], D[f[x,y], x, y], D[f[x,y], {y,2}]}
```

$$f_{xx}(x, y) = 12x^2, \quad f_{xy}(x, y) = 12, \quad f_{yy}(x, y) = 12.$$

Then we asked *Mathematica* to evaluate the second-order partial derivatives and $f(x, y)$ itself at the three critical points. The response was

$$(x_1, y_1): \quad A \approx 42.385067, \quad B = 12, \quad C = 12, \quad f(x_1, y_1) \approx -6.234422,$$
$$(x_2, y_2): \quad A \approx 1.447377, \quad B = 12, \quad C = 12, \quad f(x_2, y_2) \approx 10.680045,$$
$$(x_3, y_3): \quad A \approx 28.167556, \quad B = 12, \quad C = 12, \quad f(x_3, y_3) \approx 7.554378.$$

Hence there is a local (indeed, global) minimum at (x_1, y_1), a saddle point at (x_2, y_2), and a local minimum at (x_3, y_3).

C13S10.039: Given $f(x, y) = x^4 + 2y^4 - 12xy^2 - 20y^2$, note first that $f(x, y)$ is dominated by the terms x^4 and $2y^4$. Therefore it can have no global maximum and must have a global minimum. We continue with the aid of *Mathematica* 3.0, but as usual, we will rewrite its responses for a little more clarity.

```
{D[f[x,y],x], D[f[x,y],y]}
```

$$f_x(x, y) = 4x^3 - 12y^2, \quad f_y(x, y) = 8y^3 - 24xy - 40y.$$

```
N[Solve[ % == 0, {x,y} ], 20]
```

(We let *Mathematica* find the exact solutions, then approximate them to 20 places.)

$$x_1 = 0, \quad y_1 = 0;$$
$$x_2 \approx 3.624678956480, \quad y_2 \approx -3.984223496422;$$
$$x_3 \approx 3.624678956480, \quad y_3 \approx 3.984223496422.$$

```
{D[f[x,y], {x,2}], D[f[x,y], x, y], D[f[x,y], {y,2}]}
```

$$f_{xx}(x, y) = 12x^2, \quad f_{xy}(x, y) = -24y, \quad f_{yy}(x, y) = 24y^2 - 24x - 40.$$

Then we asked *Mathematica* to evaluate the second-order partial derivatives and $f(x, y)$ itself at the three critical points. The response was

$$(x_1, y_1): \quad A = 0, \quad B = 0, \quad C = -40, \quad f(x_1, y_1) = 0,$$
$$(x_2, y_2): \quad A \approx 157.659570, \quad B \approx 95.621364, \quad C \approx 253.984590, \quad f(x_2, y_2) \approx -331.355231,$$
$$(x_3, y_3): \quad A \approx 157.659570, \quad B \approx -95.621364, \quad C \approx 253.984590, \quad f(x_3, y_3) \approx -331.355231.$$

Hence there is a global minimum at (x_2, y_2), and at (x_3, y_3). A plot of the level curves of $f(x, y)$ using the `ContourPlot` command in *Mathematica* 3.0 indicates that the origin is an (ordinary) saddle point. You can verify that there is no extremum there by showing that $f(x, 0)$ has a local minimum at the origin whereas $f(0, y)$ has a local maximum there.

Chapter 13 Miscellaneous Problems

C13S0M.001: Using polar coordinates, we get

$$\lim_{(x,y)\to(0,0)} \frac{x^2 y^2}{x^2 + y^2} = \lim_{r\to 0} \frac{r^4 \sin^2\theta \cos^2\theta}{r^2} = \lim_{r\to 0} (r^2 \sin^2\theta \cos^2\theta) = 0$$

because $0 \le \sin^2\theta \le 1$ and $0 \le \cos^2\theta \le 1$ for all θ.

C13S0M.003: First we note that $\lim\limits_{(x,y)\to(0,0)} g(x, y) \ne 0$ because

$$\lim_{x\to 0} g(x, x) = \lim_{x\to 0} \frac{x^2}{2x^2} = \frac{1}{2}.$$

Therefore g is not continuous at $(0, 0)$.

C13S0M.005: If $f_x(x, y) = 2xy^3 + e^x \sin y$, then

$$f(x, y) = x^2 y^3 + e^x \sin y + g(y).$$

Hence

$$f_y(x, y) = 3x^2 y^2 + e^x \cos y + g'(y) = 3x^2 y^2 + e^x \cos y + 1.$$

Therefore $g'(y) = 1$, and thus $g(y) = y + C$. Thus every solution of this problem has the form

$$f(x, y) = x^2 y^3 + e^x \sin y + y + C$$

where C is a constant.

C13S0M.007: The paraboloid is a level surface of $f(x, y, z) = x^2 + y^2 - z$ with gradient $\langle 2a, 2b, -1 \rangle$ at the point $(a, b, a^2 + b^2)$. The normal line L through that point has vector equation

$$\langle x, y, z \rangle = \langle a, b, a^2 + b^2 \rangle + t\langle 2a, 2b, -1 \rangle$$

and thus parametric equations

$$x = 2at + a, \quad y = 2bt + b, \quad z = a^2 + b^2 - t.$$

Suppose that this line passes through the point $(0, 0, 1)$. Set $x = 0$, $y = 0$, and $z = 1$ in the parametric equations of L, solve the third equation for t, then substitute the result in the other two equations to find that

$$2a(a^2 + b^2) - a = 0 = 2b(a^2 + b^2) - b.$$

It now follows that $a^2 + b^2 = \frac{1}{2}$ or $a = 0 = b$. Therefore the points on the paraboloid where the normal vector points at $(0, 0, 1)$ are the origin $(0, 0, 0)$ and the points on the circle formed by the intersection of the paraboloid and the horizontal plane $z = \frac{1}{2}$.

C13S0M.009: Let $f(x, y, z) = x^2 + y^2 - z^2$. Then $\nabla f(x, y, z) = \langle 2x, 2y, -2z \rangle$ is normal to the cone at (x, y, z). So a normal at $P(a, b, c)$ is $\mathbf{n} = \langle a, b, -c \rangle$. An equation of the line through P with direction $\mathbf{n}$ is

$$\langle x, y, z \rangle = \langle a, b, c \rangle + t\langle a, b, -c \rangle;$$

that is,

$$x = a + ta, \quad y = b + tb, \quad z = c - tc.$$

When $x = y = 0$, we have $t = -1$, and thus $z = 2c$. Hence $(0, 0, 2c)$ is the point where the line meets the z-axis. Thus the normal vector $\mathbf{n}$ (extended in length, if necessary) intersects the z-axis.

C13S0M.011: If

$$u(x, y, t) = \frac{1}{4\pi kt} \exp\left(-\frac{x^2 + y^2}{4kt}\right),$$

then

$$u_t(x, y, t) = \frac{x^2 + y^2 - 4kt}{16k^2\pi t^3} \exp\left(-\frac{x^2 + y^2}{4kt}\right),$$

$$u_x(x, y, t) = -\frac{x}{8k^2\pi t^2} \exp\left(-\frac{x^2 + y^2}{4kt}\right),$$

$$u_{xx}(x, y, t) = \frac{x^2 - 2kt}{16k^3\pi t^3} \exp\left(-\frac{x^2 + y^2}{4kt}\right),$$

$$u_y(x, y, t) = -\frac{y}{8k^2\pi t^2} \exp\left(-\frac{x^2 + y^2}{4kt}\right),$$

$$u_{yy}(x, y, t) = \frac{y^2 - 2kt}{16k^3\pi t^3} \exp\left(-\frac{x^2 + y^2}{4kt}\right), \quad \text{and}$$

$$u_{xx}(x, y, t) + u_{yy}(x, y, t) = \frac{x^2 + y^2 - 4kt}{16k^3\pi t^3} \exp\left(-\frac{x^2 + y^2}{4kt}\right).$$

Therefore $u_t = k(u_{xx} + u_{yy})$.

C13S0M.013: If $\mathbf{r}(x, y) = \langle x, y, f(x, y) \rangle$, then

$$\mathbf{r}_x(x, y) = \langle 1, 0, f_x(x, y) \rangle \quad \text{and} \quad \mathbf{r}_y(x, y) = \langle 0, 1, f_y(x, y) \rangle,$$

and hence

$$\mathbf{r}_x \times \mathbf{r}_y = \begin{vmatrix} \mathbf{i} & \mathbf{j} & \mathbf{k} \\ 1 & 0 & f_x(x, y) \\ 0 & 1 & f_y(x, y) \end{vmatrix} = \langle -f_x(x, y), -f_y(x, y), 1 \rangle.$$

Let $g(x, y, z) = z - f(x, y)$. Then

$$\nabla g(x, y, z) = \langle -f_x(x, y), -f_y(x, y), 1 \rangle = \mathbf{r}_x \times \mathbf{r}_y,$$

and therefore $\mathbf{r}_x \times \mathbf{r}_y$ is normal to the surface $z = f(x, y)$.

C13S0M.015: Suppose that the bottom of the crate measures x by y (units are in feet and dollars) and that its height is z. We are to minimize its total cost $C(x, y, z) = 5xy + 2xz + 2yz$ given the constraint $V(x, y, z) = xyz - 60 = 0$. The Lagrange multiplier equations are

$$5y + 2z = \lambda yz, \quad 5x + 2z = \lambda xz, \quad \text{and} \quad 2x + 2y = \lambda xy.$$

Because x, y, and z are all positive (because of the constraint) and $\lambda \neq 0$, it follows that

$$\lambda xyz = 5xy + 2xz = 5xy + 2yz = 2xz + 2yz,$$

and thus $5x = 5y = 2z$, so that $x = y = \frac{2}{5}z$. Substitution in the constraint yields $z = 5 \cdot 3^{1/3}$, then $x = y = 2 \cdot 3^{1/3}$. The base of the shipping crate will be a square $2 \cdot 3^{1/3} \approx 2.884449914$ feet on each side and the height of the crate will be $5 \cdot 3^{1/3} \approx 7.21124785$ feet. (Its cost will be \$124.81!)

C13S0M.017: Given

$$\frac{1}{R} = \frac{1}{R_1} + \frac{1}{R_2},$$

we differentiate implicitly with respect to R_1 and thereby find that

$$-\frac{1}{R^2} \cdot \frac{\partial R}{\partial R_1} = -\frac{1}{R_1^2}, \quad \text{and thus} \quad \frac{\partial R}{\partial R_1} = \left(\frac{R}{R_1} \right)^2.$$

Similarly,

$$\frac{\partial R}{\partial R_2} = \left(\frac{R}{R_2} \right)^2.$$

Hence if the errors in measuring R_1 and R_2 are $\Delta R_1 = 3$ and $\Delta R_2 = 6$, we estimate the resulting error in computation of R as follows:

$$\Delta R \approx \frac{\partial R}{\partial R_1} \cdot \Delta R_1 + \frac{\partial R}{\partial R_2} \cdot \Delta R_2 = \left(\frac{R}{R_1} \right)^2 \Delta R_1 + \left(\frac{R}{R_2} \right)^2 \Delta R_2$$

$$= \left(\frac{200}{300} \right)^2 \cdot 3 + \left(\frac{200}{600} \right)^2 \cdot 6 = \frac{4}{9} \cdot 3 + \frac{1}{9} \cdot 6 = 2.$$

In fact, when $R_1 = 303$ and $R_2 = 606$, the value of R is exactly 202, so there is no error in this approximation.

C13S0M.019: Suppose that a, b, and c are positive. The ellipsoid with equation

$$\left(\frac{x}{a} \right)^2 + \left(\frac{y}{b} \right)^2 + \left(\frac{z}{c} \right)^2 = 1$$

and, therefore, semiaxes of lengths a, b, and c, has volume $V = \frac{4}{3} \pi abc$. Let

$$V(x, y, z) = \frac{4}{3} \pi xyz.$$

Then

$$V_x(x, y, z) = \frac{4}{3} \pi yz, \quad V_y(x, y, z) = \frac{4}{3} \pi xz, \quad \text{and} \quad V_z(x, y, z) = \frac{4}{3} \pi xy.$$

Assume that errors of at most 1% are made in measuring x, y, and z. Then the error in computing V will be at most

$$\Delta V \approx dV = V_x \, dx + V_y \, dy + V_z \, dz = \frac{4}{3} \pi yz \cdot \frac{x}{100} + \frac{4}{3} \pi xz \cdot \frac{y}{100} + \frac{4}{3} \pi xy \cdot \frac{z}{100} = \frac{1}{25} \pi xyz,$$

and hence the percentage error in computing V will be at most

772

$$100 \cdot \frac{\Delta V}{V} \approx 100 \cdot \frac{1}{25} \pi xyz \cdot \frac{3}{4\pi xyz} = 3;$$

that is, the maximum error is approximately 3%.

C13S0M.021: The ellipsoidal surface S is the level surface

$$f(x, y, z) = x^2 + 4y^2 + 9z^2 - 16 = 0,$$

and hence a vector normal to S at the point (x, y, z) is $\mathbf{n} = \nabla f = \langle 2x, 8y, 18z \rangle$. If $\mathbf{n}$ is parallel to the position vector $\langle x, y, z \rangle$, then $\mathbf{n}$ (extended if necessary) will pass through $(0, 0, 0)$. This leads to the equation

$$\langle 2x, 8y, 18z \rangle = \lambda \langle x, y, z \rangle;$$

that is,

$$2x = \lambda x, \quad 8y = \lambda y, \quad 18z = \lambda z. \tag{1}$$

Multiply both sides of the first equation by yz, the second by xz, and the third by xy to eliminate λ:

$$\lambda xyz = 2xyz = 8xyz = 18xyz.$$

We obtain a contradiction if $xyz \neq 0$, and thus at least one of x, y, and z is zero. If (say) $x \neq 0$, then multiply the second equation in (1) by z and the third by y to obtain $\lambda yz = 8yz = 18yz$. If $yz \neq 0$ we obtain a contradiction, so at least one of y and z is zero. Repeating with the other two cases, we conclude that at least two of x, y, and z are zero. Of course the third cannot be zero because of the condition $f(x, y, z) = 0$. Therefore there are six points on S at which the normal vector points toward the origin; they are $(\pm 4, 0, 0)$, $(0, \pm 2, 0)$, and $\left(0, 0, \pm\frac{4}{3}\right)$.

C13S0M.023: Given: $\mathbf{a}$, $\mathbf{b}$, and $\mathbf{c}$ are mutually perpendicular unit vectors in space and f is a function of the three variables x, y, and z. Rename the unit vectors if necessary so that $\mathbf{a}$, $\mathbf{b}$, $\mathbf{c}$ forms a right-handed triple. Then

$$\mathbf{a} \times \mathbf{b} = \mathbf{c}, \quad \mathbf{b} \times \mathbf{c} = \mathbf{a}, \quad \text{and} \quad \mathbf{c} \times \mathbf{a} = \mathbf{b}. \tag{1}$$

Write $\mathbf{a} = \langle a_1, a_2, a_3 \rangle$, $\mathbf{b} = \langle b_1, b_2, b_3 \rangle$, and $\mathbf{c} = \langle c_1, c_2, c_3 \rangle$. Then

$$\mathbf{a} D_{\mathbf{a}} f + \mathbf{b} D_{\mathbf{b}} f + \mathbf{c} D_{\mathbf{c}} f = \mathbf{a}(\nabla f) \cdot \mathbf{a} + \mathbf{b}(\nabla f) \cdot \mathbf{b} + \mathbf{c}(\nabla f) \cdot \mathbf{c}$$

$$= (a_1 f_x + a_2 f_y + a_3 f_z)\langle a_1, a_2, a_3 \rangle + (b_1 f_x + b_2 f_y + b_3 f_z)\langle b_1, b_2, b_3 \rangle + (c_1 f_x + c_2 f_y + c_3 f_z)\langle c_1, c_2, c_3 \rangle.$$

If the scalar multiplications and vector additions in the last line are carried out, the resulting vector will have first component

$$(a_1^2 + b_1^2 + c_1^2)f_x + (a_1 a_2 + b_1 b_2 + c_1 c_2)f_y + (a_1 a_3 + b_1 b_3 + c_1 c_3)f_z. \tag{2}$$

By Eq. (1),

$$c_3 = a_1 b_2 - a_2 b_1, \quad a_3 = b_1 c_2 - b_2 c_1, \quad \text{and} \quad b_3 = c_1 a_2 - c_2 a_1.$$

Hence substitution for a_3, b_3, and c_3 in Eq. (2) yields

$$(a_1^2 + b_1^2 + c_1^2)f_x + (a_1a_2 + b_1b_2 + c_1c_2)f_y + (a_1b_1c_2 - a_1b_2c_1 + b_1c_1a_2 - b_1c_2a_1 + c_1a_1b_2 - c_1a_2b_1)f_z$$

$$= (a_1^2 + b_1^2 + c_1^2)f_x + (a_1a_2 + b_1b_2 + c_1c_2)f_y + 0 \cdot f_z.$$

Also by Eq. (1),

$$a_2 = b_3c_1 - b_1c_3, \quad b_2 = c_3a_1 - c_1a_3, \quad \text{and} \quad c_2 = a_3b_1 - a_1b_3.$$

Hence substitution for a_2, b_2, and c_2 in the last expression yields

$$(a_1^2 + b_1^2 + c_1^2)f_x + (a_1a_2 + b_1b_2 + c_1c_2)f_y$$

$$= (a_1^2 + b_1^2 + c_1^2)f_x + (a_1b_3c_1 - a_1b_1c_3 + b_1c_3a_1 - b_1c_1a_3 + c_1a_3b_1 - c_1a_1b_3)f_y$$

$$= (a_1^2 + b_1^2 + c_1^2)f_x + 0 \cdot f_y = (a_1^2 + b_1^2 + c_1^2)f_x.$$

Moreover, Eq. (1) implies that

$$a_1 = b_2c_3 - b_3c_2, \quad b_1 = c_2a_3 - c_3a_2, \quad \text{and} \quad c_1 = a_2b_3 - a_3b_2.$$

Hence substitution for *only one* of each of a_1, b_1, and c_1 in the last expression yields

$$(a_1b_2c_3 - a_1b_3c_2 + b_1c_2a_3 - b_1c_3a_2 + c_1a_2b_3 - c_1a_3b_2)f_x = \mathbf{a} \cdot (\mathbf{b} \times \mathbf{c})f_x = f_x$$

by Eq. (17) of Section 11.3. Similarly, the second component of $\mathbf{a}D_\mathbf{a}f + \mathbf{b}D_\mathbf{b}f + \mathbf{c}D_\mathbf{c}f$ is f_y and its third component is f_z. Therefore $\nabla f = \mathbf{a}D_\mathbf{a}f + \mathbf{b}D_\mathbf{b}f + \mathbf{c}D_\mathbf{c}f$. ◀

C13S0M.025: If $f(x, y) = 500 - (0.003)x^2 - (0.004)y^2$, then

$$\nabla f(x, y) = \left\langle -\frac{3x}{500}, -\frac{4y}{500} \right\rangle, \quad \text{so} \quad \nabla f(-100, -100) = \left\langle \frac{3}{5}, \frac{4}{5} \right\rangle.$$

To maintain a constant altitude, you should move in a direction normal to the gradient vector; that is, you should initially move in the direction of either $\langle -4, 3 \rangle$ or $\langle 4, -3 \rangle$.

C13S0M.027: The given surface S is the level surface

$$f(x, y, z) = x^{2/3} + y^{2/3} + z^{2/3} - 1 = 0,$$

and

$$\nabla f(x, y, z) = \left\langle \frac{2}{3x^{1/3}}, \frac{2}{3y^{1/3}}, \frac{2}{3z^{1/3}} \right\rangle.$$

Hence the plane $\mathcal{P}$ tangent to S at the point $P(a, b, c)$ has normal vector

$$\mathbf{n} = \nabla f(a, b, c) = \left\langle \frac{2}{3a^{1/3}}, \frac{2}{3b^{1/3}}, \frac{2}{3c^{1/3}} \right\rangle,$$

and thus an equation of $\mathcal{P}$ is

$$\frac{x - a}{a^{1/3}} + \frac{y - b}{b^{1/3}} + \frac{z - c}{c^{1/3}} = 0.$$

Set x and y equal to zero to find the z-intercept of $\mathcal{P}$:

$$\frac{z-c}{c^{1/3}} = a^{2/3} + b^{2/3}, \quad \text{so that} \quad z = (a^{2/3} + b^{2/3} + c^{2/3})c^{1/3} = c^{1/3}.$$

Similarly, the x-intercept of $\mathcal{P}$ is $a^{1/3}$ and its y-intercept is $b^{1/3}$. So the sum of the squares of its intercepts is $a^{2/3} + b^{2/3} + c^{2/3} = 1$.

C13S0M.029: Given:

$$f(x,\,y) = \frac{x^2 y^2}{x^2 + y^2}$$

if $(x,\,y) \neq (0,\,0)$; $f(0,\,0) = 0$. Then

$$f_x(0,\,0) = \lim_{h \to 0} \frac{f(h,\,0) - f(0,\,0)}{h} = \lim_{h \to 0} \frac{0}{h^3} = 0.$$

Also $f_y(0,\,0) = 0$ by a very similar computation. If $(x,\,y) \neq (0,\,0)$, then

$$f_x(x,\,y) = \frac{2xy^4}{(x^2 + y^2)^2} \quad \text{and} \quad f_y(x,\,y) = \frac{2x^4 y}{(x^2 + y^2)^2},$$

and (using polar coordinates)

$$\lim_{(x,y) \to (0,0)} f_x(x,\,y) = \lim_{r \to 0} \frac{2r^5 \cos\theta \sin^4\theta}{r^4} = 0,$$

so that f_x is continuous at $(0,\,0)$ (as is f_y, by a similar argument). Hence both first-order partial derivatives are continuous everywhere, and therefore f is differentiable at the origin. Finally, $f(x,\,y) \geq 0 = f(0,\,0)$, so $0 = f(0,\,0)$ is a local (indeed, the global) minimum value of $f(x,\,y)$.   —C.H.E.

C13S0M.031: The absence of first-degree terms in its equation implies that the center of the given rotated ellipse $73x^2 + 72xy + 52y^2 = 100$ is the origin. Thus to find its semiaxes, we maximize and minimize $f(x,\,y) = x^2 + y^2$ subject to the constraint

$$g(x,\,y) = 73x^2 + 72xy + 52y^2 - 100 = 0.$$

The Lagrange multiplier equations are

$$2x = \lambda(146x + 72y) \quad \text{and} \quad 2y = \lambda(72x + 104y).$$

The methods of Problem 38 of Section 13.9 lead to the simultaneous equations

$$(1 - 73\lambda)x - 36\lambda y = 0,$$

$$36\lambda x + (52\lambda - 1)y = 0.$$

Because this system has a nontrivial solution, the determinant of its coefficient matrix must be zero:

$$\begin{vmatrix} 1 - 73\lambda & -36\lambda \\ 36\lambda & 52\lambda - 1 \end{vmatrix} = -2500\lambda^2 + 125\lambda - 1 = 0,$$

and hence $(100\lambda - 1)(25\lambda - 1) = 0$.

Case 1: $\lambda = \frac{1}{100}$. Then the simultaneous equations shown here lead to $3x = 4y$; then the constraint equation yields

$$(x_1,\, y_1) = \left(-\frac{4}{5},\, -\frac{3}{5}\right) \quad \text{and} \quad (x_2,\, y_2) = \left(\frac{4}{5},\, \frac{3}{5}\right).$$

Case 2: $\lambda = \frac{1}{25}$. Then the simultaneous equations shown here lead to $4x + 3y = 0$; then the constraint equation yields

$$(x_3,\, y_3) = \left(-\frac{6}{5},\, \frac{8}{5}\right) \quad \text{and} \quad (x_4,\, y_4) = \left(\frac{6}{5},\, -\frac{8}{5}\right).$$

Thus these four points are the endpoints of the semiaxes of the ellipse; the minor semiaxis (Case 1) has length 1 and the major semiaxis (Case 2) has length 2.

C13S0M.033: Let us find both the maximum and the minimum perimeter of a triangle inscribed in the unit circle. The center of the circle will be within the triangle of maximum perimeter or on its boundary by the following argument. The following figure on the left shows a triangle with the center of the circle outside it. By moving the chord AB to the new position CD shown next on the right, you increase the perimeter of the circle and now the center of the circle is within the triangle. (Note: CD and AB are parallel and have the same length.) Similarly, the center of the circle is outside the triangle of minimum perimeter, for if inside—as in the figure on the right—move the chord CD to position AB to obtain a triangle of smaller perimeter with the center of the circle now outside the triangle.

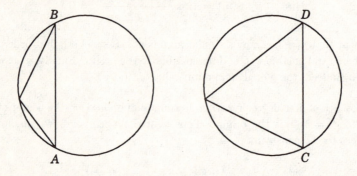

Thus to find the triangle of maximum perimeter, use the notation in Fig. 13.9.9 and maximize the perimeter

$$p(\alpha,\, \beta,\, \gamma) = \sqrt{2 - 2\cos\alpha} + \sqrt{2 - 2\cos\beta} + \sqrt{2 - 2\cos\gamma} = 2\sin\frac{\alpha}{2} + 2\sin\frac{\beta}{2} + 2\sin\frac{\gamma}{2}$$

given the constraint $g(\alpha,\, \beta,\, \gamma) = \alpha + \beta + \gamma - 2\pi = 0$. The Lagrange multiplier equations are

$$\cos\frac{\alpha}{2} = \lambda, \qquad \cos\frac{\beta}{2} = \lambda, \qquad \cos\frac{\gamma}{2} = \lambda,$$

and therefore, because $f(x) = \cos\left(\frac{1}{2}x\right)$ is single-valued on the interval $0 \leqq x \leqq 2\pi$, we may conclude that $\alpha = \beta = \gamma$. Then the constraint equation implies that $\alpha = \beta = \gamma = \frac{2}{3}\pi$, and thus we find that the triangle of maximum perimeter inscribed in the unit circle is equilateral and its perimeter is

$$p\left(\frac{2}{3}\pi,\, \frac{2}{3}\pi,\, \frac{2}{3}\pi\right) = 3\sqrt{3} \approx 5.1961524227066319.$$

Next we seek the triangle of minimum perimeter. The figure resembles the one on the left in the preceding illustration. Let C denote the third vertex of the triangle. Let α be the central angle that subtends the arc

776

BC, let β be the central angle that subtends the arc CA, and let γ be the central angle that subtends the short arc BA. Then we are to minimize the perimeter of the triangle, given by

$$p(\alpha, \beta, \gamma) = 2\sin\frac{\alpha}{2} + 2\sin\frac{\beta}{2} + 2\sin\frac{\gamma}{2}$$

(exactly as in the maximum-perimeter investigation); the constraint is now $\alpha + \beta - \gamma = 0$, and we note that $0 \leqq \gamma \leqq \pi$ as well. The Lagrange multiplier equations are

$$\cos\frac{\alpha}{2} = \lambda, \qquad \cos\frac{\beta}{2} = \lambda, \qquad \cos\frac{\gamma}{2} = -\lambda.$$

But all three of $\frac{1}{2}\alpha$, $\frac{1}{2}\beta$, and $\frac{1}{2}\gamma$ lie in the first quadrant, so these equations are impossible unless $\lambda = 0$, in which case $\alpha = \beta = \pi$ and $\gamma = 2\pi$ or, equivalently in this case, $\gamma = 0$. This implies that AC and BC are the same diameter of the circle and that AB is one endpoint of that diameter. Thus we obtain a degenerate triangle of perimeter $|AC| + |BC| = 4$. But this is not the minimum perimeter; the minimum occurs at a boundary point extremum missed by the Lagrange multiplier method.

To find the minimum, rewrite the perimeter function in the form

$$p(\alpha, \beta) = 2\sin\frac{\alpha}{2} + 2\sin\frac{\beta}{2} + 2\sin\frac{\alpha+\beta}{2}, \qquad 0 \leqq \alpha, \;\; 0 \leqq \beta, \;\; \alpha + \beta \leqq \pi.$$

There is no point in setting both partial derivatives of p equal to zero; remember, we are seeking a boundary extremum. Hence we examine the boundary of the domain of $p(\alpha, \beta)$ to find the minimum. One possibility is that $\alpha = \beta = 0$. In this case $\gamma = 0$ as well. If so, then the triangle is totally degenerate; its three vertices all coincide at a single point on the circumference of the circle, its three sides all have length zero, and its perimeter is (finally, the global minimum) zero.

C13S0M.035: First we make sure that the line and the ellipse do not intersect. Substitution of $y = 2 - x$ in the equation of the ellipse yields

$$x^2 + 2(2 - x)^2 = 1; \quad 3x^2 - 8x + 7 = 0.$$

The discriminant of the quadratic is $\Delta = 64 - 84 < 0$, so the last equation has no real solutions. Therefore the ellipse and the line do not meet.

Next, following the *Suggestion*, we let

$$f(x, y, u, v) = (x - u)^2 + (y - v)^2$$

and maximize and minimize $f(x, y, u, v)$ subject to the two constraints

$$g(x, y, u, v) = x^2 + 2y^2 - 1 = 0 \quad \text{and} \quad h(x, y, u, v) = u + v - 2 = 0.$$

The Lagrange multiplier equations are

$$2(x - u) = 2\lambda x, \quad 2(y - v) = 4\lambda y, \quad 2(u - x) = \mu, \quad \text{and} \quad 2(v - y) = \mu.$$

Elimination of μ yields $u - x = v - y$. Then elimination of λ yields

$$2y(x - u) = x(y - v) \quad \text{and} \quad u + y = v + x.$$

Finally, solving the last two equations simultaneously with the two constraint equations yields the expected number of solutions—two. They are

$$x = -\frac{1}{3}\sqrt{6}, \quad y = -\frac{1}{6}\sqrt{6}, \quad u = \frac{12 - \sqrt{6}}{6}, \quad v = \frac{12 + \sqrt{6}}{6} \quad \text{and}$$

$$x = \frac{1}{3}\sqrt{6}, \quad y = \frac{1}{6}\sqrt{6}, \quad u = \frac{12 + \sqrt{6}}{6}, \quad v = \frac{12 - \sqrt{6}}{6}.$$

Because the ellipse is "southwest" of the line, the first solution listed here gives the point (x, y) of the ellipse farthest from the line and the second gives the point closest to the line.

C13S0M.037: Let $f(x_1, x_2, \ldots, x_n) = x_1 + x_2 + \cdots + x_n$ where n is a fixed positive integer. Suppose that a is a nonnegative real number. To maximize $f(x_1, x_2, \ldots, x_n)$ subject to the constraint

$$g(x_1, x_2, \ldots, x_n) = x_1^2 + x_2^2 + \cdots + x_n^2 - a^2 = 0, \tag{1}$$

the Lagrange multiplier equations are

$$2\lambda x_i = 1, \quad 1 \leqq i \leqq n.$$

Hence $\lambda \neq 0$, and thus $x_1 = x_2 = \cdots = x_n$. Then the constraint equation implies that

$$nx_1^2 = a^2, \quad \text{so that} \quad x_1 = \pm\frac{a}{\sqrt{n}}.$$

Therefore the maximum value of f subject to the constraint in (1) is $nx_1 = a\sqrt{n}$.

It now follows that $|x_1 + x_2 + \ldots + x_n| \leqq a\sqrt{n}$ for every point $(x_1, x_2, \ldots, x_n)$ satisfying Eq. (1). Therefore

$$(x_1 + x_2 + \cdots + x_n)^2 \leqq na^2 = n(x_1^2 + x_2^2 + \cdots + x_n^2)$$

for all numbers $x_1, x_2, \ldots, x_n$ satisfying Eq. (1). But every such set of numbers satisfies Eq. (1) for some nonnegative value of a, and a is arbitrary. Therefore

$$\left(\sum_{i=1}^{n} x_i\right)^2 \leqq n \sum_{i=1}^{n} (x_i)^2 \tag{2}$$

for all n-tuples of real numbers $\{x_1, x_2, \ldots, x_n\}$. Note that equality holds *only* if $x_1 = x_2 = \cdots = x_n$; this will be important in the solution of Problem 51. Finally, divide both sides in Eq. (2) by n^2 and take square roots to obtain the result

$$\frac{x_1 + x_2 + \cdots + x_n}{n} \leqq \sqrt{\frac{x_1^2 + x_2^2 + \cdots + x_n^2}{n}}$$

in Problem 37.

C13S0M.039: We are to find the extrema of $f(x, y, z) = x^2 - yz$ subject to the constraint or side condition $g(x, y, z) = x^2 + y^2 + z^2 - 1 = 0$. The Lagrange multiplier equations are

$$2x = 2\lambda x, \quad -z = 2\lambda y, \quad \text{and} \quad -y = 2\lambda z. \tag{1}$$

Thus

$$2xyz = 2\lambda xyz, \quad -xz^2 = 2\lambda xyz, \quad \text{and} \quad -xy^2 = 2\lambda xyz,$$

778

and hence $-xy^2 = -xz^2 = 2xyz$.

Case 1: $x = 0$. Multiply the second equation in (1) by z and the third by y to obtain $-z^2 = 2\lambda yz = -y^2$, so that $y^2 = z^2$. If $y = z$, then the constraint equation yields $y^2 = \frac{1}{2}$, and we obtain the two critical points

$$A\left(0, \ \frac{1}{2}\sqrt{2}, \ \frac{1}{2}\sqrt{2}\right) \quad \text{and} \quad B\left(0, \ -\frac{1}{2}\sqrt{2}, \ -\frac{1}{2}\sqrt{2}\right).$$

If $y = -z$, then the constraint equation yields $y^2 = \frac{1}{2}$, and we obtain two more critical points,

$$C\left(0, \ \frac{1}{2}\sqrt{2}, \ -\frac{1}{2}\sqrt{2}\right) \quad \text{and} \quad D\left(0, \ -\frac{1}{2}\sqrt{2}, \ \frac{1}{2}\sqrt{2}\right).$$

Case 2: $x \ne 0$. Then $y^2 = z^2 = -2yz$. If $z = y$ then $y^2 = -2y^2$, so that $y = z = 0$. If $z = -y$ then $y^2 = 2y^2$, and again $y = z = 0$. Then the constraint equation yields $x = \pm 1$, so in this case we obtain the two critical points

$$E(1, \ 0, \ 0) \quad \text{and} \quad F(-1, \ 0, \ 0).$$

The values of $f(x, \ y, \ z)$ at these six points are

$$f(A) = -\frac{1}{2} = f(B), \quad f(C) = \frac{1}{2} = f(D), \quad \text{and} \quad f(E) = f(F) = 1.$$

Therefore the maximum value of $f(x, \ y, \ z)$ is 1 and occurs at the two critical points E and F. Its minimum value is $-\frac{1}{2}$ and occurs at the two critical points A and B. Numerical experimentation with

$$h(y, \ z) = 1 - y^2 - z^2 + yz$$

for y and z near the critical points C and D shows that there is not even a local extremum at either of these points.

C13S0M.041: If $f(x, \ y) = x^3 y - 3xy + y^2$, then when we equate both partial derivatives to zero we obtain the equations

$$3y(x^2 - 1) = 0 \quad \text{and} \quad x^3 - 3x + 2y = 0.$$

The first equation holds when $x = \pm 1$ and when $y = 0$. Then the second equation yields the critical points $P(-1, -1)$, $Q(0, 0)$, $R\left(-\sqrt{3}, 0\right)$, $S\left(\sqrt{3}, 0\right)$, and $T(1, 1)$. In the notation of Theorem 2 of Section 13.10, we find that

$$A = 6xy, \quad B = 3x^2 - 3, \quad \text{and} \quad C = 2,$$

and application of Theorem 2 yields the following results:

At $P(-1, -1)$:	$A = 6$, $B = 0$,	$C = 2$,	$\Delta = 12$,	$f(P) = -1$:	Local minimum;
At $Q(0, 0)$:	$A = 0$, $B = -3$,	$C = 2$,	$\Delta = -9$,	$f(Q) = 0$:	Saddle point;
At $R\left(-\sqrt{3}, 0\right)$:	$A = 0$, $B = 6$,	$C = 2$,	$\Delta = -36$,	$f(R) = 0$:	Saddle point;
At $S\left(\sqrt{3}, 0\right)$:	$A = 0$, $B = 6$,	$C = 2$,	$\Delta = -36$,	$f(S) = 0$:	Saddle point;
At $T(1, 1)$:	$A = 6$, $B = 0$,	$C = 2$,	$\Delta = 12$,	$f(T) = -1$:	Local minimum.

There are no global extrema; consider the behavior of $f(x, y)$ on the two lines $y = x$ and $y = -x$.

C13S0M.043: If $f(x, y) = x^3 - 6xy + y^3$, then the equations $f_x(x, y) = 0 = f_y(x, y)$ take the form

$$3x^2 - 6y = 0 \quad \text{and} \quad 3y^2 - 6x = 0.$$

Thus $2y = x^2$ and $y^2 = 2x$, so that $8x = x^4$. So the only two critical points are $(0, 0)$ and $(2, 2)$. In the notation of Section 13.10 we have $A = 6x$, $B = -6$, and $C = 6y$. Hence $\Delta = -36$ at $(0, 0)$; $A = 12$ and $\Delta = 108$ at $(2, 2)$. Therefore the graph of $z = f(x, y)$ has a saddle point at the origin and a local minimum at $(2, 2)$. There are no global extrema; examine the behavior of $f(x, 0)$.

C13S0M.045: Given $f(x, y) = x^3 y^2 (1 - x - y)$, the equations $f_x(x, y) = 0 = f_y(x, y)$ become

$$3x^2 y^2 (1 - x - y) - x^3 y^2 = 0 \quad \text{and} \quad 2x^3 y (1 - x - y) - x^3 y^2 = 0. \tag{1}$$

Clearly both partial derivatives are zero at every point where $x = 0$ (the y-axis), at every point where $y = 0$ (the x-axis). Note that $f(x, y) = 0$ at *all* such points. Moreover, there is one additional critical point; if $x \neq 0$ and $y \neq 0$, then the equations in (1) may be simplified to

$$4x + 3y = 3 \quad \text{and} \quad 2x + 3y = 2,$$

with the unique solution $(x, y) = \left(\frac{1}{2}, \frac{1}{3}\right)$. At this critical point the value of $f(x, y)$ is $\frac{1}{432}$. Because we have infinitely many critical points, we will have to deal with them by *ad hoc* methods. The next diagram shows the three lines where $f(x, y) = 0$. They divide the xy-plane into seven regions, and the sign of $f(x, y)$ is indicated on each.

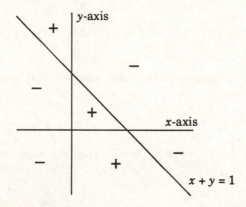

It is clear from this diagram that no point on the line $x + y = 1$ is an extremum, but that there is a saddle point at $(0, 1)$. (Theorem 2 of Section 13.10 fails at this point because $\Delta = 0$ there.) It is also clear from the diagram that there is a local maximum at every point of the negative x-axis and at every point of the x-axis for which $x > 1$; there is a local minimum at every point of the x-axis for which $0 < x < 1$. Finally, no point of the y-axis is an extremum. The fact that $f\left(\frac{1}{2}, \frac{1}{3}\right)$ is positive shows that there is a local maximum at $\left(\frac{1}{2}, \frac{1}{3}\right)$, and this conclusion is supported by Theorem 2 of Section 13.10; at that point we have $A = -\frac{1}{9}, B = -\frac{1}{12}$, and $C = -\frac{1}{8}$. Finally, because $f(x, x) = x^5(1 - 2x)$, there are no global minima; because $f(x, -2x) = 4x^5(1 + x)$, there are no global maxima.

C13S0M.047: If $f(x, y) = e^{xy} - 2xy$, then the equations $f_x(x, y) = 0 = f_y(x, y)$ take the form

$$(e^{xy} - 2)y = 0 \quad \text{and} \quad (e^{xy} - 2)x = 0.$$

780

Hence $(0, 0)$ is an isolated critical point and the points for which $e^{xy} = 2$ (the points on both branches of the hyperbola $xy = \ln 2$) are all critical points. Note that $f(0, 0) = 1$ and that $f(x, y) \equiv 2 - 2\ln 2 \approx 0.61370564$ for all points on the hyperbola $xy = \ln 2$. Next,

$$f_{xx}(x, y) = y^2 e^{xy}, \quad f_{xy}(x, y) = (xy + 1)e^{xy} - 2, \quad \text{and} \quad f_{yy}(x, y) = x^2 e^{xy}.$$

Hence Theorem 2 of Section 13.10 yields $A = 0$, $B = -1$, and $C = 0$ at the critical point $(0, 0)$, and therefore it is a saddle point. It is easy to show that the global minimum value of $g(x) = e^x - 2x$ is $2 - 2\ln 2$, and therefore every point of the hyperbola $xy = \ln 2$ is a location of the global minimum value of $f(x, y)$. There are no other extrema.

C13S0M.049: If $f(x, y) = (x - y)(xy - 1)$, then when we equate both partial derivatives to zero we obtain

$$2xy - y^2 - 1 = 0 \quad \text{and} \quad 2xy - x^2 - 1 = 0. \tag{1}$$

Therefore $y^2 = x^2$. If $y = x$ then either equation in (1) yields $x^2 = 1$, and thus we obtain the two critical points $(-1, -1)$ and $(1, 1)$. If $y = -x$ then each equation in (1) yields $3x^2 + 1 = 0$, and so there are no other critical points. When Theorem 2 of Section 13.10 is applied to the first critical point, we find that $A = -2$, $B = 0$, and $C = 2$; at the second critical point we find that $A = 2$, $B = 0$, and $C = -2$. Therefore both these points are saddle points and there are no extrema.

C13S0M.051: Given the data $\{(x_i, y_i)\}$ for $i = 1, 2, 3, \ldots, n$ (where n is a positive integer), we are to minimize

$$f(m, b) = \sum_{i=1}^{n} \left[y_i - (mx_i + b) \right]^2.$$

When we equate both partial derivatives of f to zero, we obtain the equations

$$\sum_{i=1}^{n} \left(y_i - mx_i - b \right) x_i = 0 \quad \text{and} \quad \sum_{i=1}^{n} \left(y_i - mx_i - b \right) = 0.$$

Rewrite the first of these in the form

$$b \sum_{i=1}^{n} x_i + m \sum_{i=1}^{n} (x_i)^2 = \sum_{i=1}^{n} x_i y_i \tag{1}$$

and the second in the form

$$b \sum_{i=1}^{n} 1 + m \sum_{i=1}^{n} x_i = \sum_{i=1}^{n} y_i. \tag{2}$$

Let

$$P = \sum_{i=1}^{n} x_i, \quad Q = \sum_{i=1}^{n} y_i, \quad R = \sum_{i=1}^{n} (x_i)^2, \quad \text{and} \quad S = \sum_{i=1}^{n} x_i y_i.$$

Then Eqs. (1) and (2) take the form

$$Pb + Rm = S, \qquad nb + Pm = Q. \tag{3}$$

Because P, Q, R, and S are the results of experimental observations, it is highly unlikely that the determinant of coefficients in (3) is zero, and hence these equations will normally have a unique solution (although the experimenter should be wary if the determinant is near zero). Next,

$$f_{mm}(m, b) = 2\sum_{i=1}^{n}(x_i)^2 = 2R, \qquad f_{mb}(m, b) = 2\sum_{i=1}^{n}x_i = 2P, \qquad \text{and} \qquad f_{bb}(m, b) = 2n.$$

Application of Theorem 2 of Section 13.10 yields

$$A = 2R, \quad B = 2P, \quad C = 2n, \quad \text{and} \quad \Delta = 4Rn - 4P^2 = 4(Rn - P^2).$$

Hence our sole critical point will be a local minimum (and, by the geometry of the problem, a global minimum) provided that

$$P^2 < nR; \quad \text{that is,} \quad \left(\sum_{i=1}^{n}x_i\right)^2 < n\sum_{i=1}^{n}(x_i)^2.$$

The inequality in Miscellaneous Problem 37 is strict unless all the x_i are equal, and in an experimental problem in which the least squares method is used, they will not be equal. This establishes that the critical point is a global minimum.

Section 14.1

C14S01.001: Part (a):

$$f(1, -2) \cdot 1 + f(2, -2) \cdot 1 + f(1, -1) \cdot 1 + f(2, -1) \cdot 1 + f(1, 0) \cdot 1 + f(2, 0) \cdot 1 = 198.$$

Part (b):

$$f(2, -1) \cdot 1 + f(3, -1) \cdot 1 + f(2, 0) \cdot 1 + f(3, 0) \cdot 1 + f(2, 1) \cdot 1 + f(3, 1) \cdot 1 = 480.$$

The average of the two answers is 339, fairly close to the exact value 312 of the integral.

C14S01.003: We omit Δx and Δy from the computation because each is equal to 1.

$$f\left(\tfrac{1}{2}, \tfrac{1}{2}\right) + f\left(\tfrac{3}{2}, \tfrac{1}{2}\right) + f\left(\tfrac{1}{2}, \tfrac{3}{2}\right) + f\left(\tfrac{3}{2}, \tfrac{3}{2}\right) = 8.$$

This is also the exact value of the iterated integral.

C14S01.005: The Riemann sum is

$$f(2, -1) \cdot 2 + f(4, -1) \cdot 2 + f(2, 0) \cdot 2 + f(4, 0) \cdot 2 = 88.$$

The true value of the integral is $\dfrac{416}{3} \approx 138.666666666667$.

C14S01.007: We factor out of each term in the sum the product $\Delta x \cdot \Delta y = \tfrac{1}{4}\pi^2$. The Riemann sum then takes the form

$$\frac{1}{4}\pi^2 \cdot \left[f\left(\tfrac{1}{4}\pi, \tfrac{1}{4}\pi\right) + f\left(\tfrac{3}{4}\pi, \tfrac{1}{4}\pi\right) + f\left(\tfrac{1}{4}\pi, \tfrac{3}{4}\pi\right) + f\left(\tfrac{3}{4}\pi, \tfrac{3}{4}\pi\right) \right] = \tfrac{1}{2}\pi^2 \approx 4.935.$$

The true value of the integral is 4.

C14S01.009: Because $f(x, y) = x^2 y^2$ is increasing in both the positive x-direction and the positive y-direction on $[1, 3] \times [2, 5]$, $L \leqq M \leqq U$.

C14S01.011: We integrate first with respect to x, then with respect to y:

$$\int_0^2 \int_0^4 (3x + 4y) \, dx \, dy = \int_0^2 \left[\frac{3}{2}x^2 + 4xy \right]_0^4 dy = \int_0^2 (24 + 16y) \, dy = \left[24y + 8y^2 \right]_0^2 = 80.$$

C14S01.013: We integrate first with respect to y, then with respect to x:

$$\int_{-1}^2 \int_1^3 (2x - 7y) \, dy \, dx = \int_{-1}^2 \left[2xy - \frac{7}{2}y^2 \right]_1^3 dx = \int_{-1}^2 (4x - 28) \, dx = \left[2x^2 - 28x \right]_{-1}^2 = -48 - 30 = -78.$$

C14S01.015: We integrate first with respect to x, then with respect to y:

$$\int_0^3 \int_0^3 (xy + 7x + y) \, dx \, dy = \int_0^3 \left[xy + \frac{7}{2}x^2 + \frac{1}{2}x^2 y \right]_0^3 dy$$

$$= \int_0^3 \left(\frac{3}{2}(5y + 21) \right) dy = \left[\frac{1}{4}(15y^2 + 126y) \right]_0^3 = \frac{513}{4} = 128.25.$$

C14S01.017: We integrate first with respect to y, then with respect to x:

$$\int_{-1}^{2}\int_{-1}^{2}(2xy^2 - 3x^2y)\,dy\,dx = \int_{-1}^{2}\left[\frac{2}{3}xy^3 - \frac{3}{2}x^2y^2\right]_{-1}^{2}\,dx$$

$$= \int_{-1}^{2}\frac{3}{2}(4x - 3x^2)\,dx = \left[3x^2 - \frac{3}{2}x^3\right]_{-1}^{2} = 0 - \frac{9}{2} = -\frac{9}{2} = -4.5.$$

C14S01.019: We integrate first with respect to x, then with respect to y:

$$\int_{0}^{\pi/2}\int_{0}^{\pi/2}\sin x\cos y\,dx\,dy = \int_{0}^{\pi/2}\left[-\cos x\cos y\right]_{0}^{\pi/2}\,dy$$

$$= \int_{0}^{\pi/2}\cos y\,dy = \left[\sin y\right]_{0}^{\pi/2} = 1 - 0 = 1.$$

C14S01.021: We integrate first with respect to y, then with respect to x:

$$\int_{0}^{1}\int_{0}^{1}xe^y\,dy\,dx = \int_{0}^{1}\left[xe^y\right]_{0}^{1}\,dx$$

$$= \int_{0}^{1}(ex - x)\,dx = \left[\frac{1}{2}(e-1)x^2\right]_{0}^{1} = \frac{1}{2}(e-1) \approx 0.8591409142295226.$$

C14S01.023: We integrate first with respect to y, then with respect to x:

$$\int_{0}^{1}\int_{0}^{\pi}e^x\sin y\,dy\,dx = \int_{0}^{1}\left[-e^x\cos y\right]_{0}^{\pi}\,dx$$

$$= \int_{0}^{1}2e^x\,dx = \left[2e^x\right]_{0}^{1} = 2e - 2 \approx 3.436563656918.$$

C14S01.025: We integrate first with respect to x, then with respect to y:

$$\int_{0}^{\pi}\int_{0}^{\pi}(xy + \sin x)\,dx\,dy = \int_{0}^{\pi}\left[\frac{1}{2}x^2y - \cos x\right]_{0}^{\pi}\,dy = \int_{0}^{\pi}\left(2 + \frac{1}{2}\pi^2y\right)\,dy$$

$$= \left[2y + \frac{1}{4}\pi^2y^2\right]_{0}^{\pi} = \frac{1}{4}\left(\pi^4 + 8\pi\right) \approx 30.635458065680.$$

C14S01.027: We integrate first with respect to x, then with respect to y:

$$\int_{0}^{\pi/2}\int_{1}^{e}\frac{\sin y}{x}\,dx\,dy = \int_{0}^{\pi/2}\left[(\ln x)\sin y\right]_{1}^{e}\,dy$$

$$= \int_{0}^{\pi/2}\sin y\,dy = \left[-\cos y\right]_{0}^{\pi/2} = 0 - (-1) = 1.$$

C14S01.029: We integrate first with respect to x, then with respect to y:

$$\int_0^1 \int_0^1 \left(\frac{1}{x+1} + \frac{1}{y+1}\right) dx\, dy = \int_0^1 \left[\frac{x}{y+1} + \ln(x+1)\right]_0^1 dy = \int_0^1 \left(\frac{1}{y+1} + \ln 2\right) dy$$

$$= \left[\ln(y+1) + y\ln 2\right]_0^1 = 2\ln 2 - 0 = 2\ln 2 \approx 1.3862943611198906.$$

C14S01.031: The first evaluation yields

$$\int_{-2}^2 \int_{-1}^1 (2xy - 3y^2)\, dx\, dy = \int_{-2}^2 \left[x^2 y - 3xy^2\right]_{-1}^1 dy$$

$$= \int_{-2}^2 (-6y^2)\, dy = \left[-2y^3\right]_{-2}^2 = -16 - 16 = -32.$$

The second yields

$$\int_{-1}^1 \int_{-2}^2 (2xy - 3y^2)\, dy\, dx = \int_{-1}^1 \left[xy^2 - y^3\right]_{-2}^2 dx$$

$$= \int_{-1}^1 (-16)\, dx = \left[-16x\right]_{-1}^1 = -16 - 16 = -32.$$

C14S01.033: The first evaluation yields

$$\int_1^2 \int_0^1 (x+y)^{1/2}\, dx\, dy = \int_1^2 \left[\frac{2}{3}(x+y)^{3/2}\right]_0^1 dy = \int_1^2 \left(\frac{2}{3}(y+1)^{3/2} - \frac{2}{3}y^{3/2}\right) dy$$

$$= \left[\frac{4}{15}(y+1)^{5/2} - \frac{4}{15}y^{5/2}\right]_1^2 = \frac{4}{15}\left(9\sqrt{3} - 8\sqrt{2} + 1\right) \approx 1.406599671769.$$

The second yields

$$\int_0^1 \int_1^2 (x+y)^{1/2}\, dy\, dx = \int_0^1 \left[\frac{2}{3}(x+y)^{3/2}\right]_1^2 dx = \int_0^1 \left(\frac{2}{3}(x+2)^{3/2} - \frac{2}{3}(x+1)^{3/2}\right) dx$$

$$= \left[\frac{4}{15}(x+2)^{5/2} - \frac{4}{15}(x+1)^{5/2}\right]_0^1 = \frac{4}{15}\left(9\sqrt{3} - 8\sqrt{2} + 1\right).$$

C14S01.035: We may assume that $n \geq 1$ and, if you wish, even that n is a positive integer. Then

$$\int_0^1 \int_0^1 x^n y^n\, dx\, dy = \int_0^1 \left[\frac{x^{n+1}y^n}{n+1}\right]_0^1 dy = \int_0^1 \frac{y^n}{n+1}\, dy = \left[\frac{y^{n+1}}{(n+1)^2}\right]_0^1 = \frac{1}{(n+1)^2}.$$

Therefore

$$\lim_{n \to \infty} \int_0^1 \int_0^1 x^n y^n\, dx\, dy = \lim_{n \to \infty} \frac{1}{(n+1)^2} = 0.$$

C14S01.037: Let $a(R)$ denote the area of R. If $0 \leq x \leq \pi$ and $0 \leq y \leq \pi$, then $0 \leq f(x, y) \leq \sin\frac{1}{2}\pi = 1$. Hence every Riemann sum lies between $0 \cdot a(R)$ and $1 \cdot a(R)$. Therefore

$$0 \leq \int_0^\pi \int_0^\pi \sin\sqrt{xy}\ dx\ dy \leq a(R) = \pi^2 \approx 9.869604401.$$

The exact value of the integral is

$$\int_0^\pi \int_0^\pi \sin\sqrt{xy}\ dx\ dy = 4\int_0^\pi \frac{\sin t}{t}\ dt \approx 7.4077482079298646814442134806319654533832.$$

C14S01.039: The corresponding relation among Riemann sums is

$$\sum_{i=1}^n [f(x_i^*, y_i^*) + g(x_i^*, y_i^*)] \cdot \Delta A_i = \left[\sum_{i=1}^n f(x_i^*, y_i^*) \cdot \Delta A_i\right] + \left[\sum_{i=1}^n g(x_i^*, y_i^*) \cdot \Delta A_i\right].$$

Section 14.2

C14S02.001: $\displaystyle \int_0^1 \int_0^x (1+x)\ dy\ dx = \int_0^1 \Big[y + xy\Big]_{y=0}^x\ dx = \int_0^1 (x + x^2)\ dx = \left[\frac{1}{2}x^2 + \frac{1}{3}x^3\right]_0^1 = \frac{5}{6}.$

C14S02.003: $\displaystyle \int_0^1 \int_y^1 (x+y)\ dx\ dy = \int_0^1 \left[\frac{1}{2}x^2 + xy\right]_y^1\ dy$

$$= \int_0^1 \left(\frac{1}{2} + y - \frac{3}{2}y^2\right)\ dy = \left[\frac{1}{2}(y + y^2 - y^3)\right]_0^1 = \frac{1}{2}.$$

C14S02.005: $\displaystyle \int_0^1 \int_0^{x^2} xy\ dy\ dx = \int_0^1 \left[\frac{1}{2}xy^2\right]_0^{x^2}\ dx = \int_0^1 \frac{1}{2}x^5\ dx = \left[\frac{1}{12}x^6\right]_0^1 = \frac{1}{12}.$

C14S02.007: $\displaystyle \int_0^1 \int_x^{\sqrt{x}} (2x - y)\ dy\ dx = \int_0^1 \left[2xy - \frac{1}{2}y^2\right]_x^{\sqrt{x}}\ dx$

$$= \int_0^1 \left(-\frac{1}{2}x + 2x^{3/2} - \frac{3}{2}x^2\right)\ dx = \left[-\frac{1}{4}x^2 + \frac{4}{5}x^{5/2} - \frac{1}{2}x^3\right]_0^1 = \frac{1}{20}.$$

C14S02.009: $\displaystyle \int_0^1 \int_{x^4}^x (y - x)\ dy\ dx = \int_0^1 \left[\frac{1}{2}y^2 - xy\right]_{x^4}^x\ dx$

$$= \int_0^1 \left(-\frac{1}{2}x^2 + x^5 - \frac{1}{2}x^8\right)\ dx = \left[-\frac{1}{6}x^3 + \frac{1}{6}x^6 - \frac{1}{18}x^9\right]_0^1 = -\frac{1}{18}.$$

C14S02.011: $\displaystyle \int_0^1 \int_0^{x^3} e^{y/x}\ dy\ dx = \int_0^1 \left[xe^{y/x}\right]_0^{x^3}\ dx = \int_0^1 \left(x\exp(x^2) - x\right)\ dx$

$$= \left[\frac{1}{2}\left(\exp(x^2) - x^2\right)\right]_0^1 = \frac{e-2}{2} \approx 0.3591409142295226.$$

C14S02.013: $\displaystyle \int_0^3 \int_0^y (y^2 + 16)^{1/2}\ dx\ dy = \int_0^3 \left[x(y^2 + 16)^{1/2}\right]_0^y\ dy$

$$= \int_0^3 y(y^2 + 16)^{1/2} \, dy = \left[\frac{1}{3}(y^2 + 16)^{3/2} \right]_0^3 = \frac{125}{3} - \frac{64}{3} = \frac{61}{3}.$$

C14S02.015: The following sketch of the graphs of $y = x^2$ and $y \equiv 4$ is extremely helpful in finding the limits of integration.

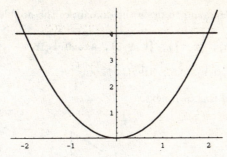

Answer:

$$\int_{-2}^{2} \int_{x^2}^{4} xy \, dy \, dx = \int_{-2}^{2} \left[\frac{1}{2} xy^2 \right]_{x^2}^{4} \, dx = \int_{-2}^{2} \left(8x - \frac{1}{2} x^5 \right) \, dx = \left[4x^2 - \frac{1}{12} x^6 \right]_{-2}^{2} = \frac{32}{3} - \frac{32}{3} = 0.$$

C14S02.017: The following diagram, showing the graphs of $y = x^2$ and $y = 8 - x^2$, is useful in finding the limits of integration. (Solve $y = x^2$ and $y = 8 - x^2$ simultaneously to find where the two curves cross.)

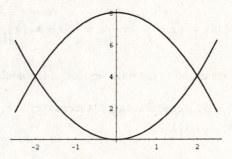

Answer:

$$\int_{-2}^{2} \int_{x^2}^{8-x^2} x \, dy \, dx = \int_{-2}^{2} \left[xy \right]_{x^2}^{8-x^2} \, dx = \int_{-2}^{2} (8x - 2x^3) \, dx = \left[4x^2 - \frac{1}{2} x^4 \right]_{-2}^{2} = 8 - 8 = 0.$$

C14S02.019: The following graph of $y = \sin x$ is helpful in determining the limits of integration.

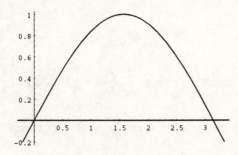

Answer:

$$\int_0^\pi \int_0^{\sin x} x \, dy \, dx = \int_0^\pi \left[xy \right]_0^{\sin x} dx = \int_0^\pi x \sin x \, dx = \left[\sin x - x \cos x \right]_0^\pi = \pi.$$

C14S02.021: A *Mathematica* 3.0 command to draw the domain of the double integral is

```
ParametricPlot[ {{t,1}, {E,t}, {t,t}}, {t,1,E}, AxesOrigin → {0,0},
         PlotRange → {{-0.5,3}, {-0.5,3}} ];
```

The graph produced by this command is shown next.

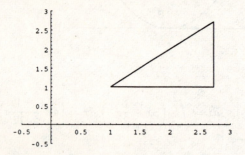

Answer:

$$\int_1^e \int_1^x \frac{1}{y} \, dy \, dx = \int_1^e \left[\ln y \right]_1^x dx = \int_1^e \ln x \, dx = \left[-x + x \ln x \right]_1^e = 0 - (-1) = 1.$$

C14S02.023: A *Mathematica* 3.0 command to draw the domain of the double integral is

```
Plot[ {x, -x/2, 1}, {x, -2, 1}, AspectRatio → Automatic,
      PlotRange → {{-2,1}, {0,1}} ];
```

the resulting figure is next.

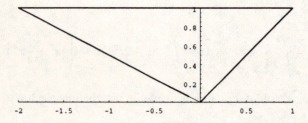

The value of the double integral is

$$\int_0^1 \int_{-2y}^y (1-x) \, dx \, dy = \int_0^1 \left[x - \frac{1}{2}x^2 \right]_{-2y}^y dy = \int_0^1 \left(3y + \frac{3}{2}y^2 \right) dy = \left[\frac{1}{2}(3y^2 + y^3) \right]_0^1 = 2 - 0 = 2.$$

C14S02.025: The domain of the double integral can be plotted by executing the following *Mathematica* 3.0 command:

```
Plot[ {x*x, 4}, {x, -2, 2} ];
```

and the resulting figure is next.

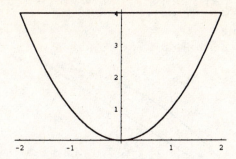

When the order of integration is reversed, we obtain

$$\int_0^4 \int_{-\sqrt{y}}^{\sqrt{y}} x^2 y \, dx \, dy = \int_0^4 \left[\frac{1}{3}x^3 y\right]_{-\sqrt{y}}^{\sqrt{y}} dy = \int_0^4 \frac{2}{3} y^{5/2} \, dy = \left[\frac{4}{21}y^{7/2}\right]_0^4 = \frac{512}{21} \approx 24.380952380952.$$

C14S02.027: The *Mathematica* 3.0 command

```
Plot[ { x*x, 2*x + 3 }, { x, -1, 3 } ];
```

produces a figure showing the domain of the double integral; it appears next.

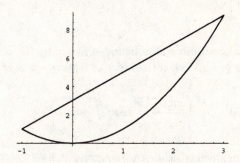

When the order of integration is reversed, two integrals are required—one for the part of the region for which $0 \leq y \leq 1$, the other for the part when $1 \leq y \leq 9$. The reason is that the lower limit of integration for x changes where $y = 1$. The first integral is

$$I_1 = \int_0^1 \int_{-\sqrt{y}}^{\sqrt{y}} x \, dx \, dy = \int_0^1 \left[\frac{1}{2}x^2\right]_{-\sqrt{y}}^{\sqrt{y}} dy = \int_0^1 0 \, dy = 0.$$

The second integral is

$$I_2 = \int_1^9 \int_{(y-3)/2}^{\sqrt{y}} x \, dx \, dy = \int_1^9 \left[\frac{1}{2}x^2\right]_{(y-3)/2}^{\sqrt{y}} dy$$

$$= \int_1^9 \frac{1}{8}(10y - 9 - y^2) \, dy = \left[\frac{1}{24}(15y^2 - 27y - y^3)\right]_1^9 = \frac{32}{3}.$$

```
```

C14S02.029: To see the domain of the double integral, execute the *Mathematica* 3.0 command

```
Plot[ { 2*x, 4*x - x*x }, { x, 0, 2 } ];
```

the result is shown next.

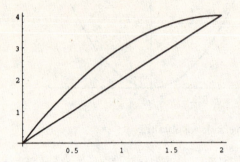

When the order of integration is reversed, the given integral becomes

$$\int_0^4 \int_{2-\sqrt{4-y}}^{y/2} 1 \; dx \; dy = \int_0^4 \left[x \right]_{2-\sqrt{4-y}}^{y/2} dy$$

$$= \int_0^4 \left(\frac{1}{2} y + (4-y)^{1/2} - 2 \right) dy = \left[\frac{1}{4} y^2 - 2y - \frac{2}{3}(4-y)^{3/2} \right]_0^4 = \frac{4}{3}.$$

C14S02.031: The domain of the given integral is bounded above by the line $y = \pi$, on the left by the y-axis, and on the right by the line $y = x$. When the order of integration is reversed, we obtain

$$\int_0^\pi \int_0^y \frac{\sin y}{y} \; dx \; dy = \int_0^\pi \left[\frac{x \sin y}{y} \right]_0^y dy = \int_0^\pi \sin y \; dy = \left[-\cos y \right]_0^\pi = 2.$$

If the improper integral is disturbing, merely define the integrand to have the value 1 at $y = 0$. Then it will be continuous and the integral will no longer be improper. Because the antiderivative

$$F(t) = \int_0^t \frac{\sin y}{y} \; dy$$

is known to be nonelementary, the only way to evaluate the given integral by hand is first to reverse the order of integration.

C14S02.033: To generate a figure showing the domain of the given integral, use the *Mathematica* 3.0 command

```
ParametricPlot[ {{t,t}, {1,t}, {t,0}}, {t,0,1}, AspectRatio → Automatic,

        PlotRange → {{-0.1, 1.1}, {-0.1, 1.1}} ];
```

The result is shown next.

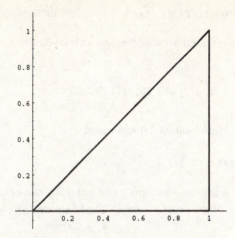

When the order of integration is reversed, we obtain

$$\int_0^1 \int_0^x \frac{1}{1+x^4} \, dy \, dx = \int_0^1 \left[\frac{y}{1+x^4} \right]_0^x dx = \int_0^1 \frac{x}{1+x^4} \, dx = \left[\frac{1}{2} \arctan x^2 \right]_0^1 = \frac{\pi}{8}.$$

The integration can be carried out in the order given in the textbook, but finding the partial fraction decomposition of the integrand is long and complex if machine aid is not available.

C14S02.035: We used *Mathematica* 3.0 in this problem. First we entered

```
Solve[ x∧3 + 1 == 3*x*x, x ]
```

and the computer returned the exact answers. Then we asked for the numerical values to 40 places, and we found that the curves intersect at the three points with approximate coordinates

$$(a, 3a^2) \approx (-0.5320888862379561, 0.8493557485738457),$$

$$(b, 3b^2) \approx (0.6527036446661393, 1.270661432813855), \quad \text{and}$$

$$(c, 3c^2) \approx (2.8793852415718168, 24.8725781081447688).$$

We then entered the command

```
Plot[ {x∧3 + 1, 3*x*x}, {x, a, b}, PlotRange → {-1, 25} ];
```

and thereby discovered that the cubic graph is above the quadratic on (a, b) but below it on (b, c). Thus we needed to compute two integrals:

```
i1 = Integrate[ Integrate[ x, {y, 3*x*x, x∧3 + 1} ], {x, a, b} ]
```

and

```
i2 = Integrate[ Integrate[ x, {y, x∧3 + 1, 3*x*x} ], {x, b, c} ]
```

Results:

791

$$I_1 \approx 0.0276702414879754 \quad \text{and} \quad I_2 \approx 7.9240769481663325.$$

All the digits are correct or correctly rounded because we used at least 32 decimal digits in every computation. The answer, therefore, is

$$I_1 + I_2 \approx 7.9517471896543079.$$

C14S02.037: We began with the *Mathematica* 3.0 command

```
Solve[ x*x - 1 == 1/(1 + x*x), x ]
```

and were rewarded with the exact solutions—two real, two complex non-real. The two real solutions are $a = -2^{1/4}$ and $b = 2^{1/4}$, so we used these exact values in the following computations. The double integral has the value

```
Integrate[ Integrate[ x, {y, x*x - 1, 1/(x*x + 1)}, {x, a, b} ]
```

$$\frac{1}{2}\left[1 - \sqrt{2} - \ln\left(1 + \sqrt{2}\right)\right] + \frac{1}{2}\left[-1 + \sqrt{2} + \ln\left(1 + \sqrt{2}\right)\right] = 0.$$

C14S02.039: We used *Mathematica* 3.0 to automate Newton's method for solving the equation $x^2 = \cos x$:

```
f[x_] := x*x - Cos[x]
g[x_] := N[x - f[x]/f'[x], 60]
```

The function g carries out the iteration of Newton's method, carrying 60 digits in its computations. A graph indicated that the positive solution of $f(x) = 0$ is close to $4/5$, hence we entered the successive commands

```
g[4/5]
g[%]
```

(Recall that % refers to the "last output.")

```
g[%]
```

After six iterations the results agreed to over 50 decimal degits, and thus we find that the graphs cross at the two points

$$(b, b^2) \approx (0.8241323123025224, 0.6791940681811024) \quad \text{and} \quad (a, a^2)$$

where $a = -b$. Hence the value of the double integral is

```
Integrate[ Integrate[ x, {y, x*x, Cos[x]}], {x, a, b} ]
```

$$0.0 \times 10^{-58}$$

A moment's thought about Riemann sums reveals that the exact value of the integral is zero.

C14S02.041: The integral is zero. Each term $f(x_i^*, y_i^*)\, \Delta A_i = x_i^*\, \Delta A_i$ in every Riemann sum is cancelled by a similar term in which x_i^* has the opposite sign.

C14S02.043: Every term of the form $f(x_i^*, y_i^*)\,\Delta A_i$ in every Riemann sum is cancelled by a similar term in which x_i^* has the opposite sign (but y_i^* is the same). Therefore the value of the integral is zero.

C14S02.045: Suppose that the rectangle R consists of those points (x, y) for which both $a \leqq x \leqq b$ and $c \leqq y \leqq d$. Suppose that k is a positive constant and that f is a function continuous on R. Then

$$I = \iint_R f(x, y)\,dA$$

exists. Suppose that $\epsilon > 0$ is given. Then there exists a number $\delta_1 > 0$ such that, for every partition $\mathcal{P} = \{R_1, R_2, \ldots R_n\}$ of R such that $|\mathcal{P}| < \delta_1$ and for every selection (x_i^*, y_i^*) in R_i ($i = 1, 2, \ldots, n$),

$$\left| \sum_{i=1}^n f(x_i^*, y_i^*)\,\Delta A_i \ - \ I \right| < \frac{\epsilon}{2k}$$

(where ΔA_i is the area $a(R_i)$ of R_i). Consequently,

$$\left| k \sum_{i=1}^n f(x_i^*, y_i^*)\,\Delta A_i \ - \ kI \right| < \frac{\epsilon}{2}. \tag{1}$$

Moreover, kf is continuous on R, and hence

$$J = \iint_R kf(x, y)\,dA$$

exists. So there exists a number $\delta_2 > 0$ such that, for every partition $\mathcal{P} = \{R_1, R_2, \ldots R_n\}$ of R such that $|\mathcal{P}| < \delta_2$ and every selection (x_i^*, y_i^*) in R_i ($i = 1, 2, \ldots n$),

$$\left| \sum_{i=1}^n kf(x_i^*, y_i^*)\,\Delta A_i \ - \ J \right| < \frac{\epsilon}{2}; \quad \text{that is,}$$

$$\left| k \sum_{i=1}^n f(x_i^*, y_i^*)\,\Delta A_i \ - \ J \right| < \frac{\epsilon}{2}.$$

Let δ be the minimum of δ_1 and δ_2. Then, for every partition $\mathcal{P}$ of R with $|\mathcal{P}| < \delta$ and every selection (x_i^*, y_i^*) in R_i ($i = 1, 2, \ldots n$), we have both

$$\left| kI \ - \ k \sum_{i=1}^n f(x_i^*, y_i^*)\,\Delta A_i \right| < \frac{\epsilon}{2} \quad \text{(by (1)) and}$$

$$\left| k \sum_{i=1}^n f(x_i^*, y_i^*)\,\Delta A_i \ - \ J \right| < \frac{\epsilon}{2}.$$

Add the last two inequalities. Then, by the triangle inequality (Theorem 1 of Appendix A, page A-2),

$$|kI \ - \ J| < \frac{\epsilon}{2} + \frac{\epsilon}{2} = \epsilon.$$

Because ϵ is an *arbitrary* positive number, this proves that $J = kI$; that is, we have shown that

$$\iint_R kf(x, y)\,dA \ = \ k \iint_R f(x, y)\,dA.$$

The proof is similar in the case $k < 0$, and if $k = 0$ there is nothing to prove. ◀

For a shorter proof, one that exploits both the continuity of f and the fact that R is a rectangle with sides parallel to the coordinate axes, choose a continuous function F such that $F_x = f$. Then choose a continuous function P such that $P_y = F$. Then

$$\iint_R kf(x, y) \, dA = \int_c^d \int_a^b kf(x, y) \, dx \, dy = \int_c^d \left[kF(x, y) \right]_a^b \, dy$$

$$= \int_c^d [kF(b, y) - kF(a, y)] \, dy = \left[kP(b, y) - kP(a, y) \right]_c^d$$

$$= kP(b, d) - kP(a, d) - kP(b, c) + kP(a, c) = k \left[P(b, d) - P(a, d) - P(b, c) + P(a, c) \right]$$

$$= k \left[P(b, y) - P(a, y) \right]_c^d = k \int_c^d [F(b, y) - F(a, y)] \, dy$$

$$= k \int_c^d \left[F(x, y) \right]_a^b \, dy = k \int_c^d \int_a^b f(x, y) \, dx \, dy = k \iint_R f(x, y) \, dA.$$

C14S02.047: Suppose that R is a plane rectangle with sides parallel to the coordinate axes, so that R consists of those points (x, y) for which both $a \leq x \leq b$ and $c \leq y \leq d$ for some numbers a, b, c, and d. Suppose that f is continuous on R and that $m \leq f(x, y) \leq M$ for all (x, y) in R.

Let $g(x, y) \equiv m$ and $h(x, y) \equiv M$ for (x, y) in R. Then $g(x, y) \leq f(x, y) \leq h(x, y)$ for all points (x, y) in R. Let $\mathcal{P} = \{R_1, R_2, \ldots, R_n\}$ be a partition of R and let $(x_i^\star, y_i^\star)$ be a selected point in R_i for $1 \leq i \leq n$. As usual, let $\Delta A_i = a(R_i)$ for $1 \leq i \leq n$. Then for each integer i, $1 \leq i \leq n$, we have

$$g(x_i^\star, y_i^\star) \leq f(x_i^\star, y_i^\star) \leq h(x_i^\star, y_i^\star);$$

thus

$$g(x_i^\star, y_i^\star) \, \Delta A_i \leq f(x_i^\star, y_i^\star) \, \Delta A_i \leq h(x_i^\star, y_i^\star) \, \Delta A_i$$

for $1 \leq i \leq n$. Add these inequalities to find that

$$\sum_{i=1}^n m \cdot \Delta A_i \leq \sum_{i=1}^n f(x_i^\star, y_i^\star) \, \Delta A_i \leq \sum_{i=1}^n M \cdot \Delta A_i.$$

Therefore

$$m \cdot \sum_{i=1}^n a(R_i) \leq \sum_{i=1}^n f(x_i^\star, y_i^\star) \, \Delta A_i \leq M \cdot \sum_{i=1}^n a(R_i);$$

that is,

$$m \cdot a(R) \leq \sum_{i=1}^n f(x_i^\star, y_i^\star) \, \Delta A_i \leq M \cdot a(R) \tag{1}$$

for every partition $\mathcal{P}$ of R and every selection $(x_i^\star, y_i^\star)$ for $\mathcal{P}$. That is, the inequalities in (1) hold for every Riemann sum for f on R. Because the double integral of f on R is the limit of such sums, we may conclude that

794

$$m \cdot a(R) \; \leqq \; \iint_R f(x, y) \, dA \; \leqq \; M \cdot a(R).$$

C14S02.049: Suppose that R is a rectangle with sides parallel to the coordinate axes, that $f(x, y) \leqq g(x, y)$ for all (x, y) in R, and that both

$$I = \iint_R f(x, y) \, dA \quad \text{and} \quad J = \iint_R g(x, y) \, dA$$

exist. Suppose by way of contradiction that $J < I$. Let $\epsilon = I - J$. Note that $\epsilon/3 > 0$. Choose $\delta > 0$ so small that if $\mathcal{P} = \{R_1, R_2, \ldots, R_n\}$ is a partition of R with $|\mathcal{P}| < \delta$ and $(x_i^\star, y_i^\star)$ is a selection for $\mathcal{P}$ with $(x_i^\star, y_i^\star)$ in R_i for $1 \leqq i \leqq n$, then

$$\left| \sum_{i=1}^n f(x_i^\star, y_i^\star) \, \Delta A_i - I \right| < \frac{\epsilon}{3} \quad \text{and} \quad \left| \sum_{i=1}^n g(x_i^\star, y_i^\star) \, \Delta A_i - J \right| < \frac{\epsilon}{3}.$$

With such a partition and such a selection, note that

$$\sum_{i=1}^n g(x_i^\star, y_i^\star) \, \Delta A_i \; < \; J + \frac{\epsilon}{3} \; < \; I - \frac{\epsilon}{3} \; < \; \sum_{i=1}^n f(x_i^\star, y_i^\star) \, \Delta A_i,$$

and thus

$$\sum_{i=1}^n [f(x_i^\star, y_i^\star) - g(x_i^\star, y_i^\star)] \, \Delta A_i \; > \; 0.$$

Because $\Delta A_i > 0$ for $1 \leqq i \leqq n$, it follows that

$$f(x_j^\star, y_j^\star) \; > \; g(x_j^\star, y_j^\star)$$

for some j, $1 \leqq j \leqq n$, contrary to hypothesis. Therefore $I \leqq J$. ◀

C14S02.051: Recall that R is the region in the first quadrant bounded by the circle $x^2 + y^2 = 1$ and the coordinate axes. Hence

$$\iint_R (x + y) \, dA = \int_{x=0}^1 \int_{y=0}^{\sqrt{1-x^2}} (x + y) \, dy \, dx$$

$$= \int_0^1 \left[xy + \frac{1}{2} y^2 \right]_0^{\sqrt{1-x^2}} dx = \int_0^1 \left(x\sqrt{1 - x^2} + \frac{1}{2}(1 - x^2) \right) dx$$

$$= \left[-\frac{1}{3}(1 - x^2)^{3/2} + \frac{1}{2}x - \frac{1}{6}x^3 \right]_0^1 = \frac{1}{2} - \frac{1}{6} + \frac{1}{3} = \frac{2}{3}.$$

C14S02.053: The domain of the integral and the partition using $n = 5$ subintervals of equal length in each direction is shown next.

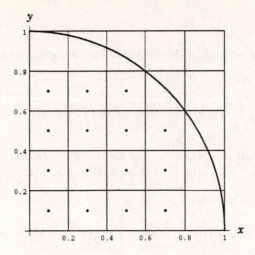

The midpoints of the subrectangles of the inner partition are indicated with "bullets" in the figure. Let $f(x, y) = xy \exp(y^2)$. Then the corresponding midpoint sum for the given integral is

$$S = \frac{1}{n^2}\Big[f(0.1, 0.1) + f(0.3, 0.1) + f(0.5, 0.1) + f(0.7, 0.1) + f(0.3, 0.1) + f(0.3, 0.3)$$

$$+ f(0.3, 0.5) + f(0.3, 0.7) + f(0.5, 0.1) + f(0.5, 0.3) + f(0.5, 0.5) + f(0.5, 0.7)$$

$$+ f(0.7, 0.1) + f(0.7, 0.3) + f(0.7, 0.5)\Big]$$

$$= \frac{1}{25}\left[\frac{4}{25}e^{0.01} + \frac{12}{25}e^{0.09} + \frac{4}{5}e^{0.25} + \frac{63}{100}e^{0.49} \right] \approx 0.109696.$$

The exact value of the integral is

$$\iint_R f(x, y)\, dA = \int_{y=0}^{1} \int_{x=0}^{\sqrt{1-y^2}} xy \exp(y^2)\, dx\, dy$$

$$= \int_0^1 \left[\frac{1}{2}x^2 y \exp(y^2) \right]_0^{\sqrt{1-y^2}} dy = \int_0^1 \frac{1}{2}(y - y^3) \exp(y^2)\, dy$$

$$= \left[\frac{2 - y^2}{4} \exp(y^2) \right]_0^1 = \frac{e - 2}{4} \approx 0.17957045711476130884007186778.$$

If you prefer the other order of integration—which avoids the integration by parts—it is

$$\iint_R f(x, y)\, dA = \int_{x=0}^{1} \int_{y=0}^{\sqrt{1-x^2}} xy \exp(y^2)\, dy\, dx$$

$$= \int_0^1 \left[\frac{1}{2}x \exp(y^2) \right]_0^{\sqrt{1-x^2}} dx = \int_0^1 \frac{1}{2}\left[x \exp(1 - x^2) - x \right] dx$$

$$= \left[-\frac{1}{4}x^2 - \frac{1}{4} \exp\left(1 - x^2\right) \right]_0^1 = \frac{e-2}{4}.$$

Section 14.3

C14S03.001: The area is

$$A = \int_{y=0}^{1} \int_{x=y^2}^{y} 1 \, dx \, dy = \int_{y=0}^{1} \left[x \right]_{x=y^2}^{y} \, dy$$

$$= \int_{y=0}^{1} (y - y^2) \, dy = \left[\frac{1}{2}y^2 - \frac{1}{3}y^3 \right]_{y=0}^{1} = \frac{1}{2} - \frac{1}{3} = \frac{1}{6}.$$

To find the limits of integration, it is very helpful to sketch the domain of the double integral. The figure is next; it was produced by *Mathematica* 3.0 via the command

```
Plot[ {x, Sqrt[x]}, {x, 0, 1}, AspectRatio→ Automatic ];
```

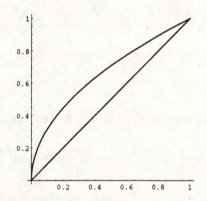

C14S03.003: The graphs cross where $x^2 = 2x + 3$; that is, where $x = -1$ and where $x = 3$. A sketch of the domain of the integral is next; it was produced by *Mathematica* 3.0 via the command

```
Plot[ {x*x, 2*x + 3}, {x, -1, 3}, AspectRatio → Automatic ];
```

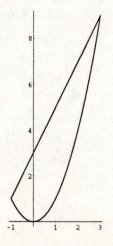

797

The area of the region is

$$A = \int_{-1}^{3} \int_{y=x^2}^{2x+3} 1 \, dy \, dx = \int_{-1}^{3} (3 + 2x - x^2) \, dx = \left[3x + x^2 - \frac{1}{3}x^3 \right]_{-1}^{3} = \frac{32}{3}.$$

C14S03.005: The graphs cross where $x^2 = 2 - x$; that is, where $x = -2$ and where $x = 1$. But the x-axis is also part of the boundary of the region in question, and hence the following figure is important to find not only the correct limits of integration, but indeed the very *region* whose area is sought. It was produced by the *Mathematica* 3.0 command

```
Plot[ {x*x, 2 - x}, {x, -2, 2}, AspectRatio → Automatic ];
```

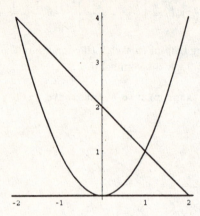

(We enhanced the result using Adobe Illustrator.) The area of the region bounded by all three graphs is

$$A = \int_{y=0}^{1} \int_{x=\sqrt{y}}^{2-y} 1 \, dx \, dy = \int_{0}^{1} (2 - \sqrt{y} - y) \, dy = \left[2y - \frac{2}{3}y^{3/2} - \frac{1}{2}y^2 \right]_{0}^{1} = \frac{5}{6}.$$

C14S03.007: The graphs cross where $x^2 + 1 = 2x^2 - 3$; that is, where $x = -2$ and where $x = 2$. The area between them is

$$A = \int_{x=-2}^{2} \int_{y=2x^2-3}^{x^2+1} 1 \, dy \, dx = \int_{-2}^{2} (4 - x^2) \, dx = \left[4x - \frac{1}{3}x^3 \right]_{-2}^{2} = \frac{32}{3}.$$

C14S03.009: The part of the region that lies in the first quadrant is shown next; the figure was generated using the *Mathematica* 3.0 command

```
Plot[ {x, 2*x, 2/x}, {x, 0, 2}, PlotRange → {0, 2.5} ];
```

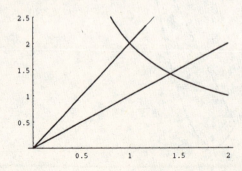

The bounding curves cross at the origin and at the points $(1, 2)$, and $(\sqrt{2}, \sqrt{2})$. The area they bound (in the first quadrant) is

$$A = \int_{x=0}^{1} \int_{y=x}^{2x} 1\, dy\, dx + \int_{x=1}^{\sqrt{2}} \int_{y=x}^{2/x} 1\, dy\, dx = \int_{0}^{1} x\, dx + \int_{1}^{\sqrt{2}} \left(\frac{2}{x} - x\right) dx$$

$$= \left[\frac{1}{2}x^2\right]_0^1 + \left[2\ln x - \frac{1}{2}x^2\right]_1^{\sqrt{2}} = \frac{1}{2} + \ln 2 - \frac{1}{2} = \ln 2 \approx 0.6931471805599453309417232.$$

The region is symmetric around the origin, so the total area is $2\ln 2$.

C14S03.011: The volume is

$$V = \int_{x=0}^{1} \int_{y=0}^{1} (1 + x + y)\, dy\, dx = \int_{x=0}^{1} \left[y + xy + \frac{1}{2}y^2\right]_{y=0}^{1} dx = \int_{0}^{1} \left(x + \frac{3}{2}\right) dx = \left[\frac{3}{2}x + \frac{1}{2}x^2\right]_0^1 = 2.$$

C14S03.013: The volume is

$$V = \int_{y=0}^{2} \int_{x=0}^{1} (y + e^x)\, dx\, dy = \int_{y=0}^{2} \left[xy + e^x\right]_{x=0}^{1} dy = \int_{0}^{2} (e + y - 1)\, dy = \left[ey + \frac{1}{2}y^2 - y\right]_0^2 = 2e.$$

C14S03.015: The domain of the integral can be drawn by using the *Mathematica* 3.0 command

```
Plot[ 1 - x, {x, 0, 1} ];
```

and the result is shown next.

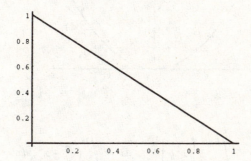

The volume is

$$V = \int_{x=0}^{1} \int_{y=0}^{1-x} (x + y)\, dy\, dx = \int_{x=0}^{1} \left[xy + \frac{1}{2}y^2\right]_{y=0}^{1-x} dx = \int_{0}^{1} \frac{1}{2}(1 - x^2)\, dx = \left[\frac{1}{2}x - \frac{1}{6}x^3\right]_0^1 = \frac{1}{3}.$$

C14S03.017: The domain of the integral can be drawn using the *Mathematica* 3.0 command

```
ParametricPlot[ {{1,t}, {t,0}, {t,t*t}}, {t,0,1}, AspectRatio → Automatic ];
```

799

and the result is shown next.

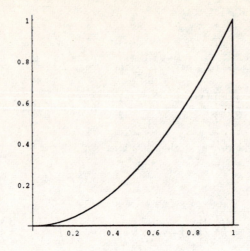

The volume is

$$V = \int_{x=0}^{1} \int_{y=0}^{x^2} (1 + x + y) \, dy \, dx = \int_{x=0}^{1} \left[y + xy + \frac{1}{2} y^2 \right]_{y=0}^{x^2} dx = \int_{0}^{1} \left(x^2 + x^3 + \frac{1}{2} x^4 \right) dx$$

$$= \left[\frac{1}{3} x^3 + \frac{1}{4} x^4 + \frac{1}{10} x^5 \right]_{0}^{1} = \frac{41}{60} \approx 0.683333333333.$$

C14S03.019: The domain of the integral is shown next.

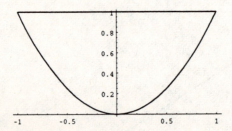

The volume of the solid is

$$V = \int_{x=-1}^{1} \int_{y=x^2}^{1} x^2 \, dy \, dx = \int_{x=-1}^{1} \left[x^2 y \right]_{y=x^2}^{1} dx = \int_{-1}^{1} (x^2 - x^4) \, dx = \left[\frac{1}{3} x^3 - \frac{1}{5} x^5 \right]_{-1}^{1} = \frac{4}{15}.$$

C14S03.021: The volume of the solid is

$$V = \int_{y=0}^{2} \int_{x=0}^{1} (x^2 + y^2) \, dx \, dy = \int_{y=0}^{2} \left[\frac{1}{3} x^3 + xy^2 \right]_{x=0}^{1} dy = \int_{0}^{2} \left(\frac{1}{3} + y^2 \right) dy = \left[\frac{1}{3} (y + y^3) \right]_{0}^{2} = \frac{10}{3}.$$

C14S03.023: The domain of the integral can be drawn by executing the *Mathematica* 3.0 command

```
ParametricPlot[ {{3, 2*t/3}, {t, 0}, {t, 2*t/3}}, {t, 0, 3} ];
```

800

and the result is shown next.

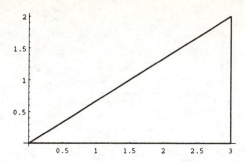

The volume of the solid is

$$V = \int_{x=0}^{3} \int_{y=0}^{2x/3} (9 - x - y)\, dy\, dx = \int_{x=0}^{3} \left[9y - xy - \frac{1}{2} y^2 \right]_{y=0}^{2x/3} dx$$

$$= \int_{0}^{3} \left(6x - \frac{8}{9} x^2 \right) dx = \left[3x^2 - \frac{8}{27} x^3 \right]_{0}^{3} = 19.$$

C14S03.025: The volume is

$$V = \int_{x=0}^{1} \int_{y=0}^{2-2x} (4x^2 + y^2)\, dy\, dx = \int_{x=0}^{1} \left[4x^2 y + \frac{1}{3} y^3 \right]_{y=0}^{2-2x} dx$$

$$= \int_{0}^{1} \frac{8}{3} (1 - 3x + 6x^2 - 4x^3)\, dx = \left[\frac{8}{3} x - 4x^2 + \frac{16}{3} x^3 - \frac{8}{3} x^4 \right]_{0}^{1} = \frac{4}{3}.$$

C14S03.027: The volume is

$$V = \int_{x=0}^{2} \int_{y=0}^{(6-3x)/2} (6 - 3x - 2y)\, dy\, dx = \int_{x=0}^{2} \left[6y - 3xy - y^2 \right]_{y=0}^{(6-3x)/2} dx$$

$$= \int_{0}^{2} \left(9 - 9x + \frac{9}{4} x^2 \right) dx = \left[9x - \frac{9}{2} x^2 + \frac{3}{4} x^3 \right]_{0}^{2} = 6.$$

C14S03.029: The triangular domain of the integral can be drawn by executing the *Mathematica* 3.0 command

```
ParametricPlot[ {{1, 2 + 2*t}, {1 + 4*t, 2}, {1 + 4*t, 4 - 2*t}},
            {t, 0, 1}, PlotRange → {{-0.5, 5.5}, {-0.5, 4.5}},
            AspectRatio → Automatic, AxesOrigin → {0, 0} ];
```

801

and the result is shown next.

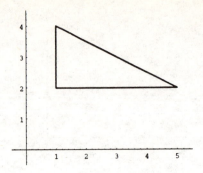

The volume of the solid is

$$V = \int_{x=1}^{5} \int_{y=2}^{(9-x)/2} xy \, dy \, dx = \int_{x=1}^{5} \left[\frac{1}{2} xy^2 \right]_{y=2}^{(9-x)/2} dx$$

$$= \int_{1}^{5} \left(\frac{65}{8} x - \frac{9}{4} x^2 + \frac{1}{8} x^3 \right) dx = \left[\frac{65}{16} x^2 - \frac{3}{4} x^3 + \frac{1}{32} x^4 \right]_{1}^{5} = 24.$$

C14S03.031: The volume is

$$V = \int_{y=-1}^{1} \int_{x=-\sqrt{1-y^2}}^{\sqrt{1-y^2}} (x+1) \, dx \, dy.$$

This integral can be evaluated exactly with a single command in *Mathematica* 3.0, but we will evaluate it one step at a time. As usual, the *Mathematica* output is rewritten slightly for more clarity.

```
Integrate[ x + 1, x ]
```

$$x + \frac{1}{2} x^2$$

```
(% /.  x → Sqrt[1 - y*y]) - (% /.  x → -Sqrt[1 - y*y])
```

$$2\sqrt{1 - y^2} + \frac{1}{2}(1 - y^2) + \frac{1}{2}(y^2 - 1)$$

```
Simplify[ % ]
```

$$2\sqrt{1 - y^2}$$

```
Integrate[ %, y ]
```

$$y\sqrt{1 - y^2} + \arcsin y$$

```
(% /.  y → 1) - (% /.  y → -1)
```

$$\pi$$

```
N[ %, 60 ]
```

802

C14S03.033: The volume is

$$V = \int_{x=-1}^{1} \int_{y=-\sqrt{1-x^2}}^{\sqrt{1-x^2}} 2\sqrt{4 - x^2 - y^2} \; dy \; dx.$$

We used *Derive* 2.56 to evaluate this integral in a step-by-step fashion much as in the solution of Problem 31. Results:

$$V = \int_{x=-1}^{1} \left[(4 - x^2) \arctan \frac{y}{\sqrt{4 - x^2 - y^2}} + y\sqrt{4 - x^2 - y^2} \right]_{y=-\sqrt{1-x^2}}^{\sqrt{1-x^2}} dx$$

$$= \int_{-1}^{1} \left[2(4 - x^2) \arctan \frac{\sqrt{3}\sqrt{1 - x^2}}{3} + 2\sqrt{3}\sqrt{1 - x^2} \right] dx$$

$$= \left[\frac{2}{3} x(12 - x^2) \arctan \frac{\sqrt{3}\sqrt{1 - x^2}}{3} + \frac{16}{3} \arctan \frac{\sqrt{3}\,(2x + 1)}{3\sqrt{1 - x^2}} \right.$$

$$\left. + \frac{16}{3} \arctan \frac{\sqrt{3}\,(2x - 1)}{3\sqrt{1 - x^2}} - 4\sqrt{3} \arcsin x + \frac{2\sqrt{3}}{3} x\sqrt{1 - x^2} \right]_{-1}^{1}$$

$$= \frac{\pi}{3} \left(32 - 12\sqrt{3} \right) \approx 11.7447292674805137.$$

In Section 14.4 we will find that this integral is quite easy to evaluate if we first convert to polar coordinates. If so, the integral takes the form

$$\int_{\theta=0}^{2\pi} \int_{r=0}^{1} 2r\sqrt{4 - r^2} \; dr \; d\theta = \int_{\theta=0}^{2\pi} \left[-\frac{2}{3}(4 - r^2)^{3/2} \right]_{r=0}^{1} d\theta = \int_{0}^{2\pi} \left(\frac{16}{3} - 2\sqrt{3} \right) d\theta = 2\pi \left(\frac{16}{3} - 2\sqrt{3} \right).$$

C14S03.035: Given: the plane with Cartesian equation

$$\frac{x}{a} + \frac{y}{b} + \frac{z}{c} = 1$$

cutting off a tetrahedron in the first octant. We set $z = 0$ and solve for

$$y = \frac{b}{a}(a - x);$$

the triangular region in the first quadrant bounded by this line and the coordinate axes is the domain of the volume integral. Hence the volume of the tetrahedron is

$$V = \int_{x=0}^{a} \int_{y=0}^{b(a-x)/a} c \left(1 - \frac{x}{a} - \frac{y}{b} \right) dy \; dx = \int_{x=0}^{a} \left[\frac{2abcy - 2bcxy - acy^2}{2ab} \right]_{y=0}^{b(a-x)/a} dx$$

$$= \int_{0}^{a} \frac{bc(a - x)^2}{2a^2} \; dx = \left[\frac{3a^2bcx - 3abcx^2 + bcx^3}{6a^2} \right]_{0}^{a} = \frac{abc}{6}.$$

C14S03.037: The volume is

$$V = \int_{y=0}^{1} \int_{x=0}^{\sqrt{1-y^2}} \sqrt{1-y^2}\, dx\, dy = \int_{y=0}^{1} \left[x\sqrt{1-y^2} \right]_{x=0}^{\sqrt{1-y^2}} dy = \int_{0}^{1} (1-y^2)\, dy = \left[y - \frac{1}{3} y^3 \right]_{0}^{1} = \frac{2}{3}.$$

As the text indicates, the other order of integration provides more difficulties. You obtain

$$V = \int_{x=0}^{1} \int_{y=0}^{\sqrt{1-x^2}} \sqrt{1-y^2}\, dy\, dx = \int_{x=0}^{1} \left[\frac{1}{2} y \sqrt{1-y^2} + \frac{1}{2} \arcsin y \right]_{y=0}^{\sqrt{1-x^2}} dx$$

$$= \int_{0}^{1} \left[\frac{1}{2} x \sqrt{1-x^2} + \frac{1}{2} \arcsin\left(\sqrt{1-x^2} \right) \right] dx$$

$$= \left[\left(\frac{1}{6} x^2 - \frac{2}{3} \right) \sqrt{1-x^2} + \frac{1}{2} x \arcsin\left(\sqrt{1-x^2} \right) \right]_{0}^{1} = \frac{2}{3}.$$

C14S03.039: We integrate to find the volume of an eighth of the sphere, then multiply by 8. Thus the volume of a sphere of radius a is

$$V = 8 \int_{x=0}^{a} \int_{y=0}^{\sqrt{a^2-x^2}} \sqrt{a^2 - x^2 - y^2}\, dy\, dx.$$

Let $y = (a^2 - x^2)^{1/2} \sin\theta$. Then $dy = (a^2 - x^2)^{1/2} \cos\theta\, d\theta$. This substitution yields

$$V = 8 \int_{x=0}^{a} \int_{\theta=0}^{\pi/2} \left[(a^2 - x^2) - (a^2 - x^2)\sin^2\theta \right]^{1/2} (a^2 - x^2)^{1/2} \cos\theta\, d\theta\, dx$$

$$= 8 \int_{x=0}^{a} \int_{\theta=0}^{\pi/2} (a^2 - x^2) \cos^2\theta\, d\theta\, dx$$

$$= 8 \int_{x=0}^{a/2} \int_{\theta=0}^{\pi/2} (a^2 - x^2) \cdot \frac{1 + \cos 2\theta}{2}\, d\theta\, dx = 8 \int_{x=0}^{a} (a^2 - x^2) \left[\frac{1}{2}\theta + \frac{1}{4} \sin 2\theta \right]_{0}^{\pi/2} dx$$

$$= 8 \int_{0}^{a} \frac{\pi}{4} (a^2 - x^2)\, dx = 2\pi \left[a^2 x - \frac{1}{3} x^3 \right]_{0}^{a} = 2\pi \cdot \frac{2}{3} a^3 = \frac{4}{3} \pi a^3.$$

C14S03.041: We integrate over the quarter-circle of radius 5 and center $(0, 0)$ in the first quadrant, then multiply by 4. Hence the volume is

$$V = 4 \int_{x=0}^{5} \int_{y=0}^{\sqrt{25-x^2}} (25 - x^2 - y^2)\, dy\, dx = 4 \int_{x=0}^{5} \left[25y - x^2 y - \frac{1}{3} y^3 \right]_{y=0}^{\sqrt{25-x^2}} dx$$

$$= \int_{0}^{5} \frac{8}{3} (25 - x^2)^{3/2}\, dx = \left[\frac{8}{3} \sqrt{25 - x^2} \left(\frac{125}{8} x - \frac{1}{4} x^3 \right) + 625 \arcsin\left(\frac{x}{5} \right) \right]_{0}^{5}$$

$$= \frac{625}{2} \pi \approx 981.7477042468103870195760057.$$

The techniques of Section 14.4 will transform this problem into one that is remarkably simple.

C14S03.043: Suppose that the cylinder is the one with equation $x^2 + z^2 = R^2$ and that the square hole is centered on the z-axis and its sides are parallel to the coordinate planes. Thus the hole meets the xy-plane

804

in the square with vertices at $(\pm\frac{1}{2}R, \pm\frac{1}{2}R)$. We will integrate over the quarter of that square that lies in the first quadrant, then multiply by 4. Hence the volume of material removed by the drill is

$$V = 4\int_{x=0}^{R/2}\int_{y=0}^{R/2} 2\sqrt{R^2 - x^2}\ dy\ dx = 4\int_{x=0}^{R/2}\left[2y\sqrt{R^2 - x^2}\right]_{y=0}^{R/2} dx$$

$$= 4\int_{0}^{R/2} R(R^2 - x^2)^{1/2}\ dx = 4\left[\frac{1}{2}Rx(R^2 - x^2)^{1/2} - \frac{1}{2}R^3 \arctan\left(\frac{x}{(R^2 - x^2)^{1/2}}\right)\right]_{0}^{R/2}$$

$$= \left[\frac{1}{2}\sqrt{3} + 2\arctan\left(\frac{1}{3}\sqrt{3}\right)\right]\cdot R^3 = \frac{3\sqrt{3} + 2\pi}{6}\cdot R^3 \approx (1.913222954981)R^3.$$

C14S03.045: The region bounded by the parabolas $y = x^2$ and $y = 8 - x^2$ in the xy-plane is a suitable domain for a double integral that gives the volume of the solid. Hence the volume of the solid is

$$V = \int_{x=-2}^{2}\int_{y=x^2}^{8-x^2}(2x^2 - x^2)\ dy\ dx = \int_{-2}^{2}\left[x^2 y\right]_{y=x^2}^{8-x^2} dx = \int_{-2}^{2}(8x^2 - 2x^4)\ dx$$

$$= \left[\frac{8}{3}x^3 - \frac{2}{5}x^5\right]_{-2}^{2} = \frac{256}{15} \approx 17.066666666667.$$

C14S03.047: We used *Mathematica* 3.0 in the usual way; the volume of the solid is

$$V = \int_{x=-\pi/2}^{\pi/2}\int_{y=-\cos x}^{\cos x}\cos y\ dy\ dx = \int_{x=-\pi/2}^{\pi/2}\left[\sin y\right]_{y=-\cos x}^{\cos x} dx$$

$$= \int_{-\pi/2}^{\pi/2} 2\sin(\cos x)\ dx \approx 3.572974963900010467337.$$

Mathematica reports that the exact value of the integral is

$$\texttt{4*HypergeometricPFQ}\left[\{1\},\ \left\{\frac{3}{2}, \frac{3}{2}\right\},\ -\frac{1}{4}\right].$$

C14S03.049: A *Mathematica* solution:

```
eq1 = z == 2*x + 3;
eq2 = z == x∧2 + y∧2;
Eliminate[ { eq1, eq2 }, z ]
    y² = -x² + 2x + 3
```

This is the circle $(x - 1)^2 + y^2 = 4$ with center $(1, 0)$ and radius 2. Therefore the volume of the solid is

```
Integrate[ 3 + 2*x − x∧2 - y∧2, { x, −1, 3 },
            { y, −Sqrt[ 3 + 2*x − x{∧2 ], Sqrt[ 3 + 2*x − x∧2 ] } ]
```

8π —C.H.E.

C14S03.051: A *Mathematica* solution:

```
eq1 = z == -16*x - 18*y;

eq2 = z == 11 - 4*x^2 - 9*y^2;

Eliminate[ { eq1, eq2 }, z ]
```

$$-9y^2 + 18y + 11 = 4x^2 - 16x$$

This is the ellipse $4(x-2)^2 + 9(y-1)^2 = 36$ with center $(2, 1)$ and semiaxes $a = 3$ and $b = 2$. Hence the volume of the solid is

```
Integrate[ 11 - 4*x^2 - 9*y^2 + 16*x + 18*y, { x, -1, 5 },
    { y, 1 - 1/3*Sqrt[ 36 - 4*(x - 2)^2 ], 1 + 1/3*Sqrt[ 36 - 4*(x - 2)^2 ] } ]
```

108π —C.H.E.

C14S03.053: First let's put the center of the sphere at the point $(-2, 0, 0)$. A *Mathematica* solution:

```
V = FullSimplify[ 2*Integrate[ Integrate[
        Sqrt[ 16 - (x + 2)^2 - y^2 ], { y, -1, 1 } ], { x, -1, 1 } ] ]
```

$$\frac{2}{3}\left[6\sqrt{6} - 2\sqrt{14} + 29\cot^{-1}\left(\sqrt{6}\right) + 41\cot^{-1}\left(\sqrt{14}\right) - 47\csc^{-1}\left(\sqrt{15}\right)\right.$$
$$+ 47\sin^{-1}\left(\sqrt{3/5}\right) + 20\tan^{-1}\left(\sqrt{3/2}\right) - 108\tan^{-1}\left(9\sqrt{3/2}\right)$$
$$\left. - 20\tan^{-1}\left(11/\sqrt{14}\right) + 108\tan^{-1}\left(19/\sqrt{14}\right)\right]$$

```
N[V]
```

26.7782

```
100*%/(4*Pi*64/3)*percent
```

9.98878 percent —C.H.E.

Section 14.4

C14S04.001: The circle with center $(0, 0)$ and radius $a > 0$ has polar description $r = a$, $0 \le \theta \le 2\pi$. Therefore its area is

$$A = \int_{\theta=0}^{2\pi} \int_{r=0}^{a} r\, dr\, d\theta = \int_{\theta=0}^{2\pi} \left[\frac{1}{2}r^2\right]_{r=0}^{a} d\theta = \int_{0}^{2\pi} \frac{1}{2}a^2\, d\theta = \left[\frac{1}{2}a^2\theta\right]_{0}^{2\pi} = \pi a^2.$$

C14S04.003: The area bounded by the cardioid with polar description $r = 1 + \cos\theta$, $0 \le \theta \le 2\pi$, is

$$A = \int_{\theta=0}^{2\pi} \int_{r=0}^{1+\cos\theta} r\, dr\, d\theta = \int_{\theta=0}^{2\pi} \left[\frac{1}{2}r^2\right]_{r=0}^{1+\cos\theta} d\theta = \int_{0}^{2\pi} \left(\frac{1}{2} + \cos\theta + \frac{1}{2}\cos^2\theta\right) d\theta$$

$$= \frac{1}{8}\left[6\theta + 8\sin\theta + \sin 2\theta\right]_{0}^{2\pi} = \frac{3}{2}\pi.$$

C14S04.005: To see the two circles, execute the *Mathematica* 3.0 command

```
ParametricPlot[ {{Cos[t], Sin[t]}, {2*Sin[t]*Cos[t], 2*Sin[t]*Sin[t]}},
              {t, 0, 2*Pi}, AspectRatio → Automatic ];
```

the result is shown next.

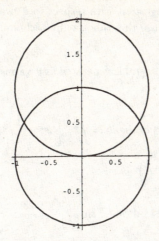

To find where the two circles meet, solve their equations simultaneously:

$$2\sin\theta = 1; \qquad \sin\theta = \frac{1}{2}; \qquad \theta = \frac{1}{6}\pi, \; \theta = \frac{5}{6}\pi.$$

To find the area between them, two integrals are required. Each is doubled because we are actually finding the area of the right half of the intersection of the circles.

$$A_1 = 2\int_{\theta=0}^{\pi/6}\int_{r=0}^{2\sin\theta} r\,dr\,d\theta = 2\int_0^{\pi/6} 2\sin^2\theta\,d\theta = \left[2\theta - \sin 2\theta\right]_0^{\pi/6} = \frac{2\pi - 3\sqrt{3}}{6};$$

$$A_2 = 2\int_{\theta=\pi/6}^{\pi/2}\int_{r=0}^{1} r\,dr\,d\theta = \int_{\theta=\pi/6}^{\pi/2} 1\,d\theta = \frac{\pi}{3}.$$

Therefore the total area enclosed by both circles is

$$A = A_1 + A_2 = \frac{4\pi - 3\sqrt{3}}{6} \approx 1.2283696986087568.$$

C14S04.007: To find where the limaçon $r = 1 - 2\sin\theta$ passes through the origin, solve $1 - 2\sin\theta = 0$ for $\theta = \frac{1}{6}\pi$, $\theta = \frac{5}{6}\pi$. The small loop of the limaçon is generated by the values of θ between these two angles, and is thus given by

$$A = \int_{\theta=\pi/6}^{5\pi/6}\int_{r=0}^{1-2\sin\theta} r\,dr\,d\theta = \int_{\theta=\pi/6}^{5\pi/6}\left[\frac{1}{2}r^2\right]_{r=0}^{1-2\sin\theta} d\theta = \int_{\pi/6}^{5\pi/6}\left[\frac{1}{2}(1 - 2\sin\theta)^2\right]d\theta$$

$$= \left[\frac{3}{2}\theta + 2\cos\theta - \frac{1}{2}\sin 2\theta\right]_{\pi/6}^{5\pi/6} = \frac{2\pi - 3\sqrt{3}}{2} \approx 0.5435164422364773.$$

C14S04.009: In polar coordinates the integral takes the form

807

$$I = \int_{\theta=0}^{2\pi} \int_{r=0}^{2} r^2 \, dr \, d\theta = 2\pi \left[\frac{1}{3} r^3 \right]_0^2 = \frac{16}{3} \pi \approx 16.7551608191455639.$$

Because the inner integral does not involve θ (either in the integrand or in the limits of integration), it is constant with respect to θ. Therefore to integrate it with respect to θ over the interval $0 \leq \theta \leq 2\pi$, simply multiply the inner integral by 2π. We will use this time-saving technique frequently and without further comment.

C14S04.011: Note that the domain of the integral is generated as θ varies from 0 to π. In polar coordinates the integral takes the form

$$I = \int_{\theta=0}^{\pi} \int_{r=0}^{\sin \theta} (10 + 2r \cos \theta + 3r \sin \theta) \cdot r \, dr \, d\theta = \int_{\theta=0}^{\pi} \left[5r^2 + \frac{1}{3} r^3 (2 \cos \theta + 3 \sin \theta) \right]_{r=0}^{\sin \theta} d\theta$$

$$= \int_0^{\pi} \left(5 \sin^2 \theta + \frac{2}{3} \sin^3 \theta \cos \theta + \sin^4 \theta \right) d\theta$$

$$= \frac{1}{96} \left[276\theta - 8 \cos 2\theta + 2 \cos 4\theta - 144 \sin 2\theta + 3 \sin 4\theta \right]_0^{\pi} = \frac{23}{8} \pi \approx 9.03207887907065556058.$$

The reduction formulas in Problems 53 and 54 of Section 7.3, and the formulas of Problem 58 there and Formula (113) of the endpapers, may be used if you prefer to avoid trigonometric identities.

C14S04.013: In polar form the given integral becomes

$$J = \int_{\theta=0}^{\pi/2} \int_{r=0}^{1} \frac{r}{1 + r^2} \, dr \, d\theta = \frac{\pi}{2} \cdot \left[\frac{1}{2} \ln(1 + r^2) \right]_0^1 = \frac{1}{4} \pi \ln 2 \approx 0.5443965225759005.$$

To evaluate the integral in Cartesian coordinates, you would first need to evaluate

$$\int_{x=0}^{\sqrt{1-y^2}} \frac{1}{1 + x^2 + y^2} \, dx = \frac{1}{\sqrt{1 + y^2}} \arctan \left(\frac{\sqrt{1 - y^2}}{\sqrt{1 + y^2}} \right). \tag{1}$$

Then you would need to antidifferentiate the last expression in (1). Neither *Mathematica* 3.0 nor *Derive* 2.56 could do so.

C14S04.015: In polar coordinates the integral becomes

$$I = \int_{\theta=0}^{\pi/2} \int_{r=0}^{2} r^4 \, dr \, d\theta = \frac{\pi}{2} \cdot \left[\frac{1}{5} r^5 \right]_0^2 = \frac{\pi}{2} \cdot \frac{32}{5} = \frac{16}{5} \pi \approx 10.0530964914873384.$$

This integral can be evaluated in Cartesian coordinates. You should obtain

$$I = \int_0^2 \int_0^{\sqrt{4-x^2}} (x^2 + y^2)^{3/2} \, dy \, dx$$

$$= \int_0^2 \left[\left(\frac{5}{8} x^2 y + \frac{1}{4} y^3 \right) \sqrt{x^2 + y^2} + \frac{3}{8} x^4 \ln \left(y + \sqrt{x^2 + y^2} \right) \right]_0^{\sqrt{4-x^2}} dx$$

$$= \int_0^2 \left[\frac{5}{4} x^2 \sqrt{4 - x^2} + \frac{1}{2} (4 - x^2)^{3/2} - \frac{3}{4} x^4 \ln x + \frac{3}{8} x^4 \ln \left(2 + \sqrt{4 - x^2} \right) \right] dx$$

$$= \left[\left(\frac{2}{5} x + \frac{3}{20} x^3 \right) \sqrt{4 - x^2} + \frac{32}{5} \arcsin \left(\frac{x}{2} \right) - \frac{3}{20} x^5 \ln x + \frac{3}{40} x^5 \ln \left(2 + \sqrt{4 - x^2} \right) \right]_0^2 = \frac{16}{5} \pi.$$

The final equality requires l'Hôpital's rule.

C14S04.017: In polar coordinates the integral becomes

$$K = \int_{\theta=0}^{\pi/2} \int_{r=0}^{1} r \sin r^2 \, dr \, d\theta = \frac{\pi}{2} \cdot \left[-\frac{1}{2} \cos r^2 \right]_0^1 d\theta = \frac{\pi}{4}(1 - \cos 1) \approx 0.3610457246892050.$$

Exact evaluation of the given integral in Cartesian coordinates may be impossible. *Mathematica 3.0* reports that

$$\int \sin(x^2 + y^2) \, dx = \sqrt{\frac{\pi}{2}} \left[(\cos y^2) \cdot \text{FresnelS}\left(x\sqrt{\frac{2}{\pi}} \right) + (\sin y^2) \cdot \text{FresnelC}\left(x\sqrt{\frac{2}{\pi}} \right) \right]$$

where

$$\text{FresnelS}(x) = \int_0^x \sin\left(\frac{\pi t^2}{2} \right) dt \quad \text{and} \quad \text{FresnelC}(x) = \int_0^x \cos\left(\frac{\pi t^2}{2} \right) dt.$$

C14S04.019: The volume is

$$V = \int_{\theta=0}^{2\pi} \int_{r=0}^{1} (2 + r\cos\theta + r\sin\theta) \cdot r \, dr \, d\theta = \int_{\theta=0}^{2\pi} \left[r^2 + \frac{1}{3} r^3 (\cos\theta + \sin\theta) \right]_{r=0}^{1} d\theta$$

$$= \int_0^{2\pi} \left(1 + \frac{1}{3} \cos\theta + \frac{1}{3} \sin\theta \right) d\theta = \left[\theta - \frac{1}{3} \cos\theta + \frac{1}{3} \sin\theta \right]_0^{2\pi} = 2\pi.$$

C14S04.021: The volume is

$$V = \int_{\theta=0}^{\pi} \int_{r=0}^{2\sin\theta} (3 + r\cos\theta + r\sin\theta) \cdot r \, dr \, d\theta = \int_{\theta=0}^{\pi} \left[\frac{3}{2} r^2 + \frac{1}{3} r^3 (\cos\theta + \sin\theta) \right]_{r=0}^{2\sin\theta} d\theta$$

$$= \int_0^{\pi} \left(6\sin^2\theta + \frac{8}{3}\sin^3\theta\cos\theta + \frac{8}{3}\sin^4\theta \right) d\theta = \frac{1}{12}\left[48\theta - 4\cos 2\theta + \cos 4\theta - 26\sin 2\theta + \sin 4\theta \right]_0^{\pi}$$

$$= \frac{1}{4} + \frac{48\pi - 3}{12} = 4\pi \approx 12.566370614359172953852.$$

C14S04.023: We will find the volume of the sphere of radius a centered at the origin:

$$V = \int_{\theta=0}^{2\pi} \int_{r=0}^{a} 2r\sqrt{a^2 - r^2} \, dr \, d\theta = 2\pi \cdot \left[-\frac{2}{3}(a^2 - r^2)^{3/2} \right]_{r=0}^{a} = 2\pi \cdot \frac{2}{3}a^3 = \frac{4}{3}\pi a^3.$$

C14S04.025: The volume of the solid is

$$V = \int_{\theta=0}^{2\pi} \int_{r=0}^{a} (h + r\cos\theta) \cdot r \, dr \, d\theta = \int_{\theta=0}^{2\pi} \left[\frac{1}{2} r^2 h + \frac{1}{3} r^2 \cos\theta \right]_{r=0}^{a} d\theta = \int_0^{2\pi} \left(\frac{1}{2} a^2 h + \frac{1}{3} a^3 \cos\theta \right) d\theta$$

$$= \frac{1}{6}\left[3a^2 h\theta + 2a^3 \sin\theta \right]_0^{2\pi} = \pi a^2 h.$$

C14S04.027: When we solve the equations of the paraboloids simultaneously, we find that one consequence is that $x^2 + y^2 = 1$. Thus the curve in which the paraboloids meet lies on that cylinder, and hence the circular disk $x^2 + y^2 \leq 1$ in the xy-plane is a suitable domain for the double integral that yields the volume between the paraboloids. That volume is therefore

$$V = \int_{\theta=0}^{2\pi} \int_{r=0}^{1} (4 - 4r^2) \cdot r \, dr \, d\theta = 2\pi \cdot \left[2r^2 - r^4 \right]_{r=0}^{1} d\theta = 2\pi \approx 6.2831853071795865.$$

C14S04.029: When the equations of the sphere and the cone are solved simultaneously, one consequence is that $x^2 + y^2 = \frac{1}{2}a^2$. Therefore the circle in which the sphere and cone meet lies on the cylinder with that equation. Thus a suitable domain for the double integral that gives the volume in question is the circle in the xy-plane with equation $x^2 + y^2 = \frac{1}{2}a^2$. Hence the volume of the "ice-cream cone" is

$$V = \int_{\theta=0}^{2\pi} \int_{r=0}^{a/\sqrt{2}} \left(\sqrt{a^2 - r^2} - r \right) \cdot r \, dr \, d\theta = 2\pi \cdot \left[-\frac{1}{3} \left((a^2 - r^2)^{3/2} + r^3 \right) \right]_{r=0}^{a/\sqrt{2}}$$

$$= 2\pi \cdot \frac{1}{12} \left(4 - 2\sqrt{2} \right) a^3 = \frac{1}{3}\pi \left(2 - \sqrt{2} \right) a^3 \approx (0.6134341230070734)a^3.$$

C14S04.031: The curve $r^2 = 2\sin\theta$ is not a lemniscate. The lemniscate was discovered in 1694 by Jacques Bernoulli (1654–1705) and has an equation of the form $r^2 = a^2 \cos(2\theta - \omega)$ or of the form $r^2 = a^2 \sin(2\theta - \omega)$ where ω and a are constants. The effect of ω is to rotate the graph through the angle ω and the effect of a is to magnify the graph. The curve given in Problem 31 has the shape shown in Figure 14.04.031A, shown next, and was generated by the *Mathematica* 3.0 command

```
ParametricPlot[ {{(Sqrt[2*Sin[t]])*Cos[t], (Sqrt[2*Sin[t]])*Sin[t]},
                 {-(Sqrt[2*Sin[t]])*Cos[t], -(Sqrt[2*Sin[t]])*Sin[t]}},
                 {t, 0, Pi}, PlotPoints -> 43, AspectRatio -> Automatic ];
```

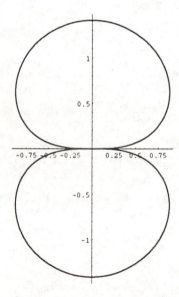

It seems likely that there is a typographical error in this problem and the equation given for the lemniscate should be $r^2 = 2\sin 2\theta$. This curve has the shape shown in Figure 14.04.031B, shown next, and was generated by the *Mathematica* 3.0 command

```
ParametricPlot[ {{(Sqrt[2*Sin[2*t]])*Cos[t], (Sqrt[2*Sin[2*t]])*Sin[t]},
                 {-(Sqrt[2*Sin[2*t]])*Cos[t], -(Sqrt[2*Sin[2*t]])*Sin[t]}},
                 {t, 0, Pi/2}, PlotPoints → 43, AspectRatio → Automatic ];
```

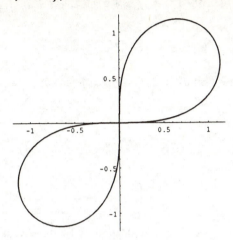

We will solve the problem both ways. First, with the curve $r^2 = 2\sin\theta$, we find the volume to be

$$V = \int_{\theta=0}^{\pi} \int_{r=0}^{\sqrt{2\sin\theta}} r^3 \, dr \, d\theta = \int_{\theta=0}^{\pi} \left[\frac{1}{4} r^4 \right]_{r=0}^{\sqrt{2\sin\theta}} d\theta = \int_{0}^{\pi} \sin^2\theta \, d\theta$$

$$= \frac{1}{4} \left[2\theta - \sin 2\theta \right]_{0}^{\pi} = \frac{1}{2}\pi \approx 1.5707963267948966.$$

Second, using the lemniscate with polar equation $r^2 = 2\sin 2\theta$, we find the volume to be

$$V = \int_{\theta=0}^{\pi/2} \int_{r=0}^{\sqrt{2\sin 2\theta}} r^3 \, dr \, d\theta = \int_{\theta=0}^{\pi/2} \left[\frac{1}{4} r^4 \right]_{r=0}^{\sqrt{2\sin 2\theta}} d\theta = \int_{0}^{\pi/2} \sin^2 2\theta \, d\theta$$

$$= \frac{1}{8} \left[4\theta - \sin 4\theta \right]_{0}^{\pi/2} = \frac{1}{4}\pi \approx 0.7853981633974483.$$

C14S04.033: Part(a): A cross section of Fig. 14.4.23 in the xz-plane reveals a right triangle with legs b and $a - h$ and hypotenuse a, and it follows immediately that $b^2 = 2ah - h^2$. Part (b): The volume of the spherical segment is

$$V = \int_{0}^{2\pi} \int_{0}^{b} r \left[\sqrt{a^2 - r^2} - (a - h) \right] dr \, d\theta = 2\pi \left[-\frac{1}{3}(a^2 - r^2)^{3/2} - \frac{1}{2}ar^2 + \frac{1}{2}hr^2 \right]_{0}^{b}$$

$$= \frac{\pi}{3} \left[3r^2 h - 3r^2 a - 2(a^2 - r^2)^{3/2} \right]_{0}^{b} = \frac{\pi}{3} \left[3b^2 h - 3b^2 a - 2(a^2 - b^2)^{3/2} + 2a^3 \right]$$

$$= \frac{\pi}{3} \left[3b^2 h - 3b^2 a - 2(a^2 - 2ah + h^2)^{3/2} + 2a^3 \right] = \frac{\pi}{3} \left[3b^2 h - 3b^2 a - 2(a - h)^3 + 2a^3 \right]$$

$$= \frac{\pi}{3}(3b^2 h - 3b^2 a + 6a^2 h - 6ah^2 + 2h^3) = \frac{\pi}{3} \left[3b^2 h - 3a(2ah - h^2) + 6a^2 h - 6ah^2 + 2h^3 \right]$$

$$= \frac{\pi}{3}(3b^2 h - 6a^2 h + 3ah^2 + 6a^2 h - 6ah^2 + 2h^3) = \frac{\pi}{3}(3b^2 h - 3ah^2 + 2h^3) = \frac{\pi}{3} h(3b^2 - 3ah + 2h^2).$$

811

Recall that $2ah = b^2 + h^2$, so we may substitute $\frac{3}{2}b^2 + \frac{3}{2}h^2$ for $3ah$ in the last expression. Therefore the volume of the spherical cap is

$$V = \frac{1}{3}\pi h\left(3b^2 - \frac{3}{2}b^2 - \frac{3}{2}h^2 + 2h^2\right) = \frac{1}{6}\pi h(6b^2 - 3b^2 - 3h^2 + 4h^2) = \frac{1}{6}\pi h(3b^2 + h^2).$$

C14S04.035: Using the *Suggestion*, we have

$$2\pi(b - x)\, dA = 2\pi(b - r\cos\theta)r\, dr\, d\theta,$$

and hence the volume of the torus is

$$V = \int_0^{2\pi}\int_0^a 2\pi r(b - r\cos\theta)\, dr\, d\theta = \int_0^{2\pi}\left[b\pi r^2 - \frac{2}{3}\pi r^3 \cos\theta\right]_0^a d\theta$$

$$= \left[\frac{1}{3}(3\pi a^2 b\theta - 2\pi a^3 \sin\theta)\right]_0^{2\pi} = 2\pi^2 a^2 b.$$

Read the *First Theorem of Pappus* in Section 14.5 and apply it to this circular disk of radius a rotated around a circle of radius b to obtain the same answer in a tenth of the time.

C14S04.037: The plane and the paraboloid meet in a curve that lies on the cylinder $x^2 + y^2 = 4$, so the circular disk $x^2 + y^2 \leq 4$ in the xy-plane is a suitable domain for the volume integral. The volume of the solid is thus

$$V = \int_0^{2\pi}\int_0^2 (4 - r^2)\cdot r\, dr\, d\theta = 2\pi\cdot\left[2r^2 - \frac{1}{4}r^4\right]_0^2 = 8\pi \approx 25.1327412287183459.$$

C14S04.039: When we solve the equations of the paraboloids simultaneously, we find that $x^2 + y^2 = 4$. Hence the curve in which the surfaces intersect lies on the cylinder $x^2 + y^2 = 4$. Thus the disk in the xy-plane bounded by the circle $x^2 + y^2 = 4$ is appropriate for the domain of the volume integral. So the volume of the solid bounded by the paraboloids is

$$V = \int_{\theta=0}^{2\pi}\int_{r=0}^2 (12 - 3r^2)\cdot r\, dr\, d\theta = 2\pi\cdot\left[6r^2 - \frac{3}{4}r^4\right]_{r=0}^2 = 24\pi \approx 75.3982236861550377.$$

C14S04.041: A *Mathematica* solution:

```
V = Simplify[ 2*Integrate[ Integrate[
        r*Sqrt[ b^2 - r^2 ], { r, 0, a } ], { θ, 0, 2*Pi } ] ]
```

$$\frac{3}{2}\left(b^3 - (b^2 - a^2)^{3/2}\right)\pi$$

```
V1 = V /. { a → 2, b → 4 }
```

$$\frac{4}{3}\left(64 - 24\sqrt{3}\right)\pi$$

Percent of material removed:

```
N[ 100*V1/(4*Pi*64/3) ]
```

812

C14S04.043: A *Mathematica* solution for the case of a square hole:

```
n = 4;
V = 4*n*Integrate[ Integrate[ r*Sqrt[ 4 - r^2 ],
            { r, 0, Cos[ Pi/n ]*Sec[ θ ] } ], { θ, 0, Pi/n } ];
```

(We suppress the output; it's not attractive.)

```
V = Chop[ N[V] ]
    7.6562
```

Percent:

```
100*V/(4*Pi*8/3)
    22.8473
```

Now for the pentagonal, hexagonal, heptagonal, and 17-sided holes. First, the pentagonal hole:

```
n = 5;
V = 4*n*Integrate[ Integrate[ r*Sqrt[ 4 - r^2 ],
            { r, 0, Cos[ Pi/n ]*Sec[ θ ] } ], { θ, 0, Pi/n } ];
V = Chop[ N[V] ]
    9.03688
```

Percent:

```
100*V/(4*Pi*8/3)
    26.9675
```

Next, the hexagonal hole:

```
n = 6;
V = 4*n*Integrate[ Integrate[ r*Sqrt[ 4 - r^2 ],
            { r, 0, Cos[ Pi/n ]*Sec[ θ ] } ], { θ, 0, Pi/n } ];
V = Chop[ N[V] ]
    9.83041
```

Percent:

```
100*V/(4*Pi*8/3)
    29.3355
```

For the heptagonal and 17-sided holes, we use *Maple* V Release 5.

```
n:=7;

V:=4*n*int(int(r*sqrt(4-r^2), r=0..cos(Pi/n)*sec(t)), t=0..Pi/n);

V:=evalf(V);
```
$$V := 10.32346688$$

Percent:
```
evalf(100*V/(4*Pi*8/3));
```
   30.80682719

```
n:=17;

V:=4*n*int(int(r*sqrt(4-r^2), r=0..cos(Pi/n)*sec(t)), t=0..Pi/n);

V:=evalf(V);
```
$$V := 11.49809060$$

Percent:
```
evalf(100*V/(4*Pi*8/3));
```
   34.31208666   —C.H.E.

Section 14.5

Note: To integrate positive integral powers of sines and cosines (and their products), there are effective techniques not illustrated in the text. They use the following identities, which are consequences of the Euler-DeMoivre formula

$$(\cos\theta + i\sin\theta)^n = \left(e^{i\theta}\right)^n = e^{in\theta} = \cos n\theta + i\sin n\theta.$$

---

1. $\sin^2\theta = \dfrac{1}{2}(1 - \cos 2\theta).$

2. $\sin^3\theta = \dfrac{1}{4}(3\sin\theta - \sin 3\theta).$

3. $\sin^4\theta = \dfrac{1}{8}(3 - 4\cos 2\theta + \cos 4\theta).$

4. $\sin^5\theta = \dfrac{1}{16}(10\sin\theta - 5\sin 3\theta + \sin 5\theta).$

5. $\sin^6\theta = \dfrac{1}{32}(10 - 15\cos 2\theta + 6\cos 4\theta - \cos 6\theta).$

6. $\sin^7\theta = \dfrac{1}{64}(35\sin\theta - 21\sin 3\theta + 7\sin 5\theta - \sin 7\theta).$

7. $\sin^8\theta = \dfrac{1}{128}(35 - 56\cos 2\theta + 28\cos 4\theta - 8\cos 6\theta + \cos 8\theta).$

814

8. $\sin^9 \theta = \dfrac{1}{256}(126 \sin \theta - 84 \sin 3\theta + 36 \sin 5\theta - 9 \sin 7\theta + \sin 9\theta)$.

9. $\sin^{10} \theta = \dfrac{1}{512}(126 - 210 \cos 2\theta + 120 \cos 4\theta - 45 \cos 6\theta + 10 \cos 8\theta - \cos 10\theta)$.

10. $\cos^2 \theta = \dfrac{1}{2}(1 + \cos 2\theta)$.

11. $\cos^3 \theta = \dfrac{1}{4}(3 \cos \theta + \cos 3\theta)$.

12. $\cos^4 \theta = \dfrac{1}{8}(3 + 4 \cos 2\theta + \cos 4\theta)$.

13. $\cos^5 \theta = \dfrac{1}{16}(10 \cos \theta + 5 \cos 3\theta + \cos 5\theta)$.

14. $\cos^6 \theta = \dfrac{1}{32}(10 + 15 \cos 2\theta + 6 \cos 4\theta + \cos 6\theta)$.

15. $\cos^7 \theta = \dfrac{1}{64}(35 \cos \theta + 21 \cos 3\theta + 7 \cos 5\theta + \cos 7\theta)$.

16. $\cos^8 \theta = \dfrac{1}{128}(35 + 56 \cos 2\theta + 28 \cos 4\theta + 8 \cos 6\theta + \cos 8\theta)$.

17. $\cos^9 \theta = \dfrac{1}{256}(126 \cos \theta + 84 \cos 3\theta + 36 \cos 5\theta + 9 \cos 7\theta + \cos 9\theta)$.

18. $\cos^{10} \theta = \dfrac{1}{512}(126 + 210 \cos 2\theta + 120 \cos 4\theta + 45 \cos 6\theta + 10 \cos 8\theta + \cos 10\theta)$.

19. $\sin^2 \theta \cos^2 \theta = \dfrac{1}{8}(1 - \cos 4\theta)$.

20. $\sin^3 \theta \cos^2 \theta = \dfrac{1}{16}(2 \sin \theta + \sin 3\theta - \sin 5\theta)$.

21. $\sin^4 \theta \cos^2 \theta = \dfrac{1}{32}(2 - \cos 2\theta - 2 \cos 4\theta + \cos 6\theta)$.

22. $\sin^5 \theta \cos^2 \theta = \dfrac{1}{64}(5 \sin \theta + \sin 3\theta - 3 \sin 5\theta + \sin 7\theta)$.

23. $\sin^6 \theta \cos^2 \theta = \dfrac{1}{128}(5 - 4 \cos 2\theta - 4 \cos 4\theta + 4 \cos 6\theta - \cos 8\theta)$.

24. $\sin^2 \theta \cos^3 \theta = \dfrac{1}{16}(2 \cos \theta - \cos 3\theta - \cos 5\theta)$.

25. $\sin^2 \theta \cos^4 \theta = \dfrac{1}{32}(2 + \cos 2\theta - 2 \cos 4\theta - \cos 6\theta)$.

26. $\sin^2 \theta \cos^5 \theta = \dfrac{1}{64}(5 \cos \theta - \cos 3\theta - 3 \cos 5\theta - \cos 7\theta)$.

27. $\sin^2 \theta \cos^6 \theta = \dfrac{1}{128}(5 + 4 \cos 2\theta - 4 \cos 4\theta - 4 \cos 6\theta - \cos 8\theta)$.

28. $\sin^3 \theta \cos^3 \theta = \dfrac{1}{32}(3\sin 2\theta - \sin 6\theta).$

29. $\sin^4 \theta \cos^3 \theta = \dfrac{1}{64}(3\cos\theta - 3\cos 3\theta - \cos 5\theta + \cos 7\theta).$

30. $\sin^5 \theta \cos^3 \theta = \dfrac{1}{128}(6\sin 2\theta - 2\sin 4\theta - 2\sin 6\theta + \sin 8\theta).$

40. $\sin^6 \theta \cos^3 \theta = \dfrac{1}{256}(6\cos\theta - 8\cos 3\theta + 3\cos 7\theta - \cos 9\theta).$

41. $\sin^4 \theta \cos^4 \theta = \dfrac{1}{128}(3 - 4\cos 4\theta + \cos 8\theta).$

42. $\sin^5 \theta \cos^4 \theta = \dfrac{1}{256}(6\sin\theta + 4\sin 3\theta - 4\sin 5\theta - \sin 7\theta + \sin 9\theta).$

43. $\sin^6 \theta \cos^4 \theta = \dfrac{1}{512}(6 - 2\cos 2\theta - 8\cos 4\theta + 3\cos 6\theta + 2\cos 8\theta - \cos 10\theta).$

44. $\sin^5 \theta \cos^5 \theta = \dfrac{1}{512}(10\sin 2\theta - 5\sin 6\theta + \sin 10\theta).$

45. $\sin^6 \theta \cos^5 \theta = \dfrac{1}{1024}(10\cos\theta - 10\cos 3\theta - 5\cos 5\theta + 5\cos 7\theta + \cos 9\theta - \cos 11\theta).$

46. $\sin^6 \theta \cos^6 \theta = \dfrac{1}{2048}(10 - 15\cos 4\theta + 6\cos 8\theta - \cos 12\theta).$

C14S05.001: By the symmetry principle, the centroid is at $(\overline{x},\, \overline{y}) = (2,\, 3)$.

C14S05.003: By the symmetry principle, the centroid is at $(\overline{x},\, \overline{y}) = (1,\, 1)$.

C14S05.005: The mass and moments are

$$m = \int_0^4 \int_0^{(4-x)/2} 1 \; dy \; dx = \int_0^4 \frac{4-x}{2} \; dx = \left[2x - \frac{1}{4}x^2 \right]_0^4 = 4,$$

$$M_y = \int_0^4 \int_0^{(4-x)/2} x \; dy \; dx = \int_0^4 \left(2x - \frac{1}{2}x^2 \right) dx = \left[x^2 - \frac{1}{6}x^3 \right]_0^4 = \frac{16}{3}, \quad \text{and}$$

$$M_x = \int_0^4 \int_0^{(4-x)/2} y \; dy \; dx = \int_0^4 \frac{1}{8}(4-x)^2 \; dx = \left[\frac{1}{24}(x-4)^3 \right]_0^4 = \frac{8}{3}.$$

Therefore the centroid is located at $(\overline{x},\, \overline{y}) = \left(\frac{4}{3},\, \frac{2}{3} \right)$.

C14S05.007: The mass and moments are

$$m = \int_0^2 \int_0^{x^2} 1 \; dy \; dx = \int_0^2 x^2 \; dx = \left[\frac{1}{3}x^3 \right]_0^2 = \frac{8}{3},$$

$$M_y = \int_0^2 \int_0^{x^2} x \; dy \; dx = \int_0^2 x^3 \; dx = \left[\frac{1}{4}x^4 \right]_0^2 = 4, \quad \text{and}$$

$$M_x = \int_0^2 \int_0^{x^2} y \; dy \; dx = \int_0^2 \frac{1}{2}x^4 \; dx = \left[\frac{1}{10}x^5 \right]_0^2 = \frac{16}{5}.$$

Therefore the centroid is $(\overline{x}, \overline{y}) = \left(\frac{3}{2}, \frac{6}{5}\right)$.

C14S05.009: By symmetry, $M_y = 0$. Next,

$$m = \int_{-2}^{2} \int_{x^2-4}^{0} 1 \, dy \, dx = \int_{-2}^{2} (4 - x^2) \, dx = \left[4x - \frac{1}{3}x^3 \right]_{-2}^{2} = \frac{32}{3} \quad \text{and}$$

$$M_x = \int_{-2}^{2} \int_{x^2-4}^{0} y \, dy \, dx = \int_{-2}^{2} -\frac{1}{2}(4 - x^2)^2 \, dx = \left[\frac{4}{3}x^3 - 8x - \frac{1}{10}x^5 \right]_{-2}^{2} = -\frac{256}{15}.$$

Therefore the centroid is at the point $\left(0, -\frac{8}{5}\right)$.

C14S05.011: The mass and moments are

$$m = \int_{0}^{1} \int_{0}^{1-x} xy \, dy \, dx = \int_{0}^{1} \frac{1}{2}(x - 2x^2 + x^3) \, dx = \frac{1}{24}\left[6x^2 - 8x^3 + 3x^4 \right]_{0}^{1} = \frac{1}{24},$$

$$M_y = \int_{0}^{1} \int_{0}^{1-x} x^2 y \, dy \, dx = \int_{0}^{1} \left[\frac{1}{2}x^2 y^2 \right]_{0}^{1-x} dx = \int_{0}^{1} \frac{1}{2}(x^2 - 2x^3 + x^4) \, dx$$

$$= \frac{1}{60}\left[10x^3 - 15x^4 + 6x^5 \right]_{0}^{1} = \frac{1}{60},$$

$$M_x = \int_{0}^{1} \int_{0}^{1-x} xy^2 \, dy \, dx = \int_{0}^{1} \left[\frac{1}{3}xy^3 \right]_{0}^{1-x} dx = \int_{0}^{1} \frac{1}{3}(x - 3x^2 + 3x^3 - x^4) \, dx$$

$$= \frac{1}{60}\left[10x^2 - 20x^3 + 15x^4 - 4x^5 \right]_{0}^{1} = \frac{1}{60}.$$

Therefore the centroid is located at the point $\left(\frac{2}{5}, \frac{2}{5}\right)$.

C14S05.013: The mass and moments of the lamina are

$$m = \int_{-2}^{2} \int_{0}^{4-x^2} y \, dy \, dx = \int_{-2}^{2} \frac{1}{2}(4 - x^2)^2 \, dx = \frac{1}{30}\left[240x - 40x^3 + 3x^5 \right]_{-2}^{2} = \frac{256}{15},$$

$$M_y = \int_{-2}^{2} \int_{0}^{4-x^2} xy \, dy \, dx = \int_{-2}^{2} \frac{1}{2}x(4 - x^2)^2 \, dx = \frac{1}{12}\left[48x^2 - 12x^4 + x^6 \right]_{-2}^{2} = 0, \quad \text{and}$$

$$M_x = \int_{-2}^{2} \int_{0}^{4-x^2} y^2 \, dy \, dx = \int_{-2}^{2} \frac{1}{3}(4 - x^2)^3 \, dx = \left[\frac{64}{3}x - \frac{16}{3}x^3 + \frac{4}{5}x^5 - \frac{1}{21}x^7 \right]_{-2}^{2} = \frac{4096}{105}.$$

Therefore the centroid of the lamina is $(\overline{x}, \overline{y}) = \left(0, \frac{16}{7}\right)$.

C14S05.015: The mass and moments of the lamina are

$$m = \int_{0}^{1} \int_{x^2}^{\sqrt{x}} xy \, dy \, dx = \int_{0}^{1} \frac{1}{2}(x^2 - x^5) \, dx = \frac{1}{12}\left[2x^3 - x^6 \right]_{0}^{1} = \frac{1}{12},$$

$$M_y = \int_{0}^{1} \int_{x^2}^{\sqrt{x}} x^2 y \, dy \, dx = \int_{0}^{1} \frac{1}{2}(x^3 - x^6) \, dx = \frac{1}{56}\left[7x^4 - 4x^7 \right]_{0}^{1} = \frac{3}{56},$$

$$M_x = \int_{0}^{1} \int_{x^2}^{\sqrt{x}} xy^2 \, dy \, dx = \int_{0}^{1} \frac{1}{3}(x^{5/2} - x^7) \, dx = \frac{1}{168}\left[16x^{7/2} - 7x^8 \right]_{0}^{1} = \frac{3}{56}.$$

817

Hence the centroid of the lamina is at $(\overline{x}, \overline{y}) = \left(\frac{9}{14}, \frac{9}{14}\right)$.

C14S05.017: The mass and moments of the lamina are

$$m = \int_{-1}^{1} \int_{x^2}^{2-x^2} y\, dy\, dx = \int_{-1}^{1} (2 - 2x^2)\, dx = \left[2x - \frac{2}{3}x^3 \right]_{-1}^{1} = \frac{8}{3},$$

$$M_y = \int_{-1}^{1} \int_{x^2}^{2-x^2} xy\, dy\, dx = \int_{-1}^{1} (2x - 2x^3)\, dx = \left[x^2 - \frac{1}{2}x^4 \right]_{-1}^{1} = 0,$$

$$M_x = \int_{-1}^{1} \int_{x^2}^{2-x^2} y^2\, dy\, dx = \int_{-1}^{1} \left[\frac{1}{3}y^3 \right]_{x^2}^{2-x^2} dx = \int_{-1}^{1} \left(\frac{1}{3}(2 - x^2)^3 - \frac{1}{3}x^6 \right) dx$$

$$= \left[\frac{8}{3}x - \frac{4}{3}x^3 + \frac{2}{5}x^5 - \frac{2}{21}x^7 \right]_{-1}^{1} = \frac{344}{105}.$$

Hence the centroid of the lamina is at $(\overline{x}, \overline{y}) = \left(0, \frac{43}{35}\right)$.

C14S05.019: The mass and moments of the lamina are

$$m = \int_{0}^{\pi} \int_{0}^{\sin x} 1\, dy\, dx = \int_{0}^{\pi} \sin x\, dx = \left[-\cos x \right]_{0}^{\pi} = 2,$$

$$M_y = \int_{0}^{\pi} \int_{0}^{\sin x} x\, dy\, dx = \int_{0}^{\pi} x \sin x\, dx = \left[\sin x - x \cos x \right]_{0}^{\pi} = \pi,$$

$$M_x = \int_{0}^{\pi} \int_{0}^{\sin x} y\, dy\, dx = \int_{0}^{\pi} \frac{1}{2}\sin^2 x\, dx = \left[\frac{1}{8}(2x - \sin 2x) \right]_{0}^{\pi} = \frac{1}{4}\pi.$$

Therefore the centroid of the lamina is located at

$$(\overline{x}, \overline{y}) = \left(\frac{\pi}{2}, \frac{\pi}{8}\right) \approx (1.5707963267948966, 0.3926990816987242).$$

C14S05.021: The mass and moments of the lamina are

$$m = \int_{0}^{a} \int_{0}^{a} (x + y)\, dy\, dx = \int_{0}^{a} \left(\frac{1}{2}a^2 + ax \right) dx = \left[\frac{1}{2}(a^2 x + ax^2) \right]_{0}^{a} = a^3,$$

$$M_y = \int_{0}^{a} \int_{0}^{a} x(x + y)\, dy\, dx = \int_{0}^{a} \left(\frac{1}{2}a^2 x + ax^2 \right) dx = \left[\frac{1}{4}a^2 x^2 + \frac{1}{3}ax^3 \right]_{0}^{a} = \frac{7}{12}a^4, \quad \text{and}$$

$$M_x = \int_{0}^{a} \int_{0}^{a} (x + y)y\, dy\, dx = \int_{0}^{a} \left(\frac{1}{3}a^3 + \frac{1}{2}a^2 x \right) dx = \left[\frac{1}{3}a^3 x + \frac{1}{4}a^2 x^2 \right]_{0}^{a} = \frac{7}{12}a^4.$$

Therefore its centroid is located at the point $(\overline{x}, \overline{y}) = \left(\frac{7}{12}a, \frac{7}{12}a\right)$.

C14S05.023: The mass and moments of the lamina are

$$m = \int_{-2}^{2} \int_{x^2}^{4} y \, dy \, dx = \int_{-2}^{2} \left(8 - \frac{1}{2} x^4 \right) dx = \left[8x - \frac{1}{10} x^5 \right]_{-2}^{2} = \frac{128}{5},$$

$$M_y = \int_{-2}^{2} \int_{x^2}^{4} xy \, dy \, dx = \int_{-2}^{2} \left(8x - \frac{1}{2} x^5 \right) dx = \left[4x^2 - \frac{1}{12} x^6 \right]_{-2}^{2} = 0, \quad \text{and}$$

$$M_x = \int_{-2}^{2} \int_{x^2}^{4} y^2 \, dy \, dx = \int_{-2}^{2} \left(\frac{64}{3} - \frac{1}{3} x^6 \right) dx = \left[\frac{64}{3} x - \frac{1}{21} x^7 \right]_{-2}^{2} = \frac{512}{7}.$$

Thus the centroid of the lamina is located at the point $(\bar{x}, \bar{y}) = \left(0, \frac{20}{7} \right)$.

C14S05.025: The mass and moments are

$$m = \int_{0}^{\pi} \int_{0}^{\sin x} x \, dy \, dx = \int_{0}^{\pi} x \sin x \, dx = \left[\sin x - x \cos x \right]_{0}^{\pi} = \pi,$$

$$M_y = \int_{0}^{\pi} \int_{0}^{\sin x} x^2 \, dy \, dx = \int_{0}^{\pi} x^2 \sin x \, dx = \left[2 \cos x - x^2 \cos x + 2x \sin x \right]_{0}^{\pi} = \pi^2 - 4, \quad \text{and}$$

$$M_x = \int_{0}^{\pi} \int_{0}^{\sin x} xy \, dy \, dx = \int_{0}^{\pi} \frac{1}{2} x \sin^2 x \, dx = \left[\frac{1}{16} \left(2x^2 - \cos 2x - 2x \sin 2x \right) \right]_{0}^{\pi} = \frac{1}{8} \pi^2.$$

Consequently the centroid of the lamina is at

$$(\bar{x}, \bar{y}) = \left(\frac{\pi^2 - 4}{\pi}, \frac{\pi}{8} \right) \approx (1.8683531088546306, \ 0.3926990816987242).$$

C14S05.027: The mass and moments are

$$m = \int_{\theta=0}^{\pi} \int_{r=0}^{a} r^2 \, dr \, d\theta = \pi \cdot \left[\frac{1}{3} r^3 \right]_{0}^{a} = \frac{1}{3} \pi a^3,$$

$$M_y = \int_{\theta=0}^{\pi} \int_{r=0}^{a} r^3 \cos \theta \, dr \, d\theta = \int_{0}^{\pi} \frac{1}{4} a^4 \cos \theta \, d\theta = \left[\frac{1}{4} a^4 \sin \theta \right]_{0}^{\pi} = 0, \quad \text{and}$$

$$M_x = \int_{\theta=0}^{\pi} \int_{r=0}^{a} r^3 \sin \theta \, dr \, d\theta = \int_{0}^{\pi} \frac{1}{4} a^4 \sin \theta \, d\theta = \left[-\frac{1}{4} a^4 \cos \theta \right]_{0}^{\pi} = \frac{1}{2} a^4.$$

Therefore the centroid of the lamina is located at the point

$$(\bar{x}, \bar{y}) = \left(0, \frac{3a}{2\pi} \right) \approx (0, \ (0.47746483) \cdot a).$$

C14S05.029: The following figure, generated by *Mathematica 3.0*, shows the two circles.

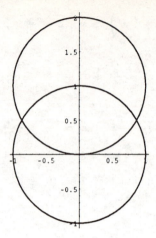

They cross where $2\sin\theta = 1$; that is, where $\theta = \frac{1}{6}\pi$ and where $\theta = \frac{5}{6}\pi$. The lamina in question is the region outside the lower circle and inside the upper circle. Its mass and moments are

$$m = \int_{\pi/6}^{5\pi/6} \int_1^{2\sin\theta} r^2 \sin\theta \, dr \, d\theta = \int_{\pi/6}^{5\pi/6} \left(\frac{8}{3}\sin^4\theta - \frac{1}{3}\sin\theta \right) d\theta$$

$$= \frac{1}{12} \left[12\theta + 4\cos\theta - 8\sin 2\theta + \sin 4\theta \right]_{\pi/6}^{5\pi/6} = \frac{8\pi + 3\sqrt{3}}{12},$$

$$M_y = \int_{\pi/6}^{5\pi/6} \int_1^{2\sin\theta} r^3 \sin\theta \cos\theta \, dr \, d\theta = \int_{\pi/6}^{5\pi/6} \left(4\sin^5\theta\cos\theta - \frac{1}{4}\sin\theta\cos\theta \right) d\theta$$

$$= \frac{1}{48} \left[6\cos 4\theta - \cos 6\theta - 12\cos 2\theta \right]_{\pi/6}^{5\pi/6} = 0,$$

$$M_x = \int_{\pi/6}^{5\pi/6} \int_1^{2\sin\theta} r^3 \sin^2\theta \, dr \, d\theta = \int_{\pi/6}^{5\pi/6} \left(4\sin^6\theta - \frac{1}{4}\sin^2\theta \right) d\theta$$

$$= \frac{1}{48} \left[54\theta - 42\sin 2\theta + 9\sin 4\theta - \sin 6\theta \right]_{\pi/6}^{5\pi/6} = \frac{12\pi + 11\sqrt{3}}{16}.$$

Thus its centroid is at the point

$$(\overline{x}, \overline{y}) = \left(0, \ \frac{36\pi + 33\sqrt{3}}{32\pi + 12\sqrt{3}} \right) \approx (0, \ 1.4034060567438982)$$

and its mass is approximately 2.5274078042854148.

C14S05.031: The polar moment of inertia of the lamina is

$$I_0 = \int_0^{2\pi} \int_0^a r^{n+3} \, dr \, d\theta = 2\pi \cdot \left[\frac{r^{n+4}}{n+4} \right]_0^a = \frac{2\pi a^{n+4}}{n+4}.$$

C14S05.033: The polar moment of inertia of the lamina is

$$I_0 = \int_{-\pi/2}^{\pi/2} \int_0^{2\cos\theta} kr^3 \, dr \, d\theta = \int_{-\pi/2}^{\pi/2} 4k \cos^4\theta \, d\theta = \frac{1}{8}k \left[12\theta + 8\sin 2\theta + \sin 4\theta \right]_{-\pi/2}^{\pi/2} = \frac{3}{2}\pi k.$$

C14S05.035: The polar moment of inertia of the lamina is

$$I_0 = \int_{-\pi/4}^{\pi/4} \int_0^{\sqrt{\cos 2\theta}} r^5 \, dr \, d\theta = \int_{-\pi/4}^{\pi/4} \frac{1}{6} \cos^3 2\theta \, d\theta = \frac{1}{144} \left[9\sin 2\theta + \sin 6\theta \right]_{-\pi/4}^{\pi/4} = \frac{1}{9}.$$

C14S05.037: In Problem 23 we found that the mass of the lamina is $m = \frac{128}{5}$. Next,

$$I_y = \int_{-2}^2 \int_{x^2}^4 x^2 y \, dy \, dx = \int_{-2}^2 \left(8x^2 - \frac{1}{2}x^6 \right) dx = \left[\frac{8}{3}x^3 - \frac{1}{14}x^7 \right]_{-2}^2 = \frac{512}{21} \quad \text{and}$$

$$I_x = \int_{-2}^2 \int_{x^2}^4 y^3 \, dy \, dx = \int_{-2}^2 \left(64 - \frac{1}{4}x^8 \right) dx = \left[64x - \frac{1}{36}x^9 \right]_{-2}^2 = \frac{2048}{9}.$$

Therefore $\hat{x} = \frac{2}{21}\sqrt{105}$ and $\hat{y} = \frac{4}{3}\sqrt{5}$.

C14S05.039: In the solution of Problem 27 we found that the mass of the lamina is $m = \frac{1}{3}\pi a^3$. Next,

$$I_y = \int_0^\pi \int_0^a r^4 \cos^2 t \, dr \, d\theta = \int_0^\pi \frac{1}{5} a^5 \cos^2 \theta \, d\theta = \left[\frac{1}{20} a^5 (2\theta + \sin 2\theta) \right]_0^\pi = \frac{1}{10}\pi a^5 \quad \text{and}$$

$$I_x = \int_0^\pi \int_0^a r^4 \sin^2 \theta \, dr \, d\theta = \int_0^\pi \frac{1}{5} a^5 \sin^2 \theta \, d\theta = \left[\frac{1}{20} a^5 (2\theta - \sin 2\theta) \right]_0^\pi = \frac{1}{10}\pi a^5.$$

Therefore $\hat{x} = \hat{y} = \frac{1}{10} a\sqrt{30}$.

C14S05.041: The quarter of the circular disk $x^2 + y^2 \leqq a^2$ that lies in the first quadrant has the polar-coordinates description $0 \leqq r \leqq a$, $0 \leqq \theta \leqq \frac{1}{2}\pi$. If we assume that it has uniform density $\delta = 1$, then its mass is $m = \frac{1}{4}\pi a^2$. Next,

$$M_y = \int_0^{\pi/2} \int_0^a r^2 \cos\theta \, dr \, d\theta = \int_0^{\pi/2} \frac{1}{3} a^3 \cos\theta \, d\theta = \left[\frac{1}{3} a^2 \sin\theta \right]_0^{\pi/2} = \frac{1}{3} a^3,$$

and $M_x = M_y$ by symmetry. Therefore the centroid is located at the point

$$(\overline{x}, \overline{y}) = \left(\frac{4a}{3\pi}, \frac{4a}{3\pi} \right) \approx ([0.4244131815783876]a, \, [0.4244131815783876]a).$$

C14S05.043: The arc $x^2 + y^2 = r^2$ can be parametrized in this way:

$$x(t) = r\cos\theta, \quad y(t) = r\sin\theta, \quad 0 \leqq \theta \leqq \frac{\pi}{2}.$$

We may also assume that it has unit density. The arc length element is

$$ds = \sqrt{[x'(\theta)]^2 + [y'(\theta)]^2} \; d\theta = r \, d\theta,$$

and hence its mass is

$$m = \int_0^{\pi/2} r \, d\theta = \frac{1}{2}\pi r.$$

Its moment around the y-axis is

$$M_y = \int_0^{\pi/2} r^2 \cos\theta \, d\theta = \left[r^2 \sin\theta \right]_0^{\pi/2} = r^2.$$

Therefore, because $\bar{y} = \bar{x}$ by symmetry,

$$(\bar{x}, \bar{y}) = \left(\frac{2r}{\pi}, \frac{2r}{\pi} \right).$$

C14S05.045: We may assume that the triangle has unit density. The hypotenuse of this right triangle has equation

$$y = f(x) = h\left(1 - \frac{x}{r}\right),$$

and hence the mass and moments of the triangle are

$$m = \int_0^r \int_0^{f(x)} 1 \, dy \, dx = \int_0^r h\left(1 - \frac{x}{r}\right) dx = \left[hx - \frac{h}{2r}x^2 \right]_0^r = \frac{1}{2}hr,$$

$$M_y = \int_0^r \int_0^{f(x)} x \, dy \, dx = \int_0^r hx\left(1 - \frac{x}{r}\right) dx = \left[\frac{h}{2}x^2 - \frac{h}{3r}x^3 \right]_0^r = \frac{1}{6}hr^2, \quad \text{and}$$

$$M_x = \int_0^r \int_0^{f(x)} y \, dy \, dx = \int_0^r \left(\frac{1}{2}h^2 - \frac{h^2}{r}x + \frac{h^2}{2r^2}x^2 \right) dx$$

$$= \left[\frac{h^2}{2}x - \frac{h^2}{2r}x^2 + \frac{h^2}{6r^2}x^3 \right]_0^r = \frac{1}{6}h^2 r.$$

Therefore the centroid of the triangle is

$$C = (\bar{x}, \bar{y}) = \left(\frac{1}{3}r, \frac{1}{3}h \right).$$

The midpoint of the hypotenuse is the point $M\left(\frac{1}{2}r, \frac{1}{2}h\right)$ and the line from the origin to M has equation $y = hx/r$, so it is clear that C lies on this line and is two-thirds of the way from the origin to M.

C14S05.047: Here we simply observe that $2\pi\dfrac{r}{2} \cdot L = \pi r L.$

C14S05.049: The diagonal side of the trapezoid has centroid at its midpoint, so that $\bar{x} = \frac{1}{2}(r_1 + r_2)$. By the second theorem of Pappus, the curved surface area of the conical frustum—which is generated by the diagonal side—is

$$A = 2\pi\bar{x} \cdot L = \pi L(r_1 + r_2).$$

C14S05.051: First we compute the mass and moments of the rectangle (we assume that it has unit density):

822

$$m = \int_{-a}^{a} \int_{0}^{b} 1 \; dy \; dx = \int_{-a}^{a} b \; dx = 2ab;$$

$$M_y = \int_{-a}^{a} \int_{0}^{b} x \; dy \; dx = \int_{-a}^{a} bx \; dx = \left[\frac{1}{2} bx^2 \right]_{-a}^{a} = 0;$$

$$M_x = \int_{-a}^{a} \int_{0}^{b} y \; dy \; dx = \int_{-a}^{a} \frac{1}{2} b^2 \; dx = ab^2.$$

Next we compute the mass and moments of the semicircle. Note that its curved boundary has equation $y = b + \sqrt{a^2 - x^2} = f(x)$. Because we have assumed unit density, its mass is $m = \frac{1}{2} \pi a^2$. By symmetry, $M_y = 0$. And

$$M_x = \int_{-a}^{a} \int_{b}^{f(x)} y \; dy \; dx = \int_{-a}^{a} \left[\frac{1}{2} \left(b + \sqrt{a^2 - x^2} \right)^2 - \frac{1}{2} b^2 \right] dx$$

$$= \int_{-a}^{a} \left[b(a^2 - x^2)^{1/2} + \frac{1}{2}(a^2 - x^2) \right] dx = b \int_{-a}^{a} (a^2 - x^2)^{1/2} \; dx + \left[\frac{1}{2} a^2 x - \frac{1}{6} x^3 \right]_{-a}^{a}$$

$$= b \cdot \frac{\pi a^2}{2} + a^3 - \frac{1}{3} a^3 = \frac{3\pi a^2 b + 4a^3}{6}.$$

Moments, like masses, are additive. Thus we find the mass and moments for the entire lamina by adding those of the rectangle and semicircle:

$$m = \frac{4ab + \pi a^2}{2}, \quad M_y = 0, \quad \text{and} \quad M_x = \frac{6ab^2 + 3\pi a^2 b + 4a^3}{6}.$$

Therefore $\bar{x} = 0$ and

$$\bar{y} = \frac{6ab^2 + 3\pi a^2 b + 4a^3}{6} \cdot \frac{2}{4ab + \pi a^2} = \frac{6b^2 + 3\pi ab + 4a^2}{12b + 3\pi a}.$$

Finally, we apply the first theorem of Pappus to find the volume of the solid generated by rotation of the lamina around the x-axis:

$$V = 2\pi \bar{y} \cdot \frac{4ab + \pi a^2}{2} = \pi \cdot \frac{6b^2 + 3\pi ab + 4a^2}{12b + 3\pi a} \cdot (4ab + \pi a^2) = \frac{\pi a}{3} \cdot (6b^2 + 3\pi ab + 4a^2).$$

To check the answer, note that it has the correct domensions of volume and that in the extreme case $b = 0$, it becomes $\frac{4}{3} \pi a^3$, the correct formula for the volume of a sphere of radius a.

C14S05.053: The plate is a rectangle of area ab, area density δ, and mass m, and hence $m = ab\delta$. The moment of inertia with respect to the y-axis is

$$I_y = \int_{-a/2}^{a/2} \int_{-b/2}^{b/2} \delta x^2 \; dy \; dx = \int_{-a/2}^{a/2} \delta b x^2 \; dx = \left[\frac{1}{3} \delta b x^3 \right]_{-a/2}^{a/2} = \frac{1}{12} \delta a^3 b = \frac{1}{12} m a^2.$$

By symmetry or by a very similar computation $I_x = \frac{1}{12} mb^2$. So, by the comment following Eq. (6) of the text,

$$I_0 = I_x + I_y = \frac{1}{12} m(a^2 + b^2).$$

823

C14S05.055: Suppose that the plane lamina L is the union of the two nonoverlapping laminae R and S. Let I_L, I_R, and I_S denote the polar moments of inertia of L, R, and S, respectively. Let $\delta(x, y)$ denote the density of L at the point (x, y). Then

$$I_L = \iint_L (x^2 + y^2)\delta(x, y)\, dA = \iint_R (x^2 + y^2)\delta(x, y)\, dA + \iint_S (x^2 + y^2)\delta(x, y)\, dA = I_R + I_S.$$

Next, let I_1 denote the polar moment of inertia of the lower rectangle in Fig. 14.5.25 and let I_2 denote the polar moment of inertia of the upper rectangle. Then

$$I_1 = \int_{-1}^{1}\int_{0}^{3}(x^2 + y^2)\cdot k\, dy\, dx = k\int_{-1}^{1}(3x^2 + 9)\, dx = k\left[x^3 + 9x\right]_{-1}^{1} = 20k;$$

$$I_2 = \int_{-4}^{4}\int_{3}^{4}(x^2 + y^2)\cdot k\, dy\, dx = k\int_{-4}^{4}\left[x^2 y + \frac{1}{3}y^3\right]_{3}^{4}dx = k\int_{-4}^{4}\left(x^2 + \frac{37}{3}\right)dx$$

$$= k\left[\frac{1}{3}x^3 + \frac{37}{3}x\right]_{-4}^{4} = k\cdot\left(\frac{128}{3} + \frac{296}{3}\right) = \frac{424}{3}k.$$

Thus, by the first result in this solution, the polar moment of inertia of the T-shaped lamina is

$$I_0 = I_1 + I_2 = \frac{484}{3}k.$$

C14S05.057: The mass and moments are

$$m = \int_{0}^{\pi}\int_{0}^{2\sin\theta} r^2 \sin\theta\, dr\, d\theta = \int_{0}^{\pi}\frac{8}{3}\sin^4\theta\, d\theta = \frac{1}{12}\left[12\theta - 8\sin 2\theta + \sin 4\theta\right]_{0}^{\pi} = \pi;$$

$$M_y = \int_{0}^{\pi}\int_{0}^{2\sin\theta} r^3 \sin\theta\cos\theta\, dr\, d\theta = \int_{0}^{\pi}4\sin^5\theta\cos\theta\, d\theta = \left[\frac{2}{3}\sin^6\theta\right]_{0}^{\pi} = 0;$$

$$M_x = \int_{0}^{\pi}\int_{0}^{2\sin\theta} r^3 \sin^2\theta\, dr\, d\theta = \int_{0}^{\pi}4\sin^6\theta\, d\theta = \frac{1}{48}\left[60\theta - 45\sin 2\theta + 9\sin 4\theta - \sin 6\theta\right]_{0}^{\pi} = \frac{5}{4}\pi.$$

Therefore the centroid is at $(\overline{x},\, \overline{y}) = \left(0,\, \dfrac{5}{4}\right)$.

C14S05.059: The mass and moments are

$$m = \int_{0}^{\pi/2}\int_{0}^{2\cos\theta} r^2 \cos\theta\, dr\, d\theta = \int_{0}^{\pi/2}\frac{8}{3}\cos^4\theta\, d\theta = \frac{1}{12}\left[12\theta + 8\sin 2\theta + \sin 4\theta\right]_{0}^{\pi/2} = \frac{1}{2}\pi;$$

$$M_y = \int_{0}^{\pi/2}\int_{0}^{2\cos\theta} r^3 \cos^2\theta\, dr\, d\theta = \int_{0}^{\pi/2}4\cos^6\theta\, d\theta = \frac{1}{48}\left[60\theta + 45\sin 2\theta + 9\sin 4\theta + \sin 6\theta\right]_{0}^{\pi/2} = \frac{5}{8}\pi;$$

$$M_x = \int_{0}^{\pi/2}\int_{0}^{2\cos\theta} r^3 \sin\theta\cos\theta\, dr\, d\theta = \int_{0}^{\pi/2}4\sin\theta\cos^5\theta\, d\theta = \left[-\frac{2}{3}\cos^6\theta\right]_{0}^{\pi/2} = \frac{2}{3}.$$

Therefore the centroid is located at $(\overline{x},\, \overline{y}) = \left(\dfrac{5}{4},\, \dfrac{4}{3\pi}\right)$.

Section 14.6

Note: For some triple integrals, there are six possible orders of integration of the corresponding iterated integrals; for many triple integrals, there are two. To save space, in this section we do not show all possibilities, but only the one that seems most natural.

C14S06.001: The value of the triple integral is

$$I = \int_{z=0}^{1} \int_{y=0}^{3} \int_{x=0}^{2} (x+y+z)\,dx\,dy\,dz = \int_{z=0}^{1} \int_{y=0}^{3} \left[\frac{1}{2}x^2 + xy + xz\right]_{x=0}^{2} dy\,dz$$

$$= \int_{z=0}^{1} \int_{y=0}^{3} (2 + 2y + 2z)\,dy\,dz = \int_{z=0}^{1} \left[2y + y^2 + 2yz\right]_{y=0}^{3} dz = \int_0^1 (15 + 6z)\,dz = \left[15z + 3z^2\right]_0^1 = 18.$$

C14S06.003: The value of the triple integral is

$$K = \int_{z=-2}^{6} \int_{y=0}^{2} \int_{x=-1}^{3} xyz\,dx\,dy\,dz = \int_{z=-2}^{6} \int_{y=0}^{2} \left[\frac{1}{2}x^2yz\right]_{x=-1}^{3} dy\,dz = \int_{z=-2}^{6} \int_{y=0}^{2} 4yz\,dy\,dz$$

$$= \int_{z=-2}^{6} \left[2y^2 z\right]_{y=0}^{2} dz = \int_{z=-2}^{6} 8z\,dz = \left[4z^2\right]_{z=-2}^{6} = 128.$$

C14S06.005: The value of the triple integral is

$$J = \int_{x=0}^{1} \int_{y=0}^{1-x} \int_{z=0}^{1-x-y} x^2\,dz\,dy\,dx = \int_{x=0}^{1} \int_{y=0}^{1-x} \left[x^2 z\right]_{z=0}^{1-x-y} dy\,dx = \int_{x=0}^{1} \int_{y=0}^{1-x} (x^2 - x^3 - x^2 y)\,dy\,dx$$

$$= \int_{x=0}^{1} \left[x^2 y - x^3 y - \frac{1}{2}x^2 y^2\right]_{y=0}^{1-x} dx = \int_0^1 \left(\frac{1}{2}x^2 - x^3 + \frac{1}{2}x^4\right) dx = \left[\frac{1}{6}x^3 - \frac{1}{4}x^4 + \frac{1}{10}x^5\right]_0^1 = \frac{1}{60}.$$

C14S06.007: The value of the triple integral is

$$I = \int_{x=-1}^{0} \int_{y=0}^{2} \int_{z=0}^{1-x^2} xyz\,dz\,dy\,dx = \int_{x=-1}^{0} \int_{y=0}^{2} \left[\frac{1}{2}xyz^2\right]_{z=0}^{1-x^2} dy\,dx$$

$$= \int_{x=-1}^{0} \int_{y=0}^{2} \left(\frac{1}{2}xy - x^3 y + \frac{1}{2}x^5 y\right) dy\,dx = \int_{x=-1}^{0} \left[\frac{1}{4}xy^2 - \frac{1}{2}x^3 y^2 + \frac{1}{4}x^5 y^2\right]_{y=0}^{2} dx$$

$$= \int_{x=-1}^{0} (x - 2x^3 + x^5)\,dx = \frac{1}{6}\left[3x^2 - 3x^4 + x^6\right]_{x=-1}^{0} = -\frac{1}{6}.$$

C14S06.009: The value of this triple integral is

$$K = \int_{-1}^{1} \int_0^3 \int_{x^2}^{2-x^2} (x+y)\,dz\,dy\,dx = \int_{-1}^{1} \int_0^3 \left[xy + xz\right]_{x^2}^{2-x^2} dy\,dx$$

$$= \int_{-1}^{1} \int_0^3 (2x - 2x^3 + 2y - 2x^2 y)\,dy\,dx = \int_{-1}^{1} \left[2xy - 2x^3 y + y^2 - x^2 y^2\right]_0^3 dx$$

$$= \int_{-1}^{1} (9 + 6x - 9x^2 - 6x^3)\,dx = \left[9x + 3x^2 - 3x^3 - \frac{3}{2}x^4\right]_{-1}^{1} = 12.$$

C14S06.011: The solid resembles the tetrahedron shown in Fig. 14.6.4 of the text. We don't have enough memory to transfer a *Mathematica*-generated graphic from Adobe Illustrator to this document, but you can see it if you execute the *Mathematica* 3.0 command

```
ParametricPlot3D[ {{x, 0, 6 - 2*x - 3*y}, {x, y, 6 - 2*x - 3*y}},
    {x, 0, 3}, {y, 0, 2}, PlotRange → {0, 3}, {0, 2}, {0, 6}},
    AspectRatio → 1.0, ViewPoint → {3.5, -1.6, 3.0},
    LightSources → {{{1., 0., 1.}, RGBColor[1., 0., 0.]},
    {{1., 1., 1.}, RGBColor[0., 1., 0.]}, {{0., 1., 1.}, RGBColor[1., 0., 1.]}} ];
```

If you change the `Viewpoint` parameters, you will see that in an effort to conserve memory, we plotted only the diagonal face of the tetrahedron and the face that lies in the xz-plane. The volume of this tetrahedron is given by

$$V = \int_{x=0}^{3} \int_{y=0}^{(6-2x)/3} \int_{z=0}^{6-2x-3y} 1 \, dz \, dy \, dx = \int_{x=0}^{3} \int_{y=0}^{(6-2x)/3} (6 - 2x - 3y) \, dy \, dx$$

$$= \int_{x=0}^{3} \left[6y - 2xy - \frac{3}{2} y^2 \right]_{y=0}^{(6-2x)/3} dx = \int_{x=0}^{3} \left(6 - 4x + \frac{2}{3} x^2 \right) dx = \left[6x - 2x^2 + \frac{2}{9} x^3 \right]_{0}^{3} = 6.$$

C14S06.013: The view of the figure next, on the left, was generated with the *Mathematica* 3.0 command

```
ParametricPlot3D[ {{u, 0, v}, {u, 4 - u*u, v}, {u, v, 4 - v}},
    {u, -2, 2}, {v, 0, 4}, AspectRatio → Automatic,
    PlotRange → {{-2.2, 2.2}, {-0.2, 4.2}, { -0.2, 4.2}}, ViewPoint → {4, 4, 5} ];
```

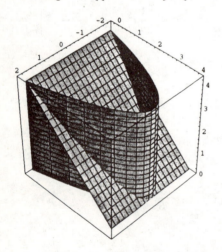 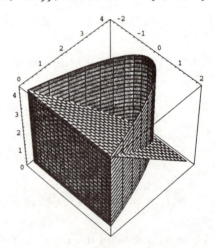

Change the `ViewPoint` command parameters to {4, -4, 7} to see the figure from another angle (this view is shown above, on the right). The volume of the solid is

$$V = \int_{x=-2}^{2} \int_{y=0}^{4-x^2} \int_{z=0}^{4-y} 1 \, dz \, dy \, dx = \int_{x=-2}^{2} \int_{y=0}^{4-x^2} (4 - y) \, dy \, dx = \int_{x=-2}^{2} \left[4y - y^2 \right]_{y=0}^{4-x^2} dx$$

$$= \int_{-2}^{2} \left(8 - \frac{1}{2} x^4 \right) dx = \left[8x - \frac{1}{10} x^5 \right]_{-2}^{2} = \frac{128}{5} = 25.6.$$

C14S06.015: It's not easy to get a clear three-dimensional plot of this solid. We got fair results with the *Mathematica 3.0* command

```
ParametricPlot3D[ {{u, u*u, v}, {u*u, u, v}, {u*Sqrt[10], v, 10 - u*u - v*v}},
    {u, 0, 1}, {v, 0, 10}, AspectRatio → 0.8,
    PlotRange → {{0, Sqrt[10]}, {0, Sqrt[10]}, {0, 10}}, ViewPoint → {6, -4, 12} ];
```

If you have the time and computer memory, experiment with changing the `AspectRatio` and `ViewPoint` parameters until you get a clear view of the solid. Its volume is

$$V = \int_{x=0}^{1} \int_{y=x^2}^{\sqrt{x}} \int_{z=0}^{10-x^2-y^2} 1 \, dz \, dy \, dx = \int_{x=0}^{1} \int_{y=x^2}^{\sqrt{x}} (10 - x^2 - y^2) \, dy \, dx$$

$$= \int_{x=0}^{1} \left[\frac{1}{3}(30y - 3x^2 y - y^3) \right]_{y=x^2}^{y=\sqrt{x}} dx = \int_{0}^{1} \left(10x^{1/2} - \frac{1}{3}x^{3/2} - 10x^2 - x^{5/2} + x^4 + \frac{1}{3}x^6 \right) dx$$

$$= \left[\frac{20}{3}x^{3/2} - \frac{2}{15}x^{5/2} - \frac{10}{3}x^3 - \frac{2}{7}x^{7/2} + \frac{1}{5}x^5 + \frac{1}{21}x^7 \right]_0^1 = \frac{332}{105} \approx 3.1619047619047619.$$

C14S06.017: The volume is

$$V = \int_{x=-2}^{2} \int_{z=x^2}^{4} \int_{y=0}^{4-z} 1 \, dy \, dz \, dx = \int_{x=-2}^{2} \int_{z=x^2}^{4} (4 - z) \, dz \, dx = \int_{x=-2}^{2} \left[4z - \frac{1}{2}z^2 \right]_{z=x^2}^{4} dx$$

$$= \int_{x=-2}^{2} \left(8 - 4x^2 + \frac{1}{2}x^4 \right) dx = \left[8x - \frac{4}{3}x^3 + \frac{1}{10}x^5 \right]_{-2}^{2} = \frac{256}{15} \approx 17.0666666666666667.$$

C14S06.019: The volume is

$$V = \int_{y=0}^{1} \int_{z=y^2}^{\sqrt{y}} \int_{x=0}^{2-y-z} 1 \, dx \, dz \, dy = \int_{y=0}^{1} \int_{z=y^2}^{\sqrt{y}} (2 - y - z) \, dz \, dy = \int_{y=0}^{1} \left[2z - yz - \frac{1}{2}z^2 \right]_{z=y^2}^{\sqrt{y}} dy$$

$$= \int_{y=0}^{1} \frac{1}{2} (4y^{1/2} - y - 2y^{3/2} - 4y^2 + 2y^3 + y^4) \, dy$$

$$= \frac{1}{60} \left[80y^{3/2} - 15y^2 - 24y^{5/2} - 40y^3 + 15y^4 + 6y^5 \right]_0^1 = \frac{11}{30} \approx 0.3666666666666667.$$

C14S06.021: Because the solid has constant density $\delta = 1$, its mass is

$$m = \int_{x=-2}^{2} \int_{y=x^2}^{4} \int_{z=0}^{y} 1 \, dz \, dy \, dx = \int_{x=-2}^{2} \int_{y=x^2}^{4} y \, dy \, dx = \int_{x=-2}^{2} \left[\frac{1}{2}y^2 \right]_{y=x^2}^{4} dx$$

$$= \int_{x=-2}^{2} \left(8 - \frac{1}{2}x^4 \right) dx = \left[8x - \frac{1}{10}x^5 \right]_{-2}^{2} = \frac{128}{5}.$$

Its moments are

$$M_{yz} = \int_{x=-2}^{2} \int_{y=x^2}^{4} \int_{z=0}^{y} x \, dz \, dy \, dx = \int_{x=-2}^{2} \int_{y=x^2}^{4} xy \, dy \, dx = \int_{x=-2}^{2} \left[\frac{1}{2}xy^2 \right]_{y=x^2}^{4} dx$$

$$= \int_{x=-2}^{2} \left(8x - \frac{1}{2}x^5\right) dx = \left[4x^2 - \frac{1}{12}x^6\right]_{-2}^{2} = 0,$$

$$M_{xz} = \int_{x=-2}^{2} \int_{y=x^2}^{4} \int_{z=0}^{y} y\, dz\, dy\, dz = \int_{x=-2}^{2} \int_{y=x^2}^{4} \frac{1}{2}y^2\, dy\, dx = \int_{x=-2}^{2} \left[\frac{1}{3}y^3\right]_{y=x^2}^{4} dx$$

$$= \int_{x=-2}^{2} \left(\frac{64}{3} - \frac{1}{3}x^6\right) dx = \left[\frac{64}{3}x - \frac{1}{21}x^7\right]_{-2}^{2} = \frac{512}{7},$$

$$M_{xy} = \int_{x=-2}^{2} \int_{y=x^2}^{4} \int_{z=0}^{y} z\, dz\, dy\, dx = \int_{x=-2}^{2} \int_{y=x^2}^{4} \left[\frac{1}{2}z^2\right]_{z=0}^{y} dy\, dx$$

$$= \frac{1}{2}\int_{x=-2}^{2} \int_{y=x^2}^{4} y^2\, dy\, dx = \frac{1}{2}\int_{x=-2}^{2} \left[\frac{1}{3}y^3\right]_{y=x^2}^{4} dx = \int_{x=-2}^{2} \left(\frac{32}{3} - \frac{1}{6}x^6\right) dx$$

$$= \left[\frac{32}{3}x - \frac{1}{42}x^7\right]_{-2}^{2} = \frac{256}{7}.$$

Therefore the centroid of the solid is located at the point $(\bar{x},\, \bar{y},\, \bar{z}) = \left(0,\, \dfrac{20}{7},\, \dfrac{10}{7}\right)$.

C14S06.023: Because the solid has unit density $\delta = 1$, its mass and moments are

$$m = \int_{x=-2}^{2} \int_{z=x^2}^{4} \int_{y=0}^{4-z} 1\, dy\, dz\, dx = \int_{x=-2}^{2} \int_{z=x^2}^{4} (4-z)\, dz\, dx = \int_{x=-2}^{2} \left[4z - \frac{1}{2}z^2\right]_{z=x^2}^{4} dx$$

$$= \int_{-2}^{2} \left(8 - 4x^2 + \frac{1}{2}x^4\right) dx = \left[8x - \frac{4}{3}x^3 + \frac{1}{10}x^5\right]_{-2}^{2} = \frac{256}{15};$$

$$M_{yz} = \int_{x=-2}^{2} \int_{z=x^2}^{4} \int_{y=0}^{4-z} x\, dy\, dz\, dx = \int_{x=-2}^{2} \int_{z=x^2}^{4} (4x - xz)\, dz\, dx = \int_{x=-2}^{2} \left[4xz - \frac{1}{2}xz^2\right]_{z=x^2}^{4} dx$$

$$= \int_{-2}^{2} \left(8x - 4x^3 + \frac{1}{2}x^5\right) dx = \left[4x^2 - x^4 + \frac{1}{12}x^6\right]_{-2}^{2} = 0;$$

$$M_{xz} = \int_{x=-2}^{2} \int_{z=x^2}^{4} \int_{y=0}^{4-z} y\, dy\, dz\, dx = \int_{x=-2}^{2} \int_{z=x^2}^{4} \frac{1}{2}(4-z)^2\, dz\, dx = \int_{x=-2}^{2} \left[8z - 2z^2 + \frac{1}{6}z^4\right]_{z=x^2}^{4} dx$$

$$= \int_{-2}^{2} \left(\frac{32}{3} - 8x^2 + 2x^4 - \frac{1}{6}x^6\right) dx = \left[\frac{32}{3}x - \frac{8}{3}x^3 + \frac{2}{5}x^5 - \frac{1}{42}x^7\right]_{-2}^{2} = \frac{2048}{105};$$

$$M_{xy} = \int_{x=-2}^{2} \int_{z=x^2}^{4} \int_{y=0}^{4-z} z\, dy\, dz\, dx = \int_{x=-2}^{2} \int_{z=x^2}^{4} (4z - z^2)\, dz\, dx = \int_{x=-2}^{2} \left[2z^2 - \frac{1}{3}z^3\right]_{z=x^2}^{4} dx$$

$$= \int_{-2}^{2} \left(\frac{32}{3} - 2x^4 + \frac{1}{3}x^6\right) dx = \left[\frac{32}{3}x - \frac{2}{5}x^5 + \frac{1}{21}x^7\right]_{-2}^{2} = \frac{1024}{35}.$$

Therefore the centroid of the solid is located at the point $(\bar{x},\, \bar{y},\, \bar{z}) = \left(0,\, \dfrac{8}{7},\, \dfrac{12}{7}\right)$.

C14S06.025: Because the solid has unit density $\delta = 1$, its mass and moments are given by

$$m = \int_{x=-\pi/2}^{\pi/2} \int_{z=0}^{\cos x} \int_{y=0}^{1-z} 1 \, dy \, dz \, dx = \int_{x=-\pi/2}^{\pi/2} \left[z - \frac{1}{2}z^2 \right]_{z=0}^{\cos x} dx$$

$$= \int_{-\pi/2}^{\pi/2} \left(\cos x - \frac{1}{2}\cos^2 x \right) dx = \int_0^{\pi/2} (2\cos x - \cos^2 x) \, dx = \frac{1}{4}\left[-2x + 8\sin x - \sin 2x \right]_0^{\pi/2} = \frac{8 - \pi}{4};$$

$M_{yz} = 0$ (by symmetry);

$$M_{xz} = \int_{x=-\pi/2}^{\pi/2} \int_{z=0}^{\cos x} \int_{y=0}^{1-z} y \, dy \, dz \, dx = \int_{x=-\pi/2}^{\pi/2} \int_{z=0}^{\cos x} \frac{1}{2}(1-z)^2 \, dz \, dx$$

$$= \int_{x=-\pi/2}^{\pi/2} \left[\frac{1}{2}z - \frac{1}{2}z^2 + \frac{1}{6}z^3 \right]_{z=0}^{\cos x} dx = \int_{-\pi/2}^{\pi/2} \left(\frac{1}{2}\cos x - \frac{1}{2}\cos^2 x + \frac{1}{6}\cos^3 x \right) dx$$

$$= \frac{1}{72}\left[-18x + 45\sin x - 9\sin 2x + \sin 3x \right]_{-\pi/2}^{\pi/2} = \frac{44 - 9\pi}{36};$$

$$M_{xy} = \int_{x=-\pi/2}^{\pi/2} \int_{z=0}^{\cos x} \int_{y=0}^{1-z} z \, dy \, dz \, dx = \int_{x=-\pi/2}^{\pi/2} \int_{z=0}^{\cos x} (z - z^2) \, dz \, dx = \int_{x=-\pi/2}^{\pi/2} \left[\frac{1}{2}z^2 - \frac{1}{3}z^3 \right]_{z=0}^{\cos x} dx$$

$$= \int_{-\pi/2}^{\pi/2} \left(\frac{1}{2}\cos^2 x - \frac{1}{3}\cos^3 x \right) dx = \frac{1}{72}\left[18x - 18\sin x + 9\sin 2x - 2\sin 3x \right]_{-\pi/2}^{\pi/2} = \frac{9\pi - 16}{36}.$$

Therefore the centroid of the solid is located at the point

$$(\overline{x}, \, \overline{y}, \, \overline{z}) = \left(0, \, \frac{44 - 9\pi}{72 - 9\pi}, \, \frac{9\pi - 16}{72 - 9\pi} \right) \approx (0, \, 0.359643831963, \, 0.280712336074).$$

C14S06.027: (See Problem 24.) The moment of inertia of the solid (with density $\delta = 1$) with respect to the y-axis is

$$I_y = \int_{x=-1}^{1} \int_{y=-1}^{1} \int_{z=0}^{1-x^2} (x^2 + z^2) \, dz \, dy \, dx = \int_{x=-1}^{1} \int_{y=-1}^{1} \left[x^2 z + \frac{1}{3}z^3 \right]_{z=0}^{1-x^2} dy \, dx$$

$$= \int_{x=-1}^{1} \int_{-1}^{1} \frac{1}{3}(1 - x^6) \, dy \, dx = \int_{-1}^{1} \frac{2}{3}(1 - x^6) \, dx = \left[\frac{2}{3}x - \frac{2}{21}x^7 \right]_{-1}^{1} = \frac{8}{7}.$$

C14S06.029: With unit density $\delta = 1$, the moment of inertia of the given tetrahedron with respect to the z-axis is

$$I_z = \int_{z=0}^{1} \int_{y=0}^{1-z} \int_{x=0}^{1-y-z} (x^2 + y^2) \, dx \, dy \, dz = \int_{z=0}^{1} \int_{y=0}^{1-z} \left[\frac{1}{3}x^3 + xy^2 \right]_{x=0}^{1-y-z} dy \, dz$$

$$= \int_{z=0}^{1} \int_{y=0}^{1-z} \left(\frac{1}{3} - y + 2y^2 - \frac{4}{3}y^3 - z + 2yz - 2y^2 z + z^2 - yz^2 - \frac{1}{3}z^3 \right) dy \, dz$$

$$= \int_{z=0}^{1} \left[-\frac{1}{3}y^4 + \frac{2}{3}y^3(1-z) - \frac{1}{2}y^2(1-z)^2 + \frac{1}{3}y(1-z)^3 \right]_{y=0}^{1-z} dz$$

829

$$= \int_0^1 \frac{1}{6}(1 - 4z + 6z^2 - 4z^3 + z^4)\, dz = \frac{1}{30}\left[5z - 10z^2 + 10z^3 - 5z^4 + z^5\right]_0^1 = \frac{1}{30}.$$

C14S06.031: It should be clear that $\bar{x} = \bar{y} = 0$ by symmetry, but there is a slight suggestion in the wording of the problem that we should *prove* this more rigorously. Hence we will compute the mass and all three moments. Assuming unit density $\delta = 1$, we have

$$m = \int_{x=-\sqrt{h}}^{\sqrt{h}} \int_{y=-\sqrt{h-x^2}}^{\sqrt{h-x^2}} \int_{z=x^2+y^2}^{h} 1\, dz\, dy\, dx = \int_{x=-\sqrt{h}}^{\sqrt{h}} \int_{y=-\sqrt{h-x^2}}^{\sqrt{h-x^2}} (h - x^2 - y^2)\, dy\, dx$$

$$= \int_{x=-\sqrt{h}}^{\sqrt{h}} \left[hy - x^2 y - \frac{1}{3}y^3\right]_{y=-\sqrt{h-x^2}}^{\sqrt{h-x^2}} dx = \int_{x=-\sqrt{h}}^{\sqrt{h}} \frac{4}{3}(h - x^2)^{3/2}\, dx.$$

The substitution $x = \sqrt{h}\,\sin\theta$, $dx = \sqrt{h}\,\cos\theta\, d\theta$ then yields

$$m = \int_{\theta=-\pi/2}^{\pi/2} \frac{4}{3}h^2 \cos^4\theta\, d\theta = \frac{1}{24}h^2\left[12\theta + 8\sin 2\theta + \sin 4\theta\right]_{-\pi/2}^{\pi/2} = \frac{1}{2}\pi h^2.$$

Next,

$$M_{yz} = \int_{x=-\sqrt{h}}^{\sqrt{h}} \int_{y=-\sqrt{h-x^2}}^{\sqrt{h-x^2}} \int_{z=x^2+y^2}^{h} x\, dz\, dy\, dx = \int_{x=-\sqrt{h}}^{\sqrt{h}} \int_{y=-\sqrt{h-x^2}}^{\sqrt{h-x^2}} (hx - x^3 - xy^2)\, dy\, dx$$

$$= \int_{x=-\sqrt{h}}^{\sqrt{h}} \left[hxy - x^3 y - \frac{1}{3}xy^3\right]_{y=-\sqrt{h-x^2}}^{\sqrt{h-x^2}} dx = \int_{-\sqrt{h}}^{\sqrt{h}} \frac{4}{3}x(h - x^2)^{3/2}\, dx$$

$$= \left[-\frac{4}{15}(h - x^2)^{5/2}\right]_{-\sqrt{h}}^{\sqrt{h}} = 0;$$

$$M_{xz} = \int_{x=-\sqrt{h}}^{\sqrt{h}} \int_{y=-\sqrt{h-x^2}}^{\sqrt{h-x^2}} \int_{z=x^2+y^2}^{h} y\, dz\, dy\, dx = \int_{x=-\sqrt{h}}^{\sqrt{h}} \int_{y=-\sqrt{h-x^2}}^{\sqrt{h-x^2}} (hy - x^2 y - y^3)\, dy\, dx$$

$$= \int_{x=-\sqrt{h}}^{\sqrt{h}} \left[\frac{1}{2}hy^2 - \frac{1}{2}x^2 y^2 - \frac{1}{4}y^4\right]_{y=-\sqrt{h-x^2}}^{\sqrt{h-x^2}} dx = \int_{-\sqrt{h}}^{\sqrt{h}} 0\, dx = 0;$$

$$M_{xy} = \int_{x=-\sqrt{h}}^{\sqrt{h}} \int_{y=-\sqrt{h-x^2}}^{\sqrt{h-x^2}} \int_{z=x^2+y^2}^{h} z\, dz\, dy\, dx = \int_{x=-\sqrt{h}}^{\sqrt{h}} \int_{y=-\sqrt{h-x^2}}^{\sqrt{h-x^2}} \left(\frac{1}{2}h^2 - \frac{1}{2}(x^2 + y^2)^2\right) dy\, dx.$$

Now rewrite the last integral in polar coordinates. Then

$$M_{xy} = \int_{\theta=0}^{2\pi} \int_{r=0}^{\sqrt{h}} \frac{1}{2}(h^2 - r^4)\cdot r\, dr\, d\theta = \int_{\theta=0}^{2\pi} \left[\frac{1}{4}h^2 r^2 - \frac{1}{12}r^6\right]_{r=0}^{\sqrt{h}} d\theta = 2\pi\cdot\frac{1}{6}h^3 = \frac{1}{3}\pi h^3.$$

Thus the centroid of the parabolic segment is located at the point

$$(\bar{x}, \bar{y}, \bar{z}) = \left(0, 0, \frac{2}{3}h\right),$$

and therefore the centroid is on the axis of symmetry of the segment, two-thirds of the way from the vertex $(0, 0, 0)$ to the base.

C14S06.033: Place the cube in the first octant with three of its faces in the coordinate planes, one vertex at $(0, 0, 0)$, and the opposite vertex at (a, a, a). With density $\delta = 1$, its moment of inertia with respect to the z-axis is

$$I_z = \int_{z=0}^{a} \int_{y=0}^{a} \int_{x=0}^{a} (x^2 + y^2)\, dx\, dy\, dz = \int_0^a \int_0^a \left[\frac{1}{3}x^3 + xy^2 \right]_0^a dy\, dz = \int_0^a \int_0^a \left(\frac{1}{3}a^3 + ay^2 \right) dy\, dz$$

$$= \int_0^a \left[\frac{1}{3}a^3 y + \frac{1}{3}ay^3 \right]_0^a dz = \int_0^a \frac{2}{3}a^4 \, dz = a \cdot \frac{2}{3}a^4 = \frac{2}{3}a^5.$$

Because the mass of the cube is $m = a^3$, we see that $I_z = \frac{2}{3}ma^2$, which is dimensionally correct.

C14S06.035: With density $\delta(x, y, z) = k(x^2 + y^2 + z^2)$ at the point (x, y, z) (k is a positive constant), the moment of inertia of the cube of Problem 34 with respect to the z-axis is

$$I_z = \int_0^a \int_0^a \int_0^a k(x^2 + y^2 + z^2)(x^2 + y^2)\, dz\, dy\, dx$$

$$= k \int_0^a \int_0^a \left[x^4 z + 2x^2 y^2 z + y^4 z + \frac{1}{3}x^2 z^3 + \frac{1}{3}y^2 z^3 \right]_0^a dy\, dx$$

$$= k \int_0^a \int_0^a \left(\frac{1}{3}a^3 x^2 + ax^4 + \frac{1}{3}a^3 y^2 + 2ax^2 y^2 + ay^4 \right) dy\, dx$$

$$= k \int_0^a \left[\frac{1}{3}a^3 x^2 y + ax^4 y + \frac{1}{9}a^3 y^3 + \frac{2}{3}ax^2 y^3 + \frac{1}{5}ay^5 \right]_0^a dx$$

$$= k \int_0^a \left(\frac{14}{45}a^6 + a^4 x^2 + a^2 x^4 \right) dx = k \left[\frac{14}{45}a^6 x + \frac{1}{3}a^4 x^3 + \frac{1}{5}a^2 x^5 \right]_0^a = \frac{38}{45}ka^7.$$

C14S06.037: The moment of inertia of the cube of Problem 36 with respect to the z-axis is

$$I_z = \int_0^1 \int_0^1 \int_0^1 k(x^2 + y^2)z\, dz\, dy\, dx = \int_0^1 \int_0^1 \frac{1}{2}k(x^2 + y^2)\, dy\, dx$$

$$= \int_0^1 \left[\frac{1}{2}kx^2 y + \frac{1}{6}ky^3 \right]_0^1 dx = \int_0^1 \left(\frac{1}{2}kx^2 + \frac{1}{6}k \right) dx = k \cdot \left[\frac{1}{6}x^3 + \frac{1}{6}x \right]_0^1 = \frac{1}{3}k.$$

C14S06.039: With constant density $\delta = 1$, the mass and moments are

$$m = \int_{z=0}^1 \int_{y=0}^{\sqrt{1-z^2}} \int_{x=0}^{\sqrt{1-z^2}} 1\, dx\, dy\, dz = \int_{z=0}^1 \int_{y=0}^{\sqrt{1-z^2}} \sqrt{1-z^2}\, dy\, dz$$

$$= \int_{z=0}^1 (1 - z^2)\, dz = \left[z - \frac{1}{3}z^3 \right]_0^1 = \frac{2}{3};$$

$$M_{yz} = \int_{z=0}^1 \int_{y=0}^{\sqrt{1-z^2}} \int_{x=0}^{\sqrt{1-z^2}} x\, dx\, dy\, dz = \int_{z=0}^1 \int_{y=0}^{\sqrt{1-z^2}} \frac{1}{2}(1 - z^2)\, dy\, dz$$

$$= \int_{z=0}^1 \frac{1}{2}(1 - z^2)^{3/2}\, dz = \left[\frac{5z - 2z^3}{16} \cdot (1 - z^2)^{1/2} + \frac{3}{16}\arcsin z \right]_0^1 = \frac{3}{32}\pi;$$

$$M_{xz} = \int_{z=0}^{1} \int_{y=0}^{\sqrt{1-z^2}} \int_{x=0}^{\sqrt{1-z^2}} y \, dx \, dy \, dz = \int_{z=0}^{1} \int_{y=0}^{\sqrt{1-z^2}} y(1-z^2)^{1/2} \, dy \, dz$$

$$= \int_{z=0}^{1} \frac{1}{2}(1-z^2)^{3/2} \, dz = \left[\frac{5z - 2z^3}{16} \cdot (1-z^2)^{1/2} + \frac{3}{16} \arcsin z \right]_0^1 = \frac{3}{32}\pi;$$

$$M_{xy} = \int_{z=0}^{1} \int_{y=0}^{\sqrt{1-z^2}} \int_{x=0}^{\sqrt{1-z^2}} z \, dx \, dy \, dz = \int_{z=0}^{1} \int_{y=0}^{\sqrt{1-z^2}} z(1-z^2)^{1/2} \, dy \, dz$$

$$= \int_{z=0}^{1} (z - z^3) \, dz = \left[\frac{1}{2}z^2 - \frac{1}{4}z^4 \right]_0^1 = \frac{1}{4}.$$

Therefore the centroid of the solid is located at the point $\left(\dfrac{9}{64}\pi, \dfrac{9}{64}\pi, \dfrac{3}{8} \right)$.

C14S06.041: The given solid projects onto the circular disk D with radius 2 and center $(0, 0)$ in the xy-plane. Hence the volume of the solid is

$$V = \iint_D \left(\int_{z=2x^2+y^2}^{12-x^2-2y^2} 1 \, dz \right) dA = \iint_D (12 - 3x^2 - 3y^2) \, dA.$$

Rewrite the last double integral as an iterated integral in polar coordinates. Thus

$$V = \int_{\theta=0}^{2\pi} \int_{r=0}^{2} (12 - 3r^2) \cdot r \, dr \, d\theta = 2\pi \cdot \left[6r^2 - \frac{3}{4}r^4 \right]_0^2 = 24\pi \approx 75.3982236861550377.$$

C14S06.043: Following the *Suggestion*, the volume is

$$V = \int_{z=0}^{1} \int_{y=-z/2}^{z/2} \int_{x=-\sqrt{z^2-4y^2}}^{\sqrt{z^2-4y^2}} 1 \, dx \, dy \, dz = \int_{z=0}^{1} \int_{y=-z/2}^{z/2} 2\sqrt{z^2-4y^2} \, dy \, dz.$$

Let $y = \frac{1}{2}z \sin u$, $dy = \frac{1}{2}z \cos u \, du$. This substitution yields

$$V = \int_{z=0}^{1} \int_{u=-\pi/2}^{\pi/2} z^2 \cos^2 u \, du \, dz = \int_{z=0}^{1} \left[\frac{1}{4}z^2(2u + \sin 2u) \right]_{u=-\pi/2}^{\pi/2} dz$$

$$= \int_0^1 \frac{1}{2}\pi z^2 \, dz = \left[\frac{1}{6}\pi z^3 \right]_0^1 = \frac{1}{6}\pi \approx 0.5235987755982989.$$

Methods of single-variable calculus also succeed here. A horizontal cross section of the solid at $z = h$ $(0 < h \leq 1)$ is an ellipse with equation $x^2 + 4y^2 = h^2$. This ellipse has major semiaxis of length h and minor semiaxis of length $\frac{1}{2}h$. Therefore its area is $\frac{1}{2}\pi h^2$. So, by the method of parallel cross sections (see Eq. (3) of Section 6.2), the volume of the solid is

$$\int_0^1 \frac{1}{2}\pi h^2 \, dh = \left[\frac{1}{6}\pi h^3 \right]_0^1 = \frac{1}{6}\pi.$$

C14S06.045: If the pyramid (tetrahedron) of Example 2 has density z at the point (x, y, z), then its mass and moments are

$$m = \int_{x=0}^{2} \int_{y=0}^{(6-3x)/2} \int_{z=0}^{6-3x-2y} z \, dz \, dy \, dx = \int_{0}^{2} \int_{0}^{(6-3x)/2} \frac{1}{2}(6 - 3x - 2y)^2 \, dy \, dx$$

$$= \int_{0}^{2} \left[18y - 18xy + \frac{9}{2}x^2y - 6y^2 + 3xy^2 + \frac{2}{3}y^3 \right]_{0}^{(6-3x)/2} dx$$

$$= \int_{0}^{2} \left[9(6 - 3x) - \frac{3}{2}(6 - 3x)^2 + \frac{1}{12}(6 - 3x)^3 - 9x(6 - 3x) + \frac{3}{4}x(6 - 3x)^2 + \frac{9}{4}x^2(6 - 3x) \right] dx$$

$$= \int_{0}^{2} \left(18 - 27x + \frac{27}{2}x^2 - \frac{9}{4}x^3 \right) dx = \left[18x - \frac{27}{2}x^2 + \frac{9}{2}x^3 - \frac{9}{16}x^4 \right]_{0}^{2} = 9;$$

$$M_{yz} = \int_{x=0}^{2} \int_{y=0}^{(6-3x)/2} \int_{z=0}^{6-3x-2y} xz \, dz \, dy \, dx = \int_{0}^{2} \int_{0}^{(6-3x)/2} \frac{1}{2}x(6 - 3x - 2y)^2 \, dy \, dx$$

$$= \int_{0}^{2} \int_{0}^{(6-3x)/2} \left(18x - 18x^2 + \frac{9}{2}x^3 - 12xy + 6x^2y + 2xy^2 \right) dy \, dx$$

$$= \int_{0}^{2} \left[\frac{9}{2}xy(x - 2)^2 + 3xy^2(x - 2) + \frac{2}{3}xy^3 \right]_{0}^{(6-3x)/2} dx$$

$$= \int_{0}^{2} \left[9x(6 - 2x) - \frac{3}{2}x(6 - 3x)^2 + \frac{1}{12}x(6 - 3x)^3 - 9x^2(6 - 3x) + \frac{3}{4}x^2(6 - 3x)^2 + \frac{9}{4}x^3(6 - 3x) \right] dx$$

$$= \int_{0}^{2} \left(18x - 27x^2 + \frac{27}{2}x^3 - \frac{9}{4}x^4 \right) dx = \left[9x^2 - 9x^3 + \frac{27}{8}x^4 - \frac{9}{20}x^5 \right]_{0}^{2} = \frac{18}{5};$$

$$M_{xz} = \int_{x=0}^{2} \int_{y=0}^{(6-3x)/2} \int_{z=0}^{6-3x-2y} yz \, dz \, dy \, dx = \int_{0}^{2} \int_{0}^{(6-3x)/2} \frac{1}{2}y(6 - 3x - 2y)^2 \, dy \, dx$$

$$= \int_{0}^{2} \int_{0}^{(6-3x)/2} \left(18y - 18xy + \frac{9}{2}x^2y - 12y^2 + 6xy^2 + 2y^3 \right) dy \, dx$$

$$= \int_{0}^{2} \left[9y^2 - 9xy^2 + \frac{9}{4}x^2y^2 - 4y^3 + 2xy^3 + \frac{1}{2}y^4 \right]_{0}^{(6-3x)/2} dx$$

$$= \int_{0}^{2} \left[\frac{9}{4}(6 - 3x)^2 - \frac{1}{2}(6 - 3x)^3 + \frac{1}{32}(6 - 3x)^4 - \frac{9}{4}x(6 - 3x)^2 + \frac{1}{4}x(6 - 3x)^3 + \frac{9}{16}x^2(6 - 3x)^2 \right] dx$$

$$= \int_{0}^{2} \left(\frac{27}{2} - 27x + \frac{81}{4}x^2 - \frac{27}{4}x^3 + \frac{27}{32}x^4 \right) dx = \left[\frac{27}{2}x - \frac{27}{2}x^2 + \frac{27}{4}x^3 - \frac{27}{16}x^4 + \frac{27}{160}x^5 \right]_{0}^{2} = \frac{27}{5};$$

$$M_{xy} = \int_{x=0}^{2} \int_{y=0}^{(6-3x)/2} \int_{z=0}^{6-3x-2y} z^2 \, dz \, dy \, dx = \int_{0}^{2} \int_{0}^{(6-3x)/2} \frac{1}{3}(6 - 3x - 2y)^3 \, dy \, dx$$

$$= \int_{0}^{2} \int_{0}^{(6-3x)/2} \left(72 - 108x + 54x^2 - 9x^3 - 72y + 72xy - 18x^2y + 24y^2 - 12xy^2 - \frac{8}{3}y^3 \right) dy \, dx$$

$$= \int_{0}^{2} \left[72y - 108xy + 54x^2y - 9x^3y - 36y^2 + 36xy^2 - 9x^2y^2 + 8y^3 - 4xy^3 - \frac{2}{3}y^4 \right]_{y=0}^{(6-3x)/2} dx$$

$$= \int_0^2 \left(54 - 108x + 81x^2 - 27x^3 + \frac{27}{8}x^4\right) dx = \left[54x - 54x^2 + 27x^3 - \frac{27}{4}x^4 + \frac{27}{40}x^5\right]_0^2 = \frac{108}{5}.$$

Therefore the centroid of the pyramid is located at the point $\left(\frac{2}{5}, \frac{3}{5}, \frac{12}{5}\right)$.

C14S06.047: First we compute

$$\int_{x=0}^2 \int_{y=0}^{(6-3x)/2} \int_{z=0}^{6-3x-2y} z \; dz \; dy \; dx = \int_0^2 \int_0^{(6-3x)/2} \frac{1}{2}(6 - 3x - 2y)^2 \; dy \; dx$$

$$= \int_0^2 \int_0^{(6-3x)/2} \left(18 - 18x + \frac{9}{2}x^2 - 12y + 6xy + 2y^2\right) dy \; dx$$

$$= \int_0^2 \left[18y - 18xy + \frac{9}{2}x^2 y - 6y^2 + 3xy^2 + \frac{2}{3}y^3\right]_0^{(6-3x)/2} dx$$

$$= \int_0^2 \left(18 - 27x + \frac{27}{2}x^2 - \frac{9}{4}x^3\right) dx = \left[18x - \frac{27}{2}x^2 + \frac{9}{2}x^3 - \frac{9}{16}x^4\right]_0^2 = 9.$$

The volume of the pyramid is 6, and hence the average value of the density function $\delta(x, y, z) = z$ on the pyramid is $\overline{\delta} = \frac{9}{6} = \frac{3}{2}$.

C14S06.049: The centroid of the cube of Problem 48 is, by symmetry, its midpoint $\left(\frac{1}{2}, \frac{1}{2}, \frac{1}{2}\right)$. Because the cube has volume 1, the average value of

$$g(x, y, z) = \left(x - \frac{1}{2}\right)^2 + \left(y - \frac{1}{2}\right)^2 + \left(z - \frac{1}{2}\right)^2$$

on the cube is

$$\overline{g} = \int_{x=0}^1 \int_{y=0}^1 \int_{z=0}^1 g(x, y, z) \; dz \; dy \; dx = \int_0^1 \int_0^1 \left[\frac{3}{4}z - xz + x^2 z - yz + y^2 z - \frac{1}{2}z^2 + \frac{1}{3}z^3\right]_0^1 dy \; dx$$

$$= \int_0^1 \int_0^1 \left(\frac{7}{12} - x + x^2 - y + y^2\right) dy \; dx = \int_0^1 \left[\frac{7}{12}y - xy + x^2 y - \frac{1}{2}y^2 + \frac{1}{3}y^3\right]_0^1 dx$$

$$= \int_0^1 \left(\frac{5}{12} - x + x^2\right) dx = \left[\frac{5}{12}x - \frac{1}{2}x^2 + \frac{1}{3}x^3\right]_0^1 = \frac{1}{4}.$$

C14S06.051: First we compute

$$J = \int_{x=0}^2 \int_{y=0}^{(6-3x)/2} \int_{z=0}^{6-3x-2y} (x^2 + y^2 + z^2) \; dz \; dy \; dx = \int_0^2 \int_0^{(6-3x)/2} \left[x^2 z + y^2 z + \frac{1}{3}z^3\right]_0^{6-3x-2y} dy \; dx$$

$$= \int_0^2 \int_0^{(6-3x)/2} \left(72 - 108x + 60x^2 - 12x^3 - 72y + 72xy - 20x^2 y + 30y^2 - 15xy^2 - \frac{14}{3}y^3\right) dy \; dx$$

$$= \int_0^2 \left[72y - 108xy + 60x^2 y - 12x^3 y - 36y^2 + 36xy^2 - 10x^2 y^2 + 10y^3 - 5xy^3 - \frac{7}{6}y^4\right]_0^{(6-3x)/2} dx$$

$$= \int_0^2 \left(\frac{135}{2} - 135x + \frac{441}{4}x^2 - \frac{171}{4}x^3 + \frac{207}{32}x^4\right) dx = \left[\frac{135}{2}x - \frac{135}{2}x^2 + \frac{147}{4}x^3 - \frac{171}{16}x^4 + \frac{207}{160}x^5\right]_0^2$$

$$= \frac{147}{5}.$$

Because the pyramid has volume $V = 6$, the average squared distance of its points from its centroid is

$$\bar{d} = \frac{J}{V} = \frac{147}{30} = \frac{49}{10} = 4.9.$$

C14S06.053: Using *Mathematica* 3.0, we find that the average distance of points of the cube of Problem 48 from the origin is

$$\bar{d} = \int_{x=0}^{1} \int_{y=1}^{1} \int_{z=0}^{1} \sqrt{x^2 + y^2 + z^2} \; dz \, dy \, dx$$

$$= \int_0^1 \int_0^1 \left[\frac{1}{2} z \sqrt{x^2 + y^2 + z^2} + \frac{1}{2}(x^2 + y^2) \ln \left(z + \sqrt{x^2 + y^2 + z^2} \right) \right]_0^1 dy \, dx$$

$$= \int_0^1 \int_0^1 \left[\frac{1}{2} \sqrt{x^2 + y^2 + 1} - \frac{1}{2}(x^2 + y^2) \ln \left(\sqrt{x^2 + y^2} \right) + \frac{1}{2}(x^2 + y^2) \ln \left(1 + \sqrt{x^2 + y^2 + 1} \right) \right] dy \, dx$$

$$= \int_0^1 \left[\frac{1}{3} y \sqrt{x^2 + y^2 + 1} - \frac{1}{3} x^2 \arctan \left(\frac{y}{x\sqrt{x^2 + y^2 + 1}} \right) - \frac{1}{6} y (3x^2 + y^2) \ln \left(\sqrt{x^2 + y^2} \right) \right.$$

$$\left. + \frac{1}{6} y (3x^2 + y^2) \ln \left(1 + \sqrt{x^2 + y^2 + 1} \right) + \frac{1}{6}(3x^2 + 1) \ln \left(y + \sqrt{x^2 + y^2 + 1} \right) \right]_0^1 dx$$

$$= \int_0^1 \left[\frac{1}{3} \sqrt{x^2 + 2} - \frac{1}{3} x^2 \arctan \left(\frac{1}{x\sqrt{x^2 + 2}} \right) - \frac{1}{3}(3x^2 + 1) \ln \left(\sqrt{x^2 + 1} \right) \right.$$

$$\left. + \frac{1}{3}(3x^2 + 1) \ln \left(1 + \sqrt{x^2 + 2} \right) \right] dx$$

$$= \left[\frac{1}{4} x \sqrt{x^2 + 2} + \frac{1}{3} \operatorname{arcsinh} \left(\frac{x}{\sqrt{2}} \right) - \frac{1}{12} x^4 \arctan \left(\frac{1}{x\sqrt{x^2 + 2}} \right) \right.$$

$$\left. - \frac{1}{6} \arctan \left(\frac{x}{\sqrt{x^2 + 2}} \right) - \frac{1}{3} x(x^2 + 1) \ln \left(\sqrt{x^2 + 1} \right) + \frac{1}{3} x(x^2 + 1) \ln \left(1 + \sqrt{x^2 + 2} \right) \right]_0^1$$

$$= \frac{1}{24} \left[6\sqrt{3} - \pi + 8 \operatorname{arcsinh} \left(\frac{\sqrt{2}}{2} \right) - 8 \ln 2 + 16 \ln \left(1 + \sqrt{3} \right) \right] \approx 0.960591956455052959425108.$$

Section 14.7

C14S07.001: The volume is

$$V = \int_{\theta=0}^{2\pi} \int_{r=0}^{2} \int_{z=r^2}^{4} r \, dz \, dr \, d\theta = 2\pi \int_{r=0}^{2} (4r - r^3) \, dr = 2\pi \left[2r^2 - \frac{1}{4} r^4 \right]_0^2 = 8\pi.$$

C14S07.003: Place the center of the sphere at the origin. Then its volume is

$$V = \int_{\theta=0}^{2\pi} \int_{r=0}^{a} \int_{z=-\sqrt{a^2-r^2}}^{\sqrt{a^2-r^2}} r \, dz \, dr \, d\theta = 2\pi \int_{r=0}^{a} 2r\sqrt{a^2 - r^2} \, dr = 2\pi \cdot \left[-\frac{2}{3}(a^2 - r^2)^{3/2} \right]_0^a = \frac{4}{3} \pi a^3.$$

C14S07.005: The volume is

$$V = \int_{\theta=0}^{2\pi} \int_{r=0}^{1} \int_{z=-\sqrt{4-r^2}}^{\sqrt{4-r^2}} r \; dz \; dr \; d\theta = 2\pi \int_{r=0}^{1} 2r(4-r^2)^{1/2} \; dr = 2\pi \cdot \left[-\frac{2}{3}(4-r^2)^{3/2} \right]_0^1$$

$$= 2\pi \left(\frac{16}{3} - 2\sqrt{3} \right) = \frac{4}{3}\pi \left(8 - 3\sqrt{3} \right) \approx 11.7447292674805137.$$

C14S07.007: The mass of the cylinder is

$$m = \int_{\theta=0}^{2\pi} \int_{r=0}^{a} \int_{z=0}^{h} rz \; dz \; dr \; d\theta = 2\pi \int_{r=0}^{a} \frac{1}{2}rh^2 \; dr = 2\pi \cdot \left[\frac{1}{4}r^2h^2 \right]_0^a = \frac{1}{2}\pi a^2 h^2.$$

C14S07.009: The moment of inertia of the cylinder of Problem 7 with respect to the z-axis is

$$I_z = \int_{\theta=0}^{2\pi} \int_{r=0}^{a} \int_{z=0}^{h} r^3 z \; dz \; dr \; d\theta = 2\pi \int_{r=0}^{a} \left[\frac{1}{2}r^3 z^2 \right]_{z=0}^{h} \; dr$$

$$= 2\pi \int_{r=0}^{a} \frac{1}{2}r^3 h^2 \; dr = 2\pi \cdot \left[\frac{1}{8}r^4 h^2 \right]_{r=0}^{a} = 2\pi \cdot \frac{1}{8}a^4 h^2 = \frac{1}{4}\pi a^4 h^2.$$

C14S07.011: The volume is

$$V = \int_{\theta=0}^{2\pi} \int_{r=0}^{3} \int_{z=0}^{9-r^2} r \; dz \; dr \; d\theta = 2\pi \int_{r=0}^{3} (9r - r^3) \; dr = 2\pi \cdot \left[\frac{9}{2}r^2 - \frac{1}{4}r^4 \right]_0^3 = \frac{81}{2}\pi.$$

By symmetry, $\overline{x} = \overline{y} = 0$. The moment of the solid with respect to the xy-plane is

$$M_{xy} = \int_{\theta=0}^{2\pi} \int_{r=0}^{3} \int_{z=0}^{9-r^2} rz \; dz \; dr \; d\theta = 2\pi \int_{r=0}^{3} \frac{1}{2}r(9-r^2)^2 \; dr$$

$$= 2\pi \int_{r=0}^{3} \left(\frac{81}{2}r - 9r^3 + \frac{1}{2}r^5 \right) \; dr = 2\pi \cdot \left[\frac{81}{4}r^2 - \frac{9}{4}r^4 + \frac{1}{12}r^6 \right]_0^3 = \frac{243}{2}\pi.$$

Therefore the z-coordinate of the centroid is $\overline{z} = 3$. *Suggestion:* Compare this answer with the answer obtained by using the result in Problem 31 of Section 14.6.

C14S07.013: The curve formed by the intersection of the paraboloids lies on the cylinder $x^2 + y^2 = 4$, and hence the solid projects vertically onto the disk D with boundary $x^2 + y^2 = 4$ in the xy-plane. Therefore the volume of the solid is

$$V = \int_{\theta=0}^{2\pi} \int_{r=0}^{2} \int_{z=r^2+r^2 \cos^2 \theta}^{12-r^2-r^2 \sin^2 \theta} r \; dz \; dr \; d\theta = \int_{\theta=0}^{2\pi} \int_{r=0}^{2} (12r - 3r^3) \; dr \; d\theta = \int_{\theta=0}^{2\pi} \left[6r^2 - \frac{3}{4}r^4 \right]_{r=0}^{2} \; d\theta$$

$$= \int_{0}^{2\pi} 12 \; d\theta = 2\pi \cdot 12 = 24\pi \approx 75.3982236861550377.$$

C14S07.015: The spherical surface $r^2 + z^2 = 2$ and the paraboloid $z = r^2$ meet in a horizontal circle that projects vertically onto the circle $x^2 + y^2 = 1$ in the xy-plane. Hence the volume between the two surfaces is

$$V = \int_{\theta=0}^{2\pi} \int_{r=0}^{1} \int_{z=r^2}^{\sqrt{2-r^2}} r\, dz\, dr\, d\theta = \int_0^{2\pi} \int_0^1 \left[r(2-r^2)^{1/2} - r^3 \right] dr\, d\theta = 2\pi \cdot \left[-\frac{1}{3}(2-r^2)^{3/2} - \frac{1}{4}r^4 \right]_0^1$$

$$= 2\pi \cdot \left(\frac{2}{3}\sqrt{2} - \frac{7}{12} \right) = \frac{1}{6}\pi \left(8\sqrt{2} - 7 \right) \approx 2.2586524883563962.$$

C14S07.017: Set up a coordinate system in which the points of the cylinder are described by

$$0 \leq r \leq a, \quad 0 \leq \theta \leq 2\pi, \quad 0 \leq z \leq h.$$

Because the cylinder has constant density δ, its mass is $m = \pi\delta a^2 h$. One diameter of its base coincides with the x-axis, so we will find the moment of inertia I of the cylinder with respect to that axis. The square of the distance of the point (x, y, z) from the x-axis is $y^2 + z^2 = z^2 + r^2 \sin^2 \theta$. Therefore

$$I = I_x = \int_{\theta=0}^{2\pi} \int_{r=0}^{a} \int_{z=0}^{h} \delta(z^2 + r^2 \sin^2 \theta) \cdot r\, dz\, dr\, d\theta = \delta \int_0^{2\pi} \int_0^a \left(\frac{1}{3}h^3 r + hr^3 \sin^2 \theta \right) dr\, d\theta$$

$$= \delta \int_0^{2\pi} \left[\frac{1}{6}h^3 r^2 + \frac{1}{4}hr^4 \sin^2 \theta \right]_0^a d\theta = \delta \int_0^{2\pi} \left(\frac{1}{6}a^2 h^3 + \frac{1}{4}a^4 h \sin^2 \theta \right) d\theta$$

$$= \delta \left[\frac{6a^4 h\theta + 8a^2 h^3 \theta - 3a^4 h \sin 2\theta}{48} \right]_0^{2\pi} = \frac{1}{12} \delta \pi a^2 h(3a^2 + 4h^2) = \frac{1}{12} m(3a^2 + 4h^2)$$

where m is the mass of the cylinder.

C14S07.019: The volume is

$$V = \int_{\theta=0}^{2\pi} \int_{r=0}^{1} \int_{z=r}^{1} r\, dz\, dr\, d\theta = 2\pi \int_0^1 (r - r^2)\, dr = 2\pi \cdot \left[\frac{1}{2}r^2 - \frac{1}{3}r^3 \right]_0^1 = \frac{1}{3}\pi.$$

C14S07.021: Without loss of generality we may assume that the hemispherical solid has density $\delta = 1$. Choose a coordinate system in which the solid is bounded above by the spherical surface $\rho = a$ and below by the xy-plane. Then its mass and moments are

$$m = \int_{\theta=0}^{2\pi} \int_{\phi=0}^{\pi/2} \int_{\rho=0}^{a} \rho^2 \sin \phi\, d\rho\, d\phi\, d\theta = 2\pi \int_{\phi=0}^{\pi/2} \frac{1}{3}a^3 \sin \phi\, d\phi = \frac{2}{3}\pi a^3 \left[-\cos \phi \right]_0^{\pi/2} = \frac{2}{3}\pi a^3;$$

$$M_{yz} = \int_{\theta=0}^{2\pi} \int_{\phi=0}^{\pi/2} \int_{\rho=0}^{a} \rho^3 \sin^2 \phi \cos \theta\, d\rho\, d\phi\, d\theta = \int_{\theta=0}^{2\pi} \int_{\phi=0}^{\pi/2} \frac{1}{4}a^4 \sin^2 \phi \cos \theta\, d\phi\, d\theta$$

$$= \int_{\theta=0}^{2\pi} \left(\int_{\phi=0}^{\pi/2} \frac{1}{4}a^2 \sin^2 \phi\, d\phi \right) \cos \theta\, d\theta = \left(\int_{\phi=0}^{\pi/2} \frac{1}{4}a^2 \sin^2 \phi\, d\phi \right) \cdot \left[\sin \theta \right]_{\theta=0}^{2\pi} = 0;$$

$$M_{xz} = \int_{\theta=0}^{2\pi} \int_{\phi=0}^{\pi/2} \int_{\rho=0}^{a} \rho^3 \sin^2 \phi \sin \theta\, d\rho\, d\phi\, d\theta = \int_{\theta=0}^{2\pi} \int_{\phi=0}^{\pi/2} \frac{1}{4}a^4 \sin^2 \phi \sin \theta\, d\phi\, d\theta$$

$$= \int_{\theta=0}^{2\pi} \left(\int_{\phi=0}^{\pi/2} \frac{1}{4}a^2 \sin^2 \phi\, d\phi \right) \sin \theta\, d\theta = \left(\int_{\phi=0}^{\pi/2} \frac{1}{4}a^2 \sin^2 \phi\, d\phi \right) \cdot \left[-\cos \theta \right]_{\theta=0}^{2\pi} = 0;$$

$$M_{yx} = \int_{\theta=0}^{2\pi} \int_{\phi=0}^{\pi/2} \int_{\rho=0}^{a} \rho^3 \sin\phi \cos\phi \, d\rho \, d\phi \, d\theta = 2\pi \int_{\phi=0}^{\pi/2} \frac{1}{4} a^4 \sin\phi \cos\phi \, d\phi$$

$$= \frac{1}{2}\pi a^4 \left[\frac{1}{2}\sin^2\phi \right]_0^{\pi/2} = \frac{1}{4}\pi a^4.$$

Therefore the centroid of the hemispherical solid is located at the point $\left(0,\, 0,\, \dfrac{3}{8}a \right)$.

C14S07.023: The plane $z = 1$ has the spherical-coordinates equation $\rho = \sec\phi$; the cone with cylindrical-coordinates equation $r = z$ has spherical-coordinates equation $\phi = \frac{1}{4}\pi$. Hence the volume bounded by the plane and the cone is

$$V = \int_{\theta=0}^{2\pi} \int_{\phi=0}^{\pi/4} \int_{\rho=0}^{\sec\phi} \rho^2 \sin\phi \, d\rho \, d\phi \, d\theta = 2\pi \int_{\phi=0}^{\pi/4} \frac{1}{3}\sec^2\phi \tan\phi \, d\phi = 2\pi \cdot \left[\frac{1}{6}\sec^2\phi \right]_{\phi=0}^{\pi/4} = \frac{1}{3}\pi.$$

C14S07.025: Assume unit density. Then the mass and the volume are numerically the same; they and the moments are

$$m = V = \int_0^{2\pi} \int_0^{\pi/4} \int_0^{a} \rho^2 \sin\phi \, d\rho \, d\phi \, d\theta = 2\pi \cdot \frac{1}{3}a^3 \left[-\cos\phi \right]_0^{\pi/4}$$

$$= \frac{2}{3}\pi a^2 \left(1 - \frac{1}{2}\sqrt{2} \right) = \frac{1}{3}\pi \left(2 - \sqrt{2} \right) a^3;$$

$$M_{yz} = \int_0^{2\pi} \int_0^{\pi/4} \int_0^{a} \rho^3 \sin^2\phi \cos\theta \, d\rho \, d\phi \, d\theta = \left(\int_0^{\pi/4} \int_0^{a} \rho^3 \sin^2\phi \, d\rho \, d\phi \right) \cdot \left[\sin\theta \right]_0^{2\pi} = 0;$$

$$M_{xz} = \int_0^{2\pi} \int_0^{\pi/4} \int_0^{a} \rho^3 \sin^2\phi \sin\theta \, d\rho \, d\phi \, d\theta = \left(\int_0^{\pi/4} \int_0^{a} \rho^3 \sin^2\phi \, d\rho \, d\phi \right) \cdot \left[-\cos\theta \right]_0^{2\pi} = 0;$$

$$M_{xy} = \int_0^{2\pi} \int_0^{\pi/4} \int_0^{a} \rho^3 \sin\phi \cos\phi \, d\rho \, d\phi \, d\theta = 2\pi \cdot \frac{1}{4}a^4 \cdot \left[\frac{1}{2}\sin^2\phi \right]_0^{\pi/4} = \frac{1}{8}\pi a^4.$$

So the z-coordinate of the centroid is

$$\bar{z} = \frac{3\pi a^4}{8\pi \left(2 - \sqrt{2} \right) a^3} = \frac{3a}{8\left(2 - \sqrt{2} \right)} = \frac{3\left(2 + \sqrt{2} \right) a}{16} = \frac{3}{16}\left(2 + \sqrt{2} \right) a;$$

clearly $\bar{x} = \bar{y} = 0$. For a plausibility check, note that $\bar{z} \approx (0.6401650429449553)a$.

C14S07.027: Set up a coordinate system so that the center of the sphere is at the point with Cartesian coordinates $(a, 0, 0)$. Then its Cartesian equation is

$$(x - a)^2 + y^2 + z^2 = a^2; \quad x^2 - 2ax + y^2 + z^2 = 0;$$

and thus it has spherical-coordinates equation $\rho = 2a \sin\phi \cos\theta$ (but note that θ ranges from $-\frac{1}{2}\pi$ to $\frac{1}{2}\pi$). We plan to find its moment of inertia with respect to the z-axis, and the square of the distance of a point of the sphere from the z-axis is

$$x^2 + y^2 = \rho^2 \sin^2 \phi \cos^2 \theta + \rho^2 \sin^2 \phi \sin^2 \theta = \rho^2 \sin^2 \phi.$$

Moreover, if the sphere has mass m and constant density δ, then we also have $m = \frac{4}{3}\pi a^3 \delta$. Finally,

$$I_z = \int_{-\pi/2}^{\pi/2} \int_0^{\pi} \int_0^{2a \sin \phi \cos \theta} \delta \rho^2 \sin^2 \phi \, d\rho \, d\phi \, d\theta = 2\delta \int_0^{\pi/2} \int_0^{\pi} \frac{1}{5}(2a \sin \phi \cos \theta)^5 \sin^3 \phi \, d\phi \, d\theta$$

$$= \frac{64}{5}\delta a^5 \int_0^{\pi/2} \int_0^{\pi} \sin^8 \phi \cos^5 \theta \, d\phi \, d\theta = \frac{128}{5}\delta a^5 \int_0^{\pi/2} \int_0^{\pi/2} \sin^8 \phi \cos^5 \theta \, d\phi \, d\theta$$

$$= \frac{128}{5}\delta a^5 \cdot \left(\int_0^{\pi/2} \sin^8 \phi \, d\phi \right) \cdot \left(\int_0^{\pi/2} \cos^5 \theta \, d\theta \right).$$

Then Formula (113) from the long table of integrals (see the endpapers) yields

$$I_z = \frac{128}{5}\delta a^5 \cdot \frac{1}{2} \cdot \frac{3}{4} \cdot \frac{5}{6} \cdot \frac{7}{8} \cdot \frac{\pi}{2} \cdot \frac{2}{3} \cdot \frac{4}{5} = \frac{28}{15}\pi \delta a^5 = \frac{4}{3}\pi \delta a^3 \cdot \frac{7}{5}a^2 = \frac{7}{5}ma^2.$$

C14S07.029: The surface with spherical-coordinates equation $\rho = 2a \sin \phi$ is generated as follows. Draw the circle in the xz-plane with center $(a, 0)$ and radius a. Rotate this circle around the z-axis. This generates the surface with the given equation. It is called a *pinched torus*—a doughnut with an infinitesimal hole. Its volume is

$$V = \int_{\theta=0}^{2\pi} \int_{\phi=0}^{\pi} \int_{\rho=0}^{2a \sin \phi} \rho^2 \sin \phi \, d\rho \, d\phi \, d\theta = 2\pi \int_{\phi=0}^{\pi} \frac{1}{3}(2a \sin \phi)^3 \sin \phi \, d\phi$$

$$= \frac{16}{3}\pi a^3 \cdot 2 \int_{\phi=0}^{\pi/2} \sin^4 \phi \, d\phi = \frac{32}{3}\pi a^3 \cdot \frac{1}{2} \cdot \frac{3}{4} \cdot \frac{\pi}{2} = 2\pi^2 a^3.$$

We evaluated the last integral with the aid of Formula (113) from the long table of integrals (see the endpapers). The volume of the pinched torus is also easy to evaluate using the first theorem of **Pappus** (Section 14.5).

C14S07.031: Assuming constant density δ, we have

$$I_x = 2 \int_0^{2\pi} \int_0^a \int_0^{\sqrt{4a^2 - r^2}} \delta(r^2 \sin^2 \theta + z^2) \cdot r \, dz \, dr \, d\theta$$

$$= 2\delta \int_0^{2\pi} \int_0^a \left(\frac{1}{3}r(4a^2 - r^2)^{3/2} + r^3(4a^2 - r^2)^{1/2} \sin^2 \theta \right) dr \, d\theta$$

$$= 2\delta \int_0^{2\pi} \frac{1}{30} \left[(4a^2 - r^2)^{3/2} \left[(8a^2 + 3r^2) \cos 2\theta - 16a^2 - r^2 \right] \right]_{r=0}^a d\theta$$

$$= 2\delta \int_0^{2\pi} \frac{1}{30}a^5 \left[128 - 51\sqrt{3} + \left(33\sqrt{3} - 64 \right) \cos 2\theta \right] d\theta$$

$$= \frac{1}{30}\delta a^5 \left[\left(128 - 5\sqrt{3} \right) \cdot 2\theta + \left(33\sqrt{3} - 64 \right) \sin 2\theta \right]_0^{2\pi} = \frac{2}{15} \left(128 - 51\sqrt{3} \right) \delta \pi a^5.$$

C14S07.033: If the density at (x, y, z) of the ice-cream cone is z, then its mass and moments are

$$m = \int_0^{2\pi} \int_0^{\pi/6} \int_0^{2a\cos\phi} \rho^3 \sin\phi \cos\phi \, d\rho \, d\phi \, d\theta = 2\pi \int_0^{\pi/6} 4a^4 \sin\phi \cos^5\phi \, d\phi$$

$$= \left[-\frac{4}{3}\pi a^4 \cos^6\phi \right]_0^{\pi/6} = \frac{37}{48}\pi a^4;$$

$$M_{yz} = \int_0^{2\pi} \left(\int_0^{\pi/6} \int_0^{2a\cos\phi} \rho^4 \sin^2\phi \cos\phi \, d\rho \, d\phi \right) \cos\theta \, d\theta$$

$$= \left(\int_0^{\pi/6} \int_0^{2a\cos\phi} \rho^4 \sin^2\phi \cos\phi \, d\rho \, d\phi \right) \cdot \left[\sin\theta \right]_0^{2\pi} = 0;$$

$$M_{xz} = 0 \quad \text{(by a similar computation)};$$

$$M_{xy} = \int_0^{2\pi} \int_0^{\pi/6} \int_0^{2a\cos\phi} \rho^4 \sin\phi \cos^2\phi \, d\rho \, d\phi \, d\theta = 2\pi \int_0^{\pi/6} \frac{32}{5} a^5 \cos^7\phi \sin\phi \, d\phi$$

$$= 2\pi \cdot \left[-\frac{4}{5}a^5 \cos^8\phi \right]_0^{\pi/6} = \frac{35}{32}\pi a^5.$$

Hence the centroid is located at the point $\left(0, 0, \dfrac{105}{74}a \right)$.

C14S07.035: The similar star with uniform density k has mass $m_2 = \frac{4}{3}k\pi a^3$. The other star has mass

$$m_1 = \int_0^{2\pi} \int_0^{\pi} \int_0^{a} k\left[1 - \left(\frac{\rho}{a} \right)^2 \right] \cdot \rho^2 \sin\phi \, d\rho \, d\phi \, d\theta = 2\pi \int_0^{\pi} \left[\frac{5ka^2\rho^3 \sin\phi - 3k\rho^5 \sin\phi}{15a^2} \right]_0^{a} d\phi$$

$$= 2\pi \int_0^{\pi} \frac{2}{15}ka^3 \sin\phi \, d\phi = \left[-\frac{4}{15}k\pi a^3 \cos\phi \right]_0^{\pi} = \frac{8}{15}k\pi a^3.$$

Finally, $\dfrac{m_1}{m_2} = \dfrac{2}{5}$.

C14S07.037: The given triple integral takes the following form:

$$\int_0^{2\pi} \int_0^{\pi} \int_0^{a} \rho^2 \exp\left(-\rho^3 \right) \sin\phi \, d\rho \, d\phi \, d\theta = 2\pi \int_0^{\pi} \left[-\frac{1}{3} \exp\left(-\rho^3 \right) \sin\phi \right]_0^{a} d\phi$$

$$= 2\pi \int_0^{\pi} \frac{1}{3}\left[1 - \exp\left(-a^3 \right) \right] \sin\phi \, d\phi = \frac{4}{3}\pi\left[1 - \exp\left(-a^3 \right) \right].$$

Clearly the value of the integral approaches $\dfrac{4}{3}\pi$ as $a \to +\infty$.

C14S07.039: Let $V = \frac{4}{3}\pi a^3$, the volume of the ball. The average distance of points of such a ball from its center is then

$$\bar{d} = \frac{1}{V} \int_0^{2\pi} \int_0^{\pi} \int_0^{a} \rho^3 \sin\phi \, d\rho \, d\phi \, d\theta = \frac{2\pi}{V} \int_0^{\pi} \frac{1}{4}a^4 \sin\phi \, d\phi = \frac{\pi a^4}{2V} \left[-\cos\phi \right]_0^{\pi} = \frac{\pi a^4}{V} = \frac{3}{4}a.$$

Note that the answer is both plausible and dimensionally correct.

C14S07.041: A *Mathematica* solution:

```
m = Integrate[ Integrate[ delta*a∧2*Sin[ phi ], { phi, 0, Pi } ],
            { theta, 0, 2*Pi } ]
```

$$4\pi a^2 \delta$$

```
r = a*Sin[ phi ];
IO = Integrate[ Integrate[ r∧2*delta*a∧2*Sin[ phi ], { phi, 0, Pi } ],
            { theta, 0, 2*Pi } ]
```

$$\frac{8}{3}\pi a^4 \delta$$

```
IO/m
```

$$\frac{2}{3}a^2$$ —C.H.E.

Second solution, by hand: The spherical surface S of radius a is described by $\rho = a$, $0 \leq \phi \leq \pi$, $0 \leq \theta \leq 2\pi$. Hence its moment of inertia with respect to the z-axis is

$$I_z = \iint_S \delta(x^2 + y^2)\, dA = \int_{\theta=0}^{2\pi} \int_{\phi=0}^{\pi} \delta a^4 \sin^3 \phi \, d\phi \, d\theta$$

$$= 2\pi\delta a^4 \int_0^\pi (1 - \cos^2 \phi) \sin \phi \, d\phi = 2\pi\delta a^4 \left[\frac{1}{3}\cos^3 \phi - \cos \phi \right]_0^\pi = \left(\frac{2}{3}a^2 \right) \cdot 4\pi\delta a^2 = \frac{2}{3}ma^2.$$

C14S07.043: A *Mathematica* solution:

```
z = Sqrt[ b∧2 - a∧2 ];
m = 2*Integrate[ Integrate[ delta*r*z, { r, a, b } ], { theta, 0, 2*Pi } ]
```

$$\frac{4}{3}\pi\delta(b^2 - a^2)^{3/2}$$

```
IO = 2*Integrate[ Integrate[ delta*r∧3*z, { r, a, b } ], { theta, 0, 2*Pi } ]
```

$$\frac{4}{15}\pi\delta(2b^4 + b^2 a^2 - 3a^4)\sqrt{b^2 - a^2}$$

```
Simplify[ IO/m ]
```

$$\frac{1}{5}(3a^2 + 2b^2)$$ —C.H.E.

Second solution, by hand: Choose a coordinate system so that the z-axis is the axis of symmetry of the sphere-with-hole. The central cross section of the solid in the xz-plane is bounded by the circle with polar (or cylindrical coordinates) equation $r^2 + z^2 = b^2$. Hence the mass of the solid is

$$m = 2 \int_{\theta=0}^{2\pi} \int_{r=a}^{b} \int_{z=0}^{\sqrt{b^2-r^2}} \delta r \, dz \, dr \, d\theta = 4\pi\delta \int_a^b r(b^2 - r^2)^{1/2} \, dr$$

841

$$= 4\pi\delta\left[-\frac{1}{3}(b^2 - r^2)^{3/2}\right]_a^b = \frac{4}{3}\pi\delta(b^2 - a^2)^{3/2}.$$

The moment of inertia of this solid with respect to the z-axis is

$$I_z = 2\int_{\theta=0}^{2\pi}\int_{r=a}^{b}\int_{z=0}^{\sqrt{b^2-r^2}} \delta r^3 \, dz \, dr \, d\theta = 4\pi\delta\int_a^b r^3(b^2 - r^2)^{1/2} \, dr.$$

Integration by parts with $u = r^2$, $dv = (b^2 - r^2)^{1/2} \, dr$, so that

$$du = 2r \, dr \quad \text{and} \quad v = -\frac{1}{3}(b^2 - r^2)^{3/2},$$

then yields

$$I_z = 4\pi\delta\left(\left[-\frac{1}{3}r^2(b^2 - r^2)^{3/2}\right]_a^b + \frac{2}{3}\int_a^b r(b^2 - r^2)^{3/2} \, dr\right)$$

$$= 4\pi\delta\left(\frac{1}{3}a^2(b^2 - a^2)^{3/2} + \frac{2}{3}\left[-\frac{1}{5}(b^2 - r^2)^{5/2}\right]_a^b\right)$$

$$= \frac{4}{15}\pi\delta\left[5a^2(b^2 - a^2)^{3/2} + 2(b^2 - a^2)(b^2 - a^2)^{3/2}\right] = \frac{4}{15}\pi\delta(b^2 - a^2)^{3/2}(5a^2 + 2b^2 - 2a^2)$$

$$= \frac{4}{3}\pi\delta(b^2 - a^2)^{3/2}\cdot\frac{1}{5}(3a^2 + 2b^2) = \frac{1}{5}m(3a^2 + 2b^2).$$

C14S07.045: First we let $g(\phi, \theta) = 6 + 3\cos 3\theta \sin 5\phi$, so that the boundary of the bumpy sphere B has spherical equation $\rho = g(\phi, \theta)$. Then the volume V of B is simply

$$V = \int_{\theta=0}^{2\pi}\int_{\phi=0}^{\pi}\int_{\rho=0}^{g(\phi,\theta)} \rho^2 \sin\phi \, d\rho \, d\phi \, d\theta.$$

To evaluate V with the aid of *Mathematica 3.0*, we write r in place of ρ, f instead of ϕ, and t for θ, as usual. Then, to find the volume V of B step-by-step, we proceed as follows:

```
Integrate[ r*r*Sin[f], r ]
```

$$\frac{1}{3}\rho^3 \sin\phi$$

```
% /. r → g[f,t]
```

$$\frac{1}{3}(\sin\phi)(6 + 3\cos 3\theta \sin 5\phi)^3$$

```
Integrate[ %, f ]
```

$$\frac{3}{78848}\,[-2247168\cos\phi - 19712\cos 9\phi + 16128\cos 11\phi - 177408\cos(\phi - 6\theta)$$

$$- 9856\cos(9\phi - 6\theta) + 8064\cos(11\phi - 6\theta) - 177408\cos(\phi + 6\theta)$$

$$- 9856\cos(9\phi + 6\theta) + 8064\cos(11\phi + 6\theta) + 2772\sin(4\phi - 9\theta)$$

$$- 1848\sin(6\phi - 9\theta) - 264\sin(14\phi - 9\theta) + 231\sin(16\phi - 9\theta)$$

$$+ 185724\sin(4\phi - 3\theta) - 123816\sin(6\phi - 3\theta) - 792\sin(14\phi - 3\theta)$$

$$+ 693\sin(16\phi - 3\theta) + 185724\sin(4\phi + 3\theta) - 123816\sin(6\phi + 3\theta)$$

$$- 792\sin(14\phi + 3\theta) + 693\sin(16\phi + 3\theta) + 2772\sin(4\phi + 9\theta) - 1848\sin(6\phi + 9\theta)$$

$$- 264\sin(14\phi + 9\theta) + 231\sin(16\phi + 9\theta)\,]$$

```
(% /.  f → Pi) - (% /.  f → 0) // FullSimplify
```

$$\frac{12\,(157 + 25\cos 6\theta)}{11}$$

```
Integrate[ %, t ]
```

$$\frac{18840\theta + 50\sin 6\theta}{11}$$

```
(% /.  t → 2*Pi) - (% /.  t → 0)
```

$$\frac{3768\pi}{11}$$

```
N[ %, 60 ]
```

1076.13828352057644750247638801792426606959031178950362484939396

Thus $V = \frac{3768}{11}\pi \approx 1076.13828352$.

C14S07.047: The following figure makes some of the equations we use easy to derive.

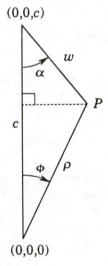

A mass element $\delta\,dV$ located at the point $P(\rho, \phi, \theta)$ of the ball exerts a force on the mass m at $(0, 0, c)$ that has vertical component

$$dF = -\frac{Gm\delta\cos\alpha}{w^2}\,dV.$$

Note the following:

$$M = \frac{4}{3}\pi\delta a^3;$$

843

$$w^2 = \rho^2 + c^2 - 2\rho c \cos\phi;$$

$$2w \, dw = 2\rho c \sin\phi \, d\phi;$$

$$w \cos\alpha + \rho \cos\phi = c;$$

$$\rho \cos\phi = \frac{\rho^2 + c^2 - w^2}{2c};$$

$$\rho \sin\phi \, d\phi = \frac{w}{c} \, dw.$$

Hence the total force exerted by the ball on m is

$$F = -\int_{\theta=0}^{2\pi} \int_{\rho=0}^{a} \int_{\phi=0}^{\pi} \frac{Gm\delta \cos\alpha}{w^2} \cdot \rho^2 \sin\phi \, d\phi \, d\rho \, d\theta = -\int_0^{2\pi} \int_0^a \int_0^\pi \frac{Gm\delta w \cos\alpha}{w^3} \cdot \rho^2 \sin\phi \, d\phi \, d\rho \, d\theta$$

$$= -2\pi \int_0^a \int_0^\pi \frac{Gm\delta}{w^3}(c - \rho\cos\phi) \cdot \rho^2 \sin\phi \, d\phi \, d\rho = -2\pi \int_0^a \int_{c-\rho}^{c+\rho} \frac{Gm\delta}{w^3}\left(c - \frac{\rho^2 + c^2 - w^2}{2c}\right) \cdot \frac{\rho w}{c} \, dw \, d\rho$$

$$= -2\pi \int_{\rho=0}^a \int_{w=c-\rho}^{c+\rho} \frac{Gm\delta}{w^3} \cdot \frac{2c^2 - \rho^2 - c^2 + w^2}{2c} \cdot \rho \cdot \frac{w}{c} \, dw \, d\rho$$

$$= -\pi Gm\delta \int_0^a \int_{c-\rho}^{c+\rho} \frac{1}{w^2} \cdot \frac{c^2 - \rho^2 + w^2}{c^2} \cdot \rho \, dw \, d\rho = -\frac{\pi Gm\delta}{c^2} \int_0^a \int_{c-\rho}^{c+\rho} \left(\frac{c^2 - \rho^2}{w^2} + 1\right) \cdot \rho \, dw \, d\rho$$

$$= -\frac{\pi Gm\delta}{c^2} \int_0^a \rho \cdot \left[\frac{\rho^2 - c^2}{w} + w\right]_{w=c-\rho}^{c+\rho} d\rho = -\frac{\pi Gm\delta}{c^2} \int_0^a \rho \cdot \left(\frac{\rho^2 - c^2}{\rho + c} + c + \rho + \frac{c^2 - \rho^2}{c - \rho} - c + \rho\right) d\rho$$

$$= -\frac{\pi Gm\delta}{c^2} \int_0^a \rho(\rho - c + \rho + c + c + \rho - c + \rho) \, d\rho = -\frac{\pi Gm\delta}{c^2} \int_0^a 4\rho^2 \, d\rho = -\frac{\pi Gm\delta}{c^2} \cdot \left[\frac{4}{3}\rho^3\right]_0^a$$

$$= -\frac{\pi Gm\delta}{c^2} \cdot \frac{4}{3} a^3 = -\frac{4}{3}\pi\delta a^3 \cdot \frac{Gm}{c^2} = -\frac{GMm}{c^2}.$$

Magnificent! You can even extend this result to show that if the density of the ball varies only as a function of ρ, then the same conclusion follows: The ball acts, for purposes of gravitational attraction of an external mass m, as if all its mass M were concentrated at its center. And note one additional item of interest: This is one of the extremely rare spherical triple integrals *not* evaluated in the order $d\rho \, d\phi \, d\theta$.

C14S07.049: A *Mathematica* solution:

```
r = 6370*1000;    k = 0.371;

d1 = 11000;    d2 = 5000;

m1 = (4/3)*Pi*d1*x^3;

m2 = (4/3)*Pi*d2*(r^3 - x^3);

m = m1 + m2;

i1 = (2/5)*m1*x^2;
```

By Problem 42, the moment of inertia of the mantle with respect to the polar axis (through the poles of the planet) is

```
i2 = (2/5)*m2*(r^5 - x^5)/(r^3 - x^3);
```

Therefore we proceed to solve for x as follows.

```
i0 = i1 + i2;

eq1 = i0 == k*m*r^2;

soln = NSolve[ eq1, x ];
```

We suppress the output, but there are five solutions. Only two are real and positive,

$$x_1 \approx 2.76447 \times 10^6 \quad \text{and} \quad x_2 \approx 5.87447 \times 10^6.$$

We are given the information that the mantle is "a few thousand kilometers thick," and x_2 does not satisfy this condition, as it implies that the mantle is less than 496 km thick. We conclude that the radius of the core is $x_1/1000$ km, and hence that the thickness of the mantle is $(r - x_1)/1000 \approx 3605.53$ km.   —C.H.E.

Section 14.8

C14S08.001: The surface area element is

$$dS = \sqrt{1 + 1^2 + 3^2} \; dA = \sqrt{11} \; dA.$$

So the area in question is

$$A = 4 \int_{x=0}^{2} \int_{y=0}^{3\sqrt{1-(x/2)^2}} \sqrt{11} \; dy \; dx = \int_0^2 6\sqrt{11} \sqrt{4 - x^2} \; dx$$

$$= \left[3x\sqrt{11} \sqrt{4 - x^2} + 12\sqrt{11} \arcsin\left(\frac{x}{2}\right) \right]_0^2 = 6\pi\sqrt{11} \approx 62.5169044565658738.$$

C14S08.003: The paraboloid meets the plane in the circle with equation $x^2 + y^2 = 4$, $z = 5$. Let D denote the circular disk $x^2 + y^2 \leqq 4$ in the xy-plane. The surface area element is

$$dS = \sqrt{1 + 4x^2 + 4y^2} \; dA,$$

so the surface area in question is

$$A = \iint_D \sqrt{1 + 4x^2 + 4y^2} \; dA = \int_{\theta=0}^{2\pi} \int_{r=0}^{2} r\sqrt{1 + 4r^2} \; dr \; d\theta = 2\pi \cdot \left[\frac{1}{12}(1 + 4r^2)^{3/2} \right]_0^2$$

$$= 2\pi \left(\frac{17\sqrt{17} - 1}{12} \right) = \frac{1}{6}\pi \left(17\sqrt{17} - 1 \right) \approx 36.1769031974114084.$$

C14S08.005: The surface area element is

$$dS = \sqrt{2 + 4y^2} \; dA,$$

845

so the area is

$$A = \int_{y=0}^{2} \int_{x=0}^{1} (2 + 4y^2)^{1/2} \, dx \, dy = \int_{y=0}^{2} (2 + 4y^2)^{1/2} \, dy = \left[\frac{1}{2} y(2 + 4y^2)^{1/2} + \frac{1}{2} \text{arcsinh} \left(y\sqrt{2} \right) \right]_{0}^{2}$$

$$= 3\sqrt{2} + \frac{1}{2} \text{arcsinh} \left(2\sqrt{2} \right) = 3\sqrt{2} + \frac{1}{2} \ln \left(3 + 2\sqrt{2} \right) \approx 5.1240142741388282.$$

C14S08.007: The surface area element is $dS = \sqrt{14} \, dA$, so the surface area is

$$A = \int_{x=0}^{3} \int_{y=0}^{(6-2x)/3} \sqrt{14} \, dy \, dx = \int_{0}^{3} \frac{1}{3} (6 - 2x)\sqrt{14} \, dx = \left[2x\sqrt{14} - \frac{1}{3} x^2 \sqrt{14} \right]_{0}^{3} = 3\sqrt{14}.$$

Alternatively, the vectors $\mathbf{u} = \langle -3, 2, 0 \rangle$ and $\mathbf{v} = \langle -3, 0, 6 \rangle$ span two adjacent sides of the triangular surface. So its area is half the magnitude of their cross product:

$$A = \frac{1}{2} |\mathbf{u} \times \mathbf{v}| = \frac{1}{2} |\langle 12, 18, 6 \rangle| = \frac{1}{2} \sqrt{504} = 3\sqrt{14} \approx 11.22497216032182415675.$$

C14S08.009: The surface area element is $dS = \sqrt{1 + x^2 + y^2} \, dA$, so the surface area is

$$A = \int_{\theta=0}^{2\pi} \int_{r=0}^{1} r(1 + r^2)^{1/2} \, dr \, d\theta = 2\pi \cdot \left[\frac{1}{3} (1 + r^2)^{3/2} \right]_{0}^{1} = \frac{2}{3} \pi \left(2\sqrt{2} - 1 \right) \approx 3.8294488151512928.$$

C14S08.011: The paraboloid meets the xy-plane in the circle with equation $x^2 + y^2 = 16$. The surface area element is

$$dS = \sqrt{1 + 4x^2 + 4y^2} \, dA,$$

so the area is

$$A = \int_{0}^{2\pi} \int_{0}^{4} r(1 + 4r^2)^{1/2} \, dr \, d\theta = 2\pi \cdot \left[\frac{1}{12} (1 + 4r^2)^{3/2} \right]_{0}^{4} = \frac{1}{6} \pi \left(65\sqrt{65} - 1 \right) \approx 273.866639786258.$$

C14S08.013: Let $\mathbf{r}(\theta, z) = \langle a \cos \theta, \, a \sin \theta, \, z \rangle$. Then

$$\mathbf{r}_\theta \times \mathbf{r}_z = \begin{vmatrix} \mathbf{i} & \mathbf{j} & \mathbf{k} \\ -a \sin \theta & a \cos \theta & 0 \\ 0 & 0 & 1 \end{vmatrix} = \langle a \cos \theta, \, a \sin \theta, \, 0 \rangle$$

and hence

$$|\mathbf{r}_\theta \times \mathbf{r}_z| = \sqrt{a^2 \cos^2 \theta + a^2 \sin^2 \theta} = a.$$

Therefore, by Eq. (8), the area of the zone is

$$A = \int_{\theta=0}^{2\pi} \int_{z=0}^{h} a \, dz \, d\theta = 2\pi \cdot \left[az \right]_{z=0}^{h} = 2\pi a h.$$

C14S08.015: Let $z(x, y) = \sqrt{a^2 - x^2}$. Then the surface area element in Cartesian coordinates is

$$dS = \frac{a}{\sqrt{a^2 - x^2}}\, dA.$$

Let D be the disk in which the vertical cylinder meets the xy-plane. Then the area of the part of the horizontal cylinder—top and bottom—that lies within the vertical cylinder is

$$A = 2 \iint_D \frac{a}{\sqrt{a^2 - x^2}}\, dA = \int_{\theta=0}^{2\pi} \int_{r=0}^{a} \frac{2a}{\sqrt{a^2 - r^2 \cos^2 \theta}} \cdot r\, dr\, d\theta$$

$$= \int_{\theta=0}^{2\pi} \left[\frac{-2a(a^2 - r^2 \cos^2 \theta)^{1/2}}{\cos^2 \theta} \right]_{r=0}^{a} d\theta = 4 \int_{\theta=0}^{\pi/2} 2a^2 (\sec^2 \theta - \sec \theta \tan \theta)\, d\theta = 8a^2 \left[\tan \theta - \sec \theta \right]_0^{\pi/2}$$

$$= 8a^2 + 8a^2 \cdot \left(\lim_{\theta \to (\pi/2)-} \left[\tan \theta - \sec \theta \right] \right) = 8a^2 + 8a^2 \cdot 0 = 8a^2.$$

The change in the limits of integration in the second line was necessary because we needed the simplification $(1 - \cos^2 \theta)^{1/2} = \sin \theta$, which is not valid if $\pi < \theta < 2\pi$; moreover, we needed to avoid the discontinuity of the following improper integral at $\theta = \pi/2$.

C14S08.017: The surface $y = f(x, z)$ is parametrized by

$$\mathbf{r}(x, z) = \langle x,\ f(x, z),\ z \rangle$$

for (x, z) in the region R in the xz-plane. Then

$$\mathbf{r}_x \times \mathbf{r}_z = \begin{vmatrix} \mathbf{i} & \mathbf{j} & \mathbf{k} \\ 1 & f_x & 0 \\ 0 & f_z & 1 \end{vmatrix} = \langle f_x,\ -1,\ f_z \rangle.$$

Therefore the area of the surface $y = f(x, z)$ lying "over" the region R is

$$A = \iint_R \sqrt{1 + \left(\frac{\partial f}{\partial x} \right)^2 + \left(\frac{\partial f}{\partial z} \right)^2}\ dx\, dz.$$

The surface $x = f(y, z)$ is parametrized by

$$\mathbf{r}(y, z) = \langle f(y, z),\ y,\ z \rangle$$

for (y, z) in the region R in the xz-plane. Then

$$\mathbf{r}_y \times \mathbf{r}_z = \begin{vmatrix} \mathbf{i} & \mathbf{j} & \mathbf{k} \\ f_y & 1 & 0 \\ f_z & 0 & 1 \end{vmatrix} = \langle 1,\ -f_y,\ -f_z \rangle.$$

Therefore the area of the surface $x = f(y, z)$ lying "over" the region R is

$$A = \iint_R \sqrt{1 + \left(\frac{\partial f}{\partial y} \right)^2 + \left(\frac{\partial f}{\partial z} \right)^2}\ dy\, dz.$$

847

C14S08.019: By Problem 50 of Section 14.2,

$$A = \int_{\theta_1}^{\theta_2} \int_{\phi_1}^{\phi_2} a^2 \sin\phi \, d\phi \, d\theta = a^2 \sin\hat{\phi} \, \Delta\phi \, \Delta\theta$$

for some $\hat{\phi}$ in (ϕ_1, ϕ_2). Therefore

$$\Delta V = \int_{\rho_1}^{\rho_2} (\rho^2 \sin\hat{\phi} \, \Delta\phi \, \Delta\theta) \, d\rho = \frac{1}{3}(\rho_2^3 - \rho_1^3)\sin\hat{\phi} \, \Delta\phi \, \Delta\theta$$

$$= \frac{1}{3} \cdot \frac{\rho_2^3 - \rho_1^3}{\rho_2 - \rho_1}\sin\hat{\phi} \, \Delta\phi \, \Delta\theta \, \Delta\rho = \frac{1}{3} \cdot 3\hat{\rho}^2 \sin\hat{\phi} \, \Delta\phi \, \Delta\theta \, \Delta\rho = \hat{\rho}^2 \sin\hat{\phi} \, \Delta\rho \, \Delta\phi \, \Delta\theta$$

for some $\hat{\rho}$ in (ρ_1, ρ_2), and this is Eq. (8) of Section 14.7.

C14S08.021: Given:

$$x = f(z)\cos\theta, \quad y = f(z)\sin\theta, \quad z = z$$

where $0 \leqq \theta \leqq 2\pi$ and $a \leqq z \leqq b$. The surface thereby generated is thereby parametrized by

$$\mathbf{r}(\theta, z) = \langle f(z)\cos\theta, \ f(z)\sin\theta, \ z \rangle,$$

and thus

$$\mathbf{r}_\theta = \langle -f(z)\sin\theta, \ f(z)\cos\theta, \ 0 \rangle \quad \text{and}$$

$$\mathbf{r}_z = \langle f'(z)\cos\theta, \ f'(z)\sin\theta, \ 1 \rangle.$$

Therefore

$$\mathbf{r}_\theta \times \mathbf{r}_z = \begin{vmatrix} \mathbf{i} & \mathbf{j} & \mathbf{k} \\ -f(z)\sin\theta & f(z)\cos\theta & 0 \\ f'(z)\cos\theta & f'(z)\sin\theta & 1 \end{vmatrix} = \langle f(z)\cos\theta, \ f(z)\sin\theta, \ -f(z)\cdot f'(z) \rangle$$

and hence

$$|\mathbf{r}_\theta \times \mathbf{r}_z| = \sqrt{[f(z)]^2 + [f(z)\cdot f'(z)]^2} = f(z)\sqrt{1 + [f'(z)]^2}.$$

Thus, by Eq. (8), the area of the surface of revolution is

$$A = \int_{\theta=0}^{2\pi} \int_{z=a}^{b} f(z)\sqrt{1 + [f'(z)]^2} \, dz \, d\theta = \int_{z=a}^{b} 2\pi f(z)\sqrt{1 + [f'(z)]^2} \, dz.$$

Compare this with Eq. (8) in Section 6.4.

C14S08.023: In the result in Problem 21, take $f(z) = r$ (the constant radius of the cylinder). Then

$$f(z)\sqrt{1 + [f'(z)]^2} = r,$$

so the curved surface area of the cylinder is

848

$$A = \int_{\theta=0}^{2\pi} \int_{z=0}^{h} r \, dz \, d\theta = 2\pi \cdot \left[rz \right]_{z=0}^{h} = 2\pi rh.$$

C14S08.025: Part (a): The *Mathematica 3.0* command

```
Plot3D[ x*x + y*y, {x, -1, 1}, {y, -1, 1}, AspectRatio → Automatic ];
```

produced the view of the surface that is shown next.

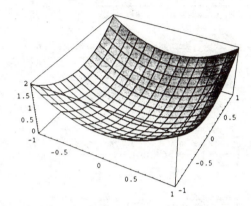

The surface area element is $dS = \sqrt{1 + 4x^2 + 4y^2} \, dA$, so the area of the surface is

$$A = \int_{x=-1}^{1} \int_{y=-1}^{1} \sqrt{1 + 4x^2 + 4y^2} \, dy \, dx$$

$$= \int_{-1}^{1} \left[\frac{1}{2} y \sqrt{1 + 4x^2 + 4y^2} + \frac{1}{4}(4x^2 + 1) \ln \left(2y + \sqrt{1 + 4x^2 + 4y^2} \right) \right]_{-1}^{1} dx$$

$$= \int_{-1}^{1} \left[\sqrt{4x^2 + 5} - \frac{1}{4}(4x^2 + 1) \ln \left(-2 + \sqrt{4x^2 + 5} \right) + \frac{1}{4}(4x^2 + 1) \ln \left(2 + \sqrt{4x^2 + 5} \right) \right] dx$$

$$= \left[\frac{2}{3} x \sqrt{4x^2 + 5} + \frac{7}{6} \operatorname{arcsinh} \left(\frac{2x}{\sqrt{5}} \right) - \frac{1}{6} \arctan \left(\frac{4x}{\sqrt{4x^2 + 5}} \right) \right.$$

$$\left. + \frac{1}{12} x(4x^2 + 3) \ln \left(-2 + \sqrt{4x^2 + 5} \right) + \frac{1}{12} x(4x^2 + 3) \ln \left(2 + \sqrt{4x^2 + 5} \right) \right]_{-1}^{1}$$

$$= 4 + \frac{7}{3} \operatorname{arcsinh} \left(\frac{2\sqrt{5}}{5} \right) - \frac{1}{3} \arctan \left(\frac{4}{3} \right) + \frac{7}{6} \ln 5 \approx 7.44625672301236346326.$$

Part (b): By symmetry, we integrate over the quarter of the square that lies in the first quadrant and multiply by 4. Thus the area is

$$A = 4 \int_{x=0}^{1} \int_{y=0}^{1-x} \sqrt{1 + 4x^2 + 4y^2} \, dy \, dx$$

$$= \int_{0}^{1} \left[2y \sqrt{1 + 4x^2 + 4y^2} + (4x^2 + 1) \ln \left(2y + \sqrt{1 + 4x^2 + 4y^2} \right) \right]_{0}^{1-x} dx$$

$$= \int_0^1 \left[2(1-x)\sqrt{1+4(1-x)^2+4x^2} - (1+4x^2)\ln\left(\sqrt{1+4x^2}\right) \right.$$

$$\left. + (1+4x^2)\ln\left(2(1-x)+\sqrt{1+4(1-x)^2+x^2}\right) \right] dx$$

$$= \left[\frac{1}{3}(4x-2x^2-1)\sqrt{8x^2-8x+5} + \frac{5\sqrt{2}}{6}\operatorname{arcsinh}\left(\frac{\sqrt{6}}{3}[2x-1]\right) \right.$$

$$- \frac{1}{6}\arctan\left(\frac{72+288x^2-25(4x+1)\sqrt{8x^2-8x+5}}{184x^2-200x-29}\right)$$

$$+ \frac{1}{6}\arctan\left(\frac{-72-288x^2+25(4x+1)\sqrt{8x^2-8x+5}}{184x^2-200x-29}\right)$$

$$\left. - \frac{1}{3}x(4x^2+3)\ln\left(\sqrt{1+4x^2}\right) + \frac{1}{3}x(4x^2+3)\ln\left(2-2x+\sqrt{8x^2-8x+5}\right) \right]_0^1$$

$$= \frac{2\sqrt{5}}{3} + \frac{5\sqrt{2}}{3}\operatorname{arcsinh}\left(\frac{\sqrt{6}}{3}\right) - \frac{1}{6}\arctan\left(\frac{72-25\sqrt{5}}{71}\right) - \frac{1}{6}\arctan\left(\frac{72-25\sqrt{5}}{29}\right)$$

$$+ \frac{1}{6}\arctan\left(\frac{25\sqrt{5}-72}{29}\right) + \frac{1}{6}\arctan\left(\frac{25\sqrt{5}-72}{71}\right) \approx 3.0046254342814410.$$

Of course it was *Mathematica* 3.0 that computed and evaluated the antiderivatives in this solution.

C14S08.027: Part (a): The following graph of the surface was generated by the *Mathematica* 3.0 command

```
Plot3D[ 1 + x*y, {x, -1, 1}, {y, -1, 1} ];
```

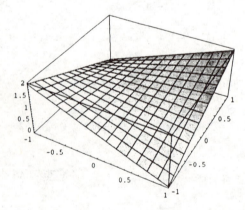

The surface area element is $dS = \sqrt{1+x^2+y^2}\, dA$. Hence the surface area is

$$\int_{x=-1}^{1}\int_{y=-1}^{1}\sqrt{1+x^2+y^2}\, dy\, dx = \int_{-1}^{1}\left[\frac{1}{2}y\sqrt{1+x^2+y^2} + \frac{1}{2}(1+x^2)\ln\left(y+\sqrt{1+x^2+y^2}\right)\right]_{-1}^{1} dx$$

$$= \int_{-1}^{1}\left[\sqrt{2+x^2} - \frac{1}{2}(1+x^2)\ln\left(-1+\sqrt{2+x^2}\right) + \frac{1}{2}(1+x^2)\ln\left(1+\sqrt{2+x^2}\right)\right] dx$$

$$= \left[\frac{2}{3}x\sqrt{2+x^2} + \frac{4}{3}\text{arcsinh}\left(\frac{x\sqrt{2}}{2}\right) - \frac{2}{3}\arctan\left(\frac{x}{\sqrt{2+x^2}}\right)\right.$$

$$\left. - \frac{1}{6}x(3+x^2)\ln\left(-1+\sqrt{2+x^2}\right) + \frac{1}{6}x(3+x^2)\ln\left(1+\sqrt{2+x^2}\right)\right]_{-1}^{1}$$

$$= \frac{4\sqrt{3}}{3} - \frac{2}{9}\pi + \frac{8}{3}\text{arcsinh}\left(\frac{\sqrt{2}}{2}\right) - \frac{4}{3}\ln\left(-1+\sqrt{3}\right) + \frac{4}{3}\ln\left(1+\sqrt{3}\right) \approx 5.123157101094.$$

Part (b): Using symmetry, we integrate over the quarter of the square in the first quadrant and multiply the answer by 4. Thus the surface area is

$$4\int_{x=0}^{1}\int_{y=0}^{1-x}\sqrt{1+x^2+y^2}\, dy\, dx = \int_{0}^{1}\left[2y\sqrt{1+x^2+y^2} + 2(1+x^2)\ln\left(y+\sqrt{1+x^2+y^2}\right)\right]_{0}^{1} dx$$

$$= \int_{0}^{1}\left[2(1-x)\sqrt{2x^2-2x+2} - 2(1+x^2)\ln\left(\sqrt{1+x^2}\right)\right.$$

$$\left. + 2(1+x^2)\ln\left(1-x+\sqrt{2x^2-2x+2}\right)\right] dx$$

$$= \left[\frac{1}{3}(4x-2x^2-1)\sqrt{2x^2-2x+2} + \frac{7\sqrt{2}}{6}\text{arcsinh}\left(\frac{2x-1}{\sqrt{3}}\right) + \frac{4}{3}\arctan\left(\frac{\sqrt{2x^2-2x+2}}{x+1}\right)\right.$$

$$\left. - \frac{2}{3}x(3+x^2)\ln\left(\sqrt{1+x^2}\right) + \frac{2}{3}x(3+x^2)\ln\left(1-x+\sqrt{2x^2-2x+2}\right)\right]_{0}^{1}$$

$$= \frac{2\sqrt{2}}{3} + \frac{7\sqrt{2}}{3}\text{arcsinh}\left(\frac{\sqrt{3}}{3}\right) + \frac{4}{3}\arctan\left(\frac{\sqrt{2}}{2}\right) - \frac{4}{3}\arctan\left(\sqrt{2}\right) \approx 2.302310960471.$$

C14S08.029: We readily verify that x, y, and z satisfy the equation

$$\frac{x^2}{a^2} + \frac{y^2}{b^2} = \frac{z}{c}$$

of an elliptic paraboloid. For a typical graph, we executed the *Mathematica* commmands

```
a = 2; b = 1; c = 3;

x = a*u*Cos[v]; y = b*u*Sin[v]; z = c*u^2;

ParametricPlot3D[ { x, y, z }, { u, 0, 1 }, { v, 0, 2*Pi } ];
```

851

and the resulting graph is next. —C.H.E.

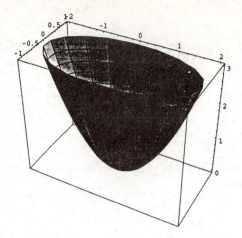

C14S08.031: We readily verify that x, y, and z satisfy the equation

$$\frac{z^2}{c^2} - \frac{x^2}{a^2} - \frac{y^2}{b^2} = 1$$

of a hyperboloid of two sheets. To see the upper half of a typical example, we executed the following *Mathematica* commands:

```
a = 2; b = 1; c = 4;

x = a*Sinh[u]*Cos[v]; y = b*Sinh[u]*Sin[v]; z = c*Cosh[u];

ParametricPlot3D[ { x, y, z }, { u, 0, 1 }, { v, 0, 2*Pi } ];
```

and the result is next. —C.H.E.

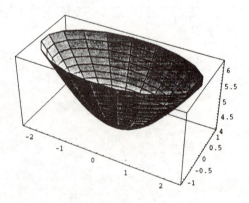

C14S08.033: The ellipsoid is parametrized via

$$\mathbf{r}(\phi,\,\theta) = \langle 4\sin\phi\cos\theta,\; 3\sin\phi\sin\theta,\; 2cos\phi\rangle.$$

Then

852

$$\mathbf{r}_\phi = \langle 4\cos\phi\cos\theta,\ 3\cos\phi\sin\theta,\ -2\sin\phi\rangle \quad \text{and}$$

$$\mathbf{r}_\theta = \langle -4\sin\phi\sin\theta,\ 3\sin\phi\cos\theta,\ 0\rangle.$$

It now follows that

$$\mathbf{r}_\phi \times \mathbf{r}_\theta = \langle 6\sin^2\phi\cos\theta,\ 8\sin^2\phi\sin\theta,\ 12\sin\phi\cos\phi\rangle,$$

so that

$$|\mathbf{r}_\phi \times \mathbf{r}_\theta| = \sqrt{36\sin^4\phi\cos^2\theta + 144\sin^2\phi\cos^2\phi + 64\sin^4\phi\sin^2\theta}\,.$$

We used the **NIntegrate** command in *Mathematica* 3.0 to approximate the surface area of the ellipsoid; it is

$$\int_{\theta=0}^{2\pi}\int_{\phi=0}^{\pi} (36\sin^4\phi\cos^2\theta + 144\sin^2\phi\cos^2\phi + 64\sin^4\phi\sin^2\theta)^{1/2}\,d\phi\,d\theta \approx 111.545774984838.$$

Section 14.9

C14S09.001: It is easy to solve the given equations for

$$x = \frac{u+v}{2}, \quad y = \frac{u-v}{2}.$$

Hence

$$\frac{\partial(x,\,y)}{\partial(u,\,v)} = \begin{vmatrix} \dfrac{1}{2} & \dfrac{1}{2} \\[2mm] \dfrac{1}{2} & -\dfrac{1}{2} \end{vmatrix} = -\frac{1}{2}.$$

C14S09.003: When we solve the equations $u = xy$ and $v = y/x$ for x and y, we find that there are two solutions:

$$x = \sqrt{\frac{u}{v}}, \quad y = \sqrt{uv} \quad \text{and} \quad x = -\sqrt{\frac{u}{v}}, \quad y = -\sqrt{uv}.$$

It doesn't matter which we choose; the value of the Jacobian will be the same. (Why?) So we choose the first solution. Then

$$\frac{\partial(x,\,y)}{\partial(u,\,v)} = \begin{vmatrix} \dfrac{1}{2u^{1/2}v^{1/2}} & -\dfrac{u^{1/2}}{2v^{3/2}} \\[3mm] \dfrac{v^{1/2}}{2u^{1/2}} & \dfrac{u^{1/2}}{2v^{1/2}} \end{vmatrix} = \frac{1}{2v}.$$

C14S09.005: When we solve the equations $u = x + 2y^2$, $v = x - 2y^2$ for x and y, we get two solutions:

$$x = \frac{u+v}{2}, \quad y = \pm\frac{\sqrt{u-v}}{2}.$$

We choose the solution for which y is nonnegative. Then

853

$$\frac{\partial(x, y)}{\partial(u, v)} = \begin{vmatrix} \dfrac{1}{2} & \dfrac{1}{2} \\[2ex] \dfrac{1}{4\sqrt{u-v}} & -\dfrac{1}{4\sqrt{u-v}} \end{vmatrix} = -\frac{1}{4\sqrt{u-v}}.$$

C14S09.007: First we solve the equations $u = x + y$ and $v = 2x - 3y$ for

$$x = \frac{3u + v}{5}, \qquad y = \frac{2u - v}{5}.$$

Substitution in the equation $x + y = 1$ then yields

$$1 = \frac{3u + v}{5} + \frac{2u - v}{5} = \frac{5u}{5} = u.$$

Similarly, $x + y = 2$ yields $u = 2$, $2x - 3y = 2$ yields $v = 2$, and $2x - 3y = 5$ yields $v = 5$. Moreover,

$$\frac{\partial(x, y)}{\partial(u, v)} = \begin{vmatrix} \dfrac{3}{5} & \dfrac{1}{5} \\[2ex] \dfrac{2}{5} & -\dfrac{1}{5} \end{vmatrix} = -\frac{1}{5}.$$

Therefore

$$\iint_R 1 \, dy \, dx = \int_{v=2}^{5} \int_{u=1}^{2} \frac{1}{5} \, du \, dv = 3 \cdot 1 \cdot \frac{1}{5} = \frac{3}{5}.$$

Note: Because R is a parallelogram with adjacent sides represented by the two vectors $\mathbf{a} = \langle \frac{3}{5}, \frac{2}{5}, 0 \rangle$ and $\mathbf{b} = \langle \frac{3}{5}, -\frac{3}{5}, 0 \rangle$, we have the following alternative method of finding the area A of R:

$$\mathbf{a} \times \mathbf{b} = \begin{vmatrix} \mathbf{i} & \mathbf{j} & \mathbf{k} \\[1ex] \dfrac{3}{5} & \dfrac{2}{5} & 0 \\[2ex] \dfrac{3}{5} & -\dfrac{3}{5} & 0 \end{vmatrix} = \left\langle 0, 0, -\frac{3}{5} \right\rangle,$$

and therefore $A = |\mathbf{a} \times \mathbf{b}| = \dfrac{3}{5}$.

C14S09.009: If $u = xy$ and $v = xy^3$, then

$$uy^2 = xy^3 = v, \qquad \text{so that} \quad y^2 = \frac{v}{u}; \quad y = \frac{v^{1/2}}{u^{1/2}}.$$

Then

$$x = \frac{u}{y} = u \cdot \frac{u^{1/2}}{v^{1/2}} = \frac{u^{3/2}}{v^{1/2}}.$$

(We do not need the solution in which x and y are negative.) Then

$$\frac{\partial(x, y)}{\partial(u, v)} = \begin{vmatrix} \dfrac{3u^{1/2}}{2v^{1/2}} & -\dfrac{u^{3/2}}{2v^{3/2}} \\[3mm] -\dfrac{v^{1/2}}{2u^{3/2}} & \dfrac{1}{2u^{1/2}v^{1/2}} \end{vmatrix} = \frac{3}{4v} - \frac{1}{4v} = \frac{1}{2v}.$$

We also find by substitution that $xy = 2$ corresponds to $u = 2$, $xy = 4$ corresponds to $u = 4$, $xy^3 = 3$ corresponds to $v = 3$, and $xy^3 = 6$ corresponds to $v = 6$. Hence the area of the region shown in Fig. 14.9.8 is

$$A = \iint_D 1 \, dx \, dy = \int_{v=3}^6 \int_{u=2}^4 \frac{1}{2v} \, du \, dv = \int_3^6 \frac{1}{v} \, dv = \ln 2 \approx 0.6931471805599453.$$

C14S09.011: Given: the region R bounded by the curves $y = x^3$, $y = 2x^3$, $x = y^3$, and $x = 4y^3$. Choose u and v so that $y = ux^3$ and $x = vy^3$. Then

$$y = uv^3 y^9; \qquad y^8 = \frac{1}{uv^3};$$

$$y = \frac{1}{u^{1/8}v^{3/8}}; \qquad x = vy^3 = \frac{v}{u^{3/8}v^{9/8}} = \frac{1}{u^{3/8}v^{1/8}}.$$

Then the curve $y = x^3$ can be written as

$$\frac{1}{u^{1/8}v^{3/8}} = \frac{1}{u^{9/8}v^{3/8}};$$

$$u^{1/8}v^{3/8} = u^{9/8}v^{3/8};$$

$$u = 1.$$

Similarly, the curve $y = 2x^3$ corresponds to $u = 2$, the curve $x = y^3$ corresponds to $v = 1$, and the curve $x = 4y^3$ corresponds to $v = 4$. Next,

$$\frac{\partial(x, y)}{\partial(u, v)} = \begin{vmatrix} -\dfrac{3}{8u^{11/8}v^{1/8}} & -\dfrac{1}{8u^{3/8}v^{9/8}} \\[3mm] -\dfrac{1}{8u^{9/8}v^{3/8}} & -\dfrac{3}{8u^{1/8}v^{11/8}} \end{vmatrix} = \frac{1}{8u^{3/2}v^{3/2}}.$$

Hence the area of R is

$$\iint_R 1 \, dx \, dy = \int_{v=1}^4 \int_{u=1}^2 \frac{1}{8u^{3/2}v^{3/2}} \, du \, dv = \int_{v=1}^4 \left[-\frac{1}{4u^{1/2}v^{3/2}} \right]_{u=1}^2 dv = \int_1^4 \left(\frac{1}{4v^{3/2}} - \frac{1}{4\sqrt{2}\, v^{3/2}} \right) dv$$

$$= \left[\frac{\sqrt{2} - 2}{4v^{1/2}} \right]_1^4 = \frac{2 - \sqrt{2}}{8} \approx 0.07322330470336311890.$$

C14S09.013: The Jacobian of the transformation $x = 3r \cos \theta$, $y = 2r \sin \theta$ is

$$\frac{\partial(x, y)}{\partial(r, \theta)} = \begin{vmatrix} 3\cos\theta & -3r\sin\theta \\ 2\sin\theta & 2r\cos\theta \end{vmatrix} = 6r\cos^2\theta + 6r\sin^2\theta = 6r.$$

The ellipse $\dfrac{x^2}{9} + \dfrac{y^2}{4} = 1$ is transformed into

$$\frac{9r^2 \cos^2 \theta}{9} + \frac{4r^2 \sin^2 \theta}{4} = 1: \quad \text{the circle} \quad r = 1.$$

The paraboloid has equation

$$z = x^2 + y^2 = 9r^2 \cos^2 \theta + 4r^2 \sin^2 \theta.$$

Therefore the volume of the solid is

$$V = \int_{\theta=0}^{2\pi} \int_{r=0}^{1} (9r^2 \cos^2 \theta + 4r^2 \sin^2 \theta) \cdot 6r \, dr \, d\theta = \int_{0}^{2\pi} \left(\frac{27}{2} \cos^2 \theta + 6 \sin^2 \theta \right) d\theta$$

$$= \int_{0}^{2\pi} \left(6 + \frac{15}{2} \cdot \frac{1 + \cos 2\theta}{2} \right) d\theta = \left[\frac{39}{4} \theta + \frac{15}{8} \sin 2\theta \right]_{0}^{2\pi} = \frac{39}{2} \pi \approx 61.26105674500096815.$$

C14S09.015: We are given the transformation $u = xy$, $v = xz$, $w = yz$. Then $uvw = x^2 y^2 z^2$. Hence

$$u^{1/2} v^{1/2} w^{1/2} = xyz = uz: \qquad z = \frac{v^{1/2} w^{1/2}}{u^{1/2}};$$

$$u^{1/2} v^{1/2} w^{1/2} = xyz = vy: \qquad y = \frac{u^{1/2} w^{1/2}}{v^{1/2}};$$

$$u^{1/2} v^{1/2} w^{1/2} = xyz = wx: \qquad x = \frac{u^{1/2} v^{1/2}}{w^{1/2}}.$$

The surface $xy = 1$ corresponds to the plane $u = 1$. Similarly, the other surfaces correspond to the planes $u = 4$, $v = 1$, $v = 9$, $w = 4$, and $w = 9$. The Jacobian of the given transformation is

$$\frac{\partial(x, y, z)}{\partial(u, v, w)} = \begin{vmatrix} \dfrac{v^{1/2}}{2u^{1/2} w^{1/2}} & \dfrac{u^{1/2}}{2v^{1/2} w^{1/2}} & -\dfrac{u^{1/2} v^{1/2}}{2w^{3/2}} \\[2ex] \dfrac{w^{1/2}}{2u^{1/2} v^{1/2}} & -\dfrac{u^{1/2} w^{1/2}}{2v^{3/2}} & \dfrac{u^{1/2}}{2v^{1/2} w^{1/2}} \\[2ex] -\dfrac{v^{1/2} w^{1/2}}{2u^{3/2}} & \dfrac{w^{1/2}}{2u^{1/2} v^{1/2}} & \dfrac{v^{1/2}}{2u^{1/2} w^{1/2}} \end{vmatrix} = -\frac{1}{2u^{1/2} v^{1/2} w^{1/2}}.$$

Therefore the volume bounded by the surfaces is

$$V = \int_{w=4}^{9} \int_{v=1}^{9} \int_{u=1}^{4} \frac{1}{2u^{1/2} v^{1/2} w^{1/2}} \, du \, dv \, dw = \int_{4}^{9} \int_{1}^{9} \left[\frac{u^{1/2}}{v^{1/2} w^{1/2}} \right]_{u=1}^{4} dv \, dw$$

$$= \int_{4}^{9} \int_{1}^{9} \frac{1}{v^{1/2} w^{1/2}} \, dv \, dw = \int_{4}^{9} \left[\frac{2v^{1/2}}{w^{1/2}} \right]_{v=1}^{9} dw = \int_{4}^{9} \frac{4}{w^{1/2}} \, dw = \left[8w^{1/2} \right]_{4}^{9} = 8.$$

C14S09.017: The substitution $x = u + v$, $y = u - v$ transforms the rotated ellipse $x^2 + xy + y^2 = 3$ into the ellipse S in "standard position," in which its axes lie on the coordinate axes. The resulting equation of S (in the uv-plane) is $3u^2 + v^2 = 3$. The Jacobian of this transformation is

$$\frac{\partial(x,\, y)}{\partial(u,\, v)} = \begin{vmatrix} 1 & 1 \\ 1 & -1 \end{vmatrix} = -2.$$

Therefore

$$I = \iint_R \exp(-x^2 - xy - y^2)\, dx\, dy = 2 \iint_S \exp(-3u^2 - v^2)\, du\, dv. \tag{1}$$

The substitution $u = r \cos\theta$, $v = r\sqrt{3}\,\sin\theta$ has Jacobian

$$\frac{\partial(u,\, v)}{\partial(r,\, \theta)} = \begin{vmatrix} \cos\theta & -r\sin\theta \\ \sqrt{3}\,\sin\theta & r\sqrt{3}\,\cos\theta \end{vmatrix} = r\sqrt{3}\,(\cos^2\theta + \sin^2\theta) = r\sqrt{3}\,.$$

This transformation, applied to the bounding ellipse of the region S, yields

$$3 = 3u^2 + v^2 = 3r^2 \cos^2\theta + 3r^2 \sin^2\theta = 3r^2,$$

and thereby transforms it into the circle with polar equation $r = 1$. Then substitution in the second integral in Eq. (1) yields

$$I = 2 \iint_S \exp(-3u^2 - u^2)\, du\, dv = 2 \int_{\theta=0}^{2\pi} \int_{r=0}^{1} r\sqrt{3}\,\exp(-3r^2)\, dr\, d\theta$$

$$= 4\pi\sqrt{3} \left[-\frac{1}{6} \exp(-3r^2) \right]_0^1 = \frac{2}{3}\pi\sqrt{3}\,(1 - e^{-3}) \approx 3.44699122256300138528.$$

C14S09.019: Suppose that k is a positive constant. First we need an integration by parts with

$$u = \rho^2 \qquad \text{and} \qquad dv = \rho\exp(-k\rho^2)\, d\rho:$$

$$du = 2\rho\, d\rho \qquad \text{and} \qquad v = -\frac{1}{2k}\exp(-k\rho^2).$$

Thus

$$\int \rho^3 \exp(-k\rho^2)\, d\rho = -\frac{1}{2k}\rho^2 \exp(-k\rho^2) + \int \frac{1}{k}\rho\exp(-k\rho^2)\, d\rho$$

$$= -\frac{1}{2k}\rho^2 \exp(-k\rho^2) - \frac{1}{2k^2}\exp(-k\rho^2) + C.$$

Then the improper triple integral given in Problem 19 will be the limit of I_a as $a \to +\infty$, where

$$I_a = \int_{\theta=0}^{2\pi} \int_{\phi=0}^{\pi} \int_{\rho=0}^{a} \rho^3 \exp(-k\rho^2) \sin\phi\, d\rho\, d\phi\, d\theta$$

$$= 2\pi \left[-\cos\phi \right]_{\phi=0}^{\pi} \left[-\frac{1}{2k}\rho^2 \exp(-k\rho^2) - \frac{1}{2k^2}\exp(-k\rho^2) \right]_{\rho=0}^{a}$$

$$= 4\pi \left[-\frac{1}{2k}a^2 \exp(-ka^2) - \frac{1}{2k^2}\exp(-ka^2) + \frac{1}{2k^2} \right].$$

Because $k > 0$, it is clear that $I_a \to \dfrac{2\pi}{k^2}$ as $a \to +\infty$.

C14S09.021: Given: The solid ellipsoid R with constant density δ and boundary surface with equation

$$\frac{x^2}{a^2} + \frac{y^2}{b^2} + \frac{z^2}{c^2} = 1 \tag{1}$$

(where a, b, and c are positive constants). The transformation

$$x = a\rho \sin\phi \cos\theta, \quad y = b\rho \sin\phi \sin\theta, \quad z = c\rho \cos\phi$$

has Jacobian

$$J = \frac{\partial(x,\,y,\,z)}{\partial(\rho,\,\phi,\,\theta)} = \begin{vmatrix} a\sin\phi\cos\theta & a\rho\cos\phi\cos\theta & -a\rho\sin\phi\sin\theta \\ b\sin\phi\sin\theta & b\rho\cos\phi\sin\theta & b\rho\sin\phi\cos\theta \\ c\cos\phi & -c\rho\sin\phi & 0 \end{vmatrix}$$

$$= abc\rho^2 \cos^2\phi \sin\phi \cos^2\theta + abc\rho^2 \sin^3\phi \cos^2\theta + abc\rho^2 \cos^2\phi \sin\phi \sin^2\theta + abc\rho^2 \sin^3\phi \sin^2\theta$$

$$= abc\rho^2 \sin\phi.$$

This transformation also transforms the ellipsoidal surface of Eq. (1) into

$$\rho^2 \sin^2\phi \cos^2\theta + \rho^2 \sin^2\phi \sin^2\theta + \rho^2 \cos^2\phi = \rho^2 \sin^2\phi + \rho^2 \cos^2\phi = \rho^2 = 1,$$

and thereby transforms R into the solid ball B of radius 1 and center at the origin. Therefore the moment of inertia of the ellipsoid with respect to the z-axis is

$$I_z = \int_{\theta=0}^{2\pi} \int_{\phi=0}^{\pi} \int_{\rho=0}^{1} (\rho^2 \sin^2\phi)(a^2 \cos^2\theta + b^2 \sin^2\theta)\delta abc\rho^2 \sin\phi \; d\rho \, d\phi \, d\theta$$

$$= \int_{0}^{2\pi} \int_{0}^{\pi} \left[\frac{1}{5}(\delta abc\rho^5 \sin^3\phi)(a^2 \cos^2\theta + b^2 \sin^2\theta) \right]_{\rho=0}^{1} d\phi \, d\theta$$

$$= \int_{0}^{2\pi} \int_{0}^{\pi} \frac{1}{5}(\delta abc \sin^3\phi)(a^2 \cos^2\theta + b^2 \cos^2\theta) \; d\phi \, d\theta$$

$$= \frac{1}{60}\delta abc \int_{0}^{2\pi} \left[(\cos 3\phi - 9\cos\phi)(a^2 \cos^2\theta + b^2 \sin^2\theta) \right]_{0}^{\pi} d\theta$$

$$= \frac{4}{15}\delta abc \int_{0}^{2\pi} (a^2 \cos^2\theta + b^2 \sin^2\theta) \; d\theta = \frac{1}{15}\delta abc \left[2a^2\theta + 2b^2\theta + a^2 \sin 2\theta - b^2 \sin 2\theta \right]_{0}^{2\pi}$$

$$= \frac{1}{15}\delta abc(4\pi a^2 + 4\pi b^2).$$

Because the mass of the sphere (found in the solution of Problem 20) is $M = \frac{4}{3}\pi \delta abc$, we see that

$$\frac{I_z}{M} = \frac{1}{5}(a^2 + b^2), \quad \text{and hence that} \quad I_z = \frac{1}{5}M(a^2 + b^2).$$

C14S09.023: If $u = xy$ and $v = xy^3$, then

858

$$uy^2 = xy^3 = v, \quad \text{so that} \quad y^2 = \frac{v}{u}; \quad y = \frac{v^{1/2}}{u^{1/2}}.$$

Then

$$x = \frac{u}{y} = u \cdot \frac{u^{1/2}}{v^{1/2}} = \frac{u^{3/2}}{v^{1/2}}.$$

(We do not need the solution in which x and y are negative.) Then

$$\frac{\partial(x, y)}{\partial(u, v)} = \begin{vmatrix} \dfrac{3u^{1/2}}{2v^{1/2}} & -\dfrac{u^{3/2}}{2v^{3/2}} \\[3mm] -\dfrac{v^{1/2}}{2u^{3/2}} & \dfrac{1}{2u^{1/2}v^{1/2}} \end{vmatrix} = \frac{3}{4v} - \frac{1}{4v} = \frac{1}{2v}.$$

We also find by substitution that $xy = 2$ corresponds to $u = 2$, $xy = 4$ corresponds to $u = 4$, $xy^3 = 3$ corresponds to $v = 3$, and $xy^3 = 6$ corresponds to $v = 6$. Hence the area of the region shown in Fig. 14.9.8 is

$$A = \iint_D 1 \; dx \; dy = \int_{v=3}^6 \int_{u=2}^4 \frac{1}{2v} \; du \; dv = \int_3^6 \frac{1}{v} \; dv = \ln 2 \approx 0.6931471805599453.$$

Its moments with respect to the coordinate axes are

$$M_y = \int_{v=3}^6 \int_{u=2}^4 \frac{1}{2v} \cdot \frac{u^{3/2}}{v^{1/2}} \; du \; dv = \int_{v=3}^6 \left[\frac{u^{5/2}}{5v^{3/2}} \right]_{u=2}^4 \; dv = \int_3^6 \frac{32 - 4\sqrt{2}}{5v^{3/2}} \; dv$$

$$= \left[\frac{8\sqrt{2} - 64}{5v^{1/2}} \right]_3^6 = \frac{72\sqrt{3} - 40\sqrt{6}}{15};$$

$$M_x = \int_{v=3}^6 \int_{u=2}^4 \frac{1}{2v} \cdot \frac{v^{1/2}}{u^{1/2}} \; du \; dv = \int_{v=3}^6 \left[\frac{u^{1/2}}{v^{1/2}} \right]_{u=2}^4 \; dv = \int_3^6 \frac{2 - \sqrt{2}}{v^{1/2}} \; dv$$

$$= \left[\left(4 - 2\sqrt{2} \right) v^{1/2} \right]_3^6 = 6\sqrt{6} - 8\sqrt{3}.$$

Therefore its centroid is located at the point

$$(\bar{x}, \bar{y}) = \left(\frac{72\sqrt{3} - 40\sqrt{6}}{15 \ln 2}, \frac{6\sqrt{6} - 8\sqrt{3}}{\ln 2} \right) \approx (2.570696785449, 1.212631342551).$$

C14S09.025: Given: The solid ellipsoid R with constant density δ and boundary surface with equation

$$\frac{x^2}{a^2} + \frac{y^2}{b^2} + \frac{z^2}{c^2} = 1 \tag{1}$$

(where a, b, and c are positive constants). The transformation

$$x = a\rho \sin\phi \cos\theta, \quad y = b\rho \sin\phi \sin\theta, \quad z = c\rho \cos\phi$$

has Jacobian

$$J = \frac{\partial(x, y, z)}{\partial(\rho, \phi, \theta)} = \begin{vmatrix} a\sin\phi\cos\theta & a\rho\cos\phi\cos\theta & -a\rho\sin\phi\sin\theta \\ b\sin\phi\sin\theta & b\rho\cos\phi\sin\theta & b\rho\sin\phi\cos\theta \\ c\cos\phi & -c\rho\sin\phi & 0 \end{vmatrix}$$

$$= abc\rho^2\cos^2\phi\sin\phi\cos^2\theta + abc\rho^2\sin^3\phi\cos^2\theta + abc\rho^2\cos^2\phi\sin\phi\sin^2\theta + abc\rho^2\sin^3\phi\sin^2\theta$$

$$= abc\rho^2\sin\phi.$$

This transformation also transforms the ellipsoidal surface of Eq. (1) into

$$\rho^2\sin^2\phi\cos^2\theta + \rho^2\sin^2\phi\sin^2\theta + \rho^2\cos^2\phi = \rho^2\sin^2\phi + \rho^2\cos^2\phi = \rho^2 = 1,$$

and thereby transforms R into the solid ball B of radius 1 and center at the origin. Note also that

$$x^2 + y^2 = a^2\rho^2\sin^2\phi\cos^2\theta + b^2r^2\sin^2\phi\sin^2\theta = (\rho^2\sin^2\phi)(a^2\cos^2\theta + b^2\sin^2\theta).$$

Assume that the solid R has constant density δ. Then its moment of inertia with respect to the z-axis is

$$I_z = \int_{\theta=0}^{2\pi}\int_{\phi=0}^{\pi}\int_{\rho=0}^{1} (\rho^2\sin^2\phi)(a^2\cos^2\theta + b^2\sin^2\theta)\delta abc\rho^2\sin\phi\,d\rho\,d\phi\,d\theta$$

$$= \int_0^{2\pi}\int_0^{\pi} \frac{1}{5}(\delta abc\sin^3\phi)(a^2\cos^2\theta + b^2\sin^2\theta)\,d\phi\,d\theta$$

$$= \int_0^{2\pi} \frac{4}{15}(\delta abc)(a^2\cos^2\theta + b^2\sin^2\theta)\,d\theta = \frac{4}{15}\pi\delta abc(a^2 + b^2) = \frac{1}{5}M(a^2 + b^2)$$

where M is the mass of the ellipsoid. By symmetry,

$$I_y = \frac{4}{15}\pi\delta abc(a^2 + c^2) = \frac{1}{5}M(a^2 + c^2) \quad \text{and} \quad I_x = \frac{4}{15}\pi\delta abc(b^2 + c^2) = \frac{1}{5}M(b^2 + c^2).$$

C14S09.027: The average distance of points of the ellipsoid from its center at (0, 0, 0) is

$$\bar{d} = \frac{1}{V}\int_0^{2\pi}\int_0^{\pi}\int_0^{1} (abc\rho^2\sin\phi)\sqrt{(a\rho\sin\phi\cos\theta)^2 + (b\rho\sin\phi\sin\theta)^2 + (c\rho\cos\phi)^2}\,d\rho\,d\phi\,d\theta$$

where $V = \frac{4}{3}\pi abc$ is the volume of the ellipsoid. In particular, if $a = 4$, $b = 3$, and $c = 2$, we find (using the NIntegrate command in *Mathematica* 3.0) that $\bar{d} \approx 2.300268522983$.

C14S09.029: Part (a): First note that

$$\int_0^1\int_0^1 \left(\frac{1}{1 - xy} - \frac{1}{1 + xy}\right)dx\,dy = \int_0^1\int_0^1 \frac{2xy}{1 - x^2y^2}\,dx\,dy.$$

The Jacobian of the substitution $u = x^2$, $v = y^2$ is

$$\frac{\partial(u, v)}{\partial(x, y)} = \begin{vmatrix} 2x & 0 \\ 0 & 2y \end{vmatrix} = 4xy,$$

860

so

$$\int_0^1 \int_0^1 \frac{2xy}{1-x^2y^2} \, dx \, dy = \frac{1}{2} \int_0^1 \int_0^1 \frac{1}{1-x^2y^2} \cdot 4xy \, dx \, dy = \frac{1}{2} \int_0^1 \int_0^1 \frac{1}{1-uv} \, du \, dv = \frac{1}{2}\zeta(2).$$

Part (b): Addition as indicated in Problem 29, and cancellation of the integrals involving $1/(1+xy)$, yields the equation

$$2 \int_0^1 \int_0^1 \frac{1}{1-xy} \, dx \, dy = \frac{1}{2}\zeta(2) + 2 \int_0^1 \int_0^1 \frac{1}{1-x^2y^2} \, dx \, dy,$$

which we readily solve for

$$\int_0^1 \int_0^1 \frac{1}{1-x^2y^2} \, dx \, dy = \int_0^1 \int_0^1 \frac{1}{1-xy} \, dx \, dy - \frac{1}{4}\zeta(2) = \frac{3}{4}\zeta(2).$$

Part (c): The Jacobian of the transformation $T : R_{uv}^2 \to R_{xy}^2$ that we define by $x = (\sin v)/(\cos u)$, $y = (\sin u)/(\cos v)$ is

$$J_T = \begin{vmatrix} \dfrac{\cos u}{\cos v} & -\dfrac{\sin u \sin v}{\cos^2 v} \\[3mm] -\dfrac{\sin u \sin v}{\cos^2 u} & \dfrac{\cos v}{\cos u} \end{vmatrix} = 1 - \frac{\sin^2 u \sin^2 v}{\cos^2 u \cos^2 v} = 1 - \tan^2 u \tan^2 v.$$

Reading the limits for the transformed integral from Fig. 14.9.10(a) in the text, we therefore find that

$$\zeta(2) = \frac{4}{3} \int_0^1 \int_0^1 \frac{1}{1-x^2y^2} \, dx \, dy$$

$$= \frac{4}{3} \int_0^{\pi/2} \int_0^{(\pi/2)-v} \left(1 - \frac{\sin^2 u \sin^2 v}{\cos^2 u \cos^2 v}\right)^{-1} \cdot (1 - \tan^2 u \tan^2 v) \, du \, dv = \frac{4}{3} \int_0^{\pi/2} \int_0^{(\pi/2)-v} 1 \, du \, dv$$

$$= \frac{4}{3} \int_0^{\pi/2} \left(\frac{\pi}{2} - v\right) dv = \frac{4}{3}\left[\frac{\pi}{2}v - \frac{1}{2}v^2\right]_0^{\pi/2} = \frac{4}{3} \cdot \frac{\pi^2}{8} = \frac{\pi^2}{6}. \qquad \text{—C.H.E.}$$

Chapter 14 Miscellaneous Problems

C14S0M.001: The domain of the given integral is bounded above by the graph of $y = x^3$, below by the x-axis, and on the right by the vertical line $x = 1$. When its order of integration is reversed, the given integral becomes

$$\int_{x=0}^1 \int_{y=0}^{x^3} \frac{1}{\sqrt{1+x^2}} \, dy \, dx = \int_0^1 \frac{x^3}{\sqrt{1+x^2}} \, dx$$

$$= \left[\frac{1}{3}(x^2 - 2)\sqrt{1+x^2}\right]_0^1 = \frac{2 - \sqrt{2}}{3} \approx 0.1952621458756350.$$

Mathematica 3.0 can evaluate the given integral without first reversing the order of integration. It obtains

861

$$\int_{y=0}^{1}\int_{x=y^{1/3}}^{1}\frac{1}{\sqrt{1+x^2}}\,dx\,dy = \int_{0}^{1}\left[\operatorname{arcsinh}(1) - \operatorname{arcsinh}(y^{1/3})\right]dy$$

$$= \left[\frac{y^{2/3}-2}{3}\sqrt{1+y^{2/3}} + y\operatorname{arcsinh}(1) - y\operatorname{arcsinh}(y^{1/3})\right]_{0}^{1} = \frac{2-\sqrt{2}}{3}.$$

C14S0M.003: The domain of the given integral is bounded above by the line $y = 1$, below and on the right by the line $y = x$, and on the left by the y-axis. When its order of integration is reversed, it becomes

$$\int_{y=0}^{1}\int_{x=0}^{y}\exp(-y^2)\,dx\,dy = \int_{0}^{1}y\exp(-y^2)\,dy = \left[-\frac{1}{2}\exp(-y^2)\right]_{0}^{1}$$

$$= \frac{1}{2} - \frac{1}{2e} = \frac{e-1}{2e} \approx 0.3160602794142788.$$

Mathematica 3.0 can evaluate the given integral without first reversing the order of integration:

$$\int_{x=0}^{1}\int_{y=x}^{1}\exp(-y^2)\,dy\,dx = \int_{x=0}^{1}\left[\frac{1}{2}\sqrt{\pi}\,\operatorname{erf}(y)\right]_{y=x}^{1}dx = \int_{0}^{1}\frac{1}{2}\sqrt{\pi}\left[\operatorname{erf}(1) - \operatorname{erf}(x)\right]dx$$

$$= \frac{1}{2}\left[\sqrt{\pi}\,x\left[\operatorname{erf}(1) - \operatorname{erf}(x)\right] - \exp(-x^2)\right]_{0}^{1} = \frac{e-1}{2e}.$$

Note: By definition, the *error function* is $\operatorname{erf}(x) = \dfrac{2}{\sqrt{\pi}}\displaystyle\int_{0}^{x}\exp(-t^2)\,dt$.

C14S0M.005: The domain of the given integral is bounded above by the graph of $y = x^2$, below by the x-axis, and on the right by the line $x = 2$. When its order of integration is reversed, it becomes

$$\int_{x=0}^{2}\int_{y=0}^{x^2}\frac{y\exp(x^2)}{x^3}\,dy\,dx = \int_{x=0}^{2}\left[\frac{y^2\exp(x^2)}{2x^3}\right]_{y=0}^{x^2}dx$$

$$= \int_{0}^{2}\frac{1}{2}x\exp(x^2)\,dx = \left[\frac{1}{4}\exp(x^2)\right]_{0}^{2} = \frac{e^4-1}{4} \approx 13.3995375082860598.$$

Mathematica 3.0 can evaluate the given integral without first reversing the order of integration, but the intermediate antiderivatives involve the *exponential integral function*

$$\operatorname{Ei}(z) = \int_{z}^{\infty}\frac{e^{-t}}{t}\,dt$$

and hence we omit the details.

C14S0M.007: The volume is

$$V = \int_{y=0}^{1}\int_{x=y}^{2-y}(x^2 + y^2)\,dx\,dy = \int_{y=0}^{1}\left[\frac{1}{3}x^3 + xy^2\right]_{x=y}^{2-y}dy$$

$$= \int_{0}^{1}\frac{4}{3}(2 - 3y + 3y^2 - 2y^3)\,dy = \left[\frac{8}{3}y - 2y^2 + \frac{4}{3}y^3 - \frac{2}{3}y^4\right]_{0}^{1} = \frac{4}{3}.$$

C14S0M.009: By symmetry, the centroid lies on the z-axis. Assume that the solid has unit density. Then its mass and volume are

$$m = V = \int_{\theta=0}^{2\pi} \int_{\phi=\pi/3}^{\pi/2} \int_{\rho=0}^{3} \rho^2 \sin\phi \, d\rho \, d\phi \, d\theta$$

$$= 2\pi \cdot \left[-\cos\phi \right]_{\pi/3}^{\pi/2} \cdot \left[\frac{1}{3}\rho^3 \right]_0^3 = 18\pi \cdot \left[0 - \left(-\frac{1}{2} \right) \right] = 9\pi \approx 28.2743338823081391.$$

The moment of the solid with respect to the xy-plane is

$$M_{xy} = \int_{\theta=0}^{2\pi} \int_{\phi=\pi/3}^{\pi/2} \int_{\rho=0}^{3} \rho^3 \sin\phi \cos\phi \, d\rho \, d\phi \, d\theta = 2\pi \int_{\phi=\pi/3}^{\pi/2} \left[\frac{1}{4}\rho^4 \sin\phi \cos\phi \right]_{\rho=0}^{3} d\phi$$

$$= 2\pi \int_{\phi=\pi/3}^{\pi/2} \frac{81}{4} \sin\phi \cos\phi \, d\phi = 2\pi \left[-\frac{81}{8} \cos^2\phi \right]_{\phi=\pi/3}^{\pi/2} = \frac{81}{16}\pi.$$

Therefore the centroid of the solid is located at the point with coordinates

$$(\overline{x}, \, \overline{y}, \, \overline{z}) = \left(0, \, 0, \, \frac{9}{16} \right).$$

C14S0M.011: First interchange y and z: We are to find the volume bounded by the paraboloid $z = x^2 + 3y^2$ and the cylinder $z = 4 - y^2$. These surfaces intersect in a curve that lies on the elliptic cylinder $x^2 + 4y^2 = 4$, bounding the region R in the xy-plane. Hence the volume is

$$V = \iint_R (4 - x^2 - 4y^2) \, dx \, dy.$$

Apply the transformation $x = 2r\cos\theta$, $y = r\sin\theta$. This transforms R into the region $0 \leq r \leq 1$, $0 \leq \theta \leq 2\pi$. Moreover, the Jacobian of this transformation is

$$\frac{\partial(x, \, y)}{\partial(r, \, \theta)} = \begin{vmatrix} 2\cos\theta & -2r\sin\theta \\ \sin\theta & r\cos\theta \end{vmatrix} = 2r.$$

Hence

$$V = \int_{\theta=0}^{2\pi} \int_{r=0}^{1} (4 - 4r^2) \cdot 2r \, dr \, d\theta = 2\pi \cdot \left[4r^2 - 2r^4 \right]_0^1 = 4\pi \approx 12.566370614359172953.85.$$

C14S0M.013: First interchange x and z: We are to find the volume enclosed by the elliptical cylinder $4x^2 + y^2 = 4$ and between the planes $z = 0$ and $z = y + 2$. Let R denote the plane region in which the elliptical cylinder meets the xy-plane. Then the volume is

$$V = \iint_R (y + 2) \, dx \, dy.$$

The transformation $x = r\cos\theta$, $y = 2r\sin\theta$ transforms R into the rectangle $0 \leq r \leq 1$, $0 \leq \theta \leq 2\pi$. The Jacobian of this transformation is

$$\frac{\partial(x, y)}{\partial(r, \theta)} = \begin{vmatrix} \cos\theta & -r\sin\theta \\ 2\sin\theta & 2r\cos\theta \end{vmatrix} = 2r.$$

Therefore

$$V = \int_{\theta=0}^{2\pi}\int_{r=0}^{1}(2 + 2r\sin\theta)\cdot 2r\,dr\,d\theta = \int_0^{2\pi}\left[2r^2 + \frac{4}{3}r^3\sin\theta\right]_{r=0}^{1}d\theta$$

$$= \int_0^{2\pi}\left(2 + \frac{4}{3}\sin\theta\right)d\theta = \left[2\theta - \frac{4}{3}\cos\theta\right]_0^{2\pi} = 4\pi \approx 12.56637061435917295385.$$

C14S0M.015: The graph of $x^4 + x^2y^2 = y^2$ in the first quadrant is the graph of

$$y = \frac{x^2}{\sqrt{1-x^2}}, \quad 0 \le x < 1.$$

This curve meets the line $y = x$ at the point $\left(\frac{1}{2}\sqrt{2}, \frac{1}{2}\sqrt{2}\right)$ and, of course, at the point $(x, y) = (0, 0)$. Conversion of the first equation into polar form yields

$$r^4\cos^4\theta + r^4\cos^2\theta\sin^2\theta = r^2\sin^2\theta;$$

$$(r^2\cos^2\theta)(\cos^2\theta + \sin^2\theta) = \sin^2\theta;$$

$$r^2\cos^2\theta = \sin^2\theta;$$

thus $r = \tan\theta$. Noting that the line $y = x$ has polar equation $\theta = \frac{1}{4}\pi$, we find that

$$\iint_R \frac{1}{(1+x^2+y^2)^2}\,dA = \int_{\theta=0}^{\pi/4}\int_{r=0}^{\tan\theta}\frac{r}{(1+r^2)^2}\,dr\,d\theta = \int_0^{\pi/4}\left[-\frac{1}{2(1+r^2)}\right]_{r=0}^{\tan\theta}d\theta$$

$$= \int_0^{\pi/4}\frac{1}{2}(1 - \cos^2\theta)\,d\theta = \frac{1}{8}\left[2\theta - \sin 2\theta\right]_0^{\pi/4} = \frac{\pi - 2}{16} \approx 0.0713495408493621.$$

C14S0M.017: The mass and and moments are

$$m = \int_{-2}^{2}\int_{2y^2}^{4+y^2}y^2\,dx\,dy = \int_{-2}^{2}\left[xy^2\right]_{2y^2}^{4+y^2}dy$$

$$= \int_{-2}^{2}(4y^2 - y^4)\,dy = \left[\frac{4}{3}y^3 - \frac{1}{5}y^5\right]_{-2}^{2} = \frac{128}{15} \approx 8.5333333333333333;$$

$$M_x = \int_{-2}^{2}\int_{2y^2}^{4+y^2}y^3\,dx\,dy = \int_{-2}^{2}\left[xy^3\right]_{2y^2}^{4+y^2}dx = \int_{-2}^{2}(4y^3 - y^5)\,dy = \left[y^4 - \frac{1}{6}y^6\right]_{-2}^{2} = 0;$$

$$M_y = \int_{-2}^{2}\int_{2y^2}^{4+y^2}xy^2\,dx\,dy = \int_{-2}^{2}\left(8y^2 + 4y^4 - \frac{3}{2}y^6\right)dy = \left[\frac{8}{3}y^3 + \frac{4}{5}y^5 - \frac{3}{14}y^7\right]_{-2}^{2} = \frac{4096}{105}.$$

Therefore the centroid of the lamina is located at the point with coordinates

$$(\overline{x}, \overline{y}) = \left(\frac{32}{7}, 0\right) \approx (4.5714285714285714, 0).$$

C14S0M.019: The mass and moments are

$$m = \int_{\theta=-\pi/2}^{\pi/2} \int_{r=0}^{2\cos\theta} kr \, dr \, d\theta = \int_{\theta=-\pi/2}^{\pi/2} \left[\frac{1}{2}kr^2\right]_{r=0}^{2\cos\theta} d\theta$$

$$= \int_{\theta=-\pi/2}^{\pi/2} 2k\cos^2\theta \, d\theta = \left[\frac{1}{2}k(2\theta + \sin 2\theta)\right]_{\theta=-\pi/2}^{\pi/2} = k\pi;$$

$$M_y = \int_{\theta=-\pi/2}^{\pi/2} \int_{r=0}^{2\cos\theta} kr^2 \cos\theta \, dr \, d\theta = \int_{\theta=-\pi/2}^{\pi/2} \left[\frac{1}{3}kr^3 \cos\theta\right]_{r=0}^{2\cos\theta} d\theta$$

$$= \int_{\theta=-\pi/2}^{\pi/2} \frac{8}{3}k\cos^4\theta \, d\theta = \left[\frac{1}{12}k(12\theta + 8\sin 2\theta + \sin 4\theta)\right]_{\theta=-\pi/2}^{\pi/2} = k\pi;$$

$$M_x = \int_{\theta=-\pi/2}^{\pi/2} \int_{r=0}^{2\cos\theta} kr^2 \sin\theta \, dr \, d\theta = \int_{\theta=-\pi/2}^{\pi/2} \left[\frac{1}{3}kr^3 \sin\theta\right]_{r=0}^{2\cos\theta} d\theta$$

$$= \int_{\theta=-\pi/2}^{\pi/2} \frac{8}{3}k\cos^3\theta \sin\theta \, d\theta = \left[-\frac{2}{3}k\cos^4\theta\right]_{\theta=-\pi/2}^{\pi/2} = 0.$$

Therefore the centroid of the lamina is located at the point $(1, 0)$.

C14S0M.021: By the first theorem of Pappus, the y-coordinate $\overline{y}$ of the centroid must satisfy the equation

$$2\pi\overline{y} \cdot \frac{1}{2}\pi ab = \frac{4}{3}\pi ab^2,$$

and it follows immediately that $\overline{y} = \dfrac{4b}{3\pi}$.

C14S0M.023: Assume that the lamina has constant density δ. By symmetry, $\overline{x} = 0$. The mass of the lamina and its moment with respect to the x-axis are

$$m = \int_{x=-2}^{2} \int_{y=0}^{4-x^2} \delta \, dy \, dx = \int_{x=-2}^{2} \delta(4 - x^2) \, dx = \delta\left[4x - \frac{1}{3}x^3\right]_{-2}^{2} = \frac{32}{3}\delta \quad \text{and}$$

$$M_x = \int_{x=-2}^{2} \int_{y=0}^{4-x^2} \delta y \, dy \, dx = \int_{x=-2}^{2} \left[\frac{1}{2}\delta y^2\right]_{y=0}^{4-x^2} dx = \frac{1}{2}\delta \int_{x=-2}^{2} (16 - 8x^2 + x^4) \, dx$$

$$= \frac{1}{2}\delta\left[16x - \frac{8}{3}x^3 + \frac{1}{5}x^5\right]_{-2}^{2} = \frac{256}{15}\delta.$$

Therefore the centroid of the lamina is at the point $\left(0, \frac{8}{5}\right)$.

C14S0M.025: The volume of the ice-cream cone is

$$V = \int_{\theta=0}^{2\pi} \int_{r=0}^{1} \int_{z=2r}^{\sqrt{5-r^2}} r \, dz \, dr \, d\theta = 2\pi \int_{r=0}^{1} \left(r\sqrt{5-r^2} - 2r^2 \right) dr$$

$$= -\frac{2}{3}\pi \cdot \left[(5-r^2)^{3/2} + 2r^3 \right]_0^1 = \frac{10}{3}\pi \left(\sqrt{5} - 2 \right) \approx 2.47209807953713305410 3626.$$

C14S0M.027: Let δ be the [constant] density of the cone and let h denote its height. Place the cone with its vertex at the origin and with its axis lying on the nonnegative z-axis. Then its mass is $M = \frac{1}{3}\pi\delta a^2 h$. The side of the cone has cylindrical equation $z = hr/a$, so the moment of inertia of the cone with respect to the z-axis is

$$I_z = \int_{\theta=0}^{2\pi} \int_{r=0}^{a} \int_{z=hr/a}^{h} \delta r^3 \, dz \, dr \, d\theta = 2\pi\delta \int_{r=0}^{a} \left(r^3 h - \frac{r^4 h}{a} \right) dr$$

$$= 2\pi\delta \left[\frac{1}{4} r^4 h - \frac{1}{5a} r^5 h \right]_{r=0}^{a} = \frac{1}{10}\pi\delta a^4 h = \frac{3}{10} M a^2.$$

Note that the answer is plausible and dimensionally correct. One of our physics teachers, Prof. J. J. Kyame of Tulane University (retired), always insisted that we express moment of inertia in terms of mass, as here, so that the answer can be inspected for plausibility and dimensional accuracy.

C14S0M.029: We are given the solid ellipsoid E with constant density $\delta = 1$ and boundary the surface with Cartesian equation

$$\frac{x^2}{a^2} + \frac{y^2}{b^2} + \frac{z^2}{c^2} = 1.$$

Its moment of inertia with respect to the x-axis is then

$$I_x = \iiint_E (y^2 + z^2) \, dV.$$

We use the transformation

$$x = a\rho\sin\phi\cos\theta, \quad y = b\rho\sin\phi\sin\theta, \quad z = c\rho\cos\phi. \tag{1}$$

Under this transformation, E is replaced with the solid B determined by

$$0 \leq \theta \leq 2\pi, \quad 0 \leq \phi \leq \pi, \quad 0 \leq \rho \leq 1.$$

The Jacobian of the transformation in (1) is

$$\frac{\partial(x, y, z)}{\partial(\rho, \phi, \theta)} = \begin{vmatrix} a\sin\phi\cos\theta & a\rho\cos\phi\cos\theta & -a\rho\sin\phi\sin\theta \\ b\sin\phi\sin\theta & b\rho\cos\phi\sin\theta & b\rho\sin\phi\cos\theta \\ c\cos\phi & -c\rho\sin\phi & 0 \end{vmatrix} = abc\rho^2 \sin\phi.$$

Therefore

$$I_x = \int_0^{2\pi} \int_0^{\pi} \int_0^{1} \left[(b\rho\sin\phi\sin\theta)^2 + (c\rho\cos\phi)^2 \right] \cdot abc\rho^2 \sin\phi \, d\rho \, d\phi \, d\theta$$

866

$$= \int_0^{2\pi} \int_0^{\pi} \int_0^1 (b^2 \sin^2 \phi \sin^2 \theta + c^2 \cos^2 \phi) \cdot abc\rho^4 \sin \phi \, d\rho \, d\phi \, d\theta$$

$$= \frac{1}{5} abc \int_0^{2\pi} \int_0^{\pi} (b^2 \sin^3 \phi \sin^2 \theta + c^2 \sin \phi \cos^2 \phi) \, d\phi \, d\rho$$

$$= \frac{1}{5} abc \int_0^{2\pi} \int_0^{\pi} \left[b^2 (1 - \cos^2 \phi) \sin \phi \sin^2 \theta + c^2 \sin \phi \cos^2 \phi \right] d\phi \, d\theta$$

$$= \frac{1}{5} abc \int_0^{2\pi} \left[\frac{1}{3} b^2 \cos^3 \phi \sin^2 \theta - b^2 \cos \phi \sin^2 \theta - \frac{1}{3} c^2 \cos^3 \phi \right]_0^{\pi} d\theta$$

$$= \frac{1}{5} abc \int_0^{2\pi} \left[-\frac{1}{3} b^2 (1 - \cos 2\theta) + b^2 (1 - \cos 2\theta) + \frac{2}{3} c^2 \right] d\theta$$

$$== \frac{1}{5} abc \left[-\frac{1}{3} b^2 \theta + \frac{1}{6} b^2 \sin 2\theta + b^2 \theta - \frac{1}{2} b^2 \sin 2\theta + \frac{2}{3} c^2 \theta \right]_0^{2\pi}$$

$$= \frac{1}{5} abc \left(-\frac{2}{3} \pi b^2 + 2\pi b^2 + \frac{4}{3} \pi c^2 \right) = \frac{4}{15} \pi abc (b^2 + c^2) = \frac{1}{5} M(b^2 + c^2)$$

where $M = \frac{4}{3} \pi abc$ is the mass of E.

C14S0M.031: The cylinder $r = 2\cos\theta$ meets the xy-plane in the circle with equation $r = 2\cos\theta$, $-\frac{1}{2}\pi \leq \theta \leq \frac{1}{2}\pi$. With density $\delta = 1$, the moment of inertia of the solid region with respect to the z-axis is

$$I_z = 2 \int_{-\pi/2}^{\pi/2} \int_0^{2\cos\theta} \int_0^{\sqrt{4-r^2}} r^3 \, dz \, dr \, d\theta = 4 \int_0^{\pi/2} \int_0^{2\cos\theta} r^3 \sqrt{4 - r^2} \, dr \, d\theta$$

$$= 4 \int_0^{\pi/2} \left[\frac{1}{15} (3r^4 - 4r^2 - 32) \sqrt{4 - r^2} \right]_0^{2\cos\theta} d\theta$$

$$= \frac{4}{15} \int_0^{\pi/2} (64 - 64 \sin\theta - 32 \sin\theta \cos^2\theta + 96 \sin\theta \cos^4\theta) \, d\theta$$

$$= \frac{8}{225} \left[480\theta + 450 \cos\theta - 25 \cos 3\theta - 9 \cos 5\theta \right]_0^{\pi/2} = \frac{128}{225}(15\pi - 26) \approx 12.0171461995217912.$$

The student who obtains the incorrect answer $\frac{128}{15}\pi$ may well have overlooked the fact that

$$\sqrt{4 - 4\cos^2\theta} = |2\sin\theta|.$$

C14S0M.033: The area element $r \, dr \, d\theta$ moves around a circle of radius $y = r\sin\theta$, and therefore of circumference $2\pi r \sin\theta$. Hence the volume swept out is

$$V = \int_{\theta=0}^{\pi} \int_{r=0}^{1+\cos\theta} 2\pi r^2 \sin\theta \, dr \, d\theta = \int_{\theta=0}^{\pi} \left[\frac{2}{3} \pi r^3 \sin\theta \right]_{r=0}^{1+\cos\theta} d\theta$$

$$= \int_0^{\pi} \frac{2}{3} \pi (1 + \cos\theta)^3 \sin\theta \, d\theta = \left[-\frac{1}{6} \pi (1 + \cos\theta)^4 \right]_0^{\pi} = \frac{8}{3} \pi \approx 8.37758040957278196923.$$

C14S0M.035: The moment of inertia of the torus of Problem 34 with respect to the line $x = -b$ (where $b \geq 0$), its natural axis of symmetry, is

$$I = \int_{\theta=0}^{2\pi} \int_{r=0}^{a} 2\pi\delta(b + r\cos\theta)^3 \cdot r \, dr \, d\theta = \int_{\theta=0}^{2\pi} \pi\delta \left[b^3 r^2 + 2b^2 r^3 \cos\theta + \frac{3}{2} br^4 \cos^2\theta + \frac{2}{5} r^5 \cos^3\theta \right]_{r=0}^{a} d\theta$$

$$= \pi\delta \int_0^{2\pi} \left(a^2 b^3 + 2a^3 b^2 \cos\theta + \frac{3}{2} a^4 b \cos^2\theta + \frac{2}{5} a^5 \cos^3\theta \right) d\theta$$

$$= \frac{1}{120} \pi\delta a^2 \left[90a^2 b\theta + 120b^3\theta + 36a^3 \sin\theta + 240ab^2 \sin\theta + 45a^2 b \sin 2\theta + 4a^3 \sin 3\theta \right]_0^{2\pi}$$

$$= \frac{1}{120} \pi\delta a^2 (180\pi a^2 b + 240\pi b^3) = \frac{1}{2} \pi^2 \delta a^2 b(3a^2 + 4b^2) = \frac{1}{4} M(3a^2 + 4b^2)$$

where $M = 2\pi^2 \delta a^2 b$ is the mass of the torus.

C14S0M.037: Use the disk bounded by the circle with polar equation $r = 2a\sin\theta$, $0 \leqq \theta \leqq \pi$. Then the origin is a point on the boundary of the disk, and the average distance of points of this disk from the origin is

$$\bar{d} = \frac{1}{\pi a^2} \int_{\theta=0}^{\pi} \int_{r=0}^{2a\sin\theta} r^2 \, dr \, d\theta = \frac{1}{\pi a^2} \int_0^{\pi} \frac{8}{3} a^3 \sin^3\theta \, d\theta$$

$$= \frac{2a^3}{9\pi a^2} \left[\cos 3\theta - 9\cos\theta \right]_0^{\pi} = \frac{32}{9\pi} a \approx (1.131768484209)a.$$

C14S0M.039: We use the ball bounded by the surface with spherical-coordinates equation $\rho = a$. Then the average distance of points of this ball from its center is

$$\bar{d} = \frac{3}{4\pi a^3} \int_{\theta=0}^{2\pi} \int_{\phi=0}^{\pi} \int_{\rho=0}^{a} \rho^3 \sin\phi \, d\rho \, d\phi \, d\theta$$

$$= \frac{3}{4\pi a^3} \int_{\theta=0}^{2\pi} \int_{\phi=0}^{\pi} \frac{1}{4} a^4 \sin\phi \, d\phi \, d\theta = \frac{3}{4\pi a^3} \cdot 2\pi \cdot \left[-\frac{1}{4} a^4 \cos\phi \right]_0^{\pi} = \frac{3}{4\pi a^3} \cdot 2\pi \cdot \frac{1}{2} a^4 = \frac{3}{4} a.$$

C14S0M.041: We will use the spheres with spherical-coordinates equations $\rho = 2\cos\phi$ and $\rho = 4\cos\phi$, which have a mutual point of tangency at the origin. Then the average distance of points outside the smaller and inside the larger sphere from the origin is

$$\bar{d} = \frac{3}{28\pi} \int_{\theta=0}^{2\pi} \int_{\phi=0}^{\pi/2} \int_{\rho=2\cos\phi}^{4\cos\phi} \rho^3 \sin\phi \, d\rho \, d\phi \, d\theta = \frac{3}{28\pi} \cdot 2\pi \cdot \int_{\phi=0}^{\pi/2} \left[\frac{1}{4} \rho^4 \sin\phi \right]_{\rho=2\cos\phi}^{4\cos\phi} d\phi$$

$$= \frac{3}{14} \int_0^{\pi/2} 60 \sin\phi \cos^4\phi \, d\phi = \frac{3}{14} \left[-12\cos^5\phi \right]_0^{\pi/2} = \frac{18}{7} \approx 2.5714285714285714.$$

C14S0M.043: The part of the paraboloid that lies between the two given planes also is the part between the cylinders $r = 2$ and $r = 3$. Let R denote the part of the xy-plane between those two cylinders. Then the surface area in question is

$$A = \iint_R \sqrt{r^2 + (rz_r)^2 + (z_\theta)^2} \, dr \, d\theta = \int_{\theta=0}^{2\pi} \int_{r=2}^{3} \sqrt{r^2 + 4r^4} \, dr \, d\theta = 2\pi \int_2^3 r(1 + 4r^2)^{1/2} \, dr$$

$$= 2\pi \left[\frac{1}{12} (1 + 4r^2)^{3/2} \right]_2^3 = \frac{1}{6} \pi \left(37\sqrt{37} - 17\sqrt{17} \right) \approx 81.1417975124065455.$$

C14S0M.045: Let R be the region in the $\phi\theta$-plane determined by the inequalities $\phi_2 \leqq \phi \leqq \phi_1$ and $0 \leqq \theta \leqq 2\pi$, where

$$\cos\phi_2 = \frac{z_2}{a} \quad \text{and} \quad \cos\phi_1 = \frac{z_1}{a}.$$

Thus the part of the sphere $\rho = a$ for which the spherical coordinates ϕ and θ satisfy these inequalities is the part of the sphere between the planes $z = z_1$ and $z = z_2$. Thus the formula in Problem 18 of Section 14.8 yields the area of this surface to be

$$A = \iint_R a^2 \sin\phi \; d\phi \; d\theta = \int_{\theta=0}^{2\pi} \int_{\phi=\phi_2}^{\phi_1} a^2 \sin\phi \; d\phi \; d\theta$$

$$= 2\pi a^2 \left[+\cos\phi \right]_{\phi_1}^{\phi_2} = 2\pi a^2 \left(\frac{z_2 - z_1}{a} \right) = 2\pi a(z_2 - z_1) = 2\pi a h$$

because $h = z_2 - z_1$.

C14S0M.047: Position the cone with its vertex at the origin and its axis on the nonnegative z-axis. The side of the cone has Cartesian equation $z = \sqrt{x^2 + y^2}$, and hence

$$dS = \sqrt{1 + (z_x)^2 + (z_y)^2} \; dx \; dy = \sqrt{1 + \frac{x^2}{x^2 + y^2} + \frac{y^2}{x^2 + y^2}} \; dx \; dy = \sqrt{2} \; dx \; dy.$$

Let S be the square with vertices at $(\pm 1, \pm 1)$. Because the area of S is 4, we see with no additional computations that the area of the part of the cone that lies directly above S is

$$\iint_S \sqrt{2} \; dx \; dy = 4\sqrt{2} \approx 5.6568542494923802.$$

C14S0M.049: Given: $x = x(t)$, $y = y(t)$, $z = z$, $a \leqq t \leqq b$, $0 \leqq z \leqq h(t)$: Let

$$\mathbf{r}(t, z) = \langle x(t), y(t), z \rangle, \quad \text{so that} \quad \mathbf{r}_t = \langle x'(t), y'(t), 0 \rangle \quad \text{and} \quad \mathbf{r}_z = \langle 0, 0, 1 \rangle.$$

Then

$$\mathbf{r}_t \times \mathbf{r}_z = \begin{vmatrix} \mathbf{i} & \mathbf{j} & \mathbf{k} \\ x'(t) & y'(t) & 0 \\ 0 & 0 & 1 \end{vmatrix} = \langle y'(t), -x'(t), 0 \rangle,$$

and hence

$$|\mathbf{r}_t \times \mathbf{r}_z| = \sqrt{[x'(t)]^2 + [y'(t)]^2}.$$

Therefore the area of the "fence" is

$$A = \int_{t=a}^{b} \int_{z=0}^{h(t)} \left([x'(t)]^2 + [y'(t)]^2 \right)^{1/2} dz \; dt.$$

C14S0M.051: We are given the region R bounded by the curves $x^2 - y^2 = 1$, $x^2 - y^2 = 4$, $xy = 1$, and $xy = 3$, of constant density δ. Its polar moment of inertia is

869

$$I_0 = \iint_R (x^2 + y^2)\, \delta\, dx\, dy.$$

The hyperbolas bounding R are u-curves and v-curves if we let $u = xy$ and $v = x^2 - y^2$. If we make this substitution, then

$$4u^2 + v^2 = 4x^2 y^2 - (x^2 - y^2)^2 = (x^2 + y^2)^2,$$

and therefore we will substitute $\sqrt{4u^2 + v^2}$ for $x^2 + y^2$ in the integral for I_0. Moreover, it is not necessary to solve for x and y in terms of u and v because of a result in Section 14.9 (see the proof in Problem 18 there). Thus

$$\frac{\partial(u, v)}{\partial(x, y)} = \begin{vmatrix} y & x \\ 2x & -2y \end{vmatrix} = -2(x^2 + y^2),$$

and therefore

$$\frac{\partial(x, y)}{\partial(u, v)} = -\frac{1}{2(x^2 + y^2)} = -\frac{1}{2\sqrt{4u^2 + v^2}}.$$

Therefore

$$I_0 = \int_{v=1}^{4} \int_{u=1}^{3} \frac{\sqrt{4u^2 + v^2}}{2\sqrt{4u^2 + v^2}}\, \delta\, du\, dv = \int_1^4 \int_1^3 \frac{1}{2}\delta\, du\, dv = 3\delta.$$

C14S0M.053: We use the transformation

$$x = a\rho \sin\phi \cos\theta, \quad y = b\rho \sin\phi \sin\theta, \quad z = c\rho \cos\phi.$$

We saw in the solution of Problem 29 that the Jacobian of this transformation is

$$\frac{\partial(x, y, z)}{\partial(\rho, \phi, \theta)} = abc\rho^2 \sin\phi.$$

Moreover, it follows from work shown in the solution of Problem 29 that the density function takes the form $\delta(\rho, \phi, \theta) = 1 - \rho^2$. Finally, the ellipsoidal surface

$$\frac{x^2}{a^2} + \frac{y^2}{b^2} + \frac{z^2}{c^2} = 1$$

is transformed into the surface $\rho = 1$, and therefore the mass of the solid ellipsoid is

$$m = \int_{\theta=0}^{2\pi} \int_{\phi=0}^{\pi} \int_{\rho=0}^{1} (1 - \rho^2)abc\rho^2 \sin\phi\, d\rho\, d\phi\, d\theta = 2\pi abc \int_{\phi=0}^{\pi} \int_{\rho=0}^{1} (\rho^2 - \rho^4)\sin\phi\, d\rho\, d\phi$$

$$= 2\pi abc \int_0^\pi \frac{2}{15}\sin\phi\, d\phi = \frac{4}{15}\pi abc \left[-\cos\phi \right]_0^\pi = \frac{8}{15}\pi abc.$$

C14S0M.055: The spherical surface with radius $\sqrt{3}$ centered at the origin has equation $x^2 + y^2 + z^2 = 3$, and hence the upper hemisphere has equation $z = \sqrt{3 - x^2 - y^2}$. Next,

$$1 + (z_x)^2 + (z_y)^2 = 1 + \frac{x^2}{3 - x^2 - y^2} + \frac{y^2}{3 - x^2 - y^2} = \frac{3}{3 - x^2 - y^2}.$$

We integrate over the unit square in the xy-plane, quadruple the result to find the area of the part of the surface above the 2-by-2 square, then double it to account for the spherical surface *below* the xy-plane. Thus the surface area is

$$A = 8 \int_{x=0}^{1} \int_{y=0}^{1} \frac{\sqrt{3}}{\sqrt{3 - x^2 - y^2}} \, dy \, dx = 8 \int_{x=0}^{1} \left[\sqrt{3} \arctan \left(\frac{y}{\sqrt{3 - x^2 - y^2}} \right) \right]_{y=0}^{1}$$

$$= 8 \int_0^1 \sqrt{3} \arctan \left(\frac{1}{\sqrt{2 - x^2}} \right) dx = \int_{x=0}^{1} 8\sqrt{3} \arcsin \left(\frac{1}{\sqrt{3 - x^2}} \right) dx.$$

Now use integration by parts with

$$u = 8\sqrt{3} \, \arcsin \frac{1}{\sqrt{3 - x^2}}, \qquad dv = dx;$$

$$du = \frac{8x\sqrt{3}}{(3 - x^2)\sqrt{2 - x^2}} \, dx, \qquad v = x.$$

Thus we find that

$$A = \left[8x\sqrt{3} \arcsin \frac{1}{\sqrt{3 - x^2}} \right]_0^1 - 8\sqrt{3} \int_0^1 \frac{x^2}{(3 - x^2)\sqrt{2 - x^2}} \, dx$$

$$= 8\sqrt{3} \arcsin \frac{1}{\sqrt{2}} - 8\sqrt{3} \int_0^1 \frac{x^2}{(3 - x^2)\sqrt{2 - x^2}} \, dx = 2\pi\sqrt{3} - 8\sqrt{3} \int_0^1 \frac{x^2}{(3 - x^2)\sqrt{2 - x^2}} \, dx.$$

Now make the substitution $x = \sqrt{2} \sin \theta$, $dx = \sqrt{2} \cos \theta \, d\theta$. This yields

$$A = 2\pi\sqrt{3} - 8\sqrt{3} \int_0^{\pi/4} \frac{2 \sin^2 \theta}{(3 - 2\sin^2 \theta)} \, d\theta$$

$$= 2\pi\sqrt{3} - 8\sqrt{3} \int_0^{\pi/4} \frac{1 - \cos 2\theta}{3 - 2\sin^2 \theta} \, d\theta = 2\pi\sqrt{3} - 4\sqrt{3} \int_{\phi=0}^{\pi/2} \frac{1 - \cos \phi}{2 + \cos \phi} \, d\phi$$

where $\phi = 2\theta$. Now substitute

$$u = \tan \frac{\phi}{2}, \qquad \sin \phi = \frac{2u}{1 + u^2}, \qquad \cos \phi = \frac{1 - u^2}{1 + u^2}, \qquad d\phi = \frac{2 \, du}{1 + u^2}$$

(see the discussion immediately following Miscellaneous Problem 134 of Chapter 8 (Chapter 7 of the "early transcendentals version")). This yields

$$A = 2\pi\sqrt{3} - 4\sqrt{3} \int_0^1 \frac{4u^2}{(u^2 + 1)(u^2 + 3)} \, du = 2\pi\sqrt{3} - 8\sqrt{3} \int_0^1 \left(\frac{3}{u^2 + 3} - \frac{1}{u^2 + 1} \right) du$$

$$= 2\pi\sqrt{3} - 8\sqrt{3} \left[\sqrt{3} \arctan \left(\frac{u}{\sqrt{3}} \right) - \arctan u \right]_0^1 = 2\pi\sqrt{3} - \left(8\sqrt{3} \right) \cdot \frac{\pi}{12} \left(2\sqrt{3} - 3 \right)$$

$$= 4\pi \left(\sqrt{3} - 1 \right) \approx 9.19922175645144125328.$$

871

C14S0M.057: We will find the volume of the part of the solid that lies in the first octant, then multiply by 8. Thus the volume of the entire solid is

$$V = 8 \int_{x=0}^{a} \int_{y=0}^{(a^{1/3}-x^{1/3})^3} (a^{1/3} - x^{1/3} - y^{1/3})^3 \, dy \, dx.$$

The substitution $y = b \sin^6 \theta$ transforms the integrand into

$$(a^{1/3} - x^{1/3} - b^{1/3} \sin^2 \theta)^3,$$

and hence will be useful provided that $b^{1/3} = a^{1/3} - x^{1/3}$. Thus we choose $b = (a^{1/3} - x^{1/3})^3$, and the substitution

$$y = (a^{1/3} - x^{1/3})^3 \sin^6 \theta, \quad dy = 6(a^{1/3} - x^{1/3})^3 \sin^5 \theta \cos \theta \, d\theta$$

then yields

$$V = 8 \int_{x=0}^{a} \int_{\theta=0}^{\pi/2} \left[(a^{1/3} - x^{1/3}) \cos^2 \theta \right]^3 \cdot 6(a^{1/3} - x^{1/3})^3 \sin^5 \theta \cos \theta \, d\theta$$

$$= 48 \int_{0}^{a} \int_{0}^{\pi/2} (a^{1/3} - x^{1/3})^6 \sin^5 \theta \cos^7 \theta \, d\theta$$

$$= 48 \int_{0}^{a} (a^{1/3} - x^{1/3})^6 \left[\frac{1}{122880} (-600 \cos 2\theta - 75 \cos 4\theta \right.$$

$$\left. \left. + 100 \cos 6\theta + 30 \cos 8\theta - 12 \cos 10\theta - 5 \cos 12\theta) \right]_{0}^{\pi/2} \, dx$$

$$= 48 \int_{0}^{a} \frac{1}{120} (a^{1/3} - x^{1/3})^6 \, dx = \frac{2}{5} \int_{0}^{a} (a^{1/3} - x^{1/3})^6 \, dx$$

$$= \frac{2}{5} \left[a^2 x - \frac{9}{2} a^{5/3} x^{4/3} + 9 a^{4/3} x^{5/3} - 10 a x^2 + \frac{45}{7} a^{2/3} x^{7/3} - \frac{9}{4} a^{1/3} x^{8/3} + \frac{1}{3} x^3 \right]_{0}^{a} = \frac{2}{5} \cdot \frac{1}{84} a^3 = \frac{1}{210} a^3.$$

C14S0M.059: Locate the cube C as shown in the next figure, with one vertex at the origin and the opposite vertex at the point $(1, 1, 1)$ in space. Let L be the line through these two points; we will rotate C around the line L to generate the solid S. We also install a coordinate system on L; it becomes the w-axis, with $w = 0$ at the origin and $w = \sqrt{3}$ at the point with Cartesian coordinates $(1, 1, 1)$. Thus distance is

872

measured on the w-axis in exactly the same way it is measured on the three Cartesian coordinate axes.

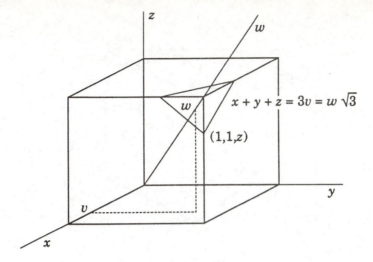

Choose a point v on the x-axis with $\frac{1}{2} \le v \le 1$. We first deal with the case $\frac{2}{3} \le v \le 1$. Then the point with Cartesian coordinates (v, v, v) determines a point $w = v\sqrt{3}$ on the w-axis, and the plane normal to the w-axis at this point intersects C in a triangle, also shown in the preceding figure. It is clear that the plane has equation $x + y + z = 3v = w\sqrt{3}$ and that it meets one edge of the cube at the point $(1, 1, z)$ for some z between 0 and 1. In fact, because $(1, 1, z)$ satisfies the equation of the plane, it follows that $z = -2 + w\sqrt{3}$.

When the cube is rotated around L, the resulting solid S meets the plane $x + y + z = 3v$ in a circular disk centered at (v, v, v), and the radius of this disk is the distance from (v, v, v) to $(1, 1, z)$, which is

$$\sqrt{(v-1)^2 + (v-1)^2 + (w\sqrt{3} - 2 - v)^2} = \sqrt{2(v-1)^2 + (3v - 2 - v)^2}$$

$$= \sqrt{6(v-1)^2} = (1-v)\sqrt{6} = \left(1 - \tfrac{1}{3}w\sqrt{3}\right) \cdot \sqrt{6}.$$

Now we turn to the case $\frac{1}{2} \le v \le \frac{2}{3}$. In this case the plane through (v, v, v) meets the surface of the cube in a semi-regular hexagon, one in which each interior angle is $2\pi/3$ and whose sides are of only two different lengths a and b, alternating as one moves around the hexagon. Such a hexagon is shown in the next figure.

One of the vertices of the hexagon is located at the point $(1, y, 0)$ on one edge of the cube. The distance from (v, v, v) to this point is the radius of the circular disk in which the plane normal to the w-axis at (v, v, v) meets the solid S. It is easy to show that $y = 3v - 1$, and it follows that the distance in question is

$$\sqrt{(v-1)^2 + (2v-1)^2 + v^2} = \sqrt{6v^2 - 6v + 2} = \sqrt{2w^2 - 2w\sqrt{3} + 2}.$$

873

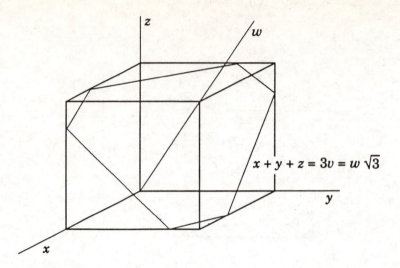

$$x + y + z = 3v = w\sqrt{3}$$

By considering only values of v for which $\frac{1}{2} \leqq v \leqq 1$, we obtain only half of the solid S, so we now can find the volume V of S as follows. We shift to coordinates on the w-axis and remember that $dw = \sqrt{3}\, dv$. The result is that

$$V = 2 \int_{v=1/2}^{2/3} \pi(2w^2 - 2w\sqrt{3} + 2)\sqrt{3}\, dw + 2 \int_{v=2/3}^{1} 6\pi \left(1 - \tfrac{1}{3}w\sqrt{3}\right)^2 \sqrt{3}\, dw.$$

The adjusted limits of integration are $w = \frac{1}{2}\sqrt{3}$ to $\frac{2}{3}\sqrt{3}$ in the first integral and $w = \frac{2}{3}\sqrt{3}$ to $\sqrt{3}$ in the second. Then *Mathematica* 3.0 promptly reports that

$$V = \frac{\pi}{\sqrt{3}} \approx 1.81379936423421785059407825764215732.$$

C15S01.001: $\mathbf{F}(x,\,y) = \langle 1,\,1 \rangle$ is a constant vector field; some vectors in this field are shown next.

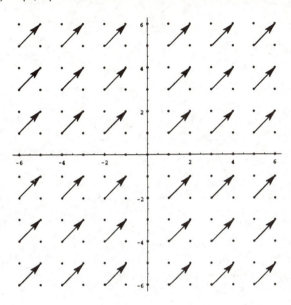

C15S01.003: Some typical vectors in the field $\mathbf{F}(x,\,y) = \langle x,\,-y \rangle$ are shown next.

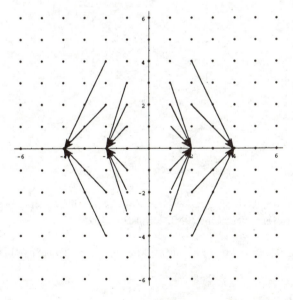

C15S01.005: Some typical vectors in the field $\mathbf{F}(x,\,y) = \langle (x^2 + y^2)^{1/2} \langle x,\,y \rangle$ are shown next. Note that the length of each vector is proportional to the square of the distance from the origin to its initial point and that each vector points directly away from the origin.

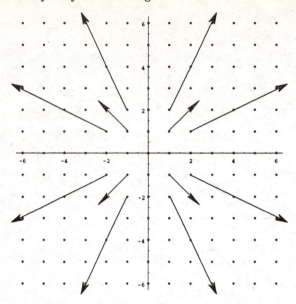

C15S01.007: The vector field $\mathbf{F}(x, y, z) = \langle 0, 1, 1 \rangle$ is a constant vector field. All vectors in this field are parallel translates of the one shown in the next figure.

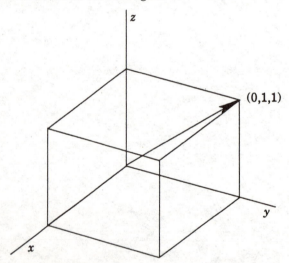

C15S01.009: Each vector in the field $\mathbf{F}(x, y, z) = \langle -x, -y \rangle$ is parallel to the xy-plane and reaches from its initial point at (x, y, z) to its terminal point $(0, 0, z)$ on the z-axis.

C15S01.011: The vector field $\mathbf{\nabla}(xy) = \langle y, x \rangle$ is shown in Fig. 15.1.8. To verify this, evaluate the gradient at $(2, 0)$.

C15S01.013: The gradient vector field

$$\mathbf{\nabla}\left(\sin \tfrac{1}{2}(x^2 + y^2)\right) = \left\langle x \cos \tfrac{1}{2}(x^2 + y^2), \; y \cos \tfrac{1}{2}(x^2 + y^2) \right\rangle$$

is shown in Fig. 15.1.10. To verify this, evaluate the gradient at $(1, 1)$ and at $(0, 1)$.

C15S01.015: If $\mathbf{F}(x, y, z) = \langle x, y, z \rangle$, then

$$\nabla \cdot \mathbf{F} = 1 + 1 + 1 = 3 \quad \text{and} \quad \nabla \times \mathbf{F} = \begin{vmatrix} \mathbf{i} & \mathbf{j} & \mathbf{k} \\ \dfrac{\partial}{\partial x} & \dfrac{\partial}{\partial y} & \dfrac{\partial}{\partial z} \\ x & y & z \end{vmatrix} = \langle 0, 0, 0 \rangle = \mathbf{0}.$$

C15S01.017: If $\mathbf{F}(x, y, z) = \langle yz, xz, xy \rangle$, then

$$\nabla \cdot \mathbf{F} = 0 + 0 + 0 = 0 \quad \text{and} \quad \nabla \times \mathbf{F} = \begin{vmatrix} \mathbf{i} & \mathbf{j} & \mathbf{k} \\ \dfrac{\partial}{\partial x} & \dfrac{\partial}{\partial y} & \dfrac{\partial}{\partial z} \\ yz & xz & xy \end{vmatrix} = \langle x - x, y - y, z - z \rangle = \mathbf{0}.$$

C15S01.019: If $\mathbf{F}(x, y, z) = \langle xy^2, yz^2, zx^2 \rangle$, then

$$\nabla \cdot \mathbf{F} = y^2 + z^2 + x^2 \quad \text{and} \quad \nabla \times \mathbf{F} = \begin{vmatrix} \mathbf{i} & \mathbf{j} & \mathbf{k} \\ \dfrac{\partial}{\partial x} & \dfrac{\partial}{\partial y} & \dfrac{\partial}{\partial z} \\ xy^2 & yz^2 & zx^2 \end{vmatrix} = \langle -2yz, -2xz, -2xy \rangle.$$

C15S01.021: If $\mathbf{F}(x, y, z) = \langle y^2 + z^2, x^2 + z^2, x^2 + y^2 \rangle$, then

$$\nabla \cdot \mathbf{F} = 0 + 0 + 0 = 0 \quad \text{and} \quad \nabla \times \mathbf{F} = \begin{vmatrix} \mathbf{i} & \mathbf{j} & \mathbf{k} \\ \dfrac{\partial}{\partial x} & \dfrac{\partial}{\partial y} & \dfrac{\partial}{\partial z} \\ y^2 + z^2 & x^2 + z^2 & x^2 + y^2 \end{vmatrix} = \langle 2y - 2z, 2z - 2x, 2x - 2y \rangle.$$

C15S01.023: If $\mathbf{F}(x, y, z) = \langle x + \sin yz, y + \sin xz, z + \sin xy \rangle$, then

$$\nabla \cdot \mathbf{F} = 1 + 1 + 1 = 3 \quad \text{and}$$

$$\nabla \times \mathbf{F} = \begin{vmatrix} \mathbf{i} & \mathbf{j} & \mathbf{k} \\ \dfrac{\partial}{\partial x} & \dfrac{\partial}{\partial y} & \dfrac{\partial}{\partial z} \\ x + \sin yz & y + \sin xz & z + \sin xy \end{vmatrix}$$

$$= \langle x \cos xy - x \cos xz, \, y \cos yz - y \cos xy, \, z \cos xz - z \cos yz \rangle.$$

C15S01.025: If a and b are constants and f and g are differentiable functions of two variables, then

$$\nabla(af + bg) = \left\langle \frac{\partial}{\partial x}(af + bg), \frac{\partial}{\partial y}(af + bg) \right\rangle = \langle af_x + bg_x, \, af_y + bg_y \rangle$$

$$= \langle af_x, af_y \rangle + \langle bg_x, bg_y \rangle = a \langle f_x, f_y \rangle + b \langle g_x, g_y \rangle = a\nabla f + b\nabla g.$$

If f and g are functions of three variables, then the proof is similar, merely longer.

C15S01.027: If a and b are constants and $\mathbf{F} = \langle P, Q, R \rangle$ and $\mathbf{G} = \langle S, T, U \rangle$ where P, Q, R, S, T, and U are each differentiable functions of three variables, then

$$\boldsymbol{\nabla} \times (a\mathbf{F} + b\mathbf{G}) = \boldsymbol{\nabla} \times \langle aP + bS,\ aQ + bT,\ aR + bU \rangle$$

$$= \begin{vmatrix} \mathbf{i} & \mathbf{j} & \mathbf{k} \\ \dfrac{\partial}{\partial x} & \dfrac{\partial}{\partial y} & \dfrac{\partial}{\partial z} \\ aP + bS & aQ + bT & aR + bU \end{vmatrix}$$

$$= \langle aR_y + bU_y - aQ_z - bT_z,\ aP_z + bS_z - aR_x - bU_x,\ aQ_x + bT_x - aP_y - bS_y \rangle$$

$$= a\langle R_y - Q_z,\ P_z - R_x,\ Q_x - P_y \rangle + b\langle U_y - T_z,\ S_z - U_x,\ T_x - S_y \rangle$$

$$= a(\boldsymbol{\nabla} \times \mathbf{F}) + b(\boldsymbol{\nabla} \times \mathbf{G}).$$

C15S01.029: Suppose that $\mathbf{G} = \langle S, T, U \rangle$ where S, T, U, and f are differentiable functions of x, y, and z. Then

$$\boldsymbol{\nabla} \times (f\mathbf{G}) = \boldsymbol{\nabla} \times \langle fS,\ fT,\ fU \rangle = \begin{vmatrix} \mathbf{i} & \mathbf{j} & \mathbf{k} \\ \dfrac{\partial}{\partial x} & \dfrac{\partial}{\partial y} & \dfrac{\partial}{\partial z} \\ fS & fT & fU \end{vmatrix}$$

$$= \langle f_y U + fU_y - f_z T - fT_z,\ f_z S + fS_z - f_x U - fU_x,\ f_x T + fT_x - f_y S - fS_y \rangle$$

$$= \langle f_y U - f_z T,\ f_z S - f_x U,\ f_x T - f_y S \rangle + \langle fU_y - fT_z,\ fS_z - fU_x,\ fT_x - fS_y \rangle$$

$$= \begin{vmatrix} \mathbf{i} & \mathbf{j} & \mathbf{k} \\ f_x & f_y & f_z \\ S & T & U \end{vmatrix} + (f) \begin{vmatrix} \mathbf{i} & \mathbf{j} & \mathbf{k} \\ \dfrac{\partial}{\partial x} & \dfrac{\partial}{\partial y} & \dfrac{\partial}{\partial z} \\ S & T & U \end{vmatrix} = (\boldsymbol{\nabla} f) \times \mathbf{G} + (f)(\boldsymbol{\nabla} \times \mathbf{G}).$$

C15S01.031: If $\mathbf{F} = \langle P, Q, R \rangle$ and $\mathbf{G} = \langle S, T, U \rangle$ where P, Q, R, S, T, and U are each differentiable functions of three variables, then

$$\boldsymbol{\nabla} \cdot (\mathbf{F} \times \mathbf{G}) = \boldsymbol{\nabla} \cdot \langle QU - RT,\ RS - PU,\ PT - QS \rangle$$

$$= Q_x U + QU_x - R_x T - RT_x + R_y S + RS_y - P_y U - PU_y + P_z T + PT_z - Q_z S - QS_z$$

and

$$\mathbf{G} \cdot (\boldsymbol{\nabla} \times \mathbf{F}) - \mathbf{F} \cdot (\boldsymbol{\nabla} \times \mathbf{G}) = \mathbf{G} \cdot \langle R_y - Q_z,\ P_z - R_x,\ Q_x - P_y \rangle - \mathbf{F} \cdot \langle U_y - T_z,\ S_z - U_x,\ T_x - S_y \rangle$$

$$= SR_y - SQ_z + TP_z - TR_x + UQ_x - UP_y - PU_y + PT_z - QS_z + QU_x - RT_x + RS_y.$$

Then comparison of the last expressions in each of the two computations reveals that

$$\nabla \cdot (\mathbf{F} \times \mathbf{G}) = \mathbf{G} \cdot (\nabla \times \mathbf{F}) - \mathbf{F} \cdot (\nabla \times \mathbf{G}).$$

C15S01.033: Suppose that f and g are twice-differentiable functions of the three variables x, y, and z. Then

$$\nabla \cdot [\nabla(fg)] = \nabla \cdot \langle f_x g + f g_x, \ f_y g + f g_y, \ f_z g + f g_z \rangle$$

$$= f_{xx} g + 2 f_x g_x + f g_{xx} + f_{yy} g + 2 f_y g_y + f g_{yy} + f_{zz} g + 2 f_z g_z + f g_{zz}$$

$$= (f)(g_{xx} + g_{yy} + g_{zz}) + (g)(f_{xx} + f_{yy} + f_{zz}) + 2(f_x g_x + f_y g_y + f_z g_z)$$

$$= f \nabla \cdot \langle g_x, g_y, g_z \rangle + g \nabla \cdot \langle f_x, f_y, f_z \rangle + 2\langle f_x, f_y, f_z \rangle \cdot \langle g_x, g_y, g_z \rangle$$

$$= f \nabla \cdot (\nabla g) + g \nabla \cdot (\nabla f) + 2(\nabla f) \cdot (\nabla g).$$

C15S01.035: If $\mathbf{r} = \langle x, y, z \rangle$, then $\nabla \cdot \mathbf{r} = 3$ and $\nabla \times \mathbf{r} = 0$ by the solution of Problem 15.

C15S01.037: By the results in Problems 28 and 35,

$$\nabla \cdot \frac{\mathbf{r}}{r^3} = \frac{1}{r^3}(\nabla \cdot \mathbf{r}) - \frac{3}{2}(x^2 + y^2 + z^2)^{-5/2}\langle 2x, 2y, 2z \rangle \cdot \mathbf{r} = \frac{3}{r^3} - 3r^{-5}(x^2 + y^2 + z^2) = \frac{3}{r^3} - \frac{3r^2}{r^5} = 0.$$

C15S01.039: If $r = |\mathbf{r}| = (x^2 + y^2 + z^2)^{1/2}$, then

$$\nabla r = \nabla(x^2 + y^2 + z^2)^{1/2} = \frac{1}{2}(x^2 + y^2 + z^2)^{-1/2}\langle 2x, 2y, 2z \rangle = \frac{\mathbf{r}}{r}.$$

C15S01.041: By the results in Problems 28, 35, and 39,

$$\nabla \cdot (r\mathbf{r}) = r\nabla \cdot \mathbf{r} + (\nabla r) \cdot \mathbf{r} = 3r + \frac{\mathbf{r} \cdot \mathbf{r}}{r} = 3r + \frac{r^2}{r} = 4r.$$

C15S01.043: Here we have

$$\nabla(\ln r) = \nabla\left(\ln\sqrt{x^2 + y^2 + z^2}\right) = \frac{1}{2}\left\langle \frac{2x}{x^2 + y^2 + z^2}, \ \frac{2y}{x^2 + y^2 + z^2}, \ \frac{2z}{x^2 + y^2 + z^2}\right\rangle$$

$$= \frac{1}{x^2 + y^2 + z^2}\langle x, y, z \rangle = \frac{\mathbf{r}}{r^2}.$$

Section 15.2

C15S02.001: If $x(t) = 4t - 1$ and $y(t) = 3t + 1$, $-1 \le t \le 1$, then

$$\int_C (x^2 + y^2)\, ds = \int_{-1}^{1}(125t^2 - 10t + 10)\, dt = \left[\frac{125}{3}t^3 - 5t^2 + 10t\right]_{-1}^{1} = \frac{310}{3},$$

$$\int_C (x^2 + y^2)\, dx = \int_{-1}^{1}(100t^2 - 8t + 8)\, dt = \left[\frac{100}{3}t^3 - 4t^2 + 8t\right]_{-1}^{1} = \frac{248}{3}, \quad \text{and}$$

$$\int_C (x^2 + y^2)\, dy = \int_{-1}^{1}(75t^2 - 6t + 6)\, dt = \left[25t^3 - 3t^2 + 6t\right]_{-1}^{1} = 62.$$

C15S02.003: If $x(t) = e^t + 1$ and $y(t) = e^t - 1$, $0 \leq t \leq \ln 2$, then

$$\int_C (x + y)\, ds = \int_0^{\ln 2} 2^{3/2} e^{2t}\, dt = \left[2^{1/2} e^{2t} \right]_0^{\ln 2} = 3\sqrt{2} \approx 4.2426406871192851,$$

$$\int_C (x + y)\, dx = \int_0^{\ln 2} 2e^{2t}\, dt = \left[e^{2t} \right]_0^{\ln 2} = 3, \quad \text{and} \quad \int_C (x + y)\, dy = \int_0^{\ln 2} 2e^{2t}\, dt = \left[e^{2t} \right]_0^{\ln 2} = 3.$$

C15S02.005: If $x(t) = 3t$ and $y(t) = t^4$, $0 \leq t \leq 1$, then

$$\int_C xy\, ds = \int_0^1 3t^5 \sqrt{9 + 16t^6}\, dt = \left[\frac{1}{48}(9 + 16t^6)^{3/2} \right]_0^1 = \frac{49}{24} \approx 2.0416666666666667,$$

$$\int_C xy\, dx = \int_0^1 9t^5\, dt = \left[\frac{3}{2} t^6 \right]_0^1 = \frac{3}{2}, \quad \text{and} \quad \int_C xy\, dy = \int_0^1 12t^8\, dt = \left[\frac{4}{3} t^9 \right]_0^1 = \frac{4}{3}.$$

C15S02.007: One parametrization of the path C is this: Let $y(t) = t$ and $x(t) = t^3$, $-1 \leq t \leq 1$. Then

$$\int_C P(x, y)\, dx + Q(x, y)\, dy = \int_C y^2\, dx + x\, dy = \int_{-1}^1 (t^3 + 3t^4)\, dt = \left[\frac{1}{4} t^4 + \frac{3}{5} t^5 \right]_{-1}^1 = \frac{6}{5}.$$

C15S02.009: Parametrize the path C in two parts. Let

$$x_1(t) = t, \ y_1(t) = 1, \ -1 \leq t \leq 2 \quad \text{and let} \quad x_2(t) = 2, \ y_2(t) = t, \ 1 \leq t \leq 5.$$

Then

$$\int_C P(x, y)\, dx + Q(x, y)\, dy = \int_C x^2 y\, dx + xy^3\, dy = \int_{-1}^2 t^2\, dt + \int_1^5 2t^3\, dt$$

$$= \left[\frac{1}{3} t^3 \right]_{-1}^2 + \left[\frac{1}{2} t^4 \right]_1^5 = 3 + 312 = 315.$$

C15S02.011: Because $\mathbf{r}(t) = \langle t, t^2, t^3 \rangle$ for $0 \leq t \leq 1$, we have

$$d\mathbf{r} = \langle 1, 2t, 3t^2 \rangle \quad \text{and} \quad \mathbf{F}(t) = \langle t^3, t, -t^2 \rangle.$$

Therefore

$$\int_C \mathbf{F} \cdot d\mathbf{r} = \int_0^1 (2t^2 + t^3 - 3t^4)\, dt = \left[\frac{2}{3} t^3 + \frac{1}{4} t^4 - \frac{3}{5} t^5 \right]_0^1 = \frac{19}{60} \approx 0.316666666667.$$

C15S02.013: The path C is parametrized by $\mathbf{r}(t) = \langle \sin t, \cos t, 2t \rangle$, $0 \leq t \leq \pi$. Hence

$$d\mathbf{r} = \langle \cos t, -\sin t, 2 \rangle \quad \text{and} \quad \mathbf{F}(t) = \langle \cos t, -\sin t, 2t \rangle,$$

and thus

$$\int_C \mathbf{F} \cdot d\mathbf{r} = \int_0^{\pi} (4t+1)\, dt = \left[2t^2 + t \right]_0^{\pi} = 2\pi^2 + \pi \approx 22.8808014557685105.$$

C15S02.015: Parametrize the path C in three sections, as follows:

$$x_1(t) = -1, \qquad y_1(t) = 2, \qquad z_1(t) = -2 + 4t, \quad 0 \leqq t \leqq 1;$$

$$x_2(t) = -1 + 2t, \quad y_2(t) = 2, \qquad z_2(t) = 2, \qquad\qquad 0 \leqq t \leqq 1;$$

$$x_3(t) = 1, \qquad y_3(t) = 2 + 3t, \quad z_3(t) = 2, \qquad\qquad 0 \leqq t \leqq 1.$$

Then

$$\int_C \mathbf{F} \cdot d\mathbf{r} \;=\; I_1 + I_2 + I_3$$

where

$$I_1 = \int_0^1 (32 - 64t)\, dt = \left[32t - 32t^2 \right]_0^1 = 0,$$

$$I_2 = \int_0^1 16\, dt = \left[16t \right]_0^1 = 16, \quad \text{and}$$

$$I_3 = \int_0^1 12\, dt = \left[12t \right]_0^1 = 12.$$

Therefore $\displaystyle\int_C \mathbf{F} \cdot d\mathbf{r} = 0 + 16 + 12 = 28.$

C15S02.017: Here we have

$$\int_C (2x + 9xy)\, ds = \int_0^1 (2t + 9t^3)(1 + 4t^2 + 9t^4)^{1/2}\, dt$$

$$= \left[\frac{1}{6}(1 + 4t^2 + 9t^4)^{3/2} \right]_0^1 = \frac{14\sqrt{14} - 1}{6} \approx 8.5638672358058632.$$

C15S02.019: Because the wire W is uniform, we may assume that its density is $\delta = 1$. Moreover, $\bar{x} = 0$ by symmetry. Parametrize the wire by

$$\mathbf{r}(t) = \langle a\cos t,\ a\sin t \rangle, \quad 0 \leqq t \leqq \pi.$$

Then $ds = a\, dt$. Also the mass of the wire is πa, so it remains only to compute the moment

$$M_x = \int_W ay\, dt = \int_0^{\pi} a^2 \sin t\, dt = \left[-a^2 \cos t \right]_0^{\pi} = 2a^2.$$

Therefore the centroid of the wire is located at the point $\left(0,\ \dfrac{2a}{\pi} \right).$

C15S02.021: First, the arc length element is

$$ds = \sqrt{9\sin^2 t + 9\cos^2 t + 16} \ dt = 5 \ dt.$$

The mass and moments of the helical wire W are

$$m = \int_W 5k \ dt = \int_0^{2\pi} 5k \ dt = 10k\pi;$$

$$M_{yz} = \int_0^{2\pi} 15k\cos t \ dt = \Big[15k\sin t \Big]_0^{2\pi} = 0;$$

$$M_{xz} = \int_0^{2\pi} 15k\sin t \ dt = \Big[-15k\cos t \Big]_0^{2\pi} = 0;$$

$$M_{xy} = \int_0^{2\pi} 20kt \ dt = \Big[10kt^2 \Big]_0^{2\pi} = 40k\pi^2.$$

Therefore the coordinates of the centroid are

$$\overline{x} = \overline{y} = 0, \quad \overline{z} = \frac{40k\pi^2}{10k\pi} = 4\pi.$$

C15S02.023: Parametrize the wire W via

$$\mathbf{r}(t) = \langle a\cos t, \ a\sin t, \ 0 \rangle, \quad 0 \le t \le \frac{\pi}{2}.$$

Then the arc-length element is $ds = |\mathbf{r}'(t)| \ dt = a \ dt$. The mass element is $dm = a \cdot k \cdot a^2 \sin t \cos t \ dt$, and hence the mass of W is

$$m = \int_0^{\pi/2} ka^3 \sin t \cos t \ dt = \Big[\frac{1}{2} ka^3 \sin^2 t \Big]_0^{\pi/2} = \frac{1}{2} ka^3.$$

Clearly the z-coordinate of the centroid is $\overline{z} = 0$, and $\overline{y} = \overline{x}$ by symmetry. The moment of the wire with respect to the y-axis is

$$M_y = \int_0^{\pi/2} ka^4 \sin t \cos^2 t \ dt = -\frac{1}{3} ka^4 \Big[\cos^3 t \Big]_0^{\pi/2} = \frac{1}{3} ka^4.$$

Therefore the x-coordinate of the centroid is

$$\overline{x} = \frac{2ka^4}{3ka^3} = \frac{2}{3} a.$$

The moments of W with respect to the coordinate axes are

$$I_x = \int_0^{\pi/2} ka^5 \sin^2 t \cos t \ dt = \frac{1}{4} ka^5 \Big[\sin^4 t \Big]_0^{\pi/2} = \frac{1}{4} ka^5 = \frac{1}{2} ma^2;$$

$I_y = I_x$ by symmetry, and $I_0 = I_x + I_y = \frac{1}{2} ka^5 = ma^2$.

C15S02.025: Using the given parametrization we find that the arc-length element is $ds = \frac{3}{2} |\sin 2t| \ dt$, and hence the polar moment of inertia of the wire is

$$I_0 = 4 \int_0^{\pi/2} k(\cos^6 t + \sin^6 t) \, ds = -\frac{3}{16} k \left[\cos 6t + 7 \cos 2t \right]_0^{\pi/2} = 3k.$$

Because the mass of the wire is

$$m = 4 \int_0^{\pi/2} k \, ds = -3k \left[\cos 2t \right]_0^{\pi/2} = 6k,$$

we can also write $I_0 = \dfrac{1}{2} m$.

C15S02.027: We are given the circle C of radius a centered at the origin and the point $(a, 0)$ on C. Suppose that (x, y) is a point of C. Let t be the angular polar coordinate of (x, y). Then, by the law of cosines, the distance w between $(a, 0)$ and (x, y) satisfies the equation

$$w^2 = a^2 + a^2 - 2a^2 \cos t = 2a^2(1 - \cos t) = 4a^2 \cdot \frac{1 - \cos t}{2} = 4a^2 \sin^2 \frac{t}{2}.$$

Because $0 \le t \le 2\pi$, it now follows that $w = 2a \sin \dfrac{t}{2}$, and hence the average value of w on C is

$$\bar{d} = \frac{1}{2\pi a} \int_0^{2\pi} 2a^2 \sin \frac{t}{2} \, dt = -\frac{4a^2}{2\pi a} \left[\cos \frac{t}{2} \right]_0^{2\pi} = \frac{8a^2}{2\pi a} = \frac{4}{\pi} a \approx (1.2732395447351627)a.$$

C15S02.029: The parametrization $x(t) = \cos^3 t$, $y(t) = \sin^3 t$, $0 \le t \le 2\pi$ of the astroid yields the arc-length element

$$ds = \sqrt{9 \cos^4 t \, \sin^2 t + 9 \cos^2 t \, \sin^4 t} \, dt = \frac{3}{2} \sqrt{\sin^2 2t} \, dt = \frac{3}{2} \sin 2t \, dt,$$

although the last equality is valid only if $\sin 2t$ is nonnegative. Hence we will find the average distance of points of the astroid in the first quadrant from the origin; by symmetry, this will be the same as the average distance of all of its points from the origin. Noting that

$$[x(t)]^2 + [y(t)]^2 = \cos^6 t + \sin^6 t$$

and noting also that the length of the first-quadrant arc of the astroid is $\frac{3}{2}$ (a consequence of the solution of Problem 30 in Section 10.5), we find that the average distance of points of the astroid from the origin is

$$\bar{d} = \frac{2}{3} \int_0^{\pi/2} \frac{3}{2} \sqrt{\cos^6 t + \sin^6 t} \, \sin 2t \, dt$$

$$= \left[\frac{\sqrt{3}}{24} \operatorname{arctanh} \left(\frac{\sqrt{6} \, \cos 2t}{\sqrt{5 + 3 \cos 4t}} \right) - \frac{\sqrt{2}}{16} (\cos 2t) \sqrt{5 + 3 \cos 4t} \right]_0^{\pi/2}$$

$$= \frac{1}{2} + \frac{\sqrt{3}}{12} \operatorname{arctanh} \frac{\sqrt{3}}{2} \approx 0.6900864990752365868827735637 2.$$

Of course we used *Mathematica 3.0* to find and simplify both the antiderivative and the value of the definite integral. By contrast, *Derive 2.56* yields the result

$$\bar{d} = \frac{2}{3} \int_0^{\pi/2} \frac{3}{2} \sqrt{\cos^6 t + \sin^6 t} \, \sin 2t \, dt$$

$$= \left[-\frac{\sqrt{3}}{24} \ln\left(2\sqrt{3}\cos^4 t - 3\cos^2 t + 1 + 2\sqrt{3}\cos^2 t - \sqrt{3}\right) - \frac{1 - 2\cos^2 t}{4} \sqrt{3\cos^4 t - 3\cos^2 t + 1} \right]_0^{\pi/2}$$

$$= \frac{6 + \sqrt{3} \ln\left(2 + \sqrt{3}\right)}{12} \approx 0.690086499075236586882773563725.$$

C15S02.031: With the given parametrization, we find that $ds = \sqrt{2}\, e^{-t}\, dt$ and that

$$\sqrt{[x(t)]^2 + [y(t)]^2} = e^{-t}.$$

The length of the spiral is

$$\int_0^\infty \sqrt{2}\, e^{-t}\, dt = \left[-\sqrt{2}\, e^{-t} \right]_0^\infty = \sqrt{2},$$

and thus the average distance of points of the spiral from the origin is

$$\bar{d} = \frac{1}{\sqrt{2}} \int_0^\infty \sqrt{2}\, e^{-2t}\, dt = \left[-\frac{1}{2} e^{-2t} \right]_0^\infty = \frac{1}{2}.$$

C15S02.033: Part (a): Parametrize the path by $x(t) = 1$, $y(t) = t$, $0 \leq t \leq 1$. Then the force is

$$\mathbf{F}(t) = \left\langle \frac{k}{1 + t^2}, \; \frac{kt}{1 + t^2} \right\rangle,$$

and so the work is

$$W = \int_0^1 \mathbf{F} \cdot d\mathbf{r} = \int_0^1 \frac{kt}{1 + t^2} \, dt = \left[\frac{1}{2} k \ln(1 + t^2) \right]_0^1 = \frac{1}{2} k \ln 2.$$

Part (b): Parametrize the path by $x(t) = 1 - t$, $y(t) = 1$, $0 \leq t \leq 1$. Then the force is

$$\mathbf{F}(t) = \left\langle \frac{k(1 - t)}{1 + (1 - t)^2}, \; \frac{k}{1 + (1 - t)^2} \right\rangle,$$

and thus the work is

$$W = \int_0^1 -\frac{k(1 - t)}{1 + (1 - t)^2} \, dt = \left[\frac{1}{2} k \ln\left(1 + (1 - t)^2\right) \right]_0^1 = -\frac{1}{2} k \ln 2.$$

C15S02.035: Parametrize the unit circle C in the usual way: $x(t) = \cos t$, $y(t) = \sin t$, $0 \leq t \leq 2\pi$. Then the force function is $\mathbf{F}(t) = \langle k\cos t, \, k\sin t \rangle$. Hence $\mathbf{F} \cdot d\mathbf{r} = 0$, and therefore

$$\int_C \mathbf{F} \cdot d\mathbf{r} = 0.$$

C15S02.037: The work done in moving along a path on the sphere is zero because $\mathbf{F}$ is normal to the sphere. Therefore $\mathbf{F} \cdot \mathbf{T}$ is identically zero on any path on the sphere.

C15S02.039: The force function is $\mathbf{F}(t) = \langle 0, -150 \rangle$ and the path may be parametrized as follows: $x(t) = 100 \sin t$, $y(t) = 100 \cos t$, $0 \leq t \leq \frac{1}{2}\pi$. Hence the work done is

$$\int_C \mathbf{F} \cdot d\mathbf{r} = \int_0^{\pi/2} 15000 \sin t \, dt = -15000 \left[\cos t \right]_0^{\pi/2} = 15000 \quad \text{(ft·lb)}.$$

C15S02.041: The force function is $\mathbf{F}(t) = \langle 0, 0, -200 \rangle$ and the path may be parametrized as follows: $x(t) = 25 \cos t$, $y(t) = 25 \sin t$, $z(t) = 100 - 100t/(10\pi)$, $0 \leq t \leq 10\pi$. Hence the work done is

$$\int_C \mathbf{F} \cdot d\mathbf{r} = \int_0^{10\pi} \frac{2000}{\pi} \, dt = \left[\frac{2000}{\pi} t \right]_0^{10\pi} = 20000 \quad \text{(ft·lb)}.$$

Section 15.3

Note: In the solutions of Problems 1 through 16, $P(x, y)$ always denotes the first component of the given vector field $\mathbf{F}(x, y)$, $Q(x, y)$ always denotes its second component, and $\phi(x, y)$ denotes a scalar potential function with gradient $\mathbf{F}(x, y)$.

C15S03.001: If $P(x, y) = 2x + 3y$ and $Q(x, y) = 3x + 2y$, then

$$\frac{\partial P}{\partial y} = 3 = \frac{\partial Q}{\partial x}.$$

Hence $\mathbf{F}$ is conservative. By inspection, $\phi(x, y) = x^2 + 3xy + y^2$.

C15S03.003: If $P(x, y) = 3x^2 + 2y^2$ and $Q(x, y) = 4xy + 6y^2$, then

$$\frac{\partial P}{\partial y} = 4y = \frac{\partial Q}{\partial x}.$$

Hence $\mathbf{F}$ is conservative. By inspection, $\phi(x, y) = x^3 + 2xy^2 + 2y^3$.

C15S03.005: If $P(x, y) = 2y + \sin 2x$ and $Q(x, y) = 3x + \cos 3y$, then

$$\frac{\partial P}{\partial y} = 2 \neq 3 = \frac{\partial Q}{\partial x}.$$

Hence $\mathbf{F}$ is not conservative.

C15S03.007: If $P(x, y) = x^3 + \frac{y}{x}$ and $Q(x, y) = y^2 + \ln x$, then

$$\frac{\partial P}{\partial y} = \frac{1}{x} = \frac{\partial Q}{\partial x}.$$

Hence $\mathbf{F}$ is conservative. By inspection, $\phi(x, y) = \frac{1}{4}x^4 + y \ln x + \frac{1}{3}y^3$.

C15S03.009: If $P(x, y) = \cos x + \ln y$ and $Q(x, y) = \frac{x}{y} + e^y$, then

$$\frac{\partial P}{\partial y} = \frac{1}{y} = \frac{\partial Q}{\partial x}.$$

Hence $\mathbf{F}$ is conservative. By inspection, $\phi(x, y) = \sin x + x \ln y + e^y$.

C15S03.011: If $P(x, y) = x\cos y + \sin y$ and $Q(x, y) = y\cos x + \sin x$, then

$$\frac{\partial P}{\partial y} = \cos y - x\sin y \neq \cos x - y\sin x = \frac{\partial Q}{\partial x}.$$

Hence $\mathbf{F}$ is not conservative.

C15S03.013: If $P(x, y) = 3x^2 y^3 + y^4$ and $Q(x, y) = 3x^3 y^2 + y^4 + 4xy^3$, then

$$\frac{\partial P}{\partial y} = 9x^2 y^2 + 4y^3 = \frac{\partial Q}{\partial x}.$$

Hence $\mathbf{F}$ is conservative. By inspection, $\phi(x, y) = x^3 y^3 + xy^4 + \frac{1}{5}y^5$.

C15S03.015: If $P(x, y) = \dfrac{2x}{y} - \dfrac{3y^2}{x^4}$ and $Q(x, y) = \dfrac{2y}{x^3} - \dfrac{x^2}{y^2} + \dfrac{1}{\sqrt{y}}$, then

$$\frac{\partial P}{\partial y} = -\frac{2x}{y^2} - \frac{6y}{x^4} = \frac{\partial Q}{\partial x}.$$

Hence $\mathbf{F}$ is conservative. By inspection, $\phi(x, y) = \dfrac{x^2}{y} + 2\sqrt{y} + \dfrac{y^2}{x^3}$.

C15S03.017: We let $x(t) = x_1 t$ and $y(t) = y_1 t$ for $0 \leqq t \leqq 1$. Also let $\mathbf{r}(t) = \langle x(t), y(t) \rangle$ and $\mathbf{F}(x, y) = \langle 3x^2 + 2y^2, 4xy + 6y^2 \rangle$. Then

$$\int_{t=0}^{1} \mathbf{F}(x(t), y(t)) \cdot \mathbf{r}'(t)\, dt = \int_0^1 3(x_1^3 + 2x_1 y_1^2 + 2y_1^3)t^2\, dt = \left[(x_1^3 + 2x_1 y_1^2 + 2y_1^3)t^3\right]_0^1 = x_1^3 + 2x_1 y_1^2 + 2y_1^3.$$

Then, as in Example 3, a scalar potential for $\mathbf{F}(x, y)$ is $\phi(x, y) = x^3 + 2xy^2 + 2y^3$.

C15S03.019: We let $x(t) = x_1 t$ and $y(t) = y_1 t$ for $0 \leqq t \leqq 1$. Also let $\mathbf{r}(t) = \langle x(t), y(t) \rangle$ and $\mathbf{F}(x, y) = \langle 3x^2 y^3 + y^4, 3x^3 y^2 + y^4 + 4xy^3 \rangle$. Then

$$\int_{t=0}^{1} \mathbf{F}(x(t), y(t)) \cdot \mathbf{r}'(t)\, dt = \int_0^1 (6x_1^3 y_1^3 t^5 + 5x_1 y_1^4 t^4 + y_1^5 t^4)\, dt$$

$$= \left[x_1^3 y_1^3 t^6 + x_1 y_1^4 t^5 + \frac{1}{5}y_1^5 t^5\right]_0^1 = x_1^3 y_1^3 + x_1 y_1^4 + \frac{1}{5}y_1^5.$$

Therefore a scalar potential for $\mathbf{F}$ is $\phi(x, y) = x^3 y^3 + xy^4 + \frac{1}{5}y^5$.

C15S03.021: Let $P(x, y) = y^2 + 2xy$ and $Q(x, y) = x^2 + 2xy$. Then

$$\frac{\partial P}{\partial y} = 2x + 2y = \frac{\partial Q}{\partial x},$$

and therefore $\mathbf{F}(x, y)$ is conservative with potential function $\phi(x, y) = x^2 y + xy^2$. Therefore

$$\int_{(0,0)}^{(1,2)} P\, dx + Q\, dy = \left[x^2 y + xy^2\right]_{(0,0)}^{(1,2)} = 6 - 0 = 6.$$

C15S03.023: Let $P(x, y) = 2xe^y$ and $Q(x, y) = x^2 e^y$. Then

$$\frac{\partial P}{\partial y} = 2xe^y = \frac{\partial Q}{\partial x},$$

and therefore $\mathbf{F}(x, y)$ is conservative with potential function $\phi(x, y) = x^2 e^y$. Consequently

$$\int_{(0,0)}^{(1,-1)} P\, dx + Q\, dy = \left[x^2 e^y \right]_{(0,0)}^{(1,-1)} = \frac{1}{e} - 0 = \frac{1}{e}.$$

C15S03.025: Let $P(x, y) = \sin y + y \cos x$ and $Q(x, y) = \sin x + x \cos y$. Then

$$\frac{\partial P}{\partial y} = \cos x + \cos y = \frac{\partial Q}{\partial x},$$

and therefore $\mathbf{F}(x, y)$ is conservative with potential function $\phi(x, y) = y \sin x + x \sin y$. Therefore

$$\int_{(\pi/2,\pi/2)}^{(\pi,\pi)} P\, dx + Q\, dy = \left[y \sin x + x \sin y \right]_{(\pi/2,\pi/2)}^{(\pi,\pi)} = 0 - \pi = -\pi.$$

C15S03.027: By inspection, $\phi(x, y, z) = xyz$. For an analytic solution, write

$$\mathbf{F}(x, y, z) = \langle\, P(x, y, z),\ Q(x, y, z),\ R(x, y, z)\,\rangle$$

where $P(x, y, z) = yz$, $Q(x, y, z) = xz$, and $R(x, y, z) = xy$. Then let

$$g(x, y, z) = \int P(x, y, z)\, dx = \int yz\, dx = xyz + h(y, z).$$

Then

$$\frac{\partial g}{\partial y} = Q(x, y, z) = xz = xz + \frac{\partial h}{\partial y} \quad \text{and} \quad \frac{\partial g}{\partial z} = R(x, y, z) = xy = xy + \frac{\partial h}{\partial z},$$

and therefore the choice $h(y, z) \equiv 0$ yields a scalar potential for $\mathbf{F}$.

C15S03.029: Write

$$\mathbf{F}(x, y, z) = \langle\, P(x, y, z),\ Q(x, y, z),\ R(x, y, z)\,\rangle$$

where $P(x, y, z) = y \cos z - yze^x$, $Q(x, y, z) = x \cos z - ze^x$, and $R(x, y, z) = -xy \sin z - ye^x$. Let

$$\phi(x, y, z) = \int P(x, y, z)\, dx = xy \cos z - yze^x + g(y, z).$$

Then

$$Q(x, y, z) = x \cos z - ze^x = \frac{\partial \phi}{\partial y} = x \cos z - ze^x + \frac{\partial g}{\partial y}.$$

Hence $g_y(x, z) = 0$, and so g is a function of z alone. Thus

$$\phi(x, y, z) = xy \cos z - yze^x + g(z),$$

and therefore

$$R(x, y, z) = -xy\sin z - ye^x = \frac{\partial \phi}{\partial z} = -xy\sin z - ye^x + g'(z).$$

Thus $g(z) = C$, a constant. So every scalar potential for $\mathbf{F}$ has the form $\phi(x, y, z) = xy\cos z - yze^x + C$. For a particular scalar potential, simply choose $C = 0$.

C15S03.031: Suppose that the force field $\mathbf{F} = \langle P, Q \rangle$ is conservative in the plane region D. Then there exists a potential function $\phi(x, y)$ for $\mathbf{F}$; that is, $\nabla \phi = \mathbf{F}$, so that

$$\frac{\partial \phi}{\partial x} = P(x, y) \quad \text{and} \quad \frac{\partial \phi}{\partial y} = Q(x, y)$$

on the interior of D. But then

$$\frac{\partial P}{\partial y} = \frac{\partial^2 \phi}{\partial y\, \partial x} = \frac{\partial^2 \phi}{\partial x\, \partial y} = \frac{\partial Q}{\partial x}$$

on the interior of D, under the assumption that the second-order mixed partial derivatives of ϕ are continuous there. Of course, continuity of P_y and Q_x on D is enough to guarantee this.

C15S03.033: The given integral is not independent of the path because

$$\frac{\partial x^2}{\partial z} = 0 \neq 2y = \frac{\partial y^2}{\partial y}.$$

C15S03.035: Part (a): If

$$f(x, y) = \arctan\left(\frac{y}{x}\right),$$

then

$$\frac{\partial f}{\partial x} = \frac{-\dfrac{y}{x^2}}{1 + \dfrac{y^2}{x^2}} = -\frac{y}{x^2 + y^2} \quad \text{and}$$

$$\frac{\partial f}{\partial y} = \frac{\dfrac{1}{x}}{1 + \dfrac{y^2}{x^2}} = \frac{x}{x^2 + y^2}.$$

Part (b): Because $\mathbf{F} = \nabla f$ has a potential on the right half-plane $x > 0$, line integrals of $\mathbf{F}$ will be independent of the path C provided that the path always remains in the right half-plane. Hence if C is such a path from $A(x_1, y_1) = (r_1, \theta_1)$ to $B(x_2, y_2) = (r_2, \theta_2)$, then

$$\int_C \mathbf{F} \cdot \mathbf{T}\, ds = \int_A^B \mathbf{F} \cdot d\mathbf{r} = \Big[f(x, y) \Big]_A^B = \theta_2 - \theta_1.$$

Part (c): Parametrize the unit circle using $x(t) = \cos t$, $y(t) = \sin t$, $-\pi \leq t \leq \pi$. It is easy to verify that

$$\mathbf{F}(x(t), y(t)) = \langle -\sin t, \cos t \rangle$$

and that $\mathbf{F}(t) \cdot \mathbf{r}'(t) = 1$. Therefore

$$\int_{C_1} \mathbf{F} \cdot d\mathbf{r} = \int_0^\pi 1 \, dt = \pi \quad \text{and} \quad \int_{C_2} \mathbf{F} \cdot d\mathbf{r} = \int_0^{-\pi} 1 \, dt = -\pi.$$

This does not contradict the fundamental theorem of calculus for line integrals because $\mathbf{F}$ does not have a scalar potential defined in a region containing both C_1 and C_2 (see the solution of Problem 30).

C15S03.037: The units are mks units throughout. We use $M = 5.97 \times 10^{24}$, $G = 6.67 \times 10^{-11}$, $m = 10000$, and let $k = GMm = 3.98199 \times 10^{18}$ in the formula in Problem 36. We also must convert r_1 and r_2 into meters: $r_1 = 9000 \cdot 1000$ and $r_2 = 11000 \cdot 1000$. Then substitution in the formula in Problem 36 yields $W = 8.04442 \times 10^{10}$ N·m.

Section 15.4

Note: As in the text, the notation

$$\oint_C P(x, y) \, dx + Q(x, y) \, dy$$

(and variations thereof) always denotes an integral around the *closed* path C with *counterclockwise* (positive) orientation.

C15S04.001: By Green's theorem,

$$\oint_C (x + y^2) \, dx + (y + x^2) \, dy = \int_{y=-1}^1 \int_{x=-1}^1 (2x - 2y) \, dx \, dy$$

$$= \int_{y=-1}^1 \left[x^2 - 2xy \right]_{x=-1}^1 dy = \int_{-1}^1 -4y \, dy = \left[-2y^2 \right]_0^1 = 0.$$

C15S04.003: By Green's theorem,

$$\oint_C (y + e^x) \, dx + (2x^2 + \cos y) \, dy = \int_0^1 \int_y^{2-y} (4x - 1) \, dx \, dy$$

$$= \int_0^1 \left[2x^2 - x \right]_y^{2-y} dy = \int_0^1 (6 - 6y) \, dy = \left[6y - 3y^2 \right]_0^1 = 3.$$

C15S04.005: By Green's theorem,

$$\oint_C \left[-y^2 + \exp(e^x) \right] dx + (\arctan y) \, dy = \int_0^1 \int_{x^2}^{\sqrt{x}} 2y \, dy \, dx = \int_0^1 \left[y^2 \right]_{x^2}^{\sqrt{x}} dx$$

$$= \int_0^1 (x - x^4) \, dx = \left[\frac{1}{2} x^2 - \frac{1}{5} x^5 \right]_0^1 = \frac{3}{10}.$$

C15S04.007: By Green's theorem,

$$\oint_C (x - y) \, dx + y \, dy = \int_0^\pi \int_0^{\sin x} 1 \, dy \, dx = \int_0^\pi \sin x \, dx = \left[-\cos x \right]_0^\pi = 2.$$

889

C15S04.009: Let D be the bounded region bounded by the curve C. Then by Green's theorem,

$$\oint_C y^2\,dx + xy\,dy = \iint_D (y - 2y)\,dA = -\iint_D y\,dA = 0$$

by symmetry.

C15S04.011: Let R denote the bounded plane region bounded by the curve C. Then by Green's theorem,

$$\oint_C xy\,dx + x^2\,dy = \iint_R (2x - x)\,dA = \int_{\theta=0}^{\pi/2} \int_{r=0}^{\sin 2\theta} r^2 \cos\theta\,dr\,d\theta = \int_0^{\pi/2} \left[\frac{1}{3}r^3 \cos\theta\right]_0^{\sin 2\theta} d\theta$$

$$= \int_0^{\pi/2} \frac{8}{3}\sin^3\theta\,\cos^4\theta\,d\theta = \frac{1}{840}\left[5\cos 7\theta + 7\cos 5\theta - 35\cos 3\theta - 105\cos\theta\right]_0^{\pi/2}$$

$$= \frac{16}{105} \approx 0.15238095238095238095.$$

C15S04.013: The given parametrization and the corollary to Green's theorem yield area

$$A = \oint_C x\,dy = \int_0^{2\pi} a^2 \cos^2 t\,dt = a^2 \left[\frac{1}{2}t + \frac{1}{2}\sin t\,\cos t\right]_0^{2\pi} = \pi a^2.$$

C15S04.015: We'll use the given parametrization, find the area of the part of the astroid in the first quadrant, then multiply by 4. The corollary to Green's theorem yields area

$$A = \oint_C x\,dy = 4\int_0^{\pi/2} 3\sin^2 t\,\cos^4 t\,dt$$

$$= \frac{1}{16}\left[12t + 3\sin 2t - 3\sin 4t - \sin 6t\right]_0^{\pi/2} = \frac{3}{8}\pi \approx 1.17809724509617246442.$$

There's no need to evaluate the line integral along the x- or y-axes because $x\,dy = 0$ there.

C15S04.017: Denote by E the bounded plane region bounded by the given curve C. Then the work is

$$W = \oint_C \mathbf{F}\cdot\mathbf{T}\,ds = \oint_C -2y\,dx + 3x\,dy = \iint_E 5\,dA = 5\cdot 6\pi = 30\pi \approx 94.2477796076937972.$$

(The area of the ellipse is 6π because it has semiaxes of lengths 2 and 3.)

C15S04.019: Denote by T the triangular region bounded by the given triangle C. Then the work done is

$$W = \oint_C \mathbf{F}\cdot\mathbf{T}\,ds = \oint_C 5x^2 y^3\,dx + 7x^3 y^2\,dy = \iint_T (21x^2 y^2 - 15x^2 y^2)\,dA = \int_{x=0}^3 \int_{y=0}^{6-2x} 6x^2 y^2\,dy\,dx$$

$$= \int_0^3 \left[2x^2 y^3\right]_{y=0}^{6-2x} dx = \int_0^3 (432x^2 - 432x^3 + 144x^4 - 16x^5)\,dx$$

$$= \left[144x^3 - 108x^4 + \frac{144}{5}x^5 - \frac{8}{3}x^6\right]_0^3 = \frac{972}{5} = 194.4.$$

C15S04.021: Denote by R the bounded plane region bounded by the given curve C. Then the outward flux of $\mathbf{F}$ across C is

$$\phi = \oint_C \mathbf{F} \cdot \mathbf{n} \, ds = \iint_R \mathbf{\nabla} \cdot \mathbf{F} \, dA = \iint_R \mathbf{\nabla} \cdot \langle 2x, \, 3y \rangle \, dA = \iint_R 5 \, dA = 30\pi \approx 94.2477796076937972.$$

C15S04.023: Denote by T the triangular region bounded by the given path C. Then the outward flux of $\mathbf{F}$ across C is

$$\phi = \oint_C \mathbf{F} \cdot \mathbf{n} \, ds = \iint_T \mathbf{\nabla} \cdot \mathbf{F} \, dA$$

$$= \iint_T \mathbf{\nabla} \cdot \langle 3x + \sqrt{1+y^2} \, , \; 2y - (1+x^4)^{1/3} \rangle \, dA = \iint_T 5 \, dA = 5 \cdot \frac{1}{2} \cdot 3 \cdot 6 = 45$$

because T is a triangle with base 3 and height 6.

C15S04.025: Given f, a twice-differentiable function of x and y, we have

$$\nabla^2 f = \mathbf{\nabla} \cdot (\nabla f) = \mathbf{\nabla} \cdot \langle f_x, \, f_y \rangle = \frac{\partial^2 f}{\partial x^2} + \frac{\partial^2 f}{\partial y^2}.$$

C15S04.027: If f and g are twice-differentiable functions of x and y, then

$$\nabla^2(fg) = \mathbf{\nabla} \cdot \left[\mathbf{\nabla}(fg) \right] = \mathbf{\nabla} \cdot \langle f_x g + f g_x, \; f_y g + f g_y \rangle$$

$$= f_{xx} g + f_x g_x + f_x g_x + f g_{xx} + f_{yy} g + f_y g_y + f_y g_y + f g_{yy}$$

$$= (f)(g_{xx} + g_{yy}) + 2(f_x g_x + f_y g_y) + (g)(f_{xx} + f_{yy})$$

$$= f\nabla^2 g + g\nabla^2 f + 2\langle f_x, \, f_y \rangle \cdot \langle g_x, \, g_y \rangle = f\nabla^2 g + g\nabla^2 f + 2\mathbf{\nabla} f \cdot \mathbf{\nabla} g.$$

Compare this with Problem 33 of Section 15.1.

C15S04.029: We may assume constant density $\delta = 1$. Then

$$A = \oint_C x \, dy = \oint_C -y \, dx.$$

Now

$$M_y = \iint_R x \, dA \quad \text{and} \quad M_x = \iint_R y \, dA.$$

Hence, by Green's theorem,

$$M_y = \oint_C \frac{1}{2} x^2 \, dy \quad \text{and} \quad M_x = -\oint_C \frac{1}{2} y^2 \, dx.$$

Therefore

$$\overline{x} = \frac{M_y}{A} = \frac{1}{2A} \oint_C x^2 \, dy \quad \text{and} \quad \overline{y} = \frac{M_x}{A} = -\frac{1}{2A} \oint_C y^2 \, dx.$$

891

C15S04.031: Suppose that the plane region R is bounded by the piecewise smooth simple closed curve C, oriented counterclockwise, and that R has constant density δ. Then, by Green's theorem,

$$I_x = \iint_R \delta y^2 \, dA = \delta \oint_C -\frac{1}{3} y^3 \, dx = -\frac{1}{3} \delta \oint_C y^3 \, dx \quad \text{and}$$

$$I_y = \iint_R \delta x^2 \, dA = \delta \oint_C \frac{1}{3} x^3 \, dy = \frac{1}{3} \delta \oint_C x^3 \, dy.$$

C15S04.033: As in Problem 30 of Section 10.4, the substitution $y = tx$ in the equation $x^3 + y^3 = 3xy$ of the folium yields $x^3 + t^3 x^3 = 3tx^2$, and thereby the parametrization

$$x(t) = \frac{3t}{1 + t^3}, \quad y(t) = \frac{3t^2}{1 + t^3}, \quad 0 \le t < +\infty$$

of the first-quadrant loop of the folium. If C is the half of its loop that stretches from $(0, 0)$ to $\left(\frac{3}{2}, \frac{3}{2}\right)$ along the lower half of the folium, then C is swept out by this parametrization as t varies from 0 to 1. Let J be the straight line segment joining $\left(\frac{3}{2}, \frac{3}{2}\right)$ with $(0, 0)$; parametrize J with $x = \frac{3}{2}(1 - t)$, $y = \frac{3}{2}(1 - t)$, $0 \le t \le 1$. Then the area of the folium is

$$A = 2 \cdot \frac{1}{2} \oint_{C \cup J} x \, dy - y \, dx = \int_C x \, dy - y \, dx + \int_J x \, dy - y \, dx.$$

The last integral is

$$\int_{t=0}^1 \left[-\frac{3}{2}(1 - t) + \frac{3}{2}(1 - t) \right] dt = 0,$$

and hence the area of the folium is

$$A = \int_0^1 \left[x(t) y'(t) - y(t) x'(t) \right] dt = \int_0^1 \left[\frac{9t(2t - t^4)}{(1 + t^3)^3} - \frac{9t^2(2t^3 - 1)}{(1 + t^3)^3} \right] dt$$

$$= \int_0^1 \frac{9t^2}{(1 + t^3)^2} \, dt = \left[-\frac{3}{1 + t^3} \right]_0^1 = \frac{3}{2}.$$

C15S04.035: We substitute $f\nabla g$ for $\mathbf{F}$ in Eq. (9),

$$\oint_C \mathbf{F} \cdot \mathbf{n} \, ds = \iint_R \nabla \cdot \mathbf{F} \, dA.$$

With the aid of the result in Problem 28 of Section 15.1, this yields

$$\oint_C f\nabla g \cdot \mathbf{n} \, ds = \iint_R \nabla \cdot (f\nabla g) \, dA = \iint_R (f\nabla \cdot \nabla g + \nabla f \cdot \nabla g) \, dA.$$

C15S04.037: It suffices to show the result in the case that $R = R_1 \cup R_1$ is the union of two regions, with C the boundary of R, $C_1 \cup C_3$ the boundary of R_1, and $-C_3 \cup C_2$ the boundary of R_2. Then $C = C_1 \cup C_2$.

Perhaps the next figure will clarify all this.

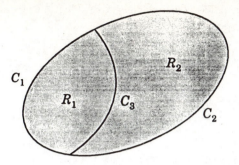

Under the assumption that Green's theorem holds for R_1 and for R_2, we have

$$\oint_{C_1 \cup C_3} P \, dx + Q \, dy = \iint_{R_1} (Q_x - P_y) \, dA \quad \text{and}$$

$$\oint_{-C_3 \cup C_2} P \, dx + Q \, dy = \iint_{R_2} (Q_x - P_y) \, dA.$$

Addition of these equations yields

$$\int_{C_1} (P \, dx + Q \, dy) + \int_{C_2} (P \, dx + Q \, dy) + \int_{C_3} (P \, dx + Q \, dy) - \int_{C_3} (P \, dx + Q \, dy) = \iint_{R} (Q_x - P_y) \, dA.$$

Therefore

$$\oint_C P \, dx + Q \, dy = \iint_R (Q_x - P_y) \, dA.$$

C15S04.039: Part (a): With $x_1 = 1$, $y_1 = 0$, $x_2 = \cos(2\pi/3)$, $y_2 = \sin(2\pi/3)$, $x_3 = \cos(4\pi/3)$, and $y_3 = \sin(4\pi/3)$, we obtain area

$$A = \frac{1}{2}(x_1 y_2 - x_2 y_1 + x_2 y_3 - x_3 y_2 + x_3 y_1 - x_1 y_3) = \frac{1}{4}\sqrt{3} + \frac{1}{4}\sqrt{3} + \frac{1}{4}\sqrt{3} = \frac{3}{4}\sqrt{3} \approx 1.29903810567666.$$

Part (b): With the points (x_i, y_i) $(1 \leqq i \leqq 5)$ chosen in a way analogous to that in part (a), we obtain area

$$A = \frac{1}{2}(x_1 y_2 - x_2 y_1 + x_2 y_3 - x_3 y_2 + x_3 y_4 - x_4 y_3 + x_4 y_5 - x_5 y_4 + x_5 y_1 - x_1 y_5)$$

$$= \frac{5}{8}\sqrt{10 + 2\sqrt{5}} \approx 2.377641290737883393029.$$

(We used a computer algebra program to evaluate the first line, then simplified the result by hand.)

C15S04.041: Using *Mathematica* 3.0, we entered the parametric functions

$$x(t) = \frac{(2n+1)t^n}{t^{2n+1} + 1} \quad \text{and} \quad y(t) = \frac{(2n+1)t^{n+1}}{t^{2n+1} + 1}.$$

Then we computed one of the integrands for area in Eq. (4):

893

```
(-y'[t]*x[t] + x[t]*y'[t])/2 // Together
```

$$\frac{(2n+1)^2 t^{2n}}{2(t^{2n+1}+1)^2}$$

Then we set $n = 1$, then 2, then 3, and found the areas of the corresponding loops as follows:

```
Integrate[ (9/2)*(t^2)/(1 + t^3)^2, t ]
```

$$-\frac{3}{2(t^3+1)}$$

```
( % /. t → Infinity ) - ( % /. t → 0 )
```

$$\frac{3}{2}$$

```
Integrate[ (25/2)*(t^4)/(1 + t^5)^2, t ]
```

$$-\frac{5}{2(t^5+1)}$$

```
( % /. t → Infinity ) - ( % /. t → 0 )
```

$$\frac{5}{2}$$

```
Integrate[ (49/2)*(t^6)/(1 + t^7)^2, t ]
```

$$-\frac{7}{2(t^7+1)}$$

```
( % /. t → Infinity ) - ( % /. t → 0 )
```

$$\frac{7}{2}$$

In fact, the integral is easy to find in the general case:

```
Integrate[ (1/2)*((2*n + 1)^2)*(t^(2*n))/(1 + t^(2*n + 1))^2, t ]
```

$$-\frac{2n+1}{2(1+t^{2n+1})}$$

Therefore

$$A_n = \left[-\frac{2n+1}{2(1+t^{2n+1})}\right]_0^\infty = 0 + \frac{2n+1}{2} = n + \frac{1}{2}.$$

To avoid the improper integral, let C denote the simple closed curve consisting of the lower half of the loop (swept out as t ranges from 0 to 1) together with the "return path" along the line $y = x$ back to the origin. On the return path P we have the parametrization $x = y = n + \frac{1}{2} - t$ as t ranges from 0 to $n + \frac{1}{2}$; but on this path, $-y\,dx + x\,dy = 0$, and hence

$$\frac{1}{2}\int_P -y\,dx + x\,dy = 0.$$

894

The area enclosed by the loop is double that enclosed by C, and thus

$$A_n = 2 \cdot \left[-\frac{2n+1}{2(1+t^{2n+1})} \right]_0^1 = -\frac{2n+1}{2} + (2n+1) = n + \frac{1}{2}.$$

The only danger in this "short cut" is that you may use one of the other two formulas in Eq. (4) of the text and forget the line integral along P.

Section 15.5

C15S05.001: Here S is the surface $z = h(x, y) = 1 - x - y$ over the plane triangle bounded by the nonnegative coordinate axes and the graph of $y = 1 - x$. So

$$dS = \sqrt{1 + (h_x)^2 + (h_y)^2} \; dx \; dy = \sqrt{3} \; dx \; dy.$$

Therefore

$$\iint_S (x + y) \; dS = \int_{x=0}^{1} \int_{y=0}^{1-x} (x + y)\sqrt{3} \; dy \; dx = \sqrt{3} \int_0^1 \left[xy + \frac{1}{2}y^2 \right]_0^{1-x} dx$$

$$= \sqrt{3} \int_0^1 \left(\frac{1}{2} - \frac{1}{2}x^2 \right) dx = \sqrt{3} \left[\frac{1}{2}x - \frac{1}{6}x^3 \right]_0^1 = \frac{1}{3}\sqrt{3} \approx 0.5773502691896258.$$

C15S05.003: First, S is the surface $z = h(x, y) = 2x + 3y$ lying over the circular disk D with center $(0, 0)$ and radius 3 in the xy-plane. Also

$$dS = \sqrt{1 + (h_x)^2 + (h_y)^2} \; dx \; dy = \sqrt{14} \; dx \; dy,$$

and thus

$$\iint_S (y + z + 3) \; dS = \iint_D (y + 2x + 3y + 3)\sqrt{14} \; dA = \int_{\theta=0}^{2\pi} \int_{r=0}^{3} (4r\sin\theta + 2r\cos\theta + 3)\left(r\sqrt{14} \right) dr \; d\theta$$

$$= \sqrt{14} \int_0^{2\pi} \left[\frac{3}{2}r^2 + \frac{2}{3}r^3(\cos\theta + 2\sin\theta) \right]_0^3 d\theta = \sqrt{14} \int_0^{2\pi} \left(\frac{27}{2} + 18\cos\theta + 36\sin\theta \right) d\theta$$

$$= \sqrt{14} \left[\frac{27}{2}\theta + 18\sin\theta - 36\cos\theta \right]_0^{2\pi} = 27\pi\sqrt{14} \approx 317.3786106805529421.$$

C15S05.005: The surface S is the part of the paraboloid $z = h(x, y) = x^2 + y^2$ that lies over the circular disk D with center $(0, 0)$ and radius 2 in the xy-plane. Also

$$dS = \sqrt{1 + (h_x)^2 + (h_y)^2} \; dx \; dy = \sqrt{1 + 4x^2 + 4y^2} \; dx \; dy,$$

and thus

$$\iint_S (xy + 1) \; dS = \iint_D (xy + 1)\sqrt{1 + 4x^2 + 4y^2} \; dx \; dy = \int_0^{2\pi} \int_0^2 (1 + r^2 \sin\theta \cos\theta) \cdot r(1 + 4r^2)^{1/2} \; dr \; d\theta$$

$$= \int_0^{2\pi} \frac{1}{240}(1 + 4r^2)^{3/2} \left[20 + (6r^2 - 1)\sin 2\theta \right]_0^2 d\theta$$

895

$$= \int_0^{2\pi} \frac{1}{240} \left[340\sqrt{17} - 20 + \left(1 + 391\sqrt{17}\right) \sin 2\theta \right] d\theta$$

$$= \frac{1}{480} \left[40 \left(17\sqrt{17} - 1\right)\theta - \left(1 + 391\sqrt{17}\right)\cos 2\theta \right]_0^{2\pi}$$

$$= \frac{1}{480} \left[1 + 391\sqrt{17} - 1 - 391\sqrt{17} + 80\pi \left(17\sqrt{17} - 1\right) \right]$$

$$= \frac{1}{6}\pi \left(-1 + 17\sqrt{17}\right) \approx 36.176903197411408364756.$$

C15S05.007: The surface S is the part of the graph of $z = h(x, y) = x + y$ that lies over the circular disk D with center $(0, 0)$ and radius 3 in the xy-plane. Also $dS = \sqrt{1 + 1 + 1}\, dx\, dy = \sqrt{3}\, dx\, dy$, and hence

$$I_z = \iint_S \delta(x^2 + y^2)\, dS = \iint_D \delta r^2 \sqrt{3}\, dA = \delta\sqrt{3} \int_0^{2\pi} \int_0^3 r^3\, dr\, d\theta = 2\pi\delta\sqrt{3}\left[\frac{1}{4}r^4\right]_0^3 = \frac{81}{2}\pi\delta\sqrt{3}\,.$$

The mass of S is

$$m = \delta\sqrt{3} \int_0^{2\pi} \int_0^3 r\, dr\, d\theta = 2\pi\delta\sqrt{3}\left[\frac{1}{2}r^2\right]_0^3 = 9\pi\delta\sqrt{3}\,,$$

and therefore $I_z = \dfrac{9}{2}m$.

C15S05.009: Suppose that (x, z) is a point in the xz-plane. Let $\mathbf{w}$ be the radius vector from the origin in the xz-plane to (x, z) and let θ be the angle that $\mathbf{w}$ makes with the nonnegative x-axis. Then points in the cylindrical surface S are described by

$$x = \cos\theta, \quad y = y, \quad z = \sin\theta, \quad -1 \leq y \leq 1, \quad 0 \leq \theta \leq 2\pi.$$

Thus the cylindrical surface S is parametrized by $\mathbf{r}(y, \theta) = \langle \cos\theta, y, \sin\theta \rangle$, for which $\mathbf{r}_y = \langle 0, 1, 0 \rangle$ and $\mathbf{r}_\theta = \langle -\sin\theta, 0, \cos\theta \rangle$. Hence

$$\mathbf{r}_y \times \mathbf{r}_\theta = \begin{vmatrix} \mathbf{i} & \mathbf{j} & \mathbf{k} \\ 0 & 1 & 0 \\ -\sin\theta & 0 & \cos\theta \end{vmatrix} = \langle \cos\theta, 0, \sin\theta \rangle.$$

Therefore $|\mathbf{r}_y \times \mathbf{r}_\theta| = 1$, so that $dS = dy\, d\theta$. Let D denote the rectangle $-1 \leq y \leq 1$, $0 \leq \theta \leq 2\pi$. Because S has constant density δ, its moment of inertia with respect to the z-axis is therefore

$$I_z = \iint_S (x^2 + y^2)\delta\, dS = \iint_D (y^2 + \cos^2\theta)\delta\, dA = \int_{\theta=0}^{2\pi} \int_{y=-1}^{1} (y^2 + \cos^2\theta)\delta\, dy\, d\theta$$

$$= \delta \int_0^{2\pi} \left[\frac{1}{3}y^3 + y\cos^2\theta\right]_{-1}^1 d\theta = \delta \int_0^{2\pi} \left(\frac{2}{3} + 2\cos^2\theta\right) d\theta = \frac{1}{6}\delta\left[10\theta + 3\sin 2\theta\right]_0^{2\pi} = \frac{10}{3}\pi\delta.$$

Because the mass m of S is the product of its surface area and its density, we have $m = 4\pi\delta$, and hence we may also express I_z in the form

896

$$I_z = \frac{5}{6}m = m \cdot (0.9128709291752769)^2.$$

C15S05.011: The surface S has the spherical-coordinates parametrization

$$\mathbf{r}(\phi,\,\theta) = \langle\, 5\sin\phi\,\cos\theta,\; 5\sin\phi\,\sin\theta,\; 5\cos\phi \,\rangle, \quad 0 \leqq \phi \leqq \arccos\left(\frac{3}{5}\right),\; 0 \leqq \theta \leqq 2\pi.$$

Therefore

$$\mathbf{r}_\phi \times \mathbf{r}_\theta = \begin{vmatrix} \mathbf{i} & \mathbf{j} & \mathbf{k} \\ 5\cos\phi\,\cos\theta & 5\cos\phi\,\sin\theta & -5\sin\phi \\ -5\sin\phi\,\sin\theta & 5\sin\phi\,\cos\theta & 0 \end{vmatrix} = \langle\, 25\sin^2\phi\,\cos\theta,\; 25\sin^2\phi\,\sin\theta,\; 25\sin\phi\,\cos\phi \,\rangle,$$

and thus

$$|\mathbf{r}_\phi \times \mathbf{r}_\theta| = \sqrt{625\sin^2\phi\,\cos^2\phi + 625\sin^4\phi\,\cos^2\theta + 625\sin^4\phi\,\sin^2\theta} = 25\sin\phi$$

(because $0 \leqq \phi \leqq \pi/2$). Hence the mass of S is

$$m = \iint_S \delta\,dS = \int_{\theta=0}^{2\pi}\int_{\phi=0}^{\arccos(3/5)} 25\delta\sin\phi\,d\phi\,d\theta = 2\pi\delta\left[\, -25\cos\phi \,\right]_0^{\arccos(3/5)}$$

$$= 20\pi\delta \approx (62.8318530717958648)\delta.$$

Next,

$$x^2 + y^2 = (5\sin\phi\,\cos\theta)^2 + (5\sin\phi\,\sin\theta)^2 = 25\sin^2\phi,$$

and hence the moment of inertia of S with respect to the z-axis is

$$I_z = \iint_S (x^2+y^2)\delta\,dS = \int_{\theta=0}^{2\pi}\int_{\phi=0}^{\arccos(3/5)} 625\delta\sin^3\phi\,d\phi\,d\theta = 2\pi\delta\left[\frac{625}{3}\cos^3\phi - 625\cos\phi\right]_0^{\arccos(3/5)}$$

$$= \frac{520}{3}\pi\delta \approx (544.5427266222308280)\delta.$$

The moment of inertia may also be expressed in the form $I_z = \dfrac{26}{3}m \approx m \cdot (2.9439202887759490)^2$.

C15S05.013: An upward unit vector normal to S is

$$\mathbf{n} = \left\langle \frac{x}{3},\; \frac{y}{3},\; \frac{z}{3} \right\rangle.$$

The surface has equation $z = h(x,y) = \sqrt{9 - x^2 - y^2}$, and therefore

$$dS = \sqrt{1 + \frac{x^2}{9 - x^2 - y^2} + \frac{y^2}{9 - x^2 - y^2}}\; dA = \frac{3}{\sqrt{9 - x^2 - y^2}}\; dA.$$

Next, $\mathbf{F} \cdot \mathbf{n} = \frac{1}{3}(x^2 + y^2)$, and S lies over the circular disk D in the plane with center $(0, 0)$ and radius 3. Therefore

$$\iint_S \mathbf{F} \cdot \mathbf{n} \, dS = \iint_D \frac{x^2 + y^2}{\sqrt{9 - x^2 - y^2}} \, dA = \int_{\theta=0}^{2\pi} \int_{r=0}^{3} \frac{r^3}{\sqrt{9 - r^2}} \, dr \, d\theta$$

$$= 2\pi \left[-\frac{1}{3}(r^2 + 18)\sqrt{9 - r^2} \right]_0^3 = 36\pi \approx 113.0973355292325566.$$

C15S05.015: An upward unit vector normal to S is

$$\mathbf{n} = \left\langle -\frac{3}{10}\sqrt{10}, \ 0, \ \frac{1}{10}\sqrt{10} \right\rangle.$$

The surface S has equation $z = h(x, y) = 3x + 2$, and therefore $dS = \sqrt{10} \, dA$. Also, S lies over the circular disk S in the xy-plane with center $(0, 0)$ and radius 2. Therefore

$$\iint_S \mathbf{F} \cdot \mathbf{n} \, dS = \iint_D 3z \, dA = \iint_D (9x + 6) \, dA = \int_{\theta=0}^{2\pi} \int_{r=0}^{2} (9r^2 \cos\theta + 6r) \, dr \, d\theta = \int_0^{2\pi} \left[3r^3 \cos\theta + 3r^2 \right]_0^2$$

$$= \int_0^{2\pi} (24\cos\theta + 12) \, d\theta = \left[24\sin\theta + 12\theta \right]_0^{2\pi} = 24\pi \approx 75.3982236861550377.$$

C15S05.017: The surface S has Cartesian equation $z = h(x, y) = \sqrt{x^2 + y^2}$, and thus has normal vector

$$\mathbf{n}_1 = \langle h_x, h_y, -1 \rangle = \left\langle \frac{x}{\sqrt{x^2 + y^2}}, \ \frac{y}{\sqrt{x^2 + y^2}}, \ -1 \right\rangle,$$

and thus (in polar coordinates) a upward-pointing vector normal to S is $\mathbf{n}_2 = \langle -\cos\theta, \ -\sin\theta, \ 1 \rangle$. Therefore an upward-pointing unit vector normal to S is

$$\mathbf{n} = \frac{\mathbf{n}_2}{|\mathbf{n}_2|} = \frac{\sqrt{2}}{2} \langle -\cos\theta, \ -\sin\theta, \ 1 \rangle.$$

Next,

$$dS = \sqrt{1 + (h_x)^2 + (h_y)^2} \, dA = \left(1 + \frac{x^2}{x^2 + y^2} + \frac{y^2}{x^2 + y^2} \right)^{1/2} dA = \sqrt{2} \, dA,$$

and in polar coordinates we have $\mathbf{F} = \langle r\sin\theta, \ -r\cos\theta, \ 0 \rangle$. But then $\mathbf{F} \cdot \mathbf{n} \, dS = 0$, so the surface integral is zero as well.

C15S05.019: On the face of the cube in the xy-plane, $z = 0$, and so

$$\mathbf{F}(x, y, z) \cdot \mathbf{n} = \langle x, 2y, 0 \rangle \cdot \langle 0, 0, -1 \rangle = 0,$$

and hence the flux of $\mathbf{F}$ across that face is zero. Similarly, the flux across the faces in the other two coordinate planes is zero. On the top face we have $z = 1$, and hence

$$\mathbf{F}(x, y, z) \cdot \mathbf{n} = \langle x, 2y, 3 \rangle \cdot \langle 0, 0, 1 \rangle = 3.$$

Similarly, the flux across the face in the plane $y = 1$ is 2 and the flux across the face in the plane $x = 1$ is 1. Hence the total flux of $\mathbf{F}$ across S is $3 + 2 + 1 = 6$.

C15S05.021: For the same reasons given in the solution of Problem 19, $\mathbf{F} \cdot \mathbf{n} = 0$ on the three faces of the pyramid in the coordinate planes. On the fourth face T a unit normal vector is

$$\mathbf{n} = \frac{1}{\sqrt{26}} \langle 3, 4, 1 \rangle,$$

and because this face is the graph of $z = h(x, y) = 12 - 3x - 4y$, we have

$$dS = \sqrt{1 + (h_x)^2 + (h_y)^2} \; dA = \sqrt{26} \; dx \, dy.$$

Therefore $\mathbf{F} \cdot \mathbf{n} \, dS = (3x - 4y) \, dy \, dx$, and consequently

$$\iint_T \mathbf{F} \cdot \mathbf{n} \, dS = \int_0^4 \int_0^{(12-3x)/4} (3x - 4y) \, dy \, dx = \int_0^4 \left[3xy - 2y^2 \right]_0^{(12-3x)/4} dx$$

$$= \int_0^4 \left(-18 + 18x - \frac{27}{8} x^2 \right) dx = \left[-18x + 9x^2 - \frac{9}{8} x^3 \right]_0^4 = 0.$$

Thus the total flux of $\mathbf{F}$ across S is zero. If you now turn two pages ahead in your textbook, you will see how the divergence theorem enables you to obtain the same result in less than two seconds and without need of pencil, paper, or computer.

C15S05.023: The paraboloids meet in the circle $x^2 + y^2 = 9$, $z = 9$, so both the upper surface and the lower surface lie over the disk D in the xy-plane with center $(0, 0)$ and radius 3. The lower surface L is the graph of $h(x, y) = x^2 + y^2$ and the upper surface U is the graph of $j(x, y) = 18 - x^2 - y^2$ for (x, y) in D. A vector normal to L is

$$\langle h_x, h_y, -1 \rangle = \langle 2x, 2y, -1 \rangle$$

and hence the outer unit vector normal to L is

$$\mathbf{n}_1 = \frac{1}{\sqrt{1 + 4x^2 + 4y^2}} \langle 2x, 2y, -1 \rangle;$$

similarly, the outer unit vector normal to U is

$$\mathbf{n}_2 = \frac{1}{\sqrt{1 + 4x^2 + 4y^2}} \langle 2x, 2y, 1 \rangle.$$

The surface area element for L is

$$dS = \sqrt{1 + (h_x)^2 + (h_y)^2} \; dA = \sqrt{1 + 4x^2 + 4y^2} \; dA$$

and the surface area element for U is the same. Next,

$$\mathbf{F} \cdot \mathbf{n}_1 \, dS = -z^2 = -(x^2 + y^2)^2 \, dA \quad \text{and} \quad \mathbf{F} \cdot \mathbf{n}_2 \, dS = z^2 = (18 - x^2 - y^2)^2 \, dA.$$

Thus

$$\iint_L \mathbf{F} \cdot \mathbf{n}_1 \, dS = -\iint_D (x^2 + y^2)^2 \, dA = -\int_0^{2\pi} \int_0^3 r^5 \, dr \, d\theta = -2\pi \left[\frac{1}{6} r^6 \right]_0^3 = -243\pi$$

899

and

$$\iint_U \mathbf{F} \cdot \mathbf{n}_2 \, dS = \iint_D (18 - x^2 - y^2)^2 \, dA = \int_0^{2\pi} \int_0^3 (18 - r^2)^2 \cdot r \, dr \, d\theta = 2\pi \left[-\frac{1}{6}(18 - r^2)^3 \right]_0^3 = 1701\pi.$$

Therefore

$$\iint_S \mathbf{F} \cdot \mathbf{n} \, dS = 1701\pi - 243\pi = 1458\pi \approx 4580.4420889339185417.$$

C15S05.025: The surface S may be parametrized by

$$x(\phi, \theta) = a \sin \phi \cos \theta, \ y(\phi, \theta) = a \sin \phi \sin \theta, \ z(\phi, \theta) = a \cos \phi, \quad 0 \le \phi \le \frac{1}{2}\pi, \ 0 \le \theta \le \frac{1}{2}\pi.$$

Then, if $\mathbf{r}(\phi, \theta) = \langle x(\phi, \theta), \ y(\phi, \theta), \ z(\phi, \theta) \rangle$, we find that

$$\mathbf{r}_\phi \times \mathbf{r}_\theta = \begin{vmatrix} \mathbf{i} & \mathbf{j} & \mathbf{k} \\ a \cos \phi \cos \theta & a \cos \phi \sin \theta & -a \sin \phi \\ -a \sin \phi \sin \theta & a \sin \phi \cos \theta & 0 \end{vmatrix} = \langle a^2 \sin^2 \phi \cos \theta, \ a^2 \sin^2 \phi \sin \theta, \ a^2 \sin \phi \cos \phi \rangle,$$

and therefore

$$|\mathbf{r}_\phi \times \mathbf{r}_\theta| = \sqrt{a^4 \cos^2 \phi \sin^2 \phi + a^4 \sin^4 \phi \cos^2 \theta + a^4 \sin^4 \phi \sin^2 \theta} = a^2 \sin \phi$$

because $0 \le \phi \le \pi$. Because the surface has unit density, its mass is

$$m = \iint_S 1 \, dS = \int_0^{\pi/2} \int_0^{\pi/2} a^2 \sin \phi \, d\phi \, d\theta = \frac{1}{2}\pi \left[-a^2 \cos \phi \right]_0^{\pi/2} = \frac{1}{2}\pi a^2.$$

The moment of S with respect to the yz-plane is

$$M_{yz} = \iint_S x(\phi, \theta) \, dS = \int_0^{\pi/2} \int_0^{\pi/2} a^3 \sin^2 \phi \cos \theta \, d\phi \, d\theta$$

$$= \int_0^{\pi/2} \left[\frac{1}{4} a^3 (\cos \theta)(2\phi - \sin 2\phi) \right]_0^{\pi/2} d\theta = \int_0^{\pi/2} \frac{1}{4}\pi a^3 \cos \theta \, d\theta = \left[\frac{1}{4}\pi a^3 \sin \theta \right]_0^{\pi/2} = \frac{1}{4}\pi a^3.$$

Therefore (by symmetry) $(\overline{x}, \overline{y}, \overline{z}) = \left(\frac{1}{2}a, \frac{1}{2}a, \frac{1}{2}a \right)$.

C15S05.027: The surface $z = r^2$, $0 \le r \le a$ is described in Cartesian coordinates by $z = h(x, y) = x^2 + y^2$, and thus

$$dS = \sqrt{1 + (h_x)^2 + (h_y)^2} \, dA = \sqrt{1 + 4x^2 + 4y^2} \, dA.$$

Because the surface S lies over the circular disk D in the xy-plane with center $(0, 0)$ and radius a and because S has constant density δ, its mass is

900

$$m = \iint_S \delta \, dS = \iint_D \delta \sqrt{1 + 4x^2 + 4y^2} \, dA = \int_0^{2\pi} \int_0^a \delta r \sqrt{1 + 4r^2} \, dr \, d\theta$$

$$= 2\pi\delta \left[\frac{1}{12}(1 + 4r^2)^{3/2} \right]_0^a = \frac{1}{6}\pi\delta \left[(1 + 4a^2)^{3/2} - 1 \right].$$

The moment of S with respect to the xy-plane is

$$M_{xy} = \iint_S \delta z \, dS = \iint_D \delta(x^2 + y^2)\sqrt{1 + 4x^2 + 4y^2} \, dA = \int_0^{2\pi} \int_0^a \delta r^3 \sqrt{1 + 4r^2} \, dr \, d\theta$$

$$= 2\pi\delta \left[\left(\frac{1}{5}r^4 + \frac{1}{60}r^2 - \frac{1}{120} \right) \sqrt{1 + 4r^2} \right]_0^a = \frac{1}{60}\pi\delta \left[(24a^4 + 2a^2 - 1)\sqrt{1 + 4a^2} + 1 \right].$$

Hence the z-coordinate of the centroid of S is

$$\overline{z} = \frac{M_{yx}}{m} = \frac{(24a^4 + 2a^2 - 1)\sqrt{1 + 4a^2} + 1}{10\left[(1 + 4a^2)^{3/2} - 1 \right]}.$$

For example, if $a = 1$, then

$$\overline{z} = \frac{1 + 25\sqrt{5}}{10\left(-1 + 5\sqrt{5} \right)} \approx 0.5589371284878981.$$

By symmetry, $\overline{x} = \overline{y} = 0$. Finally, the moment of inertia of S with respect to the z-axis is

$$I_z = \iint_S \delta(x^2 + y^2) \, dS = \iint_D \delta(x^2 + y^2)\sqrt{1 + 4x^2 + 4y^2} \, dA = \frac{1}{60}\pi\delta \left[(24a^4 + 2a^2 - 1)\sqrt{1 + 4a^2} + 1 \right]$$

(the computations are exactly the same as those in the evaluation of M_{xy}).

C15S05.029: The surface S is described by $z = h(x, y) = \sqrt{4 - x^2 - y^2}$, and

$$1 + (h_x)^2 + (h_y)^2 = 1 + \frac{x^2}{4 - x^2 - y^2} + \frac{y^2}{4 - x^2 - y^2} = \frac{4}{4 - x^2 - y^2}.$$

Thus

$$dS = \frac{2}{\sqrt{4 - x^2 - y^2}} \, dA.$$

In cylindrical coordinates, S is described by

$$z = \sqrt{4 - r^2}, \quad 0 \leq r \leq 2\cos\theta, \quad -\frac{1}{2}\pi \leq \theta \leq \frac{1}{2}\pi.$$

We may without loss of generality assume that S has constant density $\delta = 1$. Hence its mass is

$$m = \iint_S 1 \, dS = 2 \int_0^{\pi/2} \int_0^{2\cos\theta} \frac{2r}{\sqrt{4 - r^2}} \, dr \, d\theta = 2 \int_0^{\pi/2} \left[-2(4 - r^2)^{1/2} \right]_0^{2\cos\theta} d\theta$$

$$= 4 \int_0^{\pi/2} \left(2 - \sqrt{4 - 4\cos^2\theta} \right) d\theta = 8 \int_0^{\pi/2} (1 - \sin\theta) \, d\theta = 8 \left[\theta + \cos\theta \right]_0^{\pi/2} = 4\pi - 8.$$

Clearly $\bar{y} = 0$ by symmetry. The moment of S with respect to the xy-plane is

$$M_{xy} = \iint_S z\, dS = 2\int_0^{\pi/2}\int_0^{2\cos\theta} \frac{2r\sqrt{4-r^2}}{\sqrt{4-r^2}}\, dr\, d\theta = 2\int_0^{\pi/2}\left[r^2\right]_0^{2\cos\theta} d\theta$$

$$= 2\int_0^{\pi/2} 4\cos^2\theta\, d\theta = 4\int_0^{\pi/2}(1+\cos 2\theta)\, d\theta = 4\left[\theta + \frac{1}{2}\sin 2\theta\right]_0^{\pi/2} = 2\pi.$$

The moment of S with respect to the yz-plane is

$$M_{yz} = \iint_S x\, dS = 2\int_0^{\pi/2}\int_0^{2\cos\theta} \frac{2r^2\cos\theta}{\sqrt{4-r^2}}\, dr\, d\theta.$$

Let $r = 2\sin\psi$. Then $dr = 2\cos\psi\, d\psi$ and $\sqrt{4-r^2} = 2\cos\psi$. Thus

$$\int \frac{r^2}{\sqrt{4-r^2}}\, dr = \int \frac{4\sin^2\psi}{2\cos\psi}\cdot 2\cos\psi\, d\psi = 4\int \frac{1-\cos 2\psi}{2}\, d\psi$$

$$= 2\psi - \sin 2\psi + C = 2\psi - 2\sin\psi\,\cos\psi + C = 2\arcsin\left(\frac{r}{2}\right) - \frac{1}{2}r\sqrt{4-r^2} + C.$$

Therefore

$$M_{yz} = 2\int_0^{\pi/2}\left[\left(4\arcsin\frac{r}{2} - r\sqrt{4-r^2}\right)\cos\theta\right]_0^{2\cos\theta} d\theta$$

$$= 2\int_0^{\pi/2}\left[4(\cos\theta)\arcsin(\cos\theta) - 4\cos^2\theta\,\sin\theta\right] d\theta.$$

To evaluate

$$J = \int (\cos\theta)\arcsin(\cos\theta)\, d\theta,$$

we use integration by parts. Let

$$u = \arcsin(\cos\theta), \qquad dv = \cos\theta\, d\theta; \quad \text{then}$$

$$du = -\frac{\sin\theta}{\sqrt{1-\cos^2\theta}}\, d\theta, \qquad v = \sin\theta.$$

Thus

$$J = (\sin\theta)\arcsin(\cos\theta) + \int \frac{\sin^2\theta}{\sin\theta}\, d\theta = (\sin\theta)\arcsin(\cos\theta) - \cos\theta + C.$$

Consequently,

$$M_{yz} = 2\left[4(\sin\theta)\arcsin(\cos\theta) - 4\cos\theta + \frac{4}{3}\cos^3\theta\right]_0^{\pi/2} = 2\left(4 - \frac{4}{3}\right) = \frac{16}{3}.$$

Therefore

$$\bar{x} = \frac{M_{yz}}{m} = \frac{4}{3\pi - 6} \approx 1.16795892925607244080260 \quad \text{and}$$

$$\bar{z} = \frac{M_{xy}}{m} = \frac{\pi}{2\pi - 4} \approx 1.375969196942054330601955.$$

C15S05.031: The surface S is described by $h(x, y) = 4 - y^2$, and hence $dS = \sqrt{1 + 4y^2}\ dA$. Thus the moment of inertia of S with respect to the z-axis is

$$I_z = \int_{-2}^{2} \int_{-1}^{1} (x^2 + y^2)\sqrt{1 + 4y^2}\ dx\ dy \int_{-2}^{2} \left[\left(\frac{1}{3}x^3 + xy^2\right)\sqrt{1 + 4y^2} \right]_{-1}^{1} dy$$

$$= \int_{-2}^{2} \left(\frac{2}{3} + 2y^2\right)\sqrt{1 + 4y^2}\ dy = \left[\frac{24y^3 + 19y}{48}\sqrt{1 + 4y^2} + \frac{13}{96}\operatorname{arcsinh}(2y)\right]_{-2}^{2}$$

$$= \frac{460\sqrt{17} + 13\operatorname{arcsinh}(4)}{48} \approx 40.080413560385795202979817.$$

C15S05.033: By Eq. (12) in Section 15.5,

$$\cos\gamma = \frac{1}{|\mathbf{N}|} \cdot \frac{\partial(x, y)}{\partial(x, y)} = \frac{1}{|\mathbf{N}|}.$$

Therefore $|\mathbf{N}|\ dS = \sec\gamma\ dx\ dy$.

C15S05.035: The temperature within the ball is $u(x, y, z) = 4(x^2 + y^2 + z^2)$. With position vector $\mathbf{r} = \langle x, y, z \rangle$ for points of B, we find that

$$\mathbf{q} = -K\nabla u = -2 \cdot 4\langle 2x, 2y, 2z \rangle = -16\langle x, y, z \rangle = -16\mathbf{r}.$$

A unit vector normal to the concentric spherical surface S of radius 3 is $\mathbf{n} = \frac{1}{3}\mathbf{r}$, so

$$\mathbf{q} \cdot \mathbf{n} = -\frac{16}{3}(x^2 + y^2 + z^2) = -\frac{16}{3} \cdot 9 = -48.$$

Because S is a spherical surface of radius 3, its surface area is $4\pi \cdot 9 = 36\pi$. Therefore the rate of heat flow across S is

$$\iint_S \mathbf{q} \cdot \mathbf{n}\ dS = -\iint_S 48\ dS = -48 \cdot 36\pi = -1728\pi.$$

C15S05.037: The given parametrization yields $\mathbf{N} = \langle -2bu^2\cos v, -2au^2\sin v, abu \rangle$, so the area of the paraboloid is

$$A = \int_0^{2\pi} \int_0^c u\sqrt{(2au\sin v)^2 + (2bu\cos v)^2 + (ab)^2}\ du\ dv$$

$$= \int_0^{2\pi} \frac{\left[(ab)^2 + 2(ac)^2 + 2(bc)^2 - 2(a^2 - b^2)c^2\cos 2v\right]^{3/2} - (ab)^3}{6\left[a^2 + b^2 - (a^2 - b^2)\cos 2v\right]}\ dv.$$

We believe the last integral to be nonelementary (because *Mathematica* 3.0 uses elliptic functions to compute the antiderivative). With $a = 4$, $b = 3$, and $c = 2$ it reduces to

903

$$\int_0^{2\pi} \frac{-1728 + (344 - 56\cos 2v)^{3/2}}{150 - 42\cos 2v}\, dv.$$

The *Mathematica 3.0* `NIntegrate` command yields the result $A \approx 194.702812872043$. To compute the moment of inertia of the paraboloid with respect to the z-axis, we insert the factor

$$x^2 + y^2 = (au\cos v)^2 + (bu\sin v)^2 = (4u\cos v)^2 + (3u\sin v)^2$$

into the first integral, and *Mathematica* yields the result $I_z \approx 5157.168115181396$.

C15S05.039: The given parametrization yields

$$|\mathbf{N}| = (\cosh u)\sqrt{(b\cosh u\cos v)^2 + (a\cosh u\sin v)^2 + (ab\sinh u)^2}\,,$$

and hence (using $a = 4$, $b = 3$, $c = 2$, and density $\delta = 1$) we find that the hyperboloid has surface area

$$A = \int_0^{2\pi}\int_{-c}^{c} |\mathbf{N}|\, du\, dv \approx 1057.350512779488$$

and moment of inertia with respect to the z-axis

$$I_z = \int_0^{2\pi}\int_{-c}^{c} (\cosh^2 u)\big[(a\cos v)^2 + (b\sin v)^2\big] \cdot |\mathbf{N}|\, du\, dv \approx 98546.9348740325.$$

C15S05.041: We use Fig. 14.7.15 of the text and the notation there; the only change is replacement of the variable ρ with the constant radius a of the spherical surface. The spherical shell has constant density δ and total mass $M = 4\pi a^2\delta$. The "sum" of the vertical components of the gravitational forces exerted by mass elements $\delta\, dS$ of the spherical surface S on the mass m is

$$F = \iint_S \frac{Gm\delta\cos\alpha}{w^2}\, dS.$$

We saw in the solution of Problem 25 (among others) that $dS = a^2\sin\phi\, dA$. Figure 14.7.15 also shows us that

$$w\cos\alpha = c - a\cos\phi \quad\text{and}\quad w^2 = a^2 + c^2 - 2ac\cos\phi \tag{1}$$

(by the law of cosines (Appendix L, page A-49)). Note that

$$F = 2\pi Gm\delta\int_{\phi=0}^{\pi} \frac{a^2\cos\alpha\sin\phi}{w^2}\, d\phi.$$

Substitute $\cos\alpha = \dfrac{c - a\cos\phi}{w}$ to obtain

$$F = 2\pi Gm\delta\int_0^{\pi} \frac{a^2(c - a\cos\phi)\sin\phi}{w^3}\, d\phi.$$

Next note that $\phi = 0$ corresponds to $w = c - a$ and that $\phi = \pi$ corresponds to $w = c + a$. Moreover, by the second equation in (1),

$$\cos\phi = \frac{a^2 + c^2 - w^2}{2ac} \quad \text{and thus}$$

$$-\sin\phi \, d\phi = -\frac{w}{ac} \, dw.$$

These substitutions yield

$$F = 2\pi Gm\delta \int_{w=c-a}^{c+a} \frac{a^2}{w^3} \left(c - \frac{a^2 + c^2 - w^2}{2c} \right) \cdot \frac{w}{ac} \, dw$$

$$= 2\pi Gm\delta \int_{c-a}^{c+a} \frac{a}{w^2 c} \cdot \frac{1}{2c} \cdot (c^2 + w^2 - a^2) \, dw = \frac{2\pi Gm\delta a}{2c^2} \int_{c-a}^{c+a} \left(\frac{c^2 - a^2}{w^2} + 1 \right) dw$$

$$= \frac{\pi Gm\delta a}{c^2} \left[\frac{a^2 - c^2}{w} + w \right]_{c-a}^{c+a} = \frac{\pi Gm\delta a}{c^2} (a - c + a + c + c + a - c + a) = \frac{4\pi Gm\delta a^2}{c^2} = \frac{GMm}{c^2}.$$

Section 15.6

C15S06.001: The right-hand side in the divergence theorem (Eq. (1)) is

$$\iiint_B \boldsymbol{\nabla} \cdot \mathbf{F} \, dV = \iiint_B 3 \, dV = 3 \cdot \frac{4}{3}\pi \cdot 1^3 = 4\pi$$

and the left-hand side in the divergence theorem is

$$\iint_S \mathbf{F} \cdot \mathbf{n} \, dS = \iint_S \langle x, y, z \rangle \cdot \langle x, y, z \rangle \, dS = \iint_S (x^2 + y^2 + z^2) \, dS = \iint_S 1 \, dS = 1 \cdot 4\pi \cdot 1^2 = 4\pi.$$

Note that we integrate a constant function by multiplying its value by the size (length, area, or volume) of the domain of the integral. We will continue to do so without further comment.

C15S06.003: On the face F of the cube in the plane $x = 2$, a unit vector normal to F is $\mathbf{n} = \mathbf{i}$, and $\mathbf{F} \cdot \mathbf{i} = x = 2$. Hence

$$\iint_F \mathbf{F} \cdot \mathbf{n} \, dS = 2 \cdot \text{area}(F) = 8.$$

By symmetry, the same result obtains on the faces in the planes $y = 2$ and $z = 2$. On the face G of the cube in the plane $x = 0$, a unit vector normal to G is $\mathbf{n} = -\mathbf{i}$, and $\mathbf{F} \cdot (-\mathbf{i}) = -x = 0$. Hence

$$\iint_G \mathbf{F} \cdot \mathbf{n} \, dS = 0.$$

By symmetry, the same result holds on the faces in the other two coordinate planes. Hence

$$\iint_S \mathbf{F} \cdot \mathbf{n} \, dS = 3 \cdot 8 + 3 \cdot 0 = 24.$$

Let B denote the solid cube bounded by S and let V denote the volume of B. Because $\boldsymbol{\nabla} \cdot \mathbf{F} = 3$, we also have

$$\iiint_B \boldsymbol{\nabla} \cdot \mathbf{F} \, dV = 3 \cdot V = 3 \cdot 8 = 24.$$

C15S06.005: On the face F of the tetrahedron that lies in the plane $x = 0$, a unit vector normal to F is $\mathbf{n} = -\mathbf{i}$, and $\mathbf{F} \cdot \mathbf{n} = -x - y = -y$. Hence

$$\iint_F \mathbf{F} \cdot \mathbf{n} \, dS = \int_0^1 \int_0^{1-z} (-y) \, dy \, dz = \int_0^1 -\frac{1}{2}(1-z)^2 \, dz = \left[\frac{1}{6}(1-z)^3 \right]_0^1 = -\frac{1}{6}.$$

By symmetry the same result holds on the faces in the other two coordinate planes. On the fourth face G of the tetrahedron, a unit vector normal to G is

$$\mathbf{n} = \frac{\sqrt{3}}{3} \langle 1, 1, 1 \rangle,$$

and G is part of the graph of $z = h(x, y) = 1 - x - y$, so that

$$dS = \sqrt{1 + (h_x)^2 + (h_y)^2} \; dA = \sqrt{3} \; dA.$$

Therefore

$$\iint_G \mathbf{F} \cdot \mathbf{n} \, dS = \iint_G (2x + 2y + 2z) \, dA = \iint_G 2 \, dA = \int_0^1 \int_0^{1-x} 2 \, dy \, dx = \int_0^1 \left[2y \right]_0^{1-x} dx = \left[2x - x^2 \right]_0^1 = 1,$$

and therefore

$$\iint_S \mathbf{F} \cdot \mathbf{n} \, dS = 1 - 3 \cdot \frac{1}{6} = \frac{1}{2}.$$

Let B denote the solid tetrahedron itself, with volume V. Then

$$\iiint_B \nabla \cdot \mathbf{F} \, dV = \iiint_B 3 \, dV = 3 \cdot V = 3 \cdot \frac{1}{6} \cdot 1 \cdot 1 = \frac{1}{2}.$$

C15S06.007: Let B denote the solid cylinder bounded by the surface S. Then

$$\iint_S \mathbf{F} \cdot \mathbf{n} \, dS = \iiint_B \nabla \cdot \mathbf{F} \, dV = \iiint_B 3(x^2 + y^2 + z^2) \, dV = \int_0^{2\pi} \int_0^3 \int_{-1}^4 3(r^2 + z^2) \cdot r \, dz \, dr \, d\theta$$

$$= 2\pi \int_0^3 \left[3r^3 z + rz^3 \right]_{-1}^4 dr = 2\pi \int_0^3 (65r + 15r^3) \, dr = 2\pi \left[\frac{65}{2}r^2 + \frac{15}{4}r^4 \right]_0^3$$

$$= 2\pi \cdot \frac{2385}{4} = \frac{2385}{2}\pi \approx 3746.3492394058284367.$$

C15S06.009: Let B denote the tetrahedron bounded by the given surface S. Then

$$\iint_S \mathbf{F} \cdot \mathbf{n} \, dS = \iiint_B \nabla \cdot \mathbf{F} \, dV = \int_0^1 \int_0^{1-x} \int_0^{1-x-y} (2x + 1) \, dz \, dy \, dx$$

$$= \int_0^1 \int_0^{1-x} (1 + x - 2x^2 - y - 2xy) \, dy \, dx = \int_0^1 \left[(1 + x - 2x^2)y - \frac{1}{2}(2x + 1)y^2 \right]_0^{1-x} dx$$

$$= \int_0^1 \left(\frac{1}{2} - \frac{3}{2}x^2 + x^3 \right) dx = \left[\frac{1}{2}x - \frac{1}{2}x^3 + \frac{1}{4}x^4 \right]_0^1 = \frac{1}{4}.$$

C15S06.011: Let B denote the solid paraboloid bounded by the surface S. Then

$$\iint_S \mathbf{F} \cdot \mathbf{n} \, dS = \iiint_B \nabla \cdot \mathbf{F} \, dV = \iiint_B 5(x^2 + y^2 + z^2) \, dV = \int_0^{2\pi} \int_0^5 \int_0^{25-r^2} 5(r^2 + z^2) \cdot r \, dz \, dr \, d\theta$$

$$= 2\pi \int_0^5 \left[5r^3 z + \frac{5}{3} r z^3 \right]_0^{25-r^2} dr = 2\pi \int_0^5 \left(\frac{78125}{3} r - 3000 r^3 + 120 r^5 - \frac{5}{3} r^7 \right) dr$$

$$= 2\pi \left[\frac{78125}{6} r^2 - 750 r^4 + 20 r^6 - \frac{5}{24} r^8 \right]_0^5 = 2\pi \cdot \frac{703125}{8} = \frac{703125}{4} \pi \approx 552233.08363883.$$

C15S06.013: Here is a step-by-step illustration of the solution using *Mathematica* 3.0.

```
f = {x, y, 3}      (* First we define the vector function F. *)

   {x, y, 3}

D[%[[1]],x] + D[%[[2]],y] + D[%[[3]],z]       (* Then we compute div F. *)

   2

Integrate[ 2*r, z ]      (* Begin the triple integral in cylindrical coordinates.  *)

   2rz

(% /.  z → 4) - (% /.  z → r^2)      (* Substitute the limits on z.  *)

   8r - 2r³

Integrate[ %, r]
```

$$4r^2 - \frac{1}{2} r^4$$

```
(% /.  r → 2) - (% /.  r → 0)      (* Substitute the limits on r.  *)

   8

2*Pi*%      (* Integrate the constant by multiplying by 2π.  *)

   16π

N[%, 18]      (* Approximate the answer.  *)

   50.2654824574366918
```

C15S06.015: Compare this problem and its solution with Problem 25 of Section 15.4. If f is a twice-differentiable scalar function, then

$$\nabla^2 f = \nabla \cdot (\nabla f) = \nabla \cdot \langle f_x, f_y, f_z \rangle = f_{xx} + f_{yy} + f_{zz}.$$

C15S06.017: If $\nabla^2 f \equiv 0$ in the region T with boundary surface S, then by the divergence theorem and Problem 28 in Section 15.1,

$$\iint_S f \frac{\partial f}{\partial n}\, dS = \iint_S (f)(\nabla f) \cdot \mathbf{n}\, dS = \iiint_T \nabla \cdot [(f)(\nabla f)]\, dV$$

$$= \iiint_T [(f)\nabla \cdot (\nabla f) + (\nabla f) \cdot (\nabla f)]\, dV = \iiint_T [(f)\nabla^2 f + |\nabla f|^2]\, dV = \iiint_T |\nabla f|^2\, dV.$$

C15S06.019: Green's first identity states that if the space region T has surface S with a piecewise smooth parametrization, if f and g are twice-differentiable scalar functions, and if $\partial f / \partial n = (\nabla f) \cdot \mathbf{n}$ where $\mathbf{n}$ is the unit vector normal to S with outer direction, then

$$\iint_S f \frac{\partial g}{\partial n}\, dS = \iiint_T (f\nabla^2 g + \nabla f \cdot \nabla g)\, dV.$$

Interchanging the roles of f and g yields the immediate consequence

$$\iint_S g \frac{\partial f}{\partial n}\, dS = \iiint_T (g\nabla^2 f + \nabla g \cdot \nabla f)\, dV.$$

Then substraction of the second of these equations from the first yields

$$\iint_S \left(f \frac{\partial g}{\partial n} - g \frac{\partial f}{\partial n} \right) dS = \iiint_B (f\nabla^2 g + \nabla f \cdot \nabla g - g\nabla^2 f - \nabla g \cdot \nabla f)\, dV = \iiint_B (f\nabla^2 g - g\nabla^2 f)\, dV,$$

and this is Green's second identity.

C15S06.021: By the result in Problem 20, we have

$$\mathbf{B} = -\iint_S p\mathbf{n}\, dS = -\iiint_T \nabla(\delta g z)\, dV = -\iiint_T \langle 0, 0, \delta g \rangle\, dV = -\mathbf{k} \iiint_T \delta g\, dV = -W\mathbf{k}$$

because

$$\iiint_T \delta g\, dV = mg = W$$

is the weight of the fluid displaced by the body.

C15S06.023: Let B denote the region bounded by the paraboloid and the plane. Because

$$\mathbf{F}(x, y, z) = \left\langle x\sqrt{x^2 + y^2 + z^2}\, ,\ y\sqrt{x^2 + y^2 + z^2}\, ,\ z\sqrt{x^2 + y^2 + z^2}\, \right\rangle,$$

we have

$$\nabla \cdot \mathbf{F} = \frac{x^2}{\sqrt{x^2 + y^2 + z^2}} + \frac{y^2}{\sqrt{x^2 + y^2 + z^2}} + \frac{z^2}{\sqrt{x^2 + y^2 + z^2}} + 3\sqrt{x^2 + y^2 + z^2} = 4\sqrt{x^2 + y^2 + z^2}\, .$$

Thus in cylindrical coordinates, $\nabla \cdot \mathbf{F} = 4\sqrt{r^2 + z^2}$. Then, by the divergence theorem,

$$I = \iint_S \mathbf{F} \cdot \mathbf{n} \, dS = \iiint_B \nabla \cdot \mathbf{F} \, dV$$

$$= \int_0^{2\pi} \int_0^{25} \int_0^{\sqrt{25-z}} 4r\sqrt{r^2+z^2} \; dr \, dz \, d\theta = \frac{8}{3}\pi \int_0^{25} \left[(25 - z + z^2)^{3/2} - z^3 \right] dz.$$

Let

$$J = \int_0^c \left((z-a)^2 + b^2 \right)^{3/2} dz.$$

Later we will use the following values: $b = \frac{3}{2}\sqrt{11}$, $a = \frac{1}{2}$, and $c = 25 = a^2 + b^2$. The substitution $z = a + b\tan u$ yields

$$J = b^4 \int_{z=0}^c \sec^5 u \; du,$$

and the integral formulas in 37 and 28 of the endpapers of the text then yield

$$J = b^4 \int_{z=0}^c \sec^5 u \; du = b^4 \left(\left[\frac{1}{4} \sec^3 u \tan u \right]_{z=0}^c + \frac{3}{4} \int_{z=0}^c \sec^3 u \; du \right)$$

$$= b^4 \left(\left[\frac{1}{4} \sec^3 u \tan u \right]_{z=0}^c + \frac{3}{4} \left[\frac{1}{2} \sec u \tan u + \frac{1}{2} \ln|\sec u + \tan u| \right]_{z=0}^c \right).$$

The substitution $z = a + b\tan u$ also yields

$$\sec u = \frac{1}{b} \left[(z-a)^2 + b^2 \right]^{1/2} \quad \text{and} \quad \tan u = \frac{1}{b}(z-a).$$

It now follows that

$$J = b^4 \left[\frac{3(z-a)\left[(z-a)^2 + b^2 \right]^{3/2}}{4b^4} + \frac{3(z-a)\left[(z-a)^2 + b^2 \right]^{1/2}}{8b^2} + \frac{3}{8} \ln \left| \frac{z - a + \left[(z-a)^2 + b^2 \right]^{1/2}}{b} \right| \right]_{z=0}^c$$

$$= b^4 \left[\frac{3a(a^2+b^2)^{1/2}}{8b^2} + \frac{a(a^2+b^2)^{3/2}}{4b^4} + \frac{3(c-a)\left[(c-a)^2 + b^2 \right]^{1/2}}{8b^2} \right.$$

$$\left. + \frac{(c-a)\left[(c-a)^2 + b^2 \right]^{3/2}}{4b^4} - \frac{3}{8} \ln \left(\frac{-a + (a^2+b^2)^{1/2}}{b} \right) + \frac{3}{8} \ln \left(\frac{c - a + \left[(c-a)^2 + b^2 \right]^{1/2}}{b} \right) \right].$$

Then substitution of the numerical values of a, b, and c yields

$$J = \frac{9801}{16} \left[\frac{1081885}{6534} - \frac{3}{8} \ln\left(\frac{3}{\sqrt{11}} \right) + \frac{3}{8} \ln\left(3\sqrt{11} \right) \right] = \frac{3}{128} \left(4327540 + 9801 \ln 11 \right).$$

And, finally,

$$I = \frac{8}{3}\pi \left(J - \int_0^{25} z^3 \, dz \right) = \frac{8}{3}\pi \left[\frac{1}{128} \left(482620 + 29403 \ln 11 \right) \right]$$

$$= \frac{482620 + 29403 \ln 11}{48} \pi \approx 36201.96719156658969 9334774115.$$

We integrated $\sec^5 u$ using the integral formulas mentioned earlier; all the subsequent work was done by Mathematica 3.0.

C15S06.025: We begin with Gauss's law in the form

$$\iint_S \mathbf{F} \cdot \mathbf{n} \, dS = -4\pi GM.$$

Because $\mathbf{F}$ and $\mathbf{n}$ have opposite directions and $\mathbf{n}$ is a unit vector, we may in this case deduce that

$$\iint_S |\mathbf{F}| \, dS = 4\pi GM = 4\pi G \cdot 0 = 0.$$

Therefore $4\pi r^2 |\mathbf{F}| = 0$, so that $|\mathbf{F}| = 0$. Therefore $\mathbf{F} = \mathbf{0}$.

C15S06.027: Imagine a cylindrical surface of radius r and length L concentric around the wire. Because the electric field $\mathbf{E}$ is normal to the wire, there is no flux of $\mathbf{E}$ across the top and bottom of the cylinder, so the surface S of Gauss's law may be regarded as the curved side of the cylinder. It follows that

$$\iint_S \mathbf{E} \cdot \mathbf{n} \, dS = \frac{Q}{\epsilon_0} = \frac{Lq}{\epsilon_0}.$$

Because $\mathbf{E}$ and $\mathbf{n}$ are parallel and $\mathbf{n}$ is a unit vector, it now follows that

$$\iint_S |\mathbf{E}| \, dS = \frac{Lq}{\epsilon_0};$$

$$2\pi r L |\mathbf{E}| = \frac{Lq}{\epsilon_0};$$

$$|\mathbf{E}| = \frac{q}{2\pi \epsilon_0 r}.$$

Section 15.7

C15S07.001: Because $\mathbf{n}$ is to be the upper unit normal vector, we have

$$\mathbf{n} = \mathbf{n}(x, y, z) = \frac{1}{2} \langle x, y, z \rangle.$$

The boundary curve C of the hemispherical surface S has the parametrization

$$x = 2\cos\theta, \quad y = 2\sin\theta, \quad z = 0, \quad 0 \le \theta \le 2\pi.$$

Therefore

$$\iint_S (\boldsymbol{\nabla} \times \mathbf{F}) \cdot \mathbf{n} \, dS = \oint_C 3y \, dx - 2x \, dy + xyz \, dz$$

$$= \int_0^{2\pi} (-12\sin^2\theta - 8\cos^2\theta) \, d\theta = \left[-10\theta + \sin 2\theta \right]_0^{2\pi} = -20\pi.$$

C15S07.003: Parametrize the boundary curve C of the surface S as follows:

$$x = 3\cos t, \quad y = 3\sin t, \quad z = 0, \quad 0 \leq t \leq 2\pi.$$

Then

$$\iint_S (\nabla \times \mathbf{F}) \cdot \mathbf{n} \, dS = \oint_C \mathbf{F} \cdot \mathbf{T} \, ds = \int_0^{2\pi} (-6\cos t - 27 \sin^2 t \, \cos t) \, dt = \left[\frac{3}{4}(3 \sin 3t - 17 \sin t) \right]_0^{2\pi} = 0.$$

C15S07.005: Parametrize the boundary curves of the surface S as follows:

$$C_1 : \quad x = \cos t, \quad y = -\sin t, \quad z = 1, \quad 0 \leq t \leq 2\pi;$$

$$C_2 : \quad x = 3\cos t, \quad y = 3\sin t, \quad z = 3, \quad 0 \leq t \leq 2\pi.$$

Then

$$\iint_S (\nabla \times \mathbf{F}) \cdot \mathbf{n} \, dS = \oint_{C_1} \mathbf{F} \cdot \mathbf{T} \, ds + \oint_{C_2} \mathbf{F} \cdot \mathbf{T} \, ds$$

$$= \int_0^{2\pi} (\cos^2 t + \sin^2 t) \, dt + \int_0^{2\pi} -27(\cos^2 t + \sin^2 t) \, dt = \left[-26t \right]_0^{2\pi} = -52\pi.$$

C15S07.007: Parametrize S (the elliptical region bounded by C) as follows:

$$x = z = r\cos t, \quad y = r\sin t, \quad 0 \leq t \leq 2\pi.$$

Then $\mathbf{r}_r \times \mathbf{r}_t = \langle -r, 0, r \rangle$, $dS = r\sqrt{2} \, dr \, dt$, the upper unit normal for S is

$$\mathbf{n} = \frac{1}{2}\sqrt{2} \, \langle -1, 0, 1 \rangle,$$

and $\nabla \times \mathbf{F} = \langle 3, 2, 1 \rangle$. Therefore $(\nabla \times \mathbf{F}) \cdot \mathbf{n} = -\sqrt{2}$. Consequently,

$$\oint_C \mathbf{F} \cdot \mathbf{T} \, ds = \iint_S (\nabla \times \mathbf{F}) \cdot \mathbf{n} \, dS = \int_0^{2\pi} \int_0^2 (-2r) \, dr \, dt$$

$$= 2\pi \left[-r^2 \right]_0^2 = 2\pi \cdot (-4) = -8\pi \approx -25.1327412287183459.$$

C15S07.009: If $\mathbf{F}(x, y, z) = \langle y - x, \ x - z, \ x - y \rangle$, then

$$\nabla \times \mathbf{F} = \begin{vmatrix} \mathbf{i} & \mathbf{j} & \mathbf{k} \\ \dfrac{\partial}{\partial x} & \dfrac{\partial}{\partial y} & \dfrac{\partial}{\partial z} \\ y - x & x - z & x - y \end{vmatrix} = \langle 0, -1, 0 \rangle.$$

The surface S bounded by C is part of the plane with equation $x + 2y + z = 2$, so an upward normal to S is $\langle 1, 2, 1 \rangle$. Hence the unit normal vector we need is

$$\mathbf{n} = \frac{1}{6}\sqrt{6} \, \langle 1, 2, 1 \rangle.$$

Then $(\nabla \times \mathbf{F}) \cdot \mathbf{n} = -\frac{1}{3}\sqrt{6}$, a constant, so it remains only to find the area of S. Its vertices are at $A(2, 0, 0)$, $B(0, 1, 0)$, and $C(0, 0, 2)$, so we compute the "edge vectors"

$$\mathbf{u} = \overrightarrow{AB} = \langle -2, 1, 0 \rangle \quad \text{and} \quad \mathbf{v} = \overrightarrow{AC} = \langle -2, 0, 2 \rangle$$

and their cross product

$$\mathbf{u} \times \mathbf{v} = \begin{vmatrix} \mathbf{i} & \mathbf{j} & \mathbf{k} \\ -2 & 1 & 0 \\ -2 & 0 & 2 \end{vmatrix} = \langle 2, 4, 2 \rangle;$$

the area of S is then $\frac{1}{2}|\mathbf{u} \times \mathbf{v}| = \sqrt{6}$. Therefore

$$\oint_C \mathbf{F} \cdot \mathbf{T} \, ds = \iint_S (\nabla \times \mathbf{F}) \cdot \mathbf{n} \, dS = \left(-\frac{1}{3}\sqrt{6} \right) \cdot \sqrt{6} = -2.$$

C15S07.011: If $\mathbf{F}(x, y, z) = \langle 3y - 2z, 3x + z, y - 2x \rangle$, then

$$\nabla \times \mathbf{F} = \begin{vmatrix} \mathbf{i} & \mathbf{j} & \mathbf{k} \\ \dfrac{\partial}{\partial x} & \dfrac{\partial}{\partial y} & \dfrac{\partial}{\partial z} \\ 3y - 2z & 3x + z & y - 2x \end{vmatrix} = \langle 1 - 1, \, -2 + 2, \, 3 - 3 \rangle = \mathbf{0}.$$

Therefore $\mathbf{F}$ is irrotational. Next, let C be the straight line segment from $(0, 0, 0)$ to the point (u, v, w) of space. Parametrize C as follows: $x = tu$, $y = tv$, $z = tw$, $0 \leqq t \leqq 1$. Then

$$\int_C \mathbf{F} \cdot \mathbf{T} \, ds = \int_{t=0}^1 (6tuv - 4tuw + 2tvw) \, dt = \left[(3uv - 2uw + vw)t^2 \right]_0^1 = 3uv - 2uw + vw.$$

Now replace u with x, v with y, and w with z (because (u, v, w) represents an *arbitrary* point of space). This yields the potential function

$$\phi(x, y, z) = 3xy - 2xz + yz.$$

To be absolutely certain of this, verify for yourself that $\nabla \phi = \mathbf{F}$.

C15S07.013: If $\mathbf{F}(x, y, z) = \langle 3e^z - 5y \sin x, \, 5 \cos x, \, 17 + 3xe^z \rangle$, then

$$\nabla \times \mathbf{F} = \begin{vmatrix} \mathbf{i} & \mathbf{j} & \mathbf{k} \\ \dfrac{\partial}{\partial x} & \dfrac{\partial}{\partial y} & \dfrac{\partial}{\partial z} \\ 3e^z z - 5y \sin x & 5 \cos x & 17 + 3xe^z \end{vmatrix} = \langle 0, \, 3e^z - 3e^z, \, -5 \sin x + 5 \sin x \rangle = \mathbf{0}.$$

Therefore $\mathbf{F}$ is irrotational. Then the method of Example 3 yields

$$x(t) = tu, \quad y(t) = tv, \quad z(t) = tw, \quad 0 \leqq t \leqq 1,$$

$$\mathbf{r}(t) = \langle x(t), y(t), z(t) \rangle = \langle tu, tv, tw \rangle, \quad \text{and}$$

$$\mathbf{F}(t) = \langle 3 \exp(tw) - 5tv \sin tu, \, 5 \cos tu, \, 17 + 3tu \exp(tw) \rangle,$$

so that

$$\int_C \mathbf{F} \cdot d\mathbf{r} = \int_{t=0}^{1} (17w + 3tuwe^{tw} + 5v\cos tu + 3ue^{tw} - 5tuv\sin tu)\,dt$$

$$= \left[3tue^{tw} + 17tw + 5tv\cos tu\right]_0^1 = 3ue^{w} + 17w + 5v\cos u.$$

Therefore $\phi(x, y, z) = 3xe^z + 17z + 5y\cos x$.

C15S07.015: We are given $\mathbf{r} = \langle x, y, z\rangle$; suppose that $\mathbf{a} = \langle b, c, d\rangle$ is a constant vector. Part (a):

$$\mathbf{a} \times \mathbf{r} = \begin{vmatrix} \mathbf{i} & \mathbf{j} & \mathbf{k} \\ b & c & d \\ x & y & z \end{vmatrix} = \langle cz - dy,\ dx - bz,\ by - cx\rangle,$$

and hence $\nabla \cdot (\mathbf{a} \times \mathbf{r}) = 0 + 0 + 0 = 0$. Part (b): Using some of the results in part (a), we have

$$\nabla \times (\mathbf{a} \times \mathbf{r}) = \begin{vmatrix} \mathbf{i} & \mathbf{j} & \mathbf{k} \\ \dfrac{\partial}{\partial x} & \dfrac{\partial}{\partial y} & \dfrac{\partial}{\partial z} \\ cz - dy & dx - bz & by - cx \end{vmatrix} = \langle 2b,\ 2c,\ 2d\rangle = 2\mathbf{a}.$$

Part (c): First,

$$(\mathbf{r} \cdot \mathbf{r})\mathbf{a} = \langle b(x^2 + y^2 + z^2),\ c(x^2 + y^2 + z^2),\ d(x^2 + y^2 + z^2)\rangle.$$

Thus $\nabla \cdot \left[(\mathbf{r} \cdot \mathbf{r})\mathbf{a}\right] = 2bx + 2cy + 2dz = 2\mathbf{r} \cdot \mathbf{a}$. Part (d): Using some of the results in parts (a) and (c), we have

$$\nabla \times \left[(\mathbf{r} \cdot \mathbf{r})\mathbf{a}\right] = \begin{vmatrix} \mathbf{i} & \mathbf{j} & \mathbf{k} \\ \dfrac{\partial}{\partial x} & \dfrac{\partial}{\partial y} & \dfrac{\partial}{\partial z} \\ b(x^2 + y^2 + z^2) & c(x^2 + y^2 + z^2) & d(x^2 + y^2 + z^2) \end{vmatrix}$$

$$= \langle 2dy - 2cz,\ 2bz - 2dx,\ 2cx - 2by\rangle$$

$$= -2\langle cz - dy,\ dx - bz,\ by - cx\rangle = -2(\mathbf{a} \times \mathbf{r}) = 2(\mathbf{r} \times \mathbf{a}).$$

C15S07.017: Assume that S is a closed surface having a piecewise smooth parametrization, that $\mathbf{n}$ is the outer unit normal vector for S, and that $\mathbf{F}$ is continuous and has continuous first-order partial derivatives on an open region containing S and the solid T that it bounds. Part (a): By the divergence theorem,

$$\iint_S (\nabla \times \mathbf{F}) \cdot \mathbf{n}\,dS = \iiint_T \nabla \cdot (\nabla \times \mathbf{F})\,dV = \iiint_T 0\,dV = 0$$

by Problem 32 of Section 15.1. Part (b): Let C be a simple closed curve on S having a suitably differentiable parametrization and a given orientation. Then C is the common boundary of the two surfaces S_1 and S_2

into which it divides S; that is, S is the union of S_1 and S_2, S_1 and S_2 meet in the curve C, and C has positive orientation on (say) S_1 and the opposite orientation on S_2. Then

$$\iint_{S_1} (\nabla \times \mathbf{F}) \cdot \mathbf{n}\, dS = \int_C \mathbf{F} \cdot \mathbf{T}\, ds = - \iint_{S_2} (\nabla \times \mathbf{F}) \cdot \mathbf{n}\, dS.$$

Therefore

$$\iint_S (\nabla \times \mathbf{F}) \cdot \mathbf{n}\, dS = \iint_{S_1} (\nabla \times \mathbf{F}) \cdot \mathbf{n}\, dS + \iint_{S_2} (\nabla \times \mathbf{F}) \cdot \mathbf{n}\, dS = 0.$$

C15S07.019: We are given the constant vector $\mathbf{a}$ and the vector $\mathbf{r} = \langle x,\, y,\, z \rangle$. Let $\mathbf{F} = \mathbf{a} \times \mathbf{r}$. Then by Stokes' theorem,

$$\int_C (\mathbf{a} \times \mathbf{r}) \cdot \mathbf{T}\, ds = \iint_S \left[\nabla \times (\mathbf{a} \times \mathbf{r}) \right] \cdot \mathbf{n}\, dS = \iint_S 2\mathbf{a} \cdot \mathbf{n}\, dS$$

by part (b) of Problem 15. But because integration is carried out componentwise, it now follows that

$$\int_C (\mathbf{a} \times \mathbf{r}) \cdot \mathbf{T}\, ds = 2\mathbf{a} \iint_S \mathbf{n}\, dS.$$

C15S07.021: Beginning with the *Suggestion* given in the statement of Problem 21, we find that

$$\phi_x = \lim_{h \to 0} \frac{\phi(x+h,\, y,\, z) - \phi(x,\, y,\, z)}{h} = \lim_{h \to 0} \frac{1}{h} \int_{(x,y,z)}^{(x+h,y,z)} P(t,\, y,\, z)\, dt = \lim_{h \to 0} \frac{h \cdot P(x^\star,\, y,\, z)}{h} = P(x,\, y,\, z)$$

where $x^\star$ is between x and $x+h$. A similar argument shows that $\phi_y = Q$ and $\phi_z = R$. Adding these results establishes that $\nabla \phi = \mathbf{F} = \langle P,\, Q,\, R \rangle$.

Chapter 15 Miscellaneous Problems

C15S0M.001: Parametrize C: $x = t$, $y = \frac{4}{3}t$, $0 \leq t \leq 3$. Then

$$ds = \sqrt{[x'(t)]^2 + [y'(t)]^2}\, dt = \frac{5}{3}\, dt \quad \text{and} \quad [x(t)]^2 + [y(t)]^2 = \frac{25}{9}t^2.$$

Therefore

$$\int_C (x^2 + y^2)\, ds = \int_0^3 \frac{125}{27} t^2\, dt = \left[\frac{125}{81} t^3 \right]_0^3 = \frac{125}{3} \approx 41.666666666667.$$

C15S0M.003: We are given the curve C parametrized by $\mathbf{r}(t) = \langle e^{2t},\, e^t,\, e^{-t} \rangle$, $0 \leq t \leq \ln 2$. Because $\mathbf{F}(t) = \langle x,\, y,\, z \rangle$, we can also write

$$\mathbf{F}(t) = \langle e^{2t},\, e^t,\, e^{-t} \rangle,$$

and therefore

$$\int_C \mathbf{F} \cdot \mathbf{T}\, ds = \int_C \mathbf{F} \cdot d\mathbf{r} = \int_0^{\ln 2} (2e^{4t} + e^{2t} - e^{-2t})\, dt = \frac{1}{2} \left[e^{4t} + e^{2t} + e^{-2t} \right]_0^{\ln 2} = \frac{69}{8} = 8.625.$$

C15SOM.005: Given the curve C with parametrization

$$x(t) = t, \quad y(t) = t^{3/2}, \quad z(t) = t^2, \quad 0 \le t \le 4,$$

we substitute and find that

$$\int_C z^{1/2} \, dx + x^{1/2} \, dy + y^2 \, dz = \int_0^4 \left(t + \frac{3}{2}t + 2t^4 \right) dt = \left[\frac{5}{4}t^2 + \frac{2}{5}t^5 \right]_0^4 = \frac{2148}{5} = 429.6.$$

C15SOM.007: Suppose that there exists a function $\phi(x, y)$ such that $\nabla \phi = \langle x^2 y, \, xy^2 \rangle$. Then

$$\phi(x, y) = \int x^2 y \, dx = \frac{1}{3}x^3 y + g(y),$$

and hence

$$\frac{\partial \phi}{\partial y} = \frac{1}{3}x^3 + g'(y). \tag{1}$$

But there is no choice of $g(y)$ such that the last expression in Eq. (1) can equal xy^2 unless x is constant. This is not possible on any path C from $(0, 0)$ to $(1, 1)$. Thus there is no such function ϕ, and thus by Theorem 2 the given integral is not independent of the path from $(0, 0)$ to $(1, 1)$.

C15SOM.009: Parametrize the wire W using $x(t) = t$, $y(t) = \frac{1}{2}t^2$, $0 \le t \le 2$. Then

$$ds = \sqrt{[x'(t)]^2 + [y'(t)]^2} \, dt = \sqrt{1 + t^2} \, dt.$$

The density of the wire is $\delta(t) = x(t) = t$. Therefore the mass of the wire is

$$m = \int_W \delta \, ds = \int_0^2 t\sqrt{1 + t^2} \, dt = \left[\frac{1}{3}(1 + t^2)^{3/2} \right]_0^2 = \frac{5\sqrt{5} - 1}{3} \approx 3.3934466291663162.$$

Its moment of inertia with respect to the y-axis is

$$I_y = \int_W \delta x^2 \, ds = \int_0^2 t^3 (1 + t^2)^{1/2} \, dt$$

$$= \left[\frac{1}{15}(1 + t^2)^{1/2}(3t^4 + t^2 - 2) \right]_0^2 = \frac{50\sqrt{5} + 2}{15} \approx 7.5868932583326323.$$

If you prefer, $I_y \approx m \cdot (1.4952419583303542)^2$.

C15SOM.011: Let R denote the region bounded by C. Then Green's theorem yields

$$\oint_C x^2 y \, dx + xy^2 \, dy = \iint_R (y^2 - x^2) \, dA = \int_{x=-2}^2 \int_{y=x^2}^{8-x^2} (y^2 - x^2) \, dy \, dx$$

$$= \int_{-2}^2 \left[\frac{1}{3}y^3 - x^2 y \right]_{x^2}^{8-x^2} dx = \int_{-2}^2 \left(\frac{512}{3} - 72x^2 + 10x^4 - \frac{2}{3}x^6 \right) dx$$

$$= \left[\frac{512}{3}x - 24x^3 + 2x^5 - \frac{2}{21}x^7 \right]_{-2}^2 = \frac{2816}{7} \approx 402.2857142857142857.$$

C15S0M.013: Suppose that C is any positively oriented piecewise smooth simple closed curve in the xy-plane. Let $\mathbf{n}$ be the outwardly directed unit vector normal to C. We will apply the vector form of Green's theorem in Eq. (9) of Section 15.4,

$$\oint_C \mathbf{F} \cdot \mathbf{n} \, ds = \iint_R \nabla \cdot \mathbf{F} \, dA$$

where R denotes the bounded plane region with boundary C and $\mathbf{F}$ is a two-dimensional vector function of x and y with continuously differentiable component functions. If $\mathbf{F}(x, y) = \langle x^2 y, -xy^2 \rangle$, then

$$\oint_C \mathbf{F} \cdot \mathbf{n} \, ds = \iint_R (2xy - 2xy) \, dA = 0.$$

Therefore the integrals given in the statement of Problem 13 are equal because each is equal to zero. It is also possible to verify this by a direct computation. For example, using the parametrization $x = \cos t$, $y = \sin t$, $0 \leq t \leq 2\pi$ and the outer unit normal vector $\mathbf{n} = \langle \cos t, \sin t \rangle$, the first integral in Problem 13 becomes

$$\oint_C \mathbf{F} \cdot \mathbf{n} \, ds = \int_0^{2\pi} (\cos^3 t \, \sin t - \cos t \, \sin^3 t) dt = -\frac{1}{4} \left[\cos^4 t + \sin^4 t \right]_0^{2\pi} = 0 - 0 = 0.$$

C15S0M.015: Suppose that

$$\int_C P \, dx + Q \, dy$$

is independent of the path in the plane region D. By Theorem 2 of Section 15.3, $\mathbf{F} = \langle P, Q \rangle = \nabla \phi$ in D where ϕ is some differentiable scalar potential function. Suppose that C is a simple closed curve in D. Choose a point (a, b) on C. Then by Theorem 1 of Section 15.3,

$$\oint_C P \, dx + Q \, dy = \left[\phi(x, y) \right]_{(a,b)}^{(a,b)} = \phi(a, b) - \phi(a, b) = 0.$$

C15S0M.017: The surface S is described by $h(x, y) = 2 - x^2 - y^2$, and thus

$$dS = \sqrt{1 + (h_x)^2 + (h_y)^2} \, dA = \sqrt{1 + 4x^2 + 4y^2} \, dA.$$

Therefore

$$\iint_S (x^2 + y^2 + 2z) \, dS = \iint_S (4 - x^2 - y^2) \, dS = \int_{\theta=0}^{2\pi} \int_{r=0}^{\sqrt{2}} r(4 - r^2)\sqrt{1 + 4r^2} \, dr \, d\theta$$

$$= 2\pi \left[\frac{1}{120}(41 + 158r^2 - 24r^4)\sqrt{1 + 4r^2} \right]_0^{\sqrt{2}} = 2\pi \cdot \frac{371}{60} = \frac{371}{30}\pi \approx 38.8510291493937764.$$

C15S0M.019: The upper surface S_1 is described by $h(x, y) = 12 - 2x^2 - y^2$, and thus

$$dS = \sqrt{1 + (h_x)^2 + (h_y)^2} \, dA = \sqrt{1 + 16x^2 + 4y^2} \, dA.$$

An outwardly pointing vector normal to S_1 is

$$\left\langle -\frac{\partial h}{\partial x},\ -\frac{\partial h}{\partial y},\ 1 \right\rangle = \langle 4x,\ 2y,\ 1 \rangle,$$

and therefore a outwardly pointing unit vector normal to S_1 is

$$\mathbf{n} = \frac{1}{\sqrt{1 + 16x^2 + 4y^2}}\, \langle 4x,\ 2y,\ 1 \rangle.$$

Hence the outward flux of $\mathbf{F} = \langle x,\ y,\ z \rangle$ across S_1 is

$$\iint_{S_1} \mathbf{F} \cdot \mathbf{n}\ dS = \iint_{S_1} \frac{4x^2 + 2y^2 + z}{\sqrt{1 + 16x^2 + 4y^2}} \cdot \sqrt{1 + 16x^2 + 4y^2}\ dS = \iint_{S_1} (4x^2 + 2y^2 + z)\ dS$$

$$= \iint_{S} (12 + 2x^2 + y^2)\ dS = \int_{\theta=0}^{2\pi} \int_{r=0}^{2} (12 + 2r^2 \cos^2\theta + r^2 \sin^2\theta) \cdot r\ dr\ d\theta$$

$$= \int_{0}^{2\pi} \left[6r^2 + \frac{1}{8} r^4 (3 + \cos 2\theta) \right]_{0}^{2} d\theta = \int_{0}^{2\pi} (30 + 2\cos 2\theta)\ d\theta = \left[30\theta + \sin 2\theta \right]_{0}^{2\pi} = 60\pi.$$

The lower surface S_2 is described by $h(x,\ y) = x^2 + 2y^2$. By computations similar to those shown earlier in this solution, we find that

$$dS = \sqrt{1 + 4x^2 + 16y^2}\ dA$$

and that an outer unit vector normal to S_2 is

$$\mathbf{n} = \frac{1}{\sqrt{1 + 4x^2 + 16y^2}}\, \langle 2x,\ 4y,\ -1 \rangle.$$

Thus the flux of $\mathbf{F}$ across S_2 is

$$\iint_{S_2} \mathbf{F} \cdot \mathbf{n}\ dS = \iint_{S_2} (2x^2 + 4y^2 - z)\ dS = \iint_{S_2} (x^2 + 2y^2)\ dS = \int_{0}^{2\pi} \int_{0}^{2} (r^2 \cos^2\theta + 2r^2 \sin^2\theta) \cdot r\ dr\ d\theta$$

$$= \int_{0}^{2\pi} \left[\frac{1}{4} r^4 \cos^2\theta + \frac{1}{2} r^4 \sin^2\theta \right]_{0}^{2} d\theta = \int_{0}^{2\pi} (4\cos^2\theta + 8\sin^2\theta)\ d\theta = \left[6\theta - \sin 2\theta \right]_{0}^{2\pi} = 12\pi.$$

Therefore the total outward flux of $\mathbf{F}$ across the boundary of T is

$$\phi = 60\pi + 12\pi = 72\pi \approx 226.19467105846511316931.$$

C15S0M.021: We compute the three Jacobians in Eq. (17) of Section 15.5 using the parameters y and z. The result is

$$\frac{\partial(y,\, z)}{\partial(y,\, z)} = \begin{vmatrix} y_y & y_z \\ z_y & z_z \end{vmatrix} = 1,$$

$$\frac{\partial(z,\, x)}{\partial(y,\, z)} = \begin{vmatrix} z_y & z_z \\ x_y & x_z \end{vmatrix} = -\frac{\partial x}{\partial y}, \quad \text{and}$$

$$\frac{\partial(x,\, y)}{\partial(y,\, z)} = \begin{vmatrix} x_y & x_z \\ y_y & y_z \end{vmatrix} = -\frac{\partial x}{\partial z}.$$

Therefore

$$\iint_S P\, dy\; dz\; +\; Q\, dz\, dx\; +\; R\, dx\, dy = \iint_D \left(P - Q\frac{\partial x}{\partial y} - R\frac{\partial x}{\partial z} \right) dy\; dz.$$

C15S0M.023: Here we have

$$\bar{z} = \frac{1}{V}\iiint_T z\, dV = \frac{1}{V}\iiint_T \boldsymbol{\nabla}\cdot\langle 0,\, 0,\, \tfrac{1}{2}z^2 \rangle\, dV = \frac{1}{V}\iint_S \frac{1}{2}z^2\, dx\, dy = \frac{1}{2V}\iint_S z^2\, dx\, dy$$

by Eq. (4) of Section 15.6.

C15S0M.025: By Eq. (23) of Section 15.5, the heat flow across the boundary sphere S *into* B is given by

$$R = \iint_S K(\boldsymbol{\nabla}u)\cdot\mathbf{n}\, dS.$$

The divergence theorem then gives

$$R = \iiint_B \boldsymbol{\nabla}\cdot(K\boldsymbol{\nabla}u)\, dV = \iiint_B K\boldsymbol{\nabla}^2 u\, dV.$$

C15S0M.027: Problems 25 and 26 imply that

$$\iiint_B c\cdot\frac{\partial u}{\partial t}\, dV = \iiint_B K\boldsymbol{\nabla}^2 u\, dV$$

for any small ball B within the body. This can be so only if $cu_t \equiv K\boldsymbol{\nabla}^2 u$; that is, $u_t = k\boldsymbol{\nabla}^2 u$ (because $k = K/c$).

C15S0M.029: We begin with $\mathbf{r} = \langle x,\, y,\, z \rangle$ and $\phi = \phi(r)$ where $r = |\mathbf{r}| = \sqrt{x^2 + y^2 + z^2}$. Part (a):

$$\boldsymbol{\nabla}\phi(r) = \boldsymbol{\nabla}\phi\left(\sqrt{x^2 + y^2 + z^2}\right)$$

$$= \left\langle \phi'(r)\cdot\frac{x}{\sqrt{x^2 + y^2 + z^2}},\; \phi'(r)\cdot\frac{y}{\sqrt{x^2 + y^2 + z^2}},\; \phi'(r)\cdot\frac{z}{\sqrt{x^2 + y^2 + z^2}} \right\rangle$$

$$= \phi'(r)\left\langle \frac{x}{r},\, \frac{y}{r},\, \frac{z}{r} \right\rangle = \frac{\mathbf{r}}{r}\phi'(r).$$

Part (b): We use the result in part (a) and the result in Problem 28 of Section 15.1:

$$\nabla \cdot [\phi(r)\mathbf{r}] = \phi(r)(\nabla \cdot \mathbf{r}) + (\nabla\phi) \cdot \mathbf{r} = \phi(r)(1+1+1) + \frac{\mathbf{r} \cdot \mathbf{r}}{r} \cdot \frac{d\phi}{dr} = 3\phi(r) + r\frac{d\phi}{dr}.$$

Part (c): We use the result in part (a) and the results in Problems 29 and 35 of Section 15.1:

$$\nabla \times (\phi(r)\mathbf{r}) = (\phi(r))(\nabla \times \mathbf{r}) + (\nabla\phi) \times \mathbf{r} = 0 + \frac{\mathbf{r} \times \mathbf{r}}{r} \cdot \frac{d\phi}{dr} = 0.$$

C15S0M.031: Let us envision a Möbius strip M in space constructed from a long narrow rectangular strip of paper by matching its ends with a half-twist. Let the strip of paper be subdivided into smaller rectangles $R_1, R_2, \ldots, R_n$ as indicated in the following figure, with the boundary curve of each of these rectangles oriented in the positive fashion described in Section 15.7. Then the arrows cancel along any interior segment indicated in the figure, and the arrows on the ends of the strip are as indicated there—upward on the right and downward on the left.

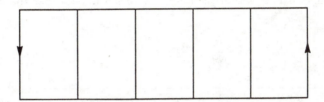

The Möbius strip is formed by matching these two ends of the rectangular paper strip, with the two end arrows matching in direction (thus providing the necessary half-twist). If we denote by S_i the ith curvilinear rectangle on the Möbius strip (corresponding to the original R_i), then Stokes' theorem gives

$$\oint_{C_i} \mathbf{F} \cdot \mathbf{T} \, ds = \iint_{S_i} (\nabla \times \mathbf{F}) \cdot \mathbf{n} \, dS$$

for each i. Taking account of cancellation of line integrals in opposite directions along segments corresponding to interior segments in the figure, summation then yields

$$\iint_M (\nabla \times \mathbf{F}) \cdot \mathbf{n} \, dS = \sum_{i=1}^n \iint_{S_i} (\nabla \times \mathbf{F}) \cdot \mathbf{n} \, dS$$

$$= \sum_{i=1}^n \oint_{C_i} \mathbf{F} \cdot \mathbf{T} \, ds = \oint_C \mathbf{F} \cdot \mathbf{T} \, ds + 2 \int_J \mathbf{F} \cdot \mathbf{T} \, ds,$$

where C denotes the boundary curve of the Möbius strip and J denotes the single interior segment along which the arrows match, so that the line integrals along J do not cancel. Because of the final term in the last equation, this calculation does not yield Stokes' theorem in the form

$$\iint_M (\nabla \times \mathbf{F}) \cdot \mathbf{n} \, dS = \oint_C \mathbf{F} \cdot \mathbf{T} \, ds$$

for the Möbius strip. Greater generality would take us too far afield, but surely you can well imagine that a similar "failure to cancel" would occur for any subdivision of the original narrow rectangular strip.

C15S0M.033: Part (a): If ΔS_i is a small piece of the boundary sphere S of the small ball B, and δ_i, $\mathbf{v}_i$, and $\mathbf{n}_i$ denote (respectively) the density, fluid flow velocity vector, and outward unit normal at time t

at a typical point of ΔS_i, then the rate of *outward* fluid flow across this area element ΔS_i is approximately $\delta_i \mathbf{v}_i \cdot \mathbf{n}_i \, \Delta S_i$. Hence the rate of flow of fluid *into* B at time t is given by

$$Q'(t) \approx -\sum_{i=1}^{n} \delta_i \mathbf{v}_i \cdot \mathbf{n}_i \, \Delta S_i \;\rightarrow\; -\iint_S \delta \mathbf{v} \cdot \mathbf{n} \, dS.$$

Part (b): Equating our two expressions for $Q'(t)$, we get

$$\iiint_B \frac{\partial \delta}{\partial t} \, dV = -\iint_S \delta \mathbf{v} \cdot \mathbf{n} \, dS = -\iiint_B \boldsymbol{\nabla} \cdot (\delta \mathbf{v}) \, dV$$

(applying the divergence theorem on the right). The fact that

$$\iiint_B \left[\frac{\partial \delta}{\partial t} + \boldsymbol{\nabla} \cdot (\delta \mathbf{v}) \right] dV = 0$$

therefore holds for *any* small ball B within the fluid flow region implies that the integrand must vanish identically; that is,

$$\frac{\partial \delta}{\partial t} + \boldsymbol{\nabla} \cdot (\delta \mathbf{v}) = 0.$$

Special thanks to the following:

Béla Bartók	César Franck
Reinhold Glière	Edvard Grieg
Percy Heath	Milt Jackson
Bob James	Connie Kay
John Lewis	Modest Mussorgsky
Sergei Rachmaninoff	Nikolai Rimsky-Korsakoff
Camille Saint-Saëns	Alexander Scriabin
Dmitri Shostakovich	Jean Sibelius

Monday, November 26, 2001

Typeset by $\mathcal{A}_{\mathcal{M}}\mathcal{S}$-TEX

Appendix A

A.001: $|3 - 17| = |-14| = -(-14) = 14.$

A.003: $\left| -0.25 - \frac{1}{4} \right| = \left| -\frac{1}{2} \right| = -\left(-\frac{1}{2} \right) = \frac{1}{2}.$

A.005: $|(-5)(4 - 9)| = |(-5)(-5)| = |25| = 25.$

A.007: $|(-3)^3| = |-27| = 27.$

A.009: According to *Mathematica* 3.0,

$$\pi \approx 3.14159265358979323846264338327950288419716939937510582097494, \quad \text{whereas}$$

$$\frac{22}{7} \approx 3.14285714285714285714285714285714285714285714285714285714286.$$

Therefore

$$\pi - \frac{22}{7} < 0, \quad \text{and thus} \quad \left| \pi - \frac{22}{7} \right| = \frac{22}{7} - \pi.$$

For a more elegant proof of the inequality, see Miscellaneous Problems 110 and 111 of Chapter 8.

A.011: If $x < 3$ then $x - 3 < 0$, so in this case $|x - 3| = -(x - 3) = 3 - x.$

A.013: $2x - 7 < -3$: $2x < 4$; $x < 2$.   Solution set: $(-\infty, 2)$.

A.015: $3x - 4 \geq 17$: $3x \geq 21$; $x \geq 7$.   Solution set: $[7, +\infty)$.

A.017: $2 - 3x < 7$: $-3x < 5$; $3x > -5$; $x > -\frac{5}{3}$.   Solution set: $\left(-\frac{5}{3}, +\infty \right)$.

A.019: $-3 < 2x + 5 < 7$: $-8 < 2x < 2$; $-4 < x < 1$.   Solution set: $(-4, 1)$.

A.021: $-6 \leq 5 - 2x < 2$: $-11 \leq -2x < -3$; $3 < 2x \leq 11$; $\frac{3}{2} < x \leq \frac{11}{2}$.   Solution set: $\left(\frac{3}{2}, \frac{11}{2} \right]$.

A.023: If $|3 - 2x| < 5$, then by the fourth property of absolute value in (3) we have $-5 < 3 - 2x < 5$. Thus $-8 < -2x < 2$, and hence $-2 < 2x < 8$. Therefore $-1 < x < 4$. So the solution set of the original inequality is $(-1, 4)$.

A.025: We will find the *complement* of the solution set—those real numbers *not* in the solution set—by solving instead the inequality $|1 - 3x| \leq 2$. By an extension of the fourth property of absolute value in (3), we then have $-2 \leq 1 - 3x \leq 2$. It follows that $-3 \leq -3x \leq 1$, and hence $-1 \leq 3x \leq 3$. So the complement of the solution set of the original equality is the closed interval $\left[-\frac{1}{3}, 1 \right]$. Therefore the solution set of the inequality $|1 - 3x| > 2$ is $(-\infty, \frac{1}{3}) \cup (1, +\infty)$.

A.027: Given: $2 \leq |4 - 5x| \leq 4$. Case 1: $4 - 5x \geq 0$. This is equivalent to the assertion that $x \leq \frac{4}{5}$; moreover, in this case the original inequality takes the form $2 \leq 4 - 5x \leq 4$. Thus $-2 \leq -5x \leq 0$, so that $0 \leq 5x \leq 2$. So the solutions we obtain in Case 1 are the real numbers in *both* of the intervals $\left[0, \frac{2}{5} \right]$ and $(-\infty, \frac{4}{5}]$. So Case 1 contributes the interval $\left[0, \frac{2}{5} \right]$ to the solution set. Case 2: $4 - 5x < 0$. This is equivalent to the assertion that $x > \frac{4}{5}$. Also, the original inequality takes the form $2 \leq 5x - 4 \leq 4$; that is, $6 \leq 5x \leq 8$, so that $\frac{6}{5} \leq x \leq \frac{8}{5}$. All these numbers satisfy the other condition that $x > \frac{4}{5}$, and hence Case

2 contributes the interval $\left[\frac{6}{5}, \frac{8}{5}\right]$ to the solution set. Therefore the solution set of the original inequality is $\left[0, \frac{2}{5}\right] \cup \left[\frac{6}{5}, \frac{8}{5}\right]$.

A.029: The given inequality

$$\frac{2}{7 - 3x} \le -5$$

implies that the denominator in the fraction is negative, so when we multiply both sides by that denominator we must reverse the inequality:

$$2 \ge 15x - 35; \qquad 15x \le 37; \qquad x \le \frac{37}{15} \approx 2.466666666667.$$

In addition, we have the condition

$$7 - 3x < 0; \qquad 3x > 7; \qquad x > \frac{7}{3} \approx 2.333333333333.$$

So the original inequality has solution set $\left(\frac{7}{3}, \frac{37}{15}\right]$.

A.031: If $x \ne \frac{1}{5}$ then $|1 - 5x| > 0$, so that

$$\frac{1}{|1 - 5x|} > 0 > -\frac{1}{3},$$

and if $x = \frac{1}{5}$ then the left-hand side of the given inequality is undefined. Consequently the solution set we seek is $\left(-\infty, \frac{1}{5}\right) \cup \left(\frac{1}{5}, +\infty\right)$.

A.033: Given $x^2 - 2x - 8 > 0$, first factor: $(x + 2)(x - 4) > 0$. Case 1: $x > -2$ and $x > 4$, so that $x > 4$. Case 2: $x < -2$ and $x < 4$, so that $x < -2$. The cases are exclusive and exhaustive, so the solution set is $(-\infty, -2) \cup (4, +\infty)$.

A.035: Given: $4x^2 - 8x + 3 \ge 0$, first factor: $(2x - 1)(2x - 3) \ge 0$. Case 1: $2x - 1 \ge 0$ and $2x - 3 \ge 0$. Then $x \ge \frac{1}{2}$ and $x \ge \frac{3}{2}$, so that $x \ge \frac{3}{2}$. Case 2: $2x - 1 \le 0$ and $2x - 3 \le 0$. Then $x \le \frac{1}{2}$ and $x \le \frac{3}{2}$, so that $x \le \frac{1}{2}$. Solution set: $\left(-\infty, \frac{1}{2}\right] \cup \left[\frac{3}{2}, +\infty\right)$.

A.037: Because $100 \le V \le 200$ and $pV = 800$, we have

$$100p \le 800 \le 200p; \qquad p \le 8 \quad \text{and} \quad 4 \le p; \qquad 4 \le p \le 8.$$

A.039: If $25 < R < 50$, then

$$25I < IR < 50I; \qquad 25I < 100 < 50I; \qquad I < 4 \quad \text{and} \quad 2 < I; \qquad 2 < I < 4.$$

A.041: Suppose that a and b are real numbers such that $0 < a$ and $0 < b$. By the second property in (1), $0 + b < a + b$; that is, $b < a + b$. But $0 < b$, so by the first property in (1), $0 < a + b$. Therefore the sum of two positive numbers is positive.

A.043: If a and b are both negative, then $a < 0$, so $a \cdot b > 0 \cdot b$ by the fourth property in (1). Thus ab is positive. If a is positive and b is negative, then $b < 0$, so $a \cdot b < a \cdot 0 = 0$ by the third property in (3). Thus $ab < 0$, and hence ab is negative.

A.045: If a, b, and c are real numbers, then

$$|a+b+c| = |(a+b)+c| \leqq |a+b| + |c| \leqq |a| + |b| + |c|.$$

That is, $|a+b+c| \leqq |a| + |b| + |c|$.

A.047: Suppose first that $|a| < b$. Then there are two cases. Case 1: $a \geqq 0$. Then $a < b$ and $b > 0$, so $-b < 0 < a < b$, and therefore $-b < a < b$. Case 2: $a < 0$. Then $-a < b$, so $b > 0$. Thus $-b < a < 0 < b$, so that $-b < a < b$. Therefore if $|a| < b$, then $-b < a < b$.

Next suppose that $-b < a < b$. Again there are two cases. Case 1: $a \geqq 0$. Then $a = |a|$, and thus $|a| < b$. Case 2: $a < 0$. Then $|a| = -a$. But from the hypothesis $-b < a < b$, it follows that

$$(-1) \cdot (-b) > (-1) \cdot a > (-1) \cdot b, \quad \text{so that} \quad -b < -a < b.$$

Hence $-b < |a| < b$, and consequently $|a| < b$.

Therefore $|a| < b$ if and only if $-b < a < b$. ◀

Appendix B

B.001: Both the segments AB and BC have the same slope 1. So the three given points do lie on a single straight line.

B.003: The segment AB has slope -2 but the segment BC has slope $-\frac{4}{3}$. Thus the three given points do not lie on a single straight line.

B.005: Both AB and CD have slope $-\frac{1}{2}$; both BC and DA have slope 2. This parallelogram is a rectangle!

B.007: AB has slope 2 and AC has slope $-\frac{1}{2}$. Hence triangle ABC has a right angle at the vertex A.

B.009: The given equation can be put in the form $y = \frac{2}{3}x$, so the line has slope $\frac{2}{3}$ and y-intercept 0.

B.011: The given equation takes the form $y = 2x + 3$, so the line has slope 2 and y-intercept 3.

B.013: The given equation can be put in the form $y = -\frac{2}{5}x + \frac{3}{5}$, and consequently the line with this equation has slope $-\frac{2}{5}$ and y-intercept $\frac{3}{5}$.

B.015: $y = -5$.

B.017: The slope of L is 2, so a point-slope equation of L is $y - 3 = 2(x - 5)$; its slope-intercept equation is $y = 2x - 7$.

B.019: The slope of L is $\tan 135° = -1$, so a point-slope equation of L is $y - 2 = -(x - 4)$; another equation of L is $x + y = 6$.

B.021: The equation $2x + y = 10$ can be written in the form $y = -2x + 10$, so the line with this equation has slope -2. So a parallel line passing through $(1, 5)$ has point-slope equation $y - 5 = -2(x - 1)$; the slope-intercept equation of this line is $y = -2x + 7$.

B.023: The given line segment has slope 2 and midpoint $(1, 6)$. Hence L has slope $-\frac{1}{2}$ and thus point-slope equation $y - 6 = -\frac{1}{2}(x - 1)$. Another equation of L is $x + 2y = 13$.

B.025: The parallel lines have slope 5, so the line $y = -\frac{1}{5}x + 1$ is perpendicular to both of them. It meets the line $y = 5x + 1$ at the point $(0, 1)$ and the line $y = 5x + 9$ at the point $\left(-\frac{20}{13}, \frac{17}{13}\right)$. The answer is the distance between the last two points, which is

$$\sqrt{\left(\frac{20}{13}\right)^2 + \left(\frac{4}{13}\right)^2} = \frac{4}{13}\sqrt{26} \approx 1.56892908110547225539.$$

B.027: With $A(-1, 6)$, $B(0, 0)$, $C(3, 1)$, and $D(2, 7)$ forming the consecutive vertices of the parallelogram, its diagonals are AC and BD. The midpoint of AC is $\left(1, \frac{7}{2}\right)$ and the midpoint of BD is the same. Therefore the diagonals of the parallelogram bisect each other.

B.029: The midpoint of AB is $P\left(\frac{5}{2}, 3\right)$ and the midpoint of BC is $Q(5, 4)$. The slope of PQ is $\frac{2}{5}$, the same as the slope of AC, and hence PQ is parallel to AC.

B.031: The slope of P_1M is

$$\frac{y_1 - \overline{y}}{x_1 - \overline{x}} = \frac{y_1 - \frac{1}{2}(y_1 + y_2)}{x_1 - \frac{1}{2}(x_1 + x_2)} = \frac{2y_1 - y_1 - y_2}{2x_1 - x_1 - x_2} = \frac{y_1 - y_2}{x_1 - x_2},$$

and the slope of MP_2 is the same.

B.033: Because F and K satisfy a linear equation, $K = mF + b$ for some constants m and b. From the data given in the statement of the problem, we find that

$$273.16 = 32m + b \qquad \text{and}$$

$$373.16 = 212m + b.$$

Subtraction of the first equation from the second yields $180m = 100$, so that $m = \dfrac{5}{9}$. Then the first equation here yields

$$273.16 = \frac{160}{9} + b, \quad \text{so that} \quad b = \frac{6829}{25} - \frac{160}{9} = \frac{57461}{225} \approx 255.382222222222.$$

Therefore $K = \dfrac{1}{225}(125F + 57461)$ provided that $F > -459.688$.

B.035: If s denotes weekly sales in gallons and p the selling price per gallon, then $s = mp + b$ for some constants m and b. From the data given in the statement of the problem, we have

$$980 = (1.69)m + b \quad \text{and} \quad 1220 = (1.49)m + b. \tag{1}$$

Subtract the second of these equations from the first to find that

$$(0.20)m = -240, \quad \text{so that} \quad m = -1200.$$

Then substitution of this value of m in the first equation in (1) yields

$$980 = -2028 + b, \quad \text{and thus} \quad b = 3008.$$

Therefore $s = s(p) = -1200p + 3008$. At \$1.56 per gallon one would expect to sell $s(1.56) = 1136$ gallons per week.

B.037: Given the simultaneous equations

$$2x + 3y = 5,$$

$$2x + 5y = 12,$$

the *Mathematica* 3.0 command

```
Solve[ { 2*x + 3*y == 5, 2*x + 5*y == 12 }, { x, y } ]
```

yields the exact solution $x = -\dfrac{11}{4}, \quad y = \dfrac{7}{2}.$

B.039: $x = \dfrac{37}{6}, \quad y = -\dfrac{1}{2}.$

B.041: $x = \dfrac{22}{5}, \quad y = -\dfrac{1}{5}.$

B.043: $x = -\dfrac{7}{4}, \quad y = \dfrac{33}{8}.$

B.045: $x = \dfrac{119}{12}, \quad y = -\dfrac{19}{4}.$

B.047: If (x_1, y_1) is a point of the plane not on the line L, then the line segment with endpoints (x_0, y_0) and (x_1, y_1) cannot have slope m. Therefore the equation $y_1 - y_0 = m(x_1 - x_0)$ cannot hold.

Appendix C

C.001: $40 \cdot \dfrac{\pi}{180} = \dfrac{2}{9}\pi \quad$ (rad).

C.003: $315 \cdot \dfrac{\pi}{180} = \dfrac{7}{4}\pi \quad$ (rad).

C.005: $-150 \cdot \dfrac{\pi}{180} = -\dfrac{5}{6}\pi \quad$ (rad).

C.007: $\dfrac{2\pi}{5} \cdot \dfrac{180}{\pi} = 72°.$

C.009: $\dfrac{15\pi}{4} \cdot \dfrac{180}{\pi} = 675°.$

C.011: If $x = -\dfrac{\pi}{3}$, then the values of the six trigonometric functions are given in the following table.

$\sin x$	$\cos x$	$\tan x$	$\sec x$	$\csc x$	$\cot x$
$-\dfrac{\sqrt{3}}{2}$	$\dfrac{1}{2}$	$-\sqrt{3}$	2	$-\dfrac{2\sqrt{3}}{3}$	$-\dfrac{\sqrt{3}}{3}$

C.013: If $x = \dfrac{7\pi}{6}$, then the values of the six trigonometric functions are given in the following table.

$\sin x$	$\cos x$	$\tan x$	$\sec x$	$\csc x$	$\cot x$
$-\dfrac{1}{2}$	$-\dfrac{\sqrt{3}}{2}$	$\dfrac{\sqrt{3}}{3}$	$-\dfrac{2\sqrt{3}}{3}$	-2	$\sqrt{3}$

C.015: Solutions: $x = n\pi$ where n is an integer.

C.017: Solutions: $x = 2n\pi - \dfrac{\pi}{2}$ where n is an integer.

C.019: Solutions: $x = 2n\pi$ where n is an integer.

C.021: Solutions: $x = n\pi$ where n is an integer.

C.023: Solutions: $x = n\pi - \dfrac{\pi}{4}$ where n is an integer.

C.025: Draw a right triangle with acute angle x and label the side opposite x as -3, the side adjacent x as 4, and the hypotenuse as 5. Then the values of the six trigonometric functions at x can be read from this triangle. The results appear in the next table.

$\sin x$	$\cos x$	$\tan x$	$\sec x$	$\csc x$	$\cot x$
$-\dfrac{3}{5}$	$\dfrac{4}{5}$	$-\dfrac{3}{4}$	$\dfrac{5}{4}$	$-\dfrac{5}{3}$	$-\dfrac{4}{3}$

C.027: Begin with the identity $\sin^2 \theta + \cos^2 \theta = 1$:

$$\frac{\sin^2 \theta}{\sin^2 \theta} + \frac{\cos^2 \theta}{\sin^2 \theta} = \frac{1}{\sin^2 \theta};$$

$$1 + \cot^2 \theta = \csc^2 \theta.$$

C.029: The sine addition formula yields

$$\sin \frac{5\pi}{6} = \sin\left(\frac{\pi}{2} + \frac{\pi}{3}\right) = \left(\sin\frac{\pi}{2}\right)\cdot\left(\cos\frac{\pi}{3}\right) + \left(\cos\frac{\pi}{2}\right)\cdot\left(\sin\frac{\pi}{3}\right) = 1\cdot\frac{1}{2} + 0\cdot\frac{\sqrt{3}}{2} = \frac{1}{2}.$$

C.031: The sine addition formula yields

$$\sin \frac{11\pi}{6} = \sin\left(2\pi - \frac{\pi}{6}\right) = (\sin 2\pi)\cdot\left(\cos\frac{\pi}{6}\right) - (\cos 2\pi)\cdot\left(\sin\frac{\pi}{6}\right) = 0\cdot\frac{\sqrt{3}}{2} - 1\cdot\frac{1}{2} = -\frac{1}{2}.$$

C.033: By the sine addition formula,

$$\sin \frac{2\pi}{3} = \sin\left(\frac{\pi}{2} + \frac{\pi}{6}\right) = \left(\sin\frac{\pi}{2}\right)\cdot\left(\cos\frac{\pi}{6}\right) + \left(\cos\frac{\pi}{2}\right)\cdot\left(\sin\frac{\pi}{6}\right) = 1\cdot\frac{\sqrt{3}}{2} + 0\cdot\frac{1}{2} = \frac{\sqrt{3}}{2}.$$

C.035: The sine addition formula yields

$$\sin \frac{5\pi}{3} = \sin\left(2\pi - \frac{\pi}{3}\right) = (\sin 2\pi)\cdot\left(\cos\frac{\pi}{3}\right) - (\cos 2\pi)\cdot\left(\sin\frac{\pi}{3}\right) = 0\cdot\frac{1}{2} - 1\cdot\frac{\sqrt{3}}{2} = -\frac{\sqrt{3}}{2}.$$

C.037: Part (a):

$$\cos\left(\frac{\pi}{2} - \theta\right) = \left(\cos\frac{\pi}{2}\right) \cdot \cos\theta + \left(\sin\frac{\pi}{2}\right) \cdot \sin\theta = 0 \cdot \cos\theta + 1 \cdot \sin\theta = \sin\theta.$$

Part (b):

$$\sin\left(\frac{\pi}{2} - \theta\right) = \left(\sin\frac{\pi}{2}\right) \cdot \cos\theta - \left(\cos\frac{\pi}{2}\right) \cdot \sin\theta = 1 \cdot \cos\theta - 0 \cdot \sin\theta = \cos\theta.$$

Part (c): We use the results in parts (a) and (b).

$$\cot\left(\frac{\pi}{2} - \theta\right) = \frac{\cos\left(\frac{\pi}{2} - \theta\right)}{\sin\left(\frac{\pi}{2} - \theta\right)} = \frac{\sin\theta}{\cos\theta} = \tan\theta.$$

C.039: The cosine addition formula yields

$$\cos(\pi \pm \theta) = \cos\pi\,\cos\theta \mp \sin\pi\,\sin\theta = (-1) \cdot \cos\theta \mp 0 \cdot \sin\theta = -\cos\theta.$$

C.041: Because AC and BD have the same length, we use the distance formula and obtain the equation

$$(\cos\phi - \cos\theta)^2 + (\sin\phi + \sin\theta)^2 = [1 - \cos(\theta + \phi)]^2 + [0 - \sin(\theta + \phi)]^2;$$

$$\cos^2\phi - 2\cos\phi\cos\theta + \cos^2\theta + \sin^2\phi + 2\sin\phi\sin\theta + \sin^2\theta$$
$$= 1 - 2\cos(\theta + \phi) + \cos^2(\theta + \phi) + \sin^2(\theta + \phi);$$

$$2 - 2\cos\phi\cos\theta + 2\sin\phi\sin\theta = 2 - 2\cos(\theta + \phi);$$

$$\cos(\theta + \phi) = \cos\phi\cos\theta - \sin\phi\sin\theta.$$

C.043: Let $u = \sin x$. Then

$$3u^2 - 1 + u^2 = 2; \qquad 4u^2 = 3; \qquad \sin x = u = \pm\frac{\sqrt{3}}{2}.$$

Thus the solutions in $[0, \pi]$ are $x = \frac{\pi}{3}$ and $x = \frac{2\pi}{3}$.

C.045: Let $u = \sin x$. Then

$$2(1 - u^2) + 3u^2 = 3; \qquad u^2 = 1; \qquad \sin x = u = \pm 1.$$

The only solution in $[0, \pi]$ is $x = \frac{\pi}{2}$.

C.047: If $8\sin^2 x \cos^2 x = 1$, then:

927

$$\sin^2 x \, \cos^2 x = \frac{1}{8}; \qquad\qquad \sin x \, \cos x = \pm\frac{\sqrt{2}}{4};$$

$$2\sin x \, \cos x = \pm\frac{\sqrt{2}}{2}; \qquad\qquad \sin 2x = \pm\frac{\sqrt{2}}{2};$$

$$2x = \frac{\pi}{4}, \frac{3\pi}{4}, \frac{5\pi}{4}, \frac{7\pi}{4}; \qquad\qquad x = \frac{\pi}{8}, \frac{3\pi}{8}, \frac{5\pi}{8}, \frac{7\pi}{8}.$$

Appendix D

D.001: Given $\epsilon > 0$, let $\delta = \epsilon$. Suppose that $0 < |x - a| < \delta$. Then $|x - a| < \epsilon$. Therefore

$$\lim_{x \to a} x = a.$$

D.003: Given $\epsilon > 0$, let $\delta = \epsilon$. Suppose that $0 < |x - 2| < \delta$. Then $|(x + 3) - 5| < \delta = \epsilon$. Therefore

$$\lim_{x \to 2} (x + 3) = 5.$$

D.005: Given $\epsilon > 0$, let δ be the minimum of 1 and $\epsilon/3$. Suppose that $0 < |x - 1| < \delta$. Then $|x - 1| < 1$; consequently,

$$-1 < x - 1 < 1;$$

$$1 < x + 1 < 3;$$

$$|x + 1| < 3.$$

Moreover, if $|x - 1| < \delta$ then $|x - 1| < \epsilon/3$. Hence

$$|x^2 - 1| = |x + 1| \cdot |x - 1| < 3 \cdot \frac{\epsilon}{3} = \epsilon.$$

Therefore, by definition, $\lim_{x \to 1} x^2 = 1$.

D.007: Given $\epsilon > 0$, let δ be the minimum of 1 and $\epsilon/6$. Suppose that $0 < |x - (-1)| < \delta$. Then $|x + 1| < \delta$, and so

$$|x + 1| < 1; \quad -1 < x + 1 < 1; \quad -3 < x - 1 < -1 < 3; \quad |x - 1| < 3.$$

Hence

$$|(2x^2 - 1) - 1| = 2|x - 1| \cdot |x + 1| < 2 \cdot 3 \cdot \frac{\epsilon}{6} = \epsilon.$$

Therefore, by definition, $\lim_{x \to -1} (2x^2 - 1) = 1$.

D.009: Case 1: $a > 0$. Let $\epsilon > 0$ be given. Choose

$$\delta = \min\left\{ \frac{a}{2}, \frac{(a^2 + 1)(a^2 + 4)}{10a} \epsilon \right\}.$$

Suppose that $0 < |x - a| < \delta$. Then

$$|x - a| < \frac{a}{2}; \qquad -\frac{a}{2} < x - a < \frac{a}{2}; \qquad \frac{a}{2} < x < \frac{3a}{2}.$$

Therefore $x > 0$ and $x + a > 0$. In addition,

$$\frac{3a}{2} < x + a < \frac{5a}{2}; \qquad |x + a| < \frac{5a}{2}.$$

Also

$$\frac{a^2}{4} < x^2 < \frac{9a^2}{4}; \qquad x^2 > \frac{a^2}{4}; \qquad x^2 + 1 > \frac{a^2 + 4}{4}; \qquad (x^2 + 1)(a^2 + 1) > \frac{(a^2 + 1)(a^2 + 4)}{4}.$$

Consequently,

$$\left| \frac{1}{x^2 + 1} - \frac{1}{a^2 + 1} \right| = \frac{|x^2 - a^2|}{|(x^2 + 1)(a^2 + 1)|} = \frac{|x - a| \cdot |x + a|}{(x^2 + 1)(a^2 + 1)} < \frac{5a}{2} \cdot \frac{4}{(a^2 + 1)(a^2 + 4)} \cdot \delta$$

$$= \frac{10a}{(a^2 + 1)(a^2 + 4)} \cdot \delta \leq \frac{10a}{(a^2 + 1)(a^2 + 4)} \cdot \frac{(a^2 + 1)(a^2 + 4)}{10a} \cdot \epsilon = \epsilon.$$

Therefore $\displaystyle \lim_{x \to a} \frac{1}{x^2 + 1} = \frac{1}{a^2 + 1}$ if $a > 0$.

Case 2: $a < 0$. The proof is similar to the one given in Case 1.

Case 3: $a = 0$. In this case we are to show that

$$\lim_{x \to 0} \frac{1}{x^2 + 1} = 1. \tag{1}$$

Given $\epsilon < 0$, let $\delta = \sqrt{\epsilon}$. Suppose that $0 < |x - 0| < \delta$. Then $x^2 < \epsilon$. Therefore

$$\left| \frac{1}{x^2 + 1} - 1 \right| = \left| \frac{x^2}{x^2 + 1} \right| = \frac{x^2}{x^2 + 1} < x^2 < \epsilon.$$

This establishes Eq. (1).

D.011: Given $\epsilon > 0$, note that $\epsilon/2$ is also positive. Choose δ_1 so that

$$|f(x) - L| < \frac{\epsilon}{2} \quad \text{if} \quad 0 < |x - a| < \delta_1;$$

choose δ_2 so that

$$|f(x) - M| < \frac{\epsilon}{2} \quad \text{if} \quad 0 < |x - a| < \delta_2.$$

Then for such x we have

$$|L - f(x)| + |f(x) - M| < \frac{\epsilon}{2} + \frac{\epsilon}{2} = \epsilon,$$

so by the triangle inequality

$$|L - M| = |L - f(x) + f(x) - M| < \epsilon.$$

929

Thus L and M differ by an amount smaller than any positive real number. Therefore $L = M$.

D.013: We treat only the case $L > 0$; the case $L < 0$ is quite similar. Let $\epsilon > 0$ be given. Choose $\delta > 0$ so small that

$$|f(x) - L| < \frac{L}{2} \quad \text{and} \quad |f(x) - L| < \frac{L^2 \epsilon}{2}$$

if $0 < |x - a| < \delta$. Then

$$-\frac{L}{2} < f(x) - L < \frac{L}{2}, \quad \text{so that} \quad \frac{L}{2} < f(x) < \frac{3L}{2};$$

moreover, $f(x) > 0$. Also

$$\frac{L^2}{2} < L \cdot f(x), \quad \text{and hence} \quad \frac{1}{|L \cdot f(x)|} < \frac{2}{L^2}.$$

Therefore, if $0 < |x - a| < \delta$, then

$$\left| \frac{1}{f(x)} - \frac{1}{L} \right| = \frac{|f(x) - L|}{|L \cdot f(x)|} < \frac{2}{L^2} \cdot \frac{L^2 \epsilon}{2} = \epsilon.$$

Consequently $\displaystyle \lim_{x \to a} \frac{1}{f(x)} = \frac{1}{L}$.

D.015: Suppose that $a > 0$ and let $\epsilon > 0$ be given. Let

$$\delta = \min \left\{ \frac{a}{2}, \ \epsilon \sqrt{2a} \right\}.$$

Suppose that $0 < |x - a| < \delta$. Then

$$|x - a| < \frac{a}{2}, \quad \text{so that} \quad \frac{a}{2} < x < \frac{3a}{2};$$

moreover, $x > 0$. Hence

$$\sqrt{\frac{a}{2}} < \sqrt{x} < \sqrt{\frac{3a}{2}};$$

$$\sqrt{\frac{a}{2}} + \sqrt{a} < \sqrt{x} + \sqrt{a} < \sqrt{\frac{3a}{2}} + \sqrt{a};$$

$$\frac{1}{\sqrt{x} + \sqrt{a}} < \frac{1}{\sqrt{\frac{a}{2}} + \sqrt{a}} < \frac{1}{\sqrt{2a}}.$$

So if $0 < |x - a| < \delta$, then

$$|\sqrt{x} - \sqrt{a}| = \frac{|x - a|}{|\sqrt{x} + \sqrt{a}|} < \frac{|x - a|}{\sqrt{2a}} \leq \frac{\epsilon \sqrt{2a}}{\sqrt{2a}} = \epsilon.$$

Therefore $\displaystyle \lim_{x \to a} \sqrt{x} = \sqrt{a}$ if $a > 0$.

Appendix G

G.001: The limit is

$$\int_0^1 x^2 \, dx = \left[\frac{1}{3} x^3 \right]_0^1 = \frac{1}{3}.$$

G.003: The limit is $\displaystyle\int_0^2 x\sqrt{4 - x^2} \, dx = \left[-\frac{1}{3}(4 - x^2)^{3/2} \right]_0^2 = \frac{8}{3}.$

G.005: The limit is $\displaystyle\int_0^{\pi/2} \sin x \cos x \, dx = \left[\frac{1}{2} \sin^2 x \right]_0^{\pi/2} = \frac{1}{2}.$

G.007: The limit is

$$\int_0^2 \sqrt{x^4 + x^7} \, dx = \int_0^2 x^2(1 + x^3)^{1/2} \, dx = \left[\frac{2}{9}(1 + x^3)^{3/2} \right]_0^2 = \frac{2}{9}(27 - 1) = \frac{52}{9} \approx 5.7777777777777778.$$

G.009: Take $G(y) = 2\pi y$ and $H(y) = \sqrt{1 + \left[g'(y) \right]^2}$ in Theorem 1. Then

$$\sum_{i=1}^n 2\pi y_i^{**} \sqrt{1 + \left[g'\left(y_i^\star\right) \right]^2} \, \Delta y \; \rightarrow \; A = \int_c^d 2\pi y \sqrt{1 + \left[g'(y) \right]^2} \, dy$$

by Theorem 1.

Typeset by $\mathcal{AMS}$-TEX